Lloyd S. Emerson and Laurence R. Paquette
Western New England College

SECOND EDITION

FUNDAMENTAL MATHEMATICS *for the* Management and Social Sciences

ALLYN AND BACON, INC.
Boston, London, Sydney, Toronto

Previous editions were published under the titles *Fundamental Mathematics for the Social and Management Sciences,* Copyright © 1975 by Allyn and Bacon, Inc., *Linear Algebra, Calculus, and Probability: Fundamental Mathematics for the Social and Management Sciences,* Copyright © 1971 by Allyn and Bacon, Inc., and *Linear Algebra, Calculus, and Probability for Business and Economics,* Copyright © 1969 by Allyn and Bacon, Inc.

Library of Congress Cataloging in Publication Data

Emerson, Lloyd S
Fundamental mathematics for the management and social sciences.

Including bibliographies and index.
1. Algebras, Linear. 2. Calculus. 3. Probabilities. 4. Social sciences—Mathematics.
I. Paquette, Laurence R., joint author.
II. Title.
QA184.E45 1978 510 77-28039

ISBN 0-205-06000-5

To our wives,
ELSIE AND BARBARA

And to our children,
DOUGLAS, EDMOND, AND JONATHAN

Contents

Preface

Many people are afraid of mathematics and avoid it whenever possible. Part of the reason for this is that many people teach mathematics for themselves and not for the students. This text has been written for the student. In addition, the focus is on mathematics as a tool for dealing with problems in the real world, which renders the text relevant and, hopefully, productive. *Fundamental Mathematics for the Management and Social Sciences* provides the ideal blend of a solid mathematical foundation as well as interesting examples of practical applications. The selected topics from the fields of linear algebra, probability, and calculus are ideal for students of the management and social sciences. It can be used at those schools having a one semester or one year mathematics course by the proper selection of topics.

The preparatory chapter should prove useful for the weak student as well as providing a review for the better prepared student. Chapter 1 is introductory in nature, beginning with a discussion of modeling. The concept of cost and revenue functions and break-even analysis are presented. Topics selected from the field of linear algebra are covered in Chapters 2, 3, and 4. Chapter 4 Determinants can be omitted without causing any disruptions in the development of future topics. Chapter 5, which treats inequalities and linear programming, is designed to provide applications that involve optimization models.

The concepts of a sequence and geometric progressions are applied to compound interest, present value, and annuities in Chapter 6. Chapters 7 and 8 contain selected topics from the field of probability.

Collection, summarization, and presentation of data, i.e., descriptive statistics, are covered in Chapter 9. Chapter 10 covers decision making under conditions of uncertainty and risk as well as game theory.

Graphing of the more important nonlinear functions is done in Chapter 11 as preparation for the calculus. Chapters 12, 13, 14, and 15 contain topics from the field of calculus dealing with elementary functions in a single variable.

To the reader who is familiar with the previous edition of this text, we would like to highlight the following distinctive features:

- Review of algebra centralized in a preparatory chapter at the beginning of text
- Reorganization of chapters so that all Finite Mathematic topics are together
- Incorporation of many new problems
- Many questions from CPA and CMA exams
- Revision and amplification of the Simplex Method
- Addition of two new chapters on Statistics and Decision Making
- Calculus toned down and more application oriented
- Theme of modeling introduced in Chapter 1 and carried throughout the text

The schematic that follows may be helpful to users of the text.

	CHAPTER*	APPLICATION
	P. Preparatory	*Ratios and Percentages*
I. Mathematical Relationships	**1. Functions and Breakeven Analysis**	• Cost and Revenue functions • Demand functions • Breakeven chart
II. Linear Algebra	**2. Introduction to Vectors and Matrices**	• Parts Explosion • Use of Inverse to Solve Systems of Equations • Input–Output Analysis
	3. Algebra of Square Matrices	
	4. Determinants	• Product Mix
	5. Inequalities and Linear Programming	• Profit Maximization • Cost Minimization
III. Functions in Finance	**6. Mathematics of Finance**	• Depreciation • Compound Interest • Present Value • Annuities • Capital Budgeting

	CHAPTER*	APPLICATION
IV. Probability, Statistics, and Applications	**7. Introduction to Probability**	• Independence of Variables
	8. Additional Topics in Probability	• Tree Diagrams • Management Information
	9. Descriptive Statistics	• Marketing Research • Brand Sharing Models
	10. Decision Making and Game Theory	• Capital Budgeting
V. Differential and Integral Calculus	**11. Pre-Calculus**	• Forecasting • Marginal Cost
	12. Introduction to Differential Calculus	• Marginal Revenue • Elasticity • Profit Maximization
	13. Additional Differentiation Techniques and Applications	• Economic Order Quantity • Inventory Control • Area Under a Normal Curve
	14. Max-Min Theory and Applications	
	15. Integral Calculus	

**Note:* Boxes with boldface lettering indicate main-track chapters.

The authors would like to thank especially William A. Thedford of Virginia Commonwealth University, reviewer of this edition, and Richard Spangler of Tacoma Community College, and reviewers of the previous edition: Professor Orlan Ohlausen of Mountain View College, Professor James E. Schaedel of Western Illinois University, Professor Kenneth L. Hankerson of the University of North Dakota, Professor Robert L. Johnson of Northern Colorado State University, Professor Paul Banks of Boston College, and Professor Alex Whitmore of McGill University.

Finally, we wish to thank the many students and professors who used the previous editions of this text for their valuable feedback, especially those at Western New England College. Special recognition is also in order for Marjorie B. Peterson, who did most of the typing for the second edition.

L. S. Emerson
L. R. Paquette

PREPARATORY CHAPTER

Algebra review

P-1 SETS, NUMBERS, AND NOTATION

Sets are used extensively in the study of modern mathematics. A *set* is a well-defined collection of objects. The objects in a set are called its elements and are said to *belong* to the set in question. Braces, { }, are normally used to enclose the elements of a set. The notation $x \in S$ is used to indicate that x is an element of the set S.

Some interesting examples of sets (collections) follow. These examples are drawn both from mathematics and from everyday experiences.

EXAMPLES OF SETS OF OBJECTS

1. The crowd of people on the corner.
2. The set of *natural numbers* $\{1, 2, 3, \ldots\}$.
3. The flock of geese flying above us.
4. The set of *whole numbers* $\{0, 1, 2, 3, \ldots\}$.
5. The herd of deer in that field.
6. The set of *integers* $\{\ldots, -3, -2, -1, 0, 1, 2, 3, \ldots\}$.
7. The set of dishes on the table.
8. The set of *rational numbers*, i.e., ratios p/q, where p and q are integers and $q \neq 0$.
9. My collection of coins.

10. The set of *irrational numbers* (numbers that cannot be expressed as a ratio of two integers).
11. The pride of lions over there.
12. The set of *real numbers.*

The sets of numbers that appear in the previous examples are shown next in an organizational chart. (Figure P-1).

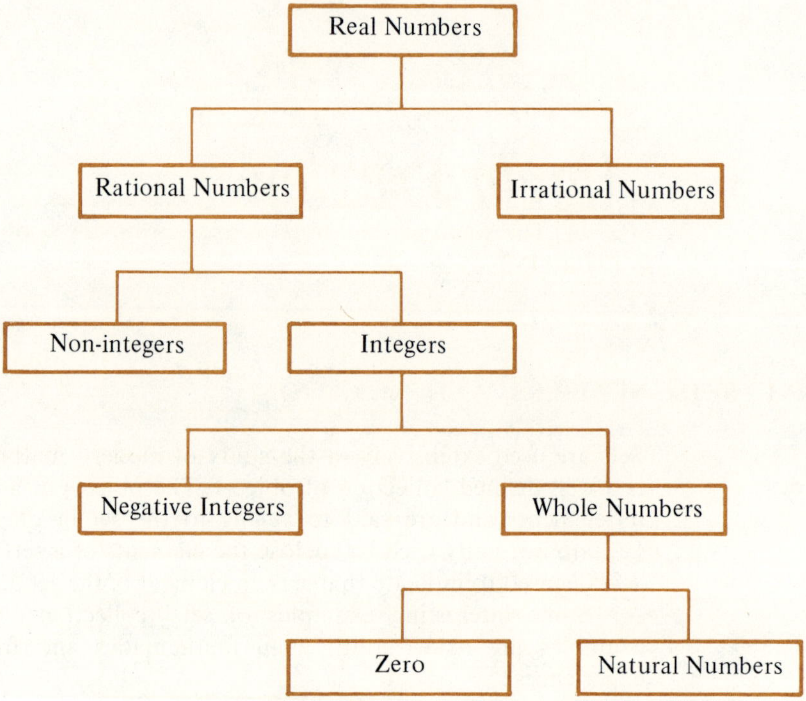

FIGURE P-1 Sets of numbers arranged in an organizational chart.

The real number system can also be represented geometrically by a number line. (Fig. P-2).

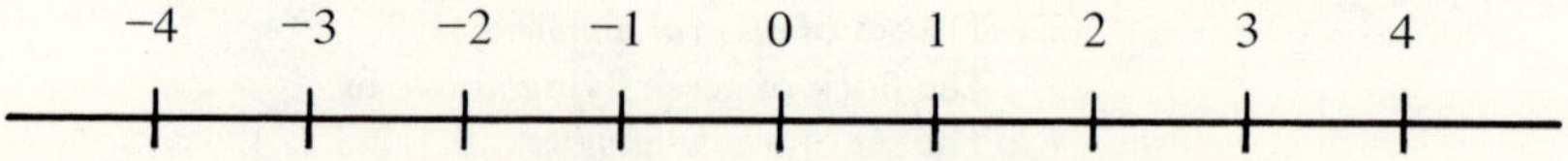

FIGURE P-2 A standard number line.

Some people find math courses difficult because of the special symbols used. Here are a few examples of standard mathematical symbols.

x, y, and z	for variables
a, b, and c	for unknown constants
$+$, $-$, $\cdot$ or $\times$, and $\div$	for operations
x^2	for the square of x
$\sqrt{x}$	for the square root of x
$=$	for equality
$<$	for less than
$>$	for greater than
$\cap$	for the intersection of sets
$\cup$	for the union of sets
$(x+y)$	for grouping x and y
$\neq$	for not equal to
$\|x\|$	for absolute value of x

Three symbols for grouping are parentheses, ( ), square brackets, [ ], and braces, { }. Braces are also used to enclose the elements of a set. The standard hierarchy for the grouping symbols is shown in Figure P-3.

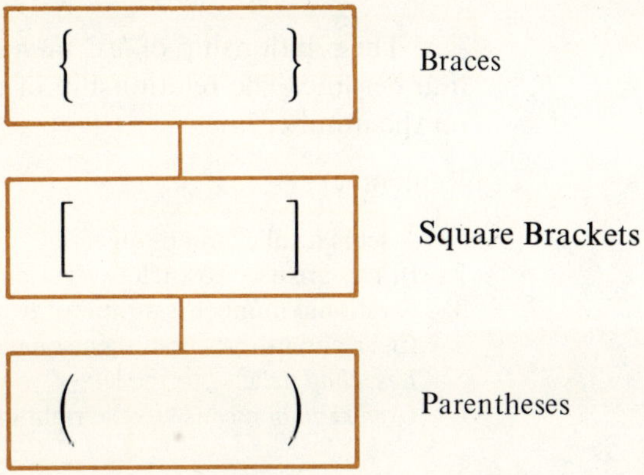

FIGURE P-3

Example 1. Clear all grouping symbols in $2\{(3+2)[4+(7-2)]\}$.

Solution:

$$\begin{aligned} 2\{(3+2)[4+(7-2)]\} &= 2\{5[4+5]\} \\ &= 2\{5[9]\} \\ &= 2\{45\} \\ &= 90 \end{aligned}$$

Notice that the parentheses are removed first, the square brackets second, and the braces last. This is the meaning of the hierarchy shown in Figure P-3.

Example 2. Compare the integers -4, 0, -1, and 2 by using the symbol $<$.

Solution:

$-4 < -1$

$-4 < 0$

$-4 < 2$

$-1 < 0$

$-1 < 2$

$0 < 2$

For $x \neq y$, there are two possibilities if x and y are real numbers:

1. $x < y$, that is, x less than y, and y greater than x

   or

2. $x > y$, that is, x greater than y, and y less than x

The relationship of *less than* means "to the left of" on a standard number line. The relationship of *greater than* means "to the right of" on the number line.

Remember:

1. A set is a collection of objects.
2. Braces are used to enclose the elements of a set.
3. A rational number is a ratio of two integers a/b, $b \neq 0$.
4. Clear parentheses first, then square brackets, then braces.
5. *Less than* means "to the left of" on the number line.
6. *Greater than* means "to the right of" on the number line.

PROBLEMS

1. Use set notation to describe the set of even natural numbers.

2. Use set notation to describe the set of odd natural numbers.

3. Make a list of eight nonintegral rational numbers.

4. The integers -4, 0, -1, and 2 are used in example 2. Indicate the points on a standard number line that would be labeled by these four numbers.

5. Compare the integers -5, 0, 3, and 10 by using the symbol $>$.

6. Clear the parentheses in the expression $2(4+3+1)$ and simplify.

7. Clear the parentheses in the expression $4(6-2+1)$ and simplify.

8. Clear all grouping symbols in the expression $3\{(1+2)[4+(7-4)]\}$ and simplify.

9. Clear all grouping symbols in the expression $(3+1)\{(4-2)[6-(8-5)]\}$ and simplify.

10. Clear the parentheses in the expression $a(b+c)$.
Answer: $ab+ac$
Remark: $a(b+c) = ab+ac$ for real numbers a, b, and c is called the distributive law.
Simplify in problems 11–14.
Sample calculations:
$4^2 = 16$
$\sqrt{81} = 9$

11. $3^2 + 4^2 + 5^2$

12. $\sqrt{100} + \sqrt{25}$

13. $\sqrt{9+16}$
Hint: the answer is *not* 7.

14. $\sqrt{38-2}$

P-2 OPERATIONS WITH SIGNED NUMBERS

The rules for multiplication and division of signed numbers are easy to remember. They are:

1. The product or quotient of two numbers with like signs is positive.
2. The product or quotient of two numbers with opposite signs is negative.

EXAMPLES OF MULTIPLYING AND DIVIDING SIGNED NUMBERS

$(3)(4) = 12$ $\qquad$ $(-10)(-12) - 120$

$(-4)(5) = -20$ $\qquad$ $(-2)(-4)(-3) = -24$

$\dfrac{-8}{-2} = 4$ $\qquad$ $\dfrac{(8)(-3)}{6} = -4$

$\dfrac{75}{-3} = -25$ $\qquad$ $\dfrac{-2}{-3} = \dfrac{2}{3}$

The *absolute value* of a real number x is denoted by $|x|$. By definition

$$|x| = \begin{cases} x & \text{if } x > 0 \text{ or } x = 0 \\ -x & \text{if } x < 0. \end{cases}$$

This means that the absolute value of any nonzero real number is positive, since the negative of a negative number is positive.

Example 1.

$|3| = 3$

$|-4| = -(-4) = 4$

$|0| = 0$

$|-0.00001| = 0.00001$

The following keys appear on most inexpensive electronic calculators.

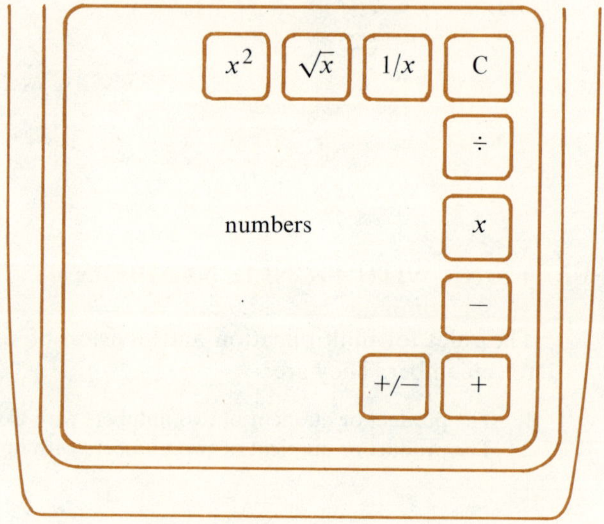

The key +/− is sometimes labeled "change sign."

The rules of addition of signed numbers are illustrated next along with a discussion of these operations on an electronic calculator. The word *press* indicates which key to use and the word *display* indicates what the user sees on the calculator display.

Example 2.

$5 + 11 = 16$

Press	*Display*
5	5
+	5
11	11
=	16

Example 3. $-5+(-11)=-16$

Press	*Display*
5	5
+/−	−5
+	−5
11	11
+/−	−11
=	−16

Rule used in examples 2 and 3: Add the absolute values and use the common sign.

Example 4. $5+(-11)=-6$

Press	*Display*
5	5
+	5
11	11
+/−	−11
=	−6

Example 5. $-5 + 11 = 6$

Press	*Display*
5	5
+/−	−5
+	−5
11	11
=	6

Rule used in examples 4 and 5: Subtract the smaller absolute value from the larger absolute value, and use the sign of the number with the larger value.

The operation of subtraction can be performed by changing the sign of the subtrahend and adding according to the rules illustrated in examples 2 through 5.

SUBTRACTION EXAMPLES

$13 - 4 = 9$	$4 - 13 = -9$
or	or
$13 - 4 = 13 + (-4)$	$4 - 13 = 4 + (-13)$
$= 9$	$= -9$
$-13 - 4 = -17$	$-13 - (-4) = -9$
or	or
$-13 - 4 = -13 + (-4)$	$-13 + 4 = -9$
$= -17$	

Remember:

1. The absolute value of any real number is nonnegative.
2. If signs are alike, add absolute values and use the common sign.
3. If signs are opposite in an addition problem, subtract absolute values and use the sign of the larger value.

4. To subtract, change the sign and add.
5. Product (quotient) is positive if signs are alike and negative if signs are opposite.

PROBLEMS

Simplify

1. $(8+3)-2$
2. $(-4-2)+6$
3. $(10-18)-7$
4. $12+(12-6)$
5. $-8-(-4-3)$
6. $-10+(-2+12)$
7. $10+6-4$
8. $-8-4+3-12$
9. $24-10+16-6$
10. $18-4-3+10$
11. $(-4)(-5)$
12. $(-1)(4)(-7)$
13. $-12/-4$
14. $27/-3$
15. $(-12)(48)/-24$
16. $(2)(-3)+(-6)(-2)$
17. $(-4)(-5)-(5)(-3)$
18. $[(12)(-4)/6]+[2+(5-8)]$
19. $(2-7)(x-y)$
20. $(x+y)[-3+(7-4)]$

P-3 OPERATIONS WITH FRACTIONS

Expressions of the form a/b, where a and b are real numbers, are called *fractions.* This definition does not apply if $b=0$. In other words, division by zero is undefined. The number a is called the *numerator*, and the number b is called the *denominator*. If a and b are both integers with $b \neq 0$, then a/b is a rational number.

Fractions are multiplied in the following manner:

$$a/b \times c/d = ac/bd$$

Fractions can be divided by inverting and multiplying as follows:

$$a/b \div c/d = a/b \times d/c$$

EXAMPLES OF MULTIPLICATION AND DIVISION OF FRACTIONS

$$\frac{1}{3}\cdot\frac{2}{5}=\frac{2}{15} \qquad \frac{1}{8}\div\frac{1}{2}=\frac{1}{4}$$

$$\frac{2}{3}\cdot\frac{-4}{7}=\frac{-8}{21} \qquad \frac{3}{4}\div\frac{6}{8}=1$$

$$\frac{-5}{4}\cdot\frac{2}{-7}=\frac{5}{14} \qquad \frac{5}{6}\div\frac{0}{3}\text{ is undefined}$$

The addition and subtraction of fractions involves the use of a *common denominator*, D, which is divisible by the denominators involved in the given problem. The smallest possible positive value of D is called the *least common denominator* (*LCD*).

Example 1.

$$\frac{1}{4}+\frac{3}{8}=\frac{2+3}{8}=\frac{5}{8}$$

$D = 8$ in this example.

In general,

$$a/b + c/d = (a^* + c^*)/D$$

where the new numerators a^* and c^* are calculated as follows:

$$a^* = (D/b) \text{ times } a$$

$$c^* = (D/d) \text{ times } c$$

Example 2.

Calculate $2/3 + 5/7$ using the preceding rule.

Solution:

$a = 2$, $c = 5$, $b = 3$, and $d = 7$

Let $D = 21$. This is the *LCD*.

$$a^* = (21/3)\times 2, \text{ and } c^* = (21/7)\times 5$$

$$a^* = 14 \text{ and } c^* = 15$$

Therefore, $2/3 + 5/7 = 29/21 = 1\frac{8}{21}$.

EXAMPLES OF ADDITION AND SUBTRACTION OF FRACTIONS

$$\frac{1}{3}+\frac{2}{10}=\frac{10+6}{30}=\frac{16}{30}=\frac{8}{15}$$

$$\frac{6}{25}-\frac{8}{75}=\frac{18-8}{75}=\frac{10}{75}=\frac{2}{15}$$

$$\frac{1}{3}-\frac{2}{9}+\frac{1}{6}=\frac{6-4+3}{18}=\frac{5}{18}$$

$$\frac{a}{6}+\frac{c}{4}=\frac{2a+3c}{12}$$

Remember:

1. Division by zero is not defined.
2. For fractions, the numerators can be added if and only if the denominators are all equal.
3. Fractions are multiplied by multiplying numerators and multiplying denominators.
4. The fraction a/b can be divided by c/d by inverting c/d and multiplying.

PROBLEMS

Perform the indicated operations and simplify.

1. $\frac{2}{5}+\frac{6}{10}$

2. $\frac{1}{8}-\frac{3}{16}$

3. $\frac{2}{3}\cdot\frac{4}{5}$

4. $2\left(\frac{1}{2}+\frac{5}{6}\right)$

5. $\frac{2}{3}\div\frac{4}{5}$

6. $\frac{6}{5}\left(\frac{1}{12}+\frac{5}{3}\right)$

7. $\frac{1}{2}+\frac{1}{3}-\frac{1}{4}$

8. $\left(\frac{1}{6}-\frac{1}{3}\right)\left(\frac{2}{5}+\frac{3}{25}\right)$

9. $\dfrac{a}{7}+\dfrac{6}{14}$

10. $\dfrac{2}{x}+\dfrac{x}{4}$

11. $\dfrac{\frac{1}{5}-\frac{2}{15}}{\frac{7}{25}}$

12. $\dfrac{\frac{7}{25}}{\frac{1}{5}+\frac{2}{15}}$

13. $\dfrac{\frac{2}{5}-\frac{1}{7}}{35}$ *Hint:* $35=\dfrac{35}{1}$

14. $\dfrac{\dfrac{x}{X+1}}{\dfrac{2x}{2x+2}}$ *where* $X \neq 0,\ x \neq -1$

15. $\dfrac{\frac{5}{6}+\frac{6}{5}}{\frac{7}{15}}$

16. $\dfrac{3}{5}\left(\dfrac{3}{8}-\dfrac{6}{16}\right)$

17. $\dfrac{0}{3}\div\dfrac{5}{6}$

18. $\dfrac{a}{b}\left(\dfrac{b}{ac}+\dfrac{b}{a}\right)$

P-4 RATIOS AND PERCENTAGES

A *ratio* a/b can be used to compare the two numbers a and b. The ratio a/b can also be written as

a to b or $a{:}b$

The next example illustrates the use of a ratio to express an economic comparison.

Example 1. The consumer price index was 150 last year, and it is 156 this year. Compare the index this year to the index of last year by using a ratio.

Solution: $156/150 = 1.04$

Note that the number following the word *to* is the denominator of the resulting fraction.

Example 2. Current assets are $420,000 and current liabilities are $150,000. Compare current assets to current liabilities by using a ratio.

Solution: 420,000/150,000 = 2.8

The ratio used in example 2 is called the *current ratio* in accounting.

Fractions are also involved in the definition and use of percentages. The ratio a to b where $b = 100$ is related to the concept of percent. For example,

$$8/100 = 0.08 = 8\%$$

The rule for changing the fraction a/b to a percent involves three simple steps:

1. Express the fraction in decimal form by dividing a by b.
2. Move the decimal point two places to the right, i.e., multiply the decimal by 100.
3. Use a percent sign, %.

Example 3. Fill in the following blanks.

	Fraction	Decimal	Percent
a.	1/4	______	______
b.	1/10	______	______
c.	______	______	50%
d.	______	0.15	______
e.	______	0.05	______
f.	______	______	100%

Solution:

a.	1/4	0.25	25%
b.	1/10	0.10	10%
c.	1/2	0.50	50%
d.	3/20	0.15	15%
e.	1/20	0.05	5%
f.	1/1	1.00	100%

Some examples of percent from the business world follow:

1. *Interest* or money paid for the use of money is expressed as a percentage of the *principal*.
2. *Discounts* are expressed as a percentage of the *list price* (base).
3. *Commissions* are expressed as a percentage of the value of the *goods sold*.

Percentage problems involve the basic relationship

$$\text{percentage} = \text{base} \times \text{rate or } (p = b \times r)$$

Example 4. Find $6\frac{1}{2}\%$ of \$330.

Solution: percentage $= \$330 \times 0.065$

$= \$21.45$

Note that the rate $6\frac{1}{2}\%$ is changed to decimal form before multiplying by the base.

It is also possible to find the rate when the base and the percentage are known. The following formula can be used.

rate = percentage/base or $r = p/b$

Example 5. Sixteen dollars is what percent of \$400?

Solution: $\text{rate} = \dfrac{\$16}{\$400} = 0.04 \text{ or } 4\%$

Note that the result of the division will always be in decimal form.

Another use of the formula is to find the *rate of increase or decrease* (often called the percent of change). To compute the rate of change simply substitute the change for the percentage and the *original* or beginning quantity for the base.

Example 6. Calculate the rate of increase of the consumer price index if it went from \$150 last year to \$156 this year.

Solution: The original value is 150 and the change is $156 - 150 = 6$. Therefore, rate of change $= 6/150$

$= 0.04$

$= 4\%$

Finally, if the rate and the percentage are given, then the following formula can be used to determine the base.

base = percentage/rate or $b = p/r$

Example 7. Calculate the principal (base) needed to yield \$14.04 interest in one year if the interest rate is $4\frac{1}{2}\%$ per year.

Solution:

$4\frac{1}{2}\% = 0.045$

base $= 14.04/0.045$

base $= 312$

Therefore, the required principal is \$312.

Example 8. The price of a certain model automobile increased by \$126. The manufacturer said that prices were increased by 3%. What was the selling price of that model last year?

Solution:

$$\text{base} = 126/0.03$$
$$= 4200$$

Therefore, the selling price increased from \$4200 last year to \$4326 this year.

Remember:

1. To change a ratio $a:b$ to a percent, divide a by b and multiply by 100.
2. Percentage = base × rate or $p = b \times r$.
3. Two other expressions equivalent to $p = b \times r$ are

   a. $r = p/b$

   b. $b = p/r$.

4. $$\text{Percentage change} = \frac{\text{change in amount}}{\text{original amount}} \times 100$$

PROBLEMS

1. Current assets for Ludlow Corporation are \$43,053,014 and current liabilities are \$26,073,313. Calculate the current ratio.

2. Net sales for Ludlow Corporation were \$220,413,000 last year and \$222,929,000 this year. Find the percentage increase in net sales.

3. Net sales for Ludlow Corporation this year were \$228,929,000 and net income was \$5,892,000. Express net income as a percentage of net sales.

4. Sales for the following product lines are as follows:

Product Line	Sales (in thousands of dollars)
Home furnishings	\$110,360
Manufactured housing	58,240
Papers and packaging	61,400

For what percent of sales does each of the product lines account?

5. Fill in the blanks in the following chart:

	Fraction	Decimal	Percent
(a)	2/8		
(b)	1/5		
(c)		0.40	
(d)		0.35	
(e)			10%
(f)			12.5%

6. Unemployment last month was 6.7% and is 6.9% this month. What is the percentage increase in unemployment?

7. The consumer price index in June was 126, in July it was 127, and in August it was 128.
(a) What was the percentage increase in July?
(b) What was the percentage increase in August?
(c) Because the answer in **(b)** is smaller than the answer in **(a)**, can we say that the consumer price index is decreasing?

8. A block of stocks was purchased on June 2, and the price per share was $10. By the end of the month the price per share increased 20 percent. What was the price per share at the end of June?

9. On August 1 the selling price of a particular radio was $25. During August the price was marked up 10 percent.
(a) What was the price at the end of August? During September it was decided to mark down the radio 10 percent.
(b) Is the selling price now $25?
(c) How can you account for this discrepancy?

10. Given the following data from a report to investors:
(a) How were the percentage gains calculated?
(b) Why would management decide to use different bases as they did?

	1974	1973	% Gain
Revenues	$2.0 billion	$1.5 billion	29%
Net Earnings	90.5 million	73.9 million	23%
Earnings per Share	2.77	2.31	20%
Dividends per Common Share	0.80	0.75	7%

11. The amount due for a fuel bill is $99.00, and there is a discount of 4 percent if the bill is paid within 10 days. For what amount should you make out a check if the bill is paid within 10 days? Calculate your answer using two different methods.

12. The following data trace the movement of two types of income—employee compensation and corporate profits. Figures represent billions of dollars.

Year	Employee Compensation	Corporate Profits
1972	$715	$92.1
1973	799	99.1
1974	875	84.8
1975	928	91.6
1976	1037	122.0

(a) Express corporate profits and employee compensation as a ratio.
(b) Do the ratios give any new insight into the growth of the two types of income?

13. Given the following data on Gross National Product (GNP) and dividend income:

Year	GNP	Dividend Income
1965	\$688	\$19.1
1968	868	21.9
1971	1063	23.0
1974	1413	30.8
1976	1708	35.6

(a) Does it appear that dividend income is growing?
(b) Express dividend income as a percentage of GNP.
(c) Would you want to modify your answer to part **(a)**?
Dollar figures are in billions.

P-5 EXPONENTS AND RADICALS

The following basic laws of exponents are discussed and illustrated in this section.

$x^0 = 1$ if $x \neq 0$	E-1	(*Definition*)
$x^m x^n = x^{m+n}$	E-2	
$x^m/x^n = x^{m-n}$	E-3	
$(x^m)^n = x^{mn}$	E-4	
$x^{-n} = 1/x^n$ if $x \neq 0$	E-5	(*Definition*)
$(xy)^n = x^n y^n$	E-6	
$x^{1/n} = \sqrt[n]{x},\ x > 0$	E-7	(*Definition*)
$x^{m/n} = \sqrt[n]{x^m},\ x > 0$	E-8	(*Definition*)

A typical electronic calculator has the following two keys.

$$x^2 \qquad\qquad \sqrt{x}$$

The 2 in x^2 is called an *exponent*, and x^2 is read *x squared*. The symbol $\sqrt{x}$ is an example of a *radical*, and it is read *the square root of x*.

By definition, if n is a positive integer, then

$$x^n = xx, \ldots, x \qquad (n \text{ factors})$$

Example 1. Illustrate the law $x^m x^n = x^{m+n}$ for the special case $m = 4$ and $n = 3$.

Solution:

$m = 4,\ n = 3$

$$
\begin{aligned}
x^4 &= xxxx \\
x^3 &= xxx \\
x^4x^3 &= (xxxx)(xxx) \\
&= xxxxxxx \qquad \textit{(7 factors)} \\
&= x^7
\end{aligned}
$$

Example 2. Illustrate the law $x^m/x^n = x^{m-n}$ for the special case $m = 5$ and $n = 2$.

Solution:

$m = 5$ and $n = 2$

$$
\begin{aligned}
x^5 &= xxxxx \\
x^2 &= xx \\
x^5/x^2 &= xxxxx/xx \\
&= xxx \qquad \textit{(3 factors)} \\
&= x^3
\end{aligned}
$$

If $x \neq 0$, then x^0 is defined to be 1. In words, any nonzero real number raised to the zero power equals one.

$$x^0 = 1 \text{ if } x \neq 0 \qquad \textit{(law E-1)}$$

The definition $x^{-n} = 1/x^n$, where n is a positive integer, allows us to work with negative integral exponents.

Example 3. Simplify 2^{-4}.

Solution:

$$
\begin{aligned}
2^{-4} &= \frac{1}{2^4} \qquad \textit{(by E-5)} \\
&= \frac{1}{2 \times 2 \times 2 \times 2} \\
&= \frac{1}{16} \\
&= 0.0625
\end{aligned}
$$

The next law deals with a product raised to a power.

$$(xy)^n = x^n y^n \qquad \textit{(law E-6)}$$

Example 4. Illustrate the law $(xy)^n = x^n y^n$ for the special case $n = 4$.

Solution:

$n = 4$

$$(xy)^4 = (xy)(xy)(xy)(xy)$$
$$= xy\,xy\,xy\,xy$$
$$= (xxxx)(yyyy)$$
$$= x^4y^4$$

It can be shown that the laws *E-2* through *E-6* are also valid if m and n are arbitrary real numbers. In particular they are valid for fractional exponents.

Example 5. Multiply $x^{1/2}$ times $x^{1/4}$.

Solution:

$$x^{1/2}x^{1/4} = x^{1/2+1/4}$$
$$= x^{3/4}$$

The number y is said to be an nth root of x ($y = \sqrt[n]{x}$) if $y^n = x$. For example

$$\sqrt[3]{125} = 5$$

and

$$\sqrt[4]{16} = 2$$

By definition $x^{1/n} = \sqrt[n]{x}$, for $x > 0$. (*E-7*)

Example 6. Simplify $32^{1/5}$.

Solution:

$$32^{1/5} = \sqrt[5]{32}$$
$$= 2$$

Finally, E-8: $x^{m/n} = \sqrt[n]{x^m}$, allows us to go back and forth at will between the exponential form and the radical form of an expression.

Example 7. Use the laws of exponents to calculate a^5a^3 and $(a^5)^3$.

Solution:

$$a^5a^3 = a^{5+3} \qquad \text{by E-2}$$
$$= a^8$$
$$(a^5)^3 = a^{5(3)} \qquad \text{by E-4}$$
$$= a^{15}$$

Example 8. Calculate a^4/a^0 two different ways, $a \neq 0$.

Solution:

$$a^4/a^0 = a^{4-0} = a^4 \qquad \text{by E-3}$$

or

$$a^4/a^0 = a^4/1 \qquad \text{by E-1}$$
$$= a^4$$

Example 9. Express $1/x$ and $\sqrt{x}$ in exponential notation.

Solution:

$$1/x = x^{-1} \qquad \text{by E-5}$$
$$\sqrt{x} = \sqrt[2]{x} \qquad \text{(understood)}$$
$$= x^{1/2} \qquad \text{by E-7}$$

Example 10. Use laws of exponents to write a^4b^2 in a different form.

Solution:

$$a^4b^2 = (a^2)^2b^2 \qquad \text{by E-4}$$
$$= (a^2b)^2 \qquad \text{by E-6}$$

Example 11. Simplify $4^{1.5}$.

Solution:

$$4^{1.5} = 4^{3/2}$$
$$= \sqrt[2]{4^3} \qquad \text{by E-8}$$
$$= \sqrt{64}$$
$$= 8$$
$$4^{3/2} = 8$$

Remember:

1. $x^0 = 1$ if $x \neq 0$
2. $x^{1/n} = \sqrt[n]{x}$, $x > 0$
3. $x^{m/n} = \sqrt[n]{x^m}$, $x > 0$

PROBLEMS

Simplify in problems 1–10.

1. x^4x^5

2. x^4x^{-3}
3. $x^{1/2}x^{3/2}$
4. $\dfrac{a^{3/2}}{a^{5/2}}$
5. $b^{1/3}b^{-4/3}$
6. 12^0
7. 2^{-3}
8. 3^{-2}
9. $\sqrt{144}$
10. $\sqrt[3]{1728}$

For problems 11–15, express in radical form.

11. $x^{3/4}$
12. $a^{5/3}$
13. $q^{0.25}$
14. $(x+y)^{1/2}$
15. $(x-y)^{-1/3}$

For problems 16–20, express in exponential form.

16. $\sqrt{x}$
17. $\sqrt[3]{x}$
18. $\sqrt{x^3}$
19. $\sqrt{(xy)^4}$
20. $\dfrac{1}{\sqrt[4]{x^2+y^2}}$

Solve for x by inspection in problems 21 32.

21. $2^x = 8$
22. $3^x = 81$
23. $4^x = 1/4$
24. $\sqrt[x]{8} = 2$
25. $\sqrt[3]{x} = 4$
26. $\sqrt{3x} = 4$
27. $10^x = 100$
28. $10^x = 1$
29. $10^x = 1000$
30. $10^x = 0.1$
31. $10^x = 0.01$
32. $10^x = 10{,}000$
33. Show that $P(1+i)^n + P(1+i)^n i = P(1+i)^{n+1}$.

P-6 LOGARITHMS

This section includes a brief introduction to logarithms. Additional information about logarithms and their applications can be found in Chapter 11. Table A-1 in appendix A is a table of common logarithms.

The logarithm (exponent) to the base, b, of a number, N, is defined next.

DEFINITION P.1. *Given the positive number b, where $b \neq 1$ and N is any positive real number, then the logarithm to the base b of N denoted by $\log_b N$ is the number that, when used as an exponent of b, gives N. In symbols, $\log_b N = x$ if $b^x = N$.*

Example 1. Use Definition P.1 to find $\log_2 8$, $\log_{16} 4$, and $\log_{1/2}(1/16)$.

Solution:

$$\log_2 8 = 3 \quad \text{since} \quad 2^3 = 8$$

$$\log_{16} 4 = 1/2 \quad \text{since} \quad 16^{1/2} = \sqrt{16} = 4$$

$$\log_{1/2}(1/16) = 4 \quad \text{since} \quad (1/2)^4 = 1/16$$

In the previous example, we used bases 2, 16, and 1/2 to illustrate Definition P.1. Although, theoretically, any positive number not equal to one could be used as a base, the number 10 is an especially important choice. The base 10 system is called the *common log system* and it can be used for computational purposes.

Example 2. Find $\log_{10} 10$, and $\log_{10} 1000$.

Solution:

$$\log_{10} 10 = x \qquad \log_{10} 1000 = x$$

$$10^x = 10 \qquad 10^x = 1000$$

$$x = 1 \qquad x = 3$$

$$\log_{10} 10 = 1 \qquad \log_{10} 1000 = 3$$

The following laws of logarithms are valid. Let M, N, and b be positive numbers with $b \neq 1$. Then

$$\log_b MN = \log_b M + \log_b N \qquad \textit{L-1}$$

$$\log_b M/N = \log_b M - \log_b N \qquad \textit{L-2}$$

$$\log_b N^P = P \log_b N \qquad \textit{L-3}$$

Proof of L-1. Let $\log_b M = x$ and $\log_b N = y$. Then, by Definition P.1, $b^x = M$ and

$$b^y = N$$

$$MN = b^x b^y$$

$$= b^{x+y} \qquad \textit{by E-2}$$

Therefore,

$$\log_b MN = x + y \qquad \textit{by Definition P.1}$$

$$\log_b MN = \log_b M + \log_b N$$

(We will omit the proofs of L-2 and L-3.)

Example 3. Illustrate numerically that $\log_2 16 = \log_2 2 + \log_2 8$.

Solution:

$$\log_2 16 = 4 \quad \text{since} \quad 2^4 = 16$$

$$\log_2 2 = 1 \quad \text{since} \quad 2^1 = 2$$

$$\log_2 8 = 3 \quad \text{since} \quad 2^3 = 8$$

Therefore,

$$\log_2 16 = 4 = 1 + 3 = \log_2 2 + \log_2 8$$

Example 4. Show that $\log_b \sqrt[n]{M} = 1/n \log_b M$.

Solution:

$$\sqrt[n]{M} = M^{1/n} \qquad \textit{by E-7}$$

$$\log_b \sqrt[n]{M} = \log_b M^{1/n}$$

$$= 1/n \log_b M \qquad \textit{by L-3}$$

Therefore,

$$\log_b \sqrt[n]{M} = 1/n \log_b M$$

Example 5. Verify L-2 for the case $\log_2 \frac{16}{4} = \log_2 16 - \log_2 4$.

Solution:

$$\log_2 \tfrac{16}{4} = \log_2 4$$

$$= 2 \quad \text{since} \quad 2^2 = 4$$

$$\log_2 16 = 4 \quad \text{and} \quad \log_2 4 = 2$$

Therefore,

$$\log_2 \tfrac{16}{4} = 2 = \log_2 16 - \log_2 4$$

Logs can be used for computational purposes (if a calculator is not available). They can be used to multiply, divide, and raise numbers to a power. The methodology needed to compute with logs is explained and illustrated in Chapter 11.

Remember:

1. A logarithm is an exponent.
2. The log of a *product* equals the *sum* of the logs of the factors.
3. The log of a *quotient* equals the *difference* of the logs of the terms in the fraction.
4. $\text{Log}\, x^n = n \log x,\ x > 0.$
5. $\text{Log}_{10} 1 = 0,\ \log_{10} 10 = 1,\ \log_{10} 100 = 2.$

PROBLEMS

1. Find $\log_3 N$ for the following values of N:

$$N = 9, 81, 27, 1, \tfrac{1}{3}, \tfrac{1}{27}, \tfrac{1}{9}$$

2. Find $\log_{1/2} N$ for the following values of N:

$$N = \tfrac{1}{4}, 1, 2, \tfrac{1}{2}, \tfrac{1}{8}, 16$$

3. Find $\log_{10} N$ for the following values of N:

$$10^2, 10^{-3}, 10^5, 10^{1/3}, \sqrt{10}, \sqrt[3]{10}, 10^{1/3}, \sqrt[3]{100}$$

4. Illustrate numerically that $\log_4 64 = \log_4 4 + \log_4 16$.

5. Illustrate numerically that $\log_2 4^3 = 3 \log_2 4$.

6. Illustrate numerically that $\log_5 \frac{625}{25} = \log_5 625 - \log_5 25$.

7. Prove L-2 and L-3.

8. Find $\log_b x$ given that $\log_b xy = 8$ and that $\log_b(x/y) = 4$.

9. Calculate or rewrite each of the following:

(a) $\log_b(b^3)$ **(c)** $\log_b(1/b)$

(b) $\log_b 1$ **(d)** $\log_b(ab^3 + b^2)$

10. Show that $\dfrac{\log_b 40 - \log_b 5}{3} = \log_b 2$.

11. Show that $\log_a N = \dfrac{\log_b N}{\log_b a}$ (change of base formula).

12. Illustrate the result in problem 11 with $N = 16$, $a = 4$, and $b = 2$.

P-7 MANIPULATION OF MONOMIALS AND BINOMIALS

Consider the expression $cx^m y^n$, where c is a real number, x and y are real variables, and m and n are positive integers. Then, the expression $cx^m y^n$ is an example of a *monomial*. Three examples of monomials are given next:

$5x^3y^7$

$3x^2$

7

The sum of two monomials (that cannot be combined) is called a *binomial*. Three examples of binomials are given next:

$x + y$

$2x^2 + 3x$

$xy^2 + x^2y$

Example 1. Clear the parentheses in the expression.

$$x^3y^4(x+y^2)$$

Solution:

$x^3y^4(x+y^2) = x^4y^4+x^3y^6$ by the *distributive law*, $a(b+c) = ab+ac$, and E-2.

Example 2. *Factor* the expression x^3y+xy^2.

Solution: xy is common to both terms of this binomial. Therefore, $x^3y+xy^2 = xy(x^2+y)$.

Example 3. Clear the parentheses in the expression $(x+3)(x+4)$.

Solution: Let $y = x+3$.

Then

$$\begin{aligned}(x+3)(x+4) &= y(x+4)\\ &= yx + 4y\\ &= xy + 4y\\ &= x(x+3) + 4(x+3)\\ &= x^2 + 3x + 4x + 12\\ &= x^2 + 7x + 12\end{aligned}$$

Therefore, $(x+3)(x+4) = x^2+7x+12$.

A short cut method for multiplying two binomials is illustrated in the next two examples.

Example 4. Multiply $(x+a)(x+b)$.

Solution: $(x+a)(x+b) = x^2 + (a+b)x + ab$.

Example 5. Multiply $(x+5)(x+6)$.

Solution: $(x+5)(x+6) = x^2 + 11x + 30.$

The answer in example 5 is an example of a *trinomial.* Now comes the hard part: factoring a trinomial into a product of two binomials.

Example 6. Factor the trinomial $x^2 - 11x + 30$.

Solution: $x^2 - 11x + 30 = (x+a)(x+b)$, where $a = -5$ and $b = -6$.

That is, $(x-5)(x-6)$ does it.

A certain amount of trial and error is often needed to factor trinomials.

Example 7. Factor the trinomial $x^2 + 12x + 36$.

Solution:

$$\begin{aligned} x^2 + 12x + 36 &= (x+6)(x+6) \\ &= (x+6)^2 \end{aligned}$$

Remember:

1. $(x+a)(x+b) = x^2 + (a+b)x + ab.$
2. $(x+a)^2 = x^2 + 2ax + a^2.$
3. $(x+a)(x-a) = x^2 - a^2.$
4. A binomial is the sum of two monomials that cannot be combined.
5. $x^m x^n = x^{m+n}.$

PROBLEMS

Clear parentheses in problems 1–10.

1. $2(x+y)$

2. $x(y+3)$

3. $x^2(x-y)$

4. $xy^2(x^2+xy+10)$

5. $(x+3)(x-2)$

6. $(x+y)(x-y)$

7. $(x+2)(x+2)$

8. $(x^2-2x)(y+3)$

9. $(x-12)(x-12)$

10. $(x-12)(x+12)$

Factor in problems 11–20

11. $xy + xz$

12. $4x^2 + 3x^3$

13. $3x^2y - 6x^2y^3$

14. $x^2 + 8x + 15$

15. $x^2 + 10x + 25$

16. $x^2 - 10x + 25$

17. $x^2 - 121$

18. $x^2 + 4x - 21$

19. $x^2 + 3x - 10$

20. $x^2 - y^2$

P-8 SOLVING SIMPLE EQUATIONS IN ONE VARIABLE

Equations that can be written in the form

$$bx + c = 0$$

where b and c are real numbers are called *linear equations*. Linear equations are very easy to solve and they have many important applications in business and economics.

EXAMPLES OF LINEAR EQUATIONS

1. $2x - 10 = 0$
2. $3x + 4 = 7$
3. $5x - 3 = 2x + 9$

The steps needed to solve the equation $bx + c = 0$ are as follows:

$$bx + c = 0$$

$$bx = -c$$

$$x = -\frac{c}{b}$$

Example 1. Solve the equation $3x + 4 = 7$ for x.

Solution:

$$3x + 4 = 7$$

$$3x = 7 - 4$$

$$3x = 3$$

$$x = 1$$

Some equations of the form

$$b^x = k$$

can be solved by inspection.

Example 2. Solve the equation $3^x = 81$ for x.

Solution: $x = 4$ by inspection

Example 3. Solve the equation

$$4(10)^x = 4000 \text{ for } x$$

Solution:

$$4(10)^x = 4000$$

$$10^x = 1000$$

$$\therefore \quad x = 3$$

Equations that can be written in the form

$$ax^2 + bx + c = 0, \qquad a \neq 0$$

are called *quadratic equations.* It is very important that you know how to solve quadratic equations. The three methods for solving a quadratic equation are:

1. Factoring
2. Completing the square, and
3. Using the quadratic formula.

EXAMPLES OF QUADRATIC EQUATIONS

1. $x^2 + 5x + 6 = 0$
2. $x^2 = 4x - 4$
3. $x^2 = 9$

These three quadratic equations are solved next by the method of factoring.

Example 4. Solve $x^2+5x+6 = 0$ for x.

Solution:

$$x^2 + 5x + 6 = 0$$

$$(x+2)(x+3) = 0$$

Now $(x+2)(x+3)$ equals zero if either factor equals zero. Therefore, we equate each factor to zero.

$$x + 2 = 0 \qquad x + 3 = 0$$

$$x = -2 \qquad x = -3$$

The two values of x that make $x^2+5x+6=0$ a true statement are $x=-2$ and $x=-3$. These values are called the *solutions* or *roots* of the given equation.

Example 5. Solve $x^2 = 4x-4$ for x.

Solution:

$$x^2 = 4x - 4$$

$$x^2 - 4x + 4 = 0$$

$$(x-2)(x-2) = 0$$

$$(x-2)^2 = 0$$

$$\therefore \quad x = 2$$

Example 6. Solve $x^2 = 9$ for x.

Solution:

$$x^2 = 9$$

$$x^2 - 9 = 0$$

$$(x+3)(x-3) = 0$$

$\therefore \quad x=-3$ and $x=3$ are the solutions.

It is possible for a quadratic equation to have no real solution. The equation $x^2+4=0$ provides an example.

$$x^2 + 4 = 0$$

is equivalent to

$$x^2 = -4$$

Since any real number squared is nonnegative, there can be no real solution for $x^2+4=0$. There are *complex* solutions, however; they are $x = 2i$ and $x = -2i$, where i denotes $\sqrt{-1}$.

We omit the method of completing the square and go next to the use of the *quadratic formula*.

QUADRATIC FORMULA

The quadratic formula for the solution(s) of the quadratic equation

$$ax^2 + bx + c = 0$$

is

$$x = \frac{-b \pm \sqrt{b^2 - 4ac}}{2a}$$

The use of the quadratic formula is illustrated in the next two examples.

Example 7. Use the quadratic formula to solve the quadratic equation $x^2 + 5x + 6 = 0$.

Solution: $a = 1$, $b = 5$, and $c = 6$

$$\frac{-5 \pm \sqrt{(5)^2 - 4(1)(6)}}{2(1)}$$

$$= \frac{-5 \pm \sqrt{25 - 24}}{2}$$

$$= \frac{-5 \pm \sqrt{1}}{2}$$

$$= \frac{-5 \pm 1}{2}$$

$$= \frac{-4}{2} \quad \text{or} \quad \frac{-6}{2}$$

$$= -2 \quad \text{or} \quad -3 \qquad \textit{(Recall example 4)}$$

Example 8. Solve $x^2 = 4x - 4$ by using the quadratic formula.

Solution:

$$x^2 = 4x - 4$$

$$x^2 - 4x + 4 = 0$$

$$a = 1, \; b = -4, \text{ and } c = 4$$

$$x = \frac{-(-4) \pm \sqrt{(-4)^2 - 4(1)(4)}}{2(1)}$$

$$= \frac{-4 \pm \sqrt{16-16}}{2}$$

$$= \frac{4 \pm \sqrt{0}}{2}$$

$$= 2 \qquad \textit{(Recall example 5)}$$

Remember:

1. All *quadratic equations* contain a squared term.
2. A quadratic equation may have no real root, one real root, or two real roots.
3. $x^3 + 3x + 7 = 0$ is *not* a quadratic equation.
4. The *quadratic formula* is

$$x = \frac{-b \pm \sqrt{b^2 - 4ac}}{2a}$$

PROBLEMS

Solve for x.

1. $2x - 6 = 0$

2. $3x + 12 = 0$

3. $-2x + 10 = 0$

4. $-3x - 12 = 0$

5. $a + bx = 0$

6. $4x + 3 = -5$

7. $x^2 - x - 6 = 0$

8. $x^2 + 4x + 4 = 0$

9. $x^2 + 7x + 12 = 0$

10. $2x^2 - 13x + 6 = 0$

11. $2x^2 + 6x = 0$

12. $x^2 - 16 = 0$

13. $x^2 + (2-a)x - 2a = 0$

14. $x^2 - 5x + 6 = 0$

15. $10x - 3 = 6x + 21$

16. $8(x-2) = (x-3)4$

17. Let $x = \dfrac{a}{P+C} - b$ be given.

Solve for P in terms of x.

18. The Ford Motor Co. said that they expect total sales of domestic and foreign autos in the United States to rise only 6% next year to about 10.6 million units. How many million units were sold this year?

19. The Agriculture Department forecast that cotton production would total 10.3 million bales this season, up 4% from its estimate of a month ago. What was the original estimate?

20. Let's look at an important question! I have an 86, 76, and 41 on each of three hour exams. What do I have to get on the final to get a 75 in the course? The final has twice the weight of an hour exam.

P-9 SOLUTION OF SYSTEMS OF LINEAR ALGEBRAIC EQUATIONS

This section explains the method of elimination to solve a system of linear algebraic equations. Our discussion is limited to systems of n equations in n unknowns. The first example involves a problem with two equations and two unknowns.

Example 1. Solve the system

$$2x + 3y = 21 \tag{1}$$

$$3x + 5y = 34 \tag{2}$$

for x and y by the method of elimination.

Solution: Multiply equation (1) by 3 to obtain

$$6x + 9y = 63 \tag{3}$$

Multiply equation (2) by 2 to obtain

$$6x + 10y = 68 \tag{4}$$

Next, subtract equation (3) from equation (4) as follows:

$$\begin{array}{r} 6x + 10y = 68 \\ 6x + \ \ 9y = 63 \\ \hline y = \ \ 5 \end{array}$$

Then, substitute $y = 5$ into either equation (1) or equation (2), say (1), to get

$$2x + 3(5) = 21$$

$$2x = 21 - 15$$

$$2x = 6$$

$$x = 3$$

Therefore, the solution of this pair of *simultaneous* equations is $x = 3$ and $y = 5$.

It is possible for a system of two linear equations in two unknowns either to fail to have a solution or to have an infinite number of solutions. These situations result if the graphs of the two equations are two different parallel lines, or if both equations represent the same straight line, respectively.

Example 2. Show that the system

$$2x + 3y - 6 = 0$$
$$4x + 6y - 10 = 0$$

has no solution, and the system

$$2x + 3y - 6 = 0$$
$$4x + 6y - 12 = 0$$

has an infinite number of solutions.

Solution: If we multiply the first equation by -2 and then add, we have

$$\begin{array}{r} -4x - 6y + 12 = 0 \\ 4x + 6y - 10 = 0 \\ \hline 0 + 0 + 2 = 0 \end{array} \quad \textit{this is a contradiction}$$

The *assumption* that the given system has a solution leads to the *contradictory statement* that $2 = 0$. Hence, this *assumption is incorrect* and the system has no solution.

Using the above method of multiplication on the second system,

$$2x + 3y - 6 = 0$$
$$4x + 6y - 12 = 0$$

leads to

$$-4x - 6y + 12 = 0$$
$$4x + 6y - 12 = 0$$

which gives, after addition, the true statement $0 = 0$. This is a reflection of the fact that both equations represent the same straight line. Any points whose coordinates satisfy the first equation also satisfy the second equation; hence the system has an infinite number of solutions.

The elimination method used here can also be used to solve systems of linear equations in more than two unknowns, as illustrated by the next example.

Example 3. Solve the system

$$3x + 2y + z = 4 \tag{1}$$
$$2x + 3y - z = 6 \tag{2}$$

$$4x + 5y + z = 2 \qquad (3)$$

for x, y, and z by the method of elimination.

Solution: (*eliminate* z)

Add equation (2) to equation (1) to obtain

$$5x + 5y = 10 \qquad (4)$$

Add equation (3) to equation (2) to obtain

$$6x + 8y = 8 \qquad (5)$$

Solve the "reduced problem"

$$5x + 5y = 10$$

$$6x + 8y = 8$$

for x and y by the method of elimination.
The reduced problem is equivalent to (has the same solution as)

$$x + y = 2 \qquad (6)$$

$$3x + 4y = 4 \qquad (7)$$

Multiply equation (6) by 3 to obtain

$$3x + 3y = 6 \qquad (8)$$

Subtract equation (8) from equation (7) to obtain

$$y = -2$$

Substitute $y = -2$ into equation (6), say, to obtain

$$x + (-2) = 2$$

$$x = 4$$

Then substitute $x = 4$ and $y = -2$ into equation (1), (2), or (3).
Choosing equation (1), we obtain

$$3(4) + 2(-2) + z = 4$$

$$z = 4 - 12 + 4$$

$$z = -4$$

Therefore, the solution is

$$x = 4, \qquad y = -2, \qquad z = -4$$

Systems of linear equations are somtimes written in the form

$$a_{11}x_1 + a_{12}x_2 + \cdots + a_{1n}x_n = b_1$$

$$a_{21}x_1 + a_{22}x_2 + \cdots + a_{2n}x_n = b_2$$

$$a_{i1}x_1 + a_{i2}x_2 + \cdots + a_{ij}x_j + \cdots + a_{in}x_n = b_i$$

$$a_{n1}x_1 + a_{n2}x_2 + \cdots + a_{nn}x_n = b_n$$

where the n unknowns are denoted by the *subscripted* variables $x_1, x_2, \ldots, x_n$.

Note also the use of *double subscripts* to label the coefficients in this system of linear equations.

Example 4. Solve the following system for x_1, x_2, and x_3 by the method of elimination.

$$x_1 + 2x_2 + 3x_3 = 1 \quad (1)$$

$$x_1 + 3x_2 + 5x_3 = 2 \quad (2)$$

$$2x_1 + 5x_2 + 9x_3 = 3 \quad (3)$$

Solution: (*eliminate* x_1)

Subtract equation (2) from equation (1) to get

$$-x_2 - 2x_3 = -1 \quad (4)$$

Multiply equation (2) by 2 and subtract the resulting equation from equation (3) to get

$$-x_2 - x_3 = -1 \quad (5)$$

Form a reduced problem

$$x_2 + 2x_3 = 1 \quad (6)$$

$$x_2 + x_3 = 1 \quad (7)$$

and solve it for x_3 by subtracting equation (7) from equation (6).
This gives $x_3 = 0$.
From $x_3 = 0$ and equation (7), we obtain $x_2 = 1$.
Finally, substitution of $x_2 = 1$ and $x_3 = 0$ into equation (1) gives

$$x_1 + 2(1) + 3(0) = 1$$

$$x_1 = -1$$

Therefore, the solution is

$$x_1 = -1,\ x_2 = 1, \text{ and } x_3 = 0$$

Remember: The solution of a system of linear algebraic equations by the method of elimination involves the following steps:

1. Select an appropriate variable to eliminate from the system.
2. Eliminate the chosen variable to obtain the reduced problem in the remaining variables.
3. Solve the *reduced problem* by the method of elimination.
4. Substitute the values of the variables found in step 3 into one of the equations of the original system to obtain the value of the variable eliminated in step 2.

Note: Any result of the form "0 = 0" resulting from the method of elimination indicates that the sytem has an *infinite number of solutions.* Or, any result of the form "$k = 0$", where k is a nonzero number, indicates that the system has *no solution.*

PROBLEMS

Solve for x and y in problems 1–11.

1. $y = 2x - 3$
$y = 3x - 4$

2. $2x + 3y - 4 = 0$
$x - 2y + 5 = 0$

3. $y = -x + 4$
$y = 3x - 2$

4. $2x + 3y + 4 = 0$
$y = -4/6x - 5/6$

5. $2x + 7y + 6 = 0$
$6x + 21y - 4 = 0$

6. $y = -2x + 3$
$-4y = 8x + 6$

7. $2y = 4x + 2$
$3y = 4x + 2$

8. $6y = -2x + 3$
$6y = x - 2$

9. $2x + 3y = 4$
$3x + 6y = 5$

10. $3x + 2y = 3$
$2x + y = 5$

11. $2x + 3y = 5$
$x + 2y = 7$

Solve for x, y, and z in problems 12–21.

12. $x + 2y + 3z = 3$
$x + 3y + 5z = 4$
$2x + 5y + 9z = 5$

13. $x + y + z = 4$
$2x + 2y + 2z = 7$
$x - y + z = 2$

14. $2x + 3y + 5z = 26$
$4x + 6y + 6z = 32$
$3x + y + 2z = 15$

15. $2x - y + z = 0$
$3x - 2y - 4z = 16$
$4x - 3y + z = 2$

16. $x + y - 2z = 9$
$2x - y + z = 6$
$3x + y + 4z = -1$

17. $\begin{aligned} -2x + y + 3z &= 1 \\ x - 2y + z &= -2 \\ 4x + 2y + z &= 4 \end{aligned}$

18. $\begin{aligned} 2x - y + z &= 4 \\ 3x + y + z &= -3 \\ x + 2y - z &= 5 \end{aligned}$

19. $\begin{aligned} x + y - 2z &= -2 \\ 2x - y + z &= 9 \\ 5x + y + 6z &= 5 \end{aligned}$

20. $\begin{aligned} 2x - y - 3z &= -2 \\ x - 2y + z &= -1 \\ 4x + 2y + z &= 3 \end{aligned}$

21. $\begin{aligned} x + 2y + 3z &= 1 \\ x + 3y + 5z &= 2 \\ 2x + 5y + 9z &= 3 \end{aligned}$

P-10 SUMMATION NOTATION

The symbol $\sum$, the Greek letter sigma, is used in mathematics to indicate summation. For example, the expression

$$x_1 + x_2 + \ldots + x_9$$

can be written more concisely as

$$\sum_{k=1}^{9} x_k$$

This expression could be read as summation of x sub k for $k = 1$ to 9. In general,

$$\sum_{k=1}^{n} x_k = x_1 + x_2 + \cdots + x_n$$

The next three examples involve summation notation.

EXAMPLES OF SUMMATION (SIGMA) NOTATION

$$\sum_{i=1}^{4} i = 1 + 2 + 3 + 4 = 10$$

$$\sum_{j=1}^{3} j^2 = 1^2 + 2^2 + 3^2 = 1 + 4 + 9 = 14$$

$$\sum_{k=3}^{7} (k+1) = 4 + 5 + 6 + 7 + 8 = 30$$

Example 1. Use summation notation to rewrite the linear equation

$$a_1 x_1 + \cdots + a_N x_N = a_0$$

Solution: Using the summation symbol, we have

$$\sum_{k=1}^{N} a_k x_k = a_0$$

Example 2. Write $\sum_{k=1}^{6} c_k x^k$ in expanded form.

Solution:

$$\sum_{k=1}^{6} c_k x^k = c_1 x + c_2 x^2 + c_3 x^3 + c_4 x^4 + c_5 x^5 + c_6 x^6$$

Example 3. Compute $f(3)$ if $f(x) = \sum_{k=1}^{4} 2^k x^k$.

Solution:

$$f(x) = 2x + 4x^2 + 8x^3 + 16x^4$$

$$f(3) = 2(3) + 4(9) + 8(27) + 16(81)$$

$$= 6 + 36 + 216 + 1296$$

$$= 1554$$

The definition of the average or *arithmetic mean* of a set of data can be stated in terms of the summation symbol as follows.

$$\bar{x} = \frac{1}{n} \sum_{i=1}^{n} x_i \qquad (mean)$$

where $\{x_1, x_2, \ldots, x_n\}$ is a set of observations.

Note that a mean, $\bar{x}$, is calculated in two steps as follows:

1. Determine the sum of the observations in the data set.
2. Divide the sum in step 1 by the number of observations, n, in the data set.

Example 4. Calculate the mean of the five numbers 2, 4, 7, 10, 12.

Solution:

$$\bar{x} = \frac{2+4+7+10+12}{5}$$

$$\bar{x} = 7$$

PROBLEMS

In problems 1–7 write the given expression in expanded form and simplify.

1. $\sum_{j=1}^{5} j$
2. $\sum_{i=2}^{5} \frac{i}{2}$
3. $\sum_{i=2}^{5} \frac{2}{i}$
4. $\sum_{k=2}^{5} (10+k)$
5. $\sum_{j=2}^{5} (j-1)^{j-2}$
6. $\sum_{j=2}^{6} (j-1)^{j-2}$
7. $\sum_{j=2}^{5} 2^j$
8. Given $a_1 = 2$, $a_2 = 5$, $a_3 = 1$, and $a_4 = 6$, compute the following:
   (a) $\sum_{i=1}^{4} a_i$
   (b) $\sum_{i=1}^{4} a_i^2$
9. Let c_{ij} be defined as $c_{ij} = \sum_{k=1}^{3} a_{ik} b_{kj}$
   (a) Express c_{13} in expanded form.
   (b) Express c_{31} in expanded form.
10. Use double subscript notation to indicate a typical system of three linear equations in four unknowns.
11. Show that $\sum_{i=1}^{2} x_i y_i$ is not equal to $(\sum_{i=1}^{2} x_i)(\sum_{i=1}^{2} y_i)$, in general.
12. Write $\sum_{i=1}^{3} x_i f(x_i)$ in expanded form.
13. $f(x)$ is defined to be $f(x) = \sum_{k=1}^{3} 3^k x^{k+1}$
   (a) Compute $f(2)$.
   (b) Compute $f(1)$.

Problems 14 through 18 are designed to illustrate the use of summation and subscript notation and to give the reader some computational experience with the mean.

14. Compute the mean of the following test scores:

i	1	2	3	4	5	6	7	8
x_i	80	75	65	84	90	95	68	75

15. The sales figures for the Go Get Em Company for the past six years are listed in the following table. Compute the six-year average sales, x, for this company. Sales figures are given in thousands of dollars.

19X1	19X2	19X3	19X4	19X5	19X6
\$1200	\$1320	\$1450	\$1300	\$1500	\$1600

16. Your company employs five types of workers, each at a different wage rate. Assumptions: (1) all employees work 40 hours per week, (2) each category has an equal number of workers. Compute the average wage rate for this company given the following data:

Type	1	2	3	4	5
Wage Rate	\$2.25	\$2.50	\$3.00	\$4.00	\$5.00

17. Discuss the implications of relaxing the assumptions in problem 16.

18. For the following data, where x_i denotes the number of lines in the ith invoice, calculate $\bar{x}$.

i	1	2	3	4	5
x_i	6	4	3	1	11

19. Show that

$$\sum_{i=1}^{n} (x_i - \bar{x})^2 = \sum_{i=1}^{n} x_i^2 - n\bar{x}^2. \qquad \textit{Hint} \quad \sum_{i=1}^{n} x_i = n\bar{x}.$$

20. Show that

$$\sum_{i=1}^{n} (x_i - \bar{x})^2 = \sum_{i-1}^{n} x_i^2 - \frac{(\sum_{i-1}^{n} x_i)^2}{n}$$

21. Calculate $\sum_{i=1}^{n} x_i$, $\sum_{i=1}^{n} y_i$, $\sum_{i=1}^{n} x_i^2$, $\sum_{i=1}^{n} x_i y_i$, and $(\sum_{i=1}^{n} x_i)^2$ for the following x_i and y_i values.

i	1	2	3	4	5
x_i	6	8	10	12	14
y_i	10	12	13	16	18

REFERENCES

Brett, William F. and Sentlowitz, Michael *Elementary Algebra by Example.* Boston: Houghton Mifflin Company, 1977.

Dubisch, Roy and Hood, Vernon R. *Elementary Algebra.* Reading, Mass.: W. A. Benjamin, Inc., 1977.

Groza, Vivian Shaw and Shelley, Susanne M. *Modern Elementary Algebra for College Students,* Third Edition. New York: Holt, Rinehart and Winston, 1977.

Leithold, Louis *College Algebra.* New York: Macmillan Publishing Co., Inc., 1975.

Setek, William M. Jr. *Algebra: A Fundamental Approach.* Philadelphia: W. B. Saunders Company, 1977.

Wooton, William and Drooyan, Irving *Intermediate Algebra,* Fourth Edition California: Wadsworth Publishing Company, Inc., 1977.

Zuckerman, Martin M. *Intermediate Algebra.* New York: W. W. Norton & Company, Inc., 1976.

ONE

Functions and breakeven analysis

1-1 MATHEMATICAL MODELS AND FUNCTIONS

There are numerous situations in business and economics where changes in one quantity result in changes in some related quantity. For example, a decrease in price generally results in an increase in sales. Or, an increase in the volume of output of goods by a firm results in an increase in the total operating cost of the firm. Relationships between quantities can be described and studied in a variety of ways. Three methods of analysis are illustrated in this section. They are denoted by the terms *verbal*, *tabular*, and *mathematical*.

The verbal method is illustrated first with an example of computer rental costs. The relationship between annual cost and the number of hours of computer usage is described.

VERBAL DESCRIPTION OF A RELATIONSHIP

The annual cost of renting the central processing unit (CPU) of a certain computer is \$250,000 plus \$50 per hour for the first 2,000 hours of use. No payment is required for usage beyond 2,000 hours. Furthermore, the maximum usage is not expected to exceed 4,000 hours per year.

The relationship of cost to usage is also displayed in tabular form in Table 1-1.

TABLE 1-1 TABULAR DESCRIPTION OF A RELATIONSHIP

Hours used (x)	Rental cost (y), in dollars
0	250,000
500	275,000
1,000	300,000
1,500	325,000
2,000	350,000
3,000	350,000
4,000	350,000

To express this relationship mathematically, we must define two *variables*. Let x denote or represent the number of hours used, and let y represent the resulting rental cost. Translating the given verbal description into mathematical form results in the following *mathematical model.*

MATHEMATICAL MODEL OF A RELATIONSHIP

$$y = 250{,}000 + 50x, \quad \text{if} \quad 0 \leqslant x \leqslant 2{,}000$$

and

$$y = 350{,}000, \quad \text{if} \quad 2{,}000 \leqslant x \leqslant 4{,}000$$

where x = hours used
y = rental cost in dollars

Note that there are two quantities in the formula $y = 250{,}000 + 50x$ that are free to vary or change. However, the value of y is dependent on the value of x. The dependence of a quantity (variable) on the value of another variable is what the mathematical concept of a *function* is all about.

The terms *dependent variable* and *independent variable* are used in connection with functions. The most common symbol for a function is

$$y = f(x)$$

where x is the independent variable
y is the dependent variable.

Returning to the computer example, it would be perfectly correct to say that y, the rental cost, *is a function of* the number of hours used, x.

A dependent variable y is said to be a function of an independent variable x, i.e., $y = f(x)$, if and only if a unique value of y *corresponds* to a given value of x. Putting it another way, a function can be thought of as a *rule* that assigns a unique y to each x. It is also true that a function can be thought of as a set of *ordered pairs* (x, y) with the property that, if (x, y) and (x, y^*) are in the set, then $y = y^*$.

This discussion of relationships between variables should help you to appreciate the next two definitions.

DEFINITION 1.1 *A symbol x that can represent any number in a set of real numbers is called a real variable.*

Other common symbols for a real variable are y, z, t, and w.

DEFINITION 1.2 *A real variable y is said to be a function of the real variable x if one and only one value of y corresponds to each value of x.*

Also some attention needs to be given to the question of which values x may assume, and to the resulting set of y values. The terms *domain* and *range* are used to denote these two sets, respectively.

The most common symbol for a function is $y = f(x)$. Note well, though, that the meaning of f changes from problem to problem. In this example, $f(x) = 250{,}000 + 50x$.

It is important to become familiar with *functional notation.* Specifically, the symbol $f(a)$, which denotes the value of y when $x = a$.

For example, in the computer example,

$$\begin{aligned} f(600) &= 250{,}000 + 50(600) \\ &= 250{,}000 + 30{,}000 \\ &= 280{,}000 \end{aligned}$$

Other common symbols for functions are $y = F(x)$, $w = g(z)$, $y = h(x)$.

Additional insights into a relationship between variables can often be obtained by graphing the mathematical model that is used to describe it. The graph of the relationship between rental cost and hours used for the computer example is shown in Figure 1-1.

In the preceding definition, both the domain and range are sets of real numbers. This is by no means the most general definition of a function. For example, the domain might consist of ordered pairs of real numbers. In this case the function could be written in the form $z = f(x, y)$.

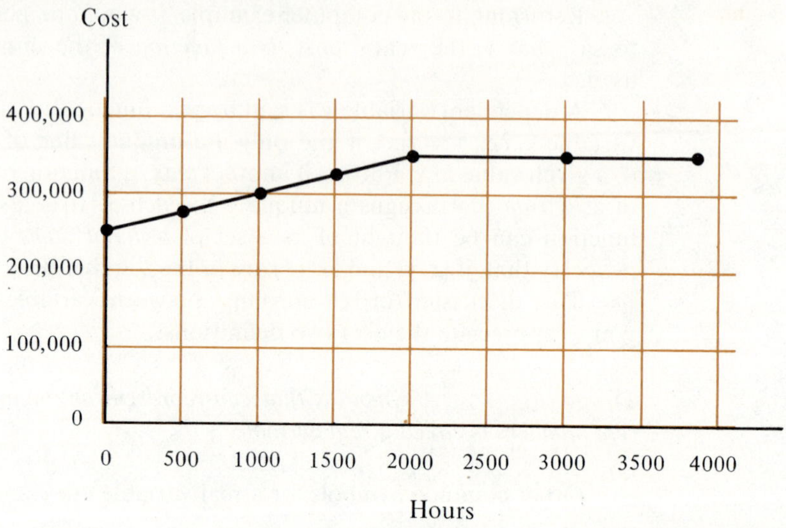

FIGURE 1-1

Example 1. Discuss the function $y = f(x)$ defined as follows: $y = 2x - 10$ for each x between 5 and 10, inclusive.

Solution: The domain of $y = f(x)$ is the set of real numbers between $x = 5$ and $x = 10$, inclusive.

x	5	6	7.2	8.5	9	10
y	0	2	4.4	7	8	10

We see that $f(5) = 0$, $f(6) = 2$, $f(7.2) = 4.4$, and so on.

The range of this function is the set of real numbers between $y = 0$ and $y = 10$, inclusive.

A function $y = f(x)$, where x and y are real variables, can be represented graphically on a rectangular coordinate system.

Example 2. Represent the function in Example 1 graphically.

Solution:

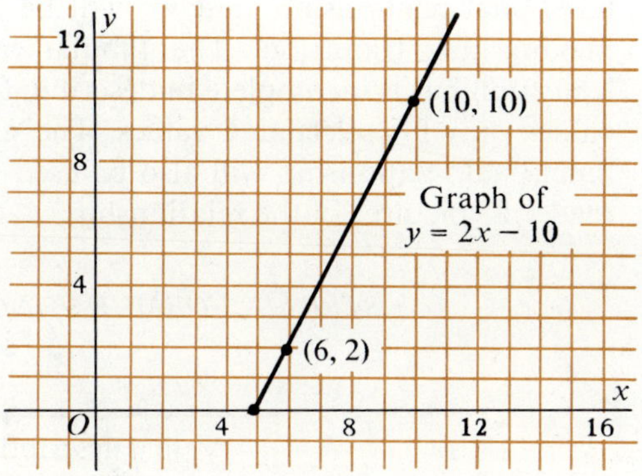

FIGURE 1-2

Example 3. Discuss and graph the function $y = x^2 - 2x + 1$ for x between -2 and 3, inclusive.

Solution: For convenience, we denote $x^2 - 2x + 1$ by $g(x)$. The domain of $y = g(x)$ is the set of x values between $x = -2$ and $x = 3$, inclusive. The range of $y = g(x)$ is the set of y values between $y = 0$ and $y = 9$, inclusive.

x	-2	-1	0	0.5	1	2	3
$g(x)$	9	4	1	0.25	0	1	4

$g(x) = x^2 - 2x + 1$

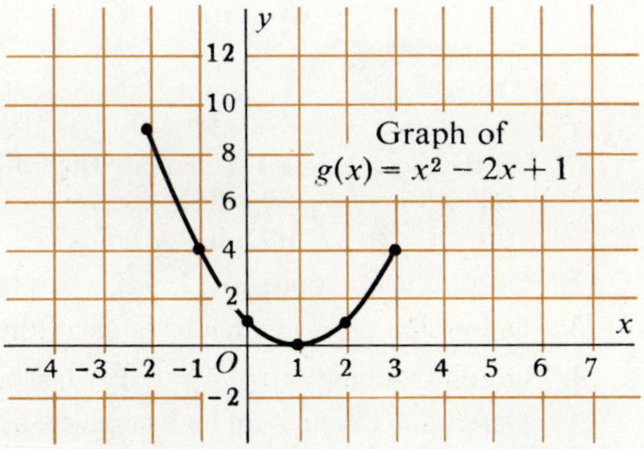

FIGURE 1-3

The function $y = x^2 - 2x + 1$ is an example of a quadratic function.

Verbal descriptions can often be cumbersome and may lack precision. We are aware that students often have trouble in making the transition from the verbal description to the *algebraic description* of a functional relationship, and we will be as explicit as possible when making this transition. The tabular approach often aids in the transition but is incomplete in describing a situation, in that it gives y values only for selected x values. The graphical representation of a functional relationship will also be used often in this text to give the reader a "picture" of the relationship.

STEPS IN FORMULATING A MODEL

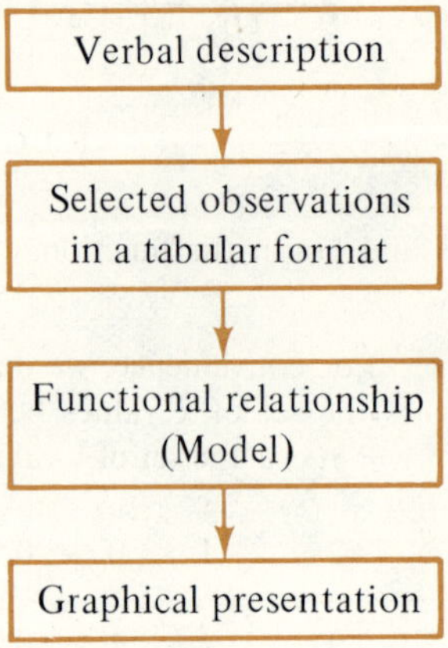

FIGURE 1-3a Steps in formulating a model.

PROBLEMS

1. Let $f(x) = x^2 + 2x + 1$. Calculate the following function values:
   (a) $f(0)$ **(c)** $f(3)$
   (b) $f(-2)$ **(d)** $f(2+h)$
2. Graph the function: $y = 3x - 2$, x lies between 4 and 9, inclusive.
3. Determine the domain and range of the function given in Prob. 2.
4. Graph the function: $y = x^2 - 1$, x lies between -1 and 2, inclusive.
5. Determine the domain and range of the function given in Prob. 4.
6. Graph the functions $y = 5x + 1$ and $y = 3x + 2$ on the same coordinate system.
7. Restate Definition 1.1 for a real variable u as a function of the real variable t.
8. Is $y = 350{,}000$ for $2000 \leqslant x \leqslant 4000$, a function of x?

9. What is the domain and the range of the function in Prob. 8?

10. If $y = 250{,}000 + 50x$ for $0 \leqslant x \leqslant 2000$, what is the domain and range?

11. Given that taxes are computed as 20 percent of taxable income, when taxable income is between \$0 and \$20,000.
- **(a)** Formulate a mathematical model to describe this relationship.
- **(b)** What are the variables in your model? Which is the dependent variable?
- **(c)** Specify the domain and range for the function used in part **a**.

1-2 LINEAR FUNCTIONS

Cost functions and rate structures can often be modeled or represented by linear functions. Applications of linear functions will be used extensively throughout the text; however, in this section the objective is simply to lay the foundation for these applications. More specifically, we will look at linear demand functions and linear cost functions in Sections 4 and 5 of this chapter.

Mathematical functions or models of the form

$$y = mx + b \tag{1.1}$$

or

$$y = a + bx \tag{1.2}$$

are treated in this section. Both of these forms represent a linear relationship between x and y.

The function $y = mx + b$ graphs as a straight line. If $x = 0$ in equation (1.1), then $y = b$. Therefore, the ordered pair $(0, b)$ represents the point of intersection of the y axis and the graph of $y = mx + b$. The number b gives the y intercept.

Example 1. Find the y intercept of the straight line with equation $2x - 3y - 12 = 0$.

Solution:

$$2x - 3y - 12 = 0$$
$$-3y = -2x + 12$$
$$y = (2/3)x - 4$$
$$= (2/3)x + (-4) \qquad (y = mx + b \text{ form})$$
$$b = -4$$

$\therefore$ the y intercept $= -4$

Next, we seek to interpret the geometric meaning of the coefficient m in the equation $y = mx + b$. If (x_1, y_1) is a point in the graph of (1.1) and (x_2, y_2) is second point on this graph, where $x_1 \neq x_2$, then the ratio $(y_2 - y_1)/(x_2 - x_1)$ is called the *slope* of the line in question. To see that m equals the slope of the graph of $y = mx + b$, let

$$y_1 = mx_1 + b \qquad \text{and} \qquad y_2 = mx_2 + b$$

Then

$$\begin{aligned} y_2 - y_1 &= mx_2 - mx_1 + (b - b) \\ &= m(x_2 - x_1) \end{aligned}$$

Therefore,

$$m = (y_2 - y_1)/(x_2 - x_1) \tag{1.3}$$

Example 2. Calculate the slope of the line with equation $y = 4x + 10$, using $x_1 = -2$ and $x_2 = 4$.

Solution:

$$y_2 = 4(4) + 10 \qquad y_1 = 4(-2) + 10$$

$$y_2 = 26 \qquad y_1 = 2$$

$$y_2 - y_1 = 24, \; x_2 - x_1 = 4 - (-2) = 6$$

$$\text{slope} = (y_2 - y_1)/(x_2 - x_1) = 24/6 = 4$$

Remark: This answer comes as no surprise, since $m = 4$ in $y = 4x + 10$ and m gives the slope of the graph of $y = mx + b$.

Geometrically speaking, the slope of a straight line is defined as the ratio of the *rise to the run*, i.e.

$$\text{slope} = \frac{\text{rise}}{\text{run}} \tag{1.4}$$

The relationship (1.4) is illustrated in Figure 1-4.

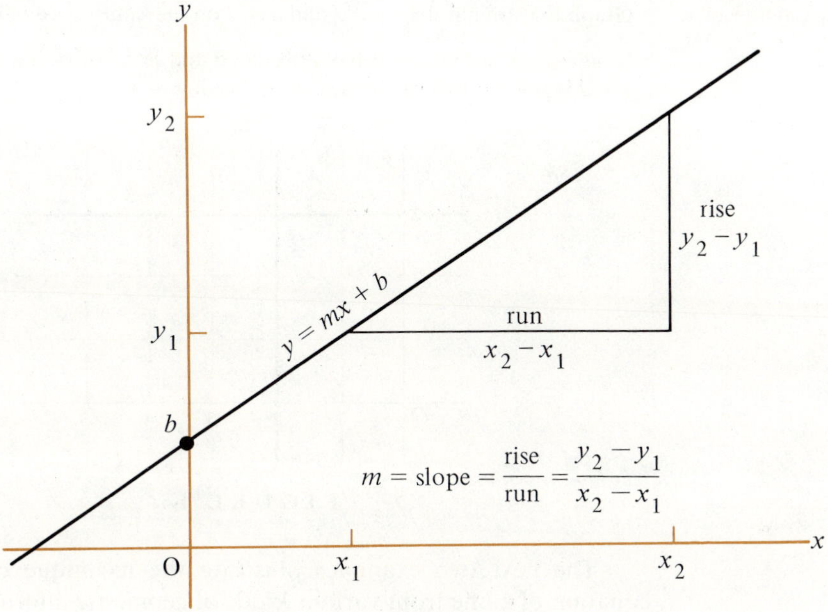

FIGURE 1-4

Example 3. Sketch the graphs of $y = 3x + 2$ and $y = 2x + 2$ on the same coordinate system.

Solution: Both lines have y intercept equal to 2. The slope of $y = 3x + 2$ is 3 and the slope of $y = 2x + 2$ is 2. This means that the first line will increase from 2 to 5 and the second line will increase from 2 to 4 if x is increased from 0 to 1. In other words, the first line passes through the points (0, 2) and (1, 5), and the second line passes through the points (0, 2) and (1, 4).

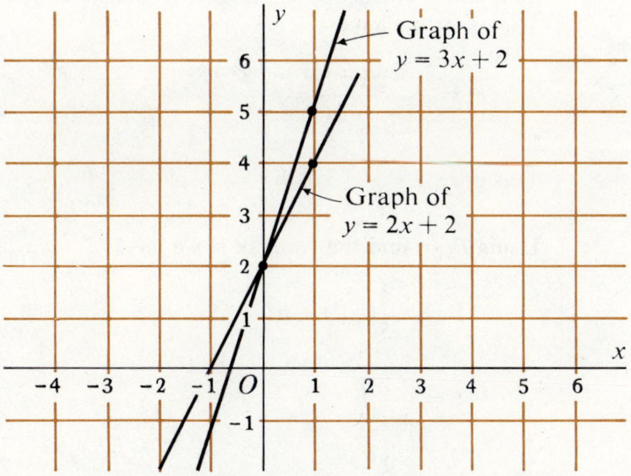

FIGURE 1-5

Example 4. Graph the straight lines $y = 3$ and $x = 2$ on the same coordinate system.

Solution: The line $y = 3$ has zero slope and is a horizontal line. The line $x = 2$ is vertical and its slope is not defined.

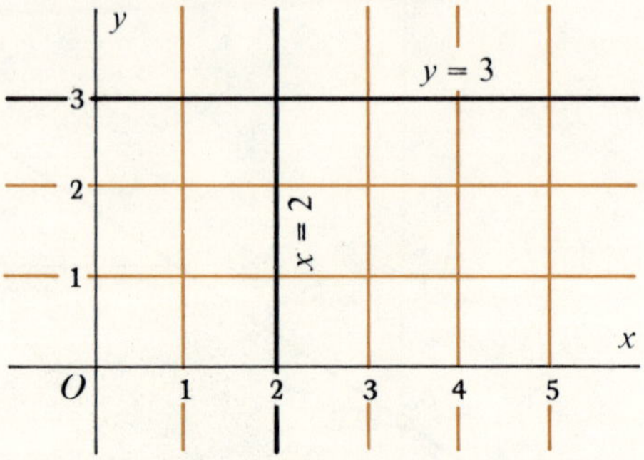

FIGURE 1-6

The next two examples illustrate the technique of finding an equation of a line from various kinds of geometric information.

Example 5. Find an equation of the line whose slope is 5 and whose y intercept is -2.

Solution: Start with the form $y = mx + b$; this means we have $m = 5$ and $b = -2$. Therefore, $y = 5x - 2$ is an equation for this line.

Example 6. Find an equation of the straight line that passes through the points $P_1\,(-2, -1)$ and $P_2\,(3, -4)$.

Solution: The slope of a line is measured by the ratio $y_2 - y_1/x_2 - x_1$. Therefore, we have

$$m = \frac{-4-(-1)}{3-(-2)} = -\frac{3}{5}$$

This gives $y = -\frac{3}{5}x + b$.

Using P_1 to find the value of b, we have

$$-1 = -\frac{3}{5}(-2) + b$$

$$-1 = \frac{6}{5} + b$$

$$b = -\frac{11}{5}$$

Hence

$$y = -\frac{3}{5}x - \frac{11}{5}$$

This can be rewritten in the form

$$3x + 5y + 11 = 0$$

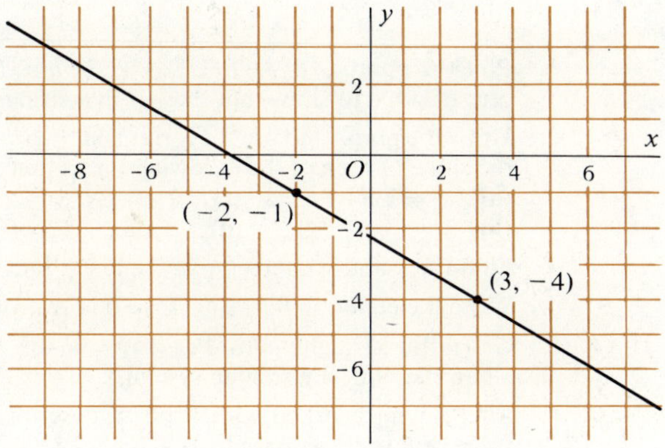

FIGURE 1-7

The equation, $y = mx + b$ (or $y = a + bx$), is expressed in the *slope-intercept* form. Another useful form for the equation of a straight line is

$$y - y_1 = m(x - x_1) \tag{1.5}$$

where m denotes the known slope and
(x_1, y_1) is a given point on the line whose equation is being sought.

This is called the *point-slope* form.

Example 7. Find an equation of the line with slope -3 that passes through the point $(-2, -1)$.

Solution: Start with the form

$$y - y_1 = m(x - x_1) \qquad \text{(point-slope form)}$$

$$x_1 = -2,\ y_1 = -1, \text{ and } m = -3$$

$$y - (-1) = -3[x - (-2)]$$

$$y + 1 = -3(x + 2)$$

or

$$y = -3x - 6 - 1$$

$$y = -3x - 7 \qquad \text{(slope-intercept form)}$$

The coefficient of x in $y = -3x - 7$ is -3, as expected from $m = -3$. The y intercept $= -7$.

PROBLEMS

1. Plot the points $A(0, 8)$, $B(1, 5)$, $C(2, 2)$, and $D(3, -1)$, and use the concept of slope to show that they lie in a straight line.

2. Draw the graph and use the formula $m = (y_2 - y_1)/(x_2 - x_1)$ to determine the slope of each of the following linear functions:
(a) $y = 2x + 3$ **(b)** $y = 3x - 3$ **(c)** $y = -2x + 5$ **(d)** $y - 2x - 6 = 0$ **(e)** $x - 2y - 3 = 0$

3. Find the x and y intercepts of the linear functions in Prob. 2.

4. Write the equation $9x + 3y + 15 = 0$ in the form $y = mx + b$.

5. Graph the straight lines $4x + 4.5y = P$ for $P = 900$, 1600, 1800, 3600, all on the same coordinate system.

6. After having worked with Prob. 5, describe the effect of changing the value of P in the equation $Ax + By = P$.

7. Find an equation for the line in Prob. 1.

8. Find an equation for each of the following straight lines:
(a) The slope of the line is 3 and its y intercept is -5.
The line passes through the point $(2, 5)$ and has slope 2.
(c) The line passes through the point $(-2, 3)$ and has y intercept 2.
(d) The line has slope 4 and x intercept 3.
(e) The line passes through the two points $(0, 3)$ and $(1, -2)$.
(f) The line passes through $(10, -3)$ and $(2, 5)$.

9. Let $A(-2, 4)$ and $B(3, -6)$ be two points in the xy-plane.
(a) Plot A and B.
(b) Draw the line L passing through A and B.
(c) Determine the slope of the line L.
(d) Find an equation for the line L.

10. Find an equation for each of the following straight lines:
(a) The line has slope 5 and x intercept -2.
(b) The line passes through the two points $(-3, 4)$ and $(2, -6)$.

11. We know that two straight lines in a plane are parallel if their slopes are equal. It can also be shown that two straight lines in a plane are perpendicular (at right angles) if the product of their slopes is -1. Given this information, determine which of the following pairs of lines are (1) parallel; (2) perpendicular; or (3) neither:
(a) $2x + 3y = 7$, $4x + 6y = 15$
(b) $x + 3y = 12$, $-3x + y = 5$
(c) $4x + 5y - 6 = 0$, $3x + 2y + 4 = 0$
(d) $2x + y = 4$, $x + 3y = 6$
(e) $2x + 6y = 4$, $-3x + y = -2$
(f) $-2x + 5y = 8$, $y = 0.4x + 3$

12. Graph each pair of lines in Prob. 11 on a separate coordinate system.

13. The formula for the distance between two points in a plane is developed next. Let $A(x_2, y_2)$ and $B(x_1, y_1)$ be given. Form the supplementary point $D(x_2, y_1)$ as, for example, in Figure 1-8. Applying the Pythagorean theorem of plane geometry to triangle ABD gives $d^2 = a^2 + b^2$, where $a = x_2 - x_1$ and $b = y_2 - y_1$; therefore, $d = \sqrt{(x_2 - x_1)^2 + (y_2 - y_1)^2}$, where d is the distance between A and B. Use this *distance formula* to compute the distance between the following pairs of points:

(a) (2, 1) and (5, 5)
(b) $(-1, 2)$ and (3, 5)
(c) (0, 1) and $(4, -2)$
(d) (2, 0) and $(-4, 1)$
(e) $(-1, -2)$ and (3, 4)
(f) (0, 0) and (a, b)

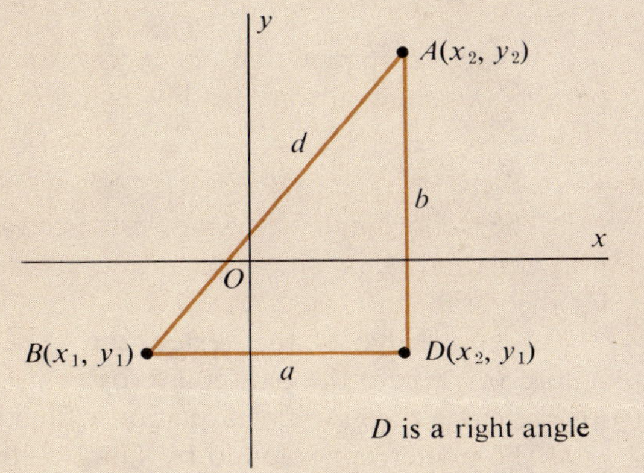

FIGURE 1-8

14. Given the points $A(-1, 15)$, $B(1, 1)$, and $C(5, 3)$;
(a) Plot the points and draw the triangle ABC.
(b) Compute the slopes of BC and AC.
(c) Is triangle ABC a right triangle?
(d) Verify the Pythagorean theorem by applying the distance formula to the sides of this triangle.

15. Given the points $A(1, 2)$, $B(2, 3)$, and $C(3, 5)$:
(a) Plot the points.
(b) Use the results of Prob. 11 to determine if ABC is a right triangle.
(c) Use the distance formula and the Pythagorean theorem to determine if ABC is a right triangle. *Hint:* ABC is a right triangle if and only if $c^2 = a^2 + b^2$, where C is the right angle.

1-3 QUADRATIC FUNCTIONS

Many business and economic relationships can be described exactly or approximately by a linear function of the form $y = mx + b$. Others require a nonlinear function to describe them adequately. A class of nonlinear models called *quadratic functions* is discussed in this section. The mathematical and graphical properties of quadratic functions are stressed here; the next section shows an application.

A *quadratic function* can be written in the form

$$y = ax^2 + bx + c \tag{1.6}$$

where $a \neq 0$.

A function of the form $y = ax^2 + bx + c$ graphs as a *parabola* that opens upward or downward. A parabola that opens upward or downward has a vertical *axis of symmetry* and a turning point called a *vertex*. Typical graphs of parabolas are shown in Figures 1-9 and 1-10.

It can be shown that the x coordinate, h, of the vertex of a parabola opening upward or downward is given by

$$h = -b/2a \tag{1.7}$$

The y coordinate, k, of the vertex can be found by substituting h into equation (1.6). The vertical line, $x = h$, is the axis of symmetry for the parabola.

A knowledge of the vertex, the axis of symmetry, and points (intercepts) where the parabola crosses or touches the axes is useful for sketching the graph of a quadratic function.

The y intercept is found by simply setting $x = 0$ and calculating the corresponding y value.

Example 1. Find the vertex and the y intercept of the parabola $y = x^2 - 9$.

Solution: $a = 1$, $b = 0$, $c = -9$

The x coordinate of the vertex is

$$x = -b/2a = -0/2(1) = 0$$

If $x = 0$, then

$$y = (0)^2 - 9$$

$$y = -9$$

$\therefore$ the vertex is $(0, -9)$.
The y intercept is also -9 in this example.

To find the points where the graph crosses or touches the x axis we must find those values of x that will make $y = 0$. These values of x are called the *zeros* of the quadratic function. The formula for finding the values of x that make $ax^2 + bx + c = 0$ a true statement is called the quadratic formula and it is discussed in the preparatory chapter.

The quadratic formula for the solution of $ax^2+bx+c=0$ is

$$x = \frac{-b \pm \sqrt{b^2-4ac}}{2a}$$

Example 2. Solve the equation $x^2-9=0$ for x, i.e., find the x intercept.

Solution: $a=1$, $b=0$, $c=-9$

$$x = \frac{-0 \pm \sqrt{0^2-4(1)(-9)}}{2(1)}$$

$$x = \frac{\pm\sqrt{36}}{2} = \frac{\pm 6}{2} = \pm 3$$

In Example 2 there are two values of x that make $y=0$. They are $x=+3$ and $x=-3$. These values of x are the *zeros* of the function $f(x)=x^2-9$. Alternatively, they are the *roots* of the quadratic equation $x^2-9=0$.

Any quadratic function of the form $y=ax^2+bx+c$ has either *two zeros, one zero, or no zero.* The three possibilities are illustrated graphically in Figure 1-9.

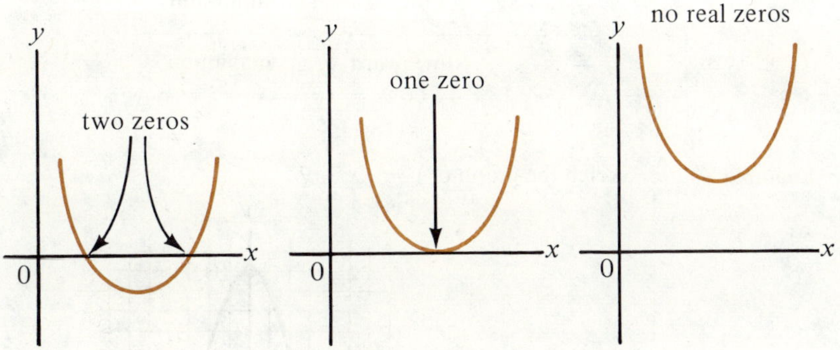

FIGURE 1-9

Example 3. Using the information obtained in Examples 1 and 2, sketch the graph of $y=x^2-9$.

Solution: Vertex at (0,9), y intercept $=-9$, axis of symmetry: $x=0$, x intercepts $=\pm 3$.

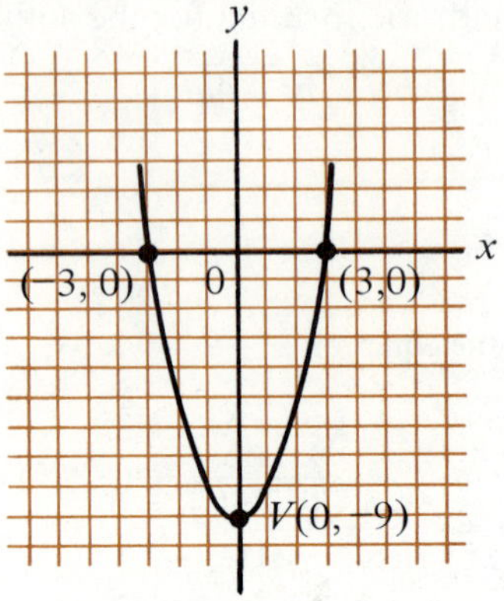

FIGURE 1-10

For a parabola opening upward or downward the vertex represents a *minimum* or *maximum* value, respectively, of the dependent variable, y. The relationship between the sign of a in the quadratic function $y = ax^2 + bx + c$ and the geometric properties of its graph are shown in Table 1-2.

TABLE 1-2

Sign of a	Parabola opens	Vertex is a
$+$	upward	minimum
$-$	downward	maximum

Example 4. Sketch the graph of $y = -x^2 + 9$.

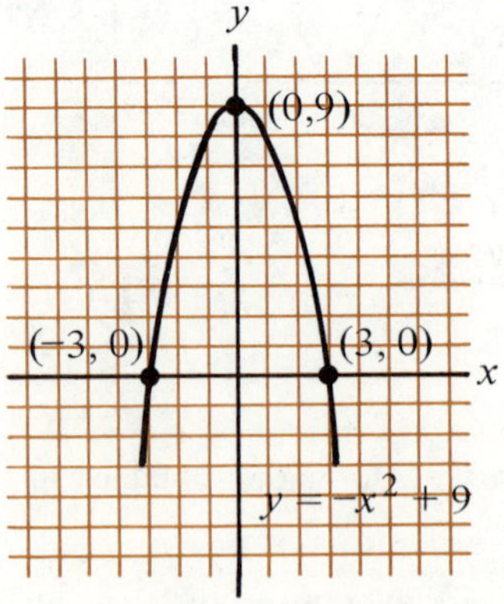

FIGURE 1-11

Solution: $a = -1$, therefore, the parabola opens downward.

The vertex is at (0, 9).
The x intercepts can be found by setting $9 - x^2$ equal to 0 and solving for x. The solutions are $x = \pm 3$.

Example 5. Sketch the graph of the following parabola:

$$y = x^2 - 6x + 9$$

Solution: Determination of vertex.

$$x = \frac{-(-6)}{2} = 3$$

$$y = 3^2 - 6(3) + 9$$

$$y = 9 - 18 + 9$$

$$y = 0$$

Vertex at (3, 0)

Determination of y intercept.

$$y = 0^2 - 6(0) + 9$$

$$y = 9$$

Determination of x intercept.

$$x = \frac{-(-6) \pm \sqrt{-6^2 - 4(1)(9)}}{2(1)}$$

$$x = \frac{6 \pm \sqrt{36 - 36}}{2}$$

$$x = 3$$

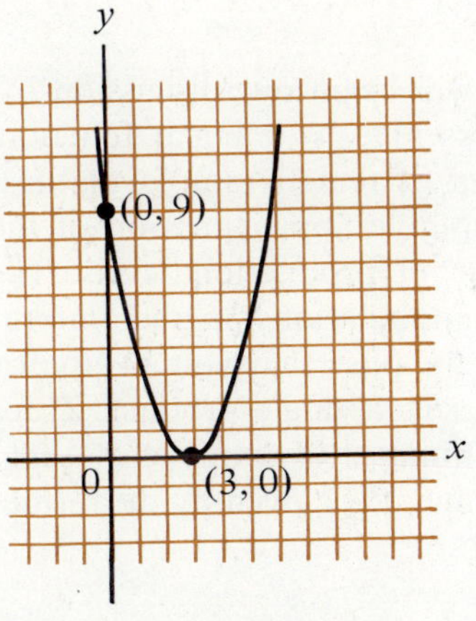

FIGURE 1-12

PROBLEMS

1. Is $y = x^2+3x+7$ a quadratic function of x?
2. Is $y = (x+3)(x-2)$ a quadratic function of x?
3. Is $y = 7x+3$ a quadratic function of x?
4. Is $y = 4-x^2$ a quadratic function of x?
5. Is $y = 2x^3+x-3$ a quadratic function of x?
6. Find the coordinates of the vertex of the parabola whose equation is $y = 4x^2+3x-5$.
7. Find the coordinates of the vertex of the parabola whose equation is $y = x^2+8x+2$.
8. Find the coordinates of the vertex of the parabola whose equation is $y = 2x^2+6x-3$.
9. Find the coordinates of the vertex of the parabola whose equation is $2y = 6x^2-4x+2$.
10. Find the coordinates of the vertex of the parabola whose equation is $y-x^2+10x-2 = 0$.
11. Graph the function $y = x^2+3x+7$.
12. Graph the function $y = (x+3)(x-2)$.
13. Graph the function $y = (x-3)^2$.
14. Graph the function $y = x^2-6x+12$.
15. Graph the function $y = x^2+6x+3$.
16. Find the maximum value of y if $y = 4-x^2$.
17. Find the maximum value of y if $y = -x^2+6x-2$.
18. Find the minimum value of y if $y = x^2-4$.
19. Find the minimum value of y if $y = 2x^2+12x-10$.
20. Find the minimum value of y if $3y = 9x^2-12x+15$.

1-4 DEMAND AND REVENUE FUNCTIONS

We will begin our discussion by assuming that total revenue can be represented by a linear function. This assumption of linearity leads to an extremely simple mathematical analysis that often gives an acceptable approximation of the real world situation for a limited range of production levels. Then, we will look at the underlying demand relationship and a more realistic revenue function.

Suppose that you are the owner of a small business that makes and sells a single product, Z, for a constant price, p. If x represents the number of units of Z that are made and sold, then the total revenue, R, is given by the formula:

$$R = px$$

where R denotes total revenue,
p denotes a constant price per unit of Z, and
x denotes the number of units of Z that are made and sold.

To be more specific, let's assume that $p = \$2.00$. Then $R = \$2.00x$. Now there are two quantities in the formula $R = \$2.00x$ that are free to change. However, one quantity is always dependent on the current value of the other quantity. The dependence of one quantity (variable) on the value of one or more variables is what the mathematical concept of a function is all about.

Example 1. You plan to sell grinders during the lunch period. Set up the weekly revenue function for your proposed enterprise assuming a selling price of $1.40 each, and calculate the weekly revenue if 400 grinders are sold.

Solution:

$R = px$

$R = \$1.40x$ (*Revenue function*)

$R = \$1.40(400)$ ($x = 400$)

$R = \$560.00$

Example 2. Graph the revenue function for the grinder enterprise described in the preceding example.

Solution:

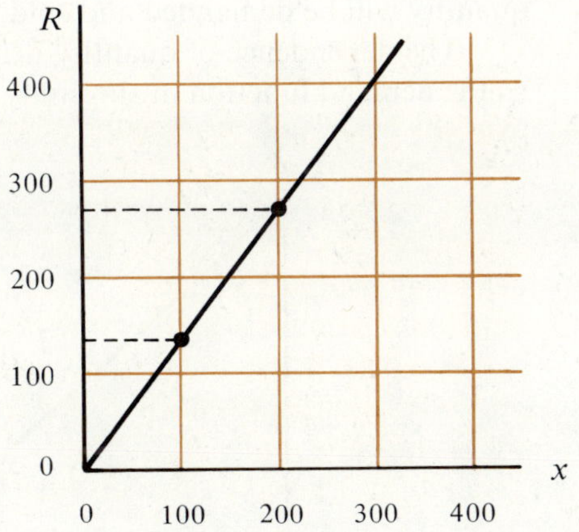

FIGURE 1-13

The revenue function is a linear function with slope, p, and R intercept, 0. Note that the higher the price, p, the steeper the graph of $R = px$ will be. Also, it should be clear that the function shown as Figure 1-13 is applicable only when $x \geqslant 0$ (i.e., one cannot sell a negative amount of grinders).

To gain further insight on the topic of total revenue we will look at the underlying demand relationship. The demand relationship for the grinder enterprise discussed in Examples 1 and 2 can be illustrated graphically as follows:

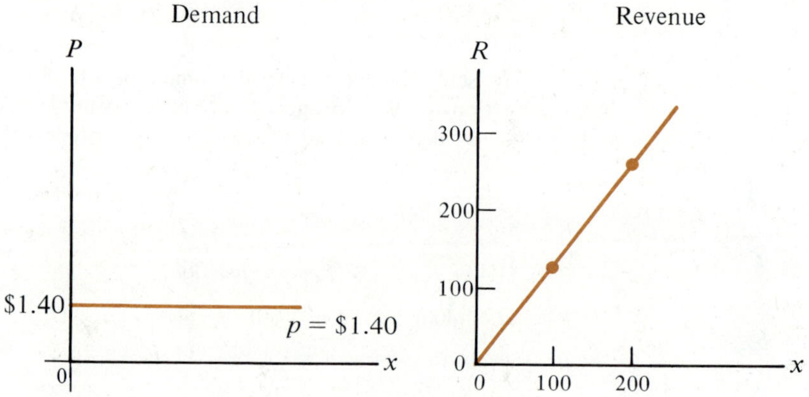

FIGURE 1-14

Note that to find total revenue one simply multiplies the price (p) by the quantity x to obtain $R = px$.

More realistically, one would expect demand to graph as in Figure 1-15, where one can see that as the price is lowered a greater quantity will be demanded and sold.

The dependence of quantity demanded (x) on price is referred to as the demand function in the study of economics. The reader should

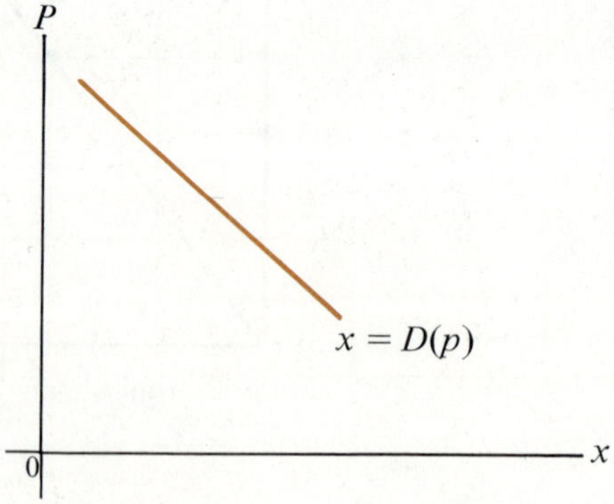

note that in Figure 1-15, the horizontal axis is used for the dependent variable quantity demanded and sold, and the vertical axis for the independent variable price. The purpose is to be consistent with the convention that economists use. The reader is reminded that this is the reverse of the convention used in mathematics!

The relationship described in Figure 1-15 can also be expressed as follows:

$$p = ax + b$$

where a and b are constants.

To obtain total revenue one would simply multiply price (p) by the quantity sold (x).

$$R = px$$

$$R = (ax+b)x, \qquad p = ax + b$$

$$R = ax^2 + bx$$

We saw in the previous section that the mathematical relationship $y = ax^2+bx+c$ is called a quadratic function and its graph is a parabola. The value of c is zero in this particular application. Note that the total revenue function derived is a quadratic function, and, therefore, graphs as a parabola.

Example 3. The demand for a particular item is represented by the following equation:

$$p = -2x + 10$$

Find the resulting revenue function, and graph the demand and revenue functions for $0 \leqslant x \leqslant 5$.

Solution:

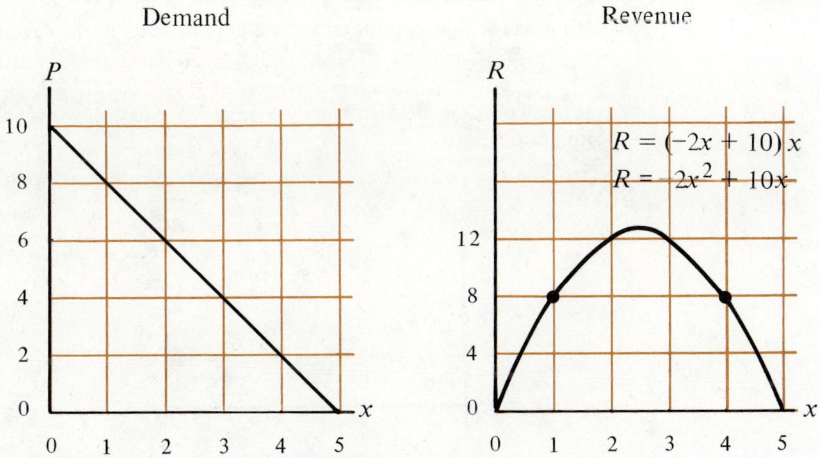

FIGURE 1-16

PROBLEMS

1. Given the following sales (i.e. revenue) data for a particular publication:
(a) What is the price per copy?
(b) What is the revenue function?

Number of copies	1000	1200	1300
Revenue	\$1500	\$1800	\$1950

2. Keeping in mind the demand relationship, $p = ax + b$, what condition must be imposed on the constant a in the revenue function $R = ax^2 + bx + c$? Does this have any bearing on whether or not the parabola opens upward or downward?

3. An appliance manufacturer believes that he will be able to sell 100,000 blenders at \$40 this Christmas season, but that 500,000 can be sold if the price is lowered to \$20. It can be assumed that the demand function is linear.
(a) What is the demand function?
(b) Graph the demand function.

4. What is the revenue function for the manufacturer described in Problem 3?
(a) Graph the revenue function.

5. Given that total revenue is equal to $-2x^2 + 120x$ where x is units sold.
(a) Graph the revenue function for $10 \leqslant x \leqslant 50$.
(b) At what quantity is revenue maximized?

6. If total revenue is known to be $R = -2x^2 + 120x$ as it is in Prob. 5, how does the quantity sold (x) depend on p? In other words, determine the linear demand function that is implied by the given total revenue function. *Hint*: $R = px$.

7. Given that total revenue is: $R = 90x - 3x^2$, where x is units sold.
(a) Graph the revenue function for those x values where R is positive.
(b) At what quantity is revenue maximized?

8. What is the demand function that corresponds to the revenue function?

9. Why would it be incorrect to refer to the following as a demand function?

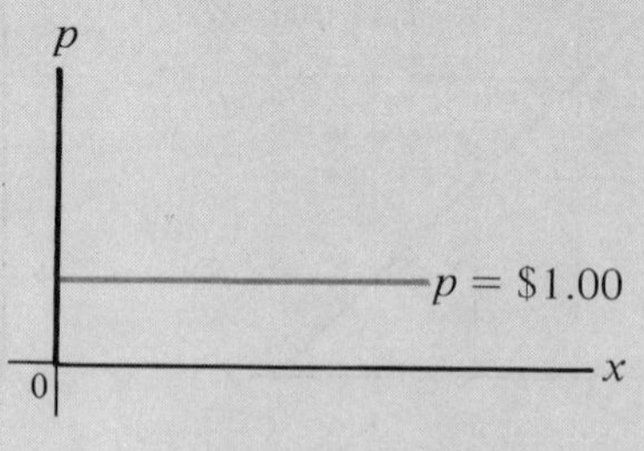

Hint: Recall that the dependent variable is on the horizontal axis.

10. Given the figures in Table 1-3.
(a) Plot the data with price plotted on the y axis and quantity demanded plotted on the x axis.
(b) What is the equation of the demand curve?
(c) What is the revenue function?
(d) At what quantity will total revenue be maximized?

TABLE 1-3

Total Quantity Demanded per Week	Price
2,000	$5
4,000	4
6,000	3
8,000	2
10,000	1

1-5 COST FUNCTIONS

In the previous section we looked at revenue functions and the underlying demand relationships. One might argue that it is easier to study and model the dollar outflows in a company or the cost of running a business. Cost behavior patterns are functional relationships between cost and level of production. The independent variable, production level, x, is on the horizontal axis. The dependent variable, cost, C, is on the vertical axis.

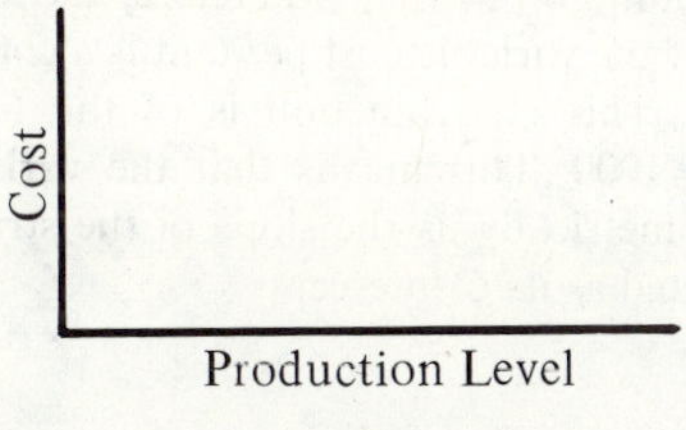

FIGURE 1-17

The ability to model cost allows one to control day-to-day operations as well as to facilitate long range planning and decision making.

Costs that remain relatively constant regardless of the level of activity are called *fixed costs.* Costs that are proportional to output are called *variable costs.* The sum of the variable cost and the fixed cost that results from producing x units is called the *total cost.*

$$\text{Total Cost} = \text{Variable Cost} + \text{Fixed Cost}$$

$$C = VC + F$$

More specifically, variable costs are obtained by multiplying *variable cost* per unit (c) by the number of units produced (x).

$$C = cx + F$$

Note that in the preceding equation for total cost, we are assuming that total cost can be represented by a linear function. This is often the case for a limited range of production.

Example 1. The total operating cost, C, of a certain firm is given as a function of the production level, x, by the linear function $C = 0.72x + 1000$. C is in dollars, and x is in units produced. Discuss the variable cost and the fixed cost for this firm. Interpret this situation geometrically.

Solution: We see from the definitions that the fixed cost is \$1000. Also, the variable cost is $0.72x$. The variable cost per unit can be obtained by dividing the variable cost by the number of units produced. This involves dividing $0.72x$ by x; therefore, the variable cost per unit is 0.72. We note that the variable cost per unit is a constant in this case.

This cost function is of the form $C = mx + b$, where $m = 0.72$ and $b = 1000$. This means that the variable cost per unit can be interpreted geometrically as the slope of the straight line, and fixed cost can be interpreted as its C intercept.

Example 2. Assume that you are in charge of a plant that has a fixed cost of \$1000, and a cost of \$3.10 is incurred for each additional unit of production. Express total cost, C, as a linear function of x, the number of units produced.

Solution: $C = cx + F$, where c is the cost per unit and F is fixed cost.

$$C = (3.10)x + 1000$$

Example 3. Graph the cost function for the plant described in Example 2.

Solution:

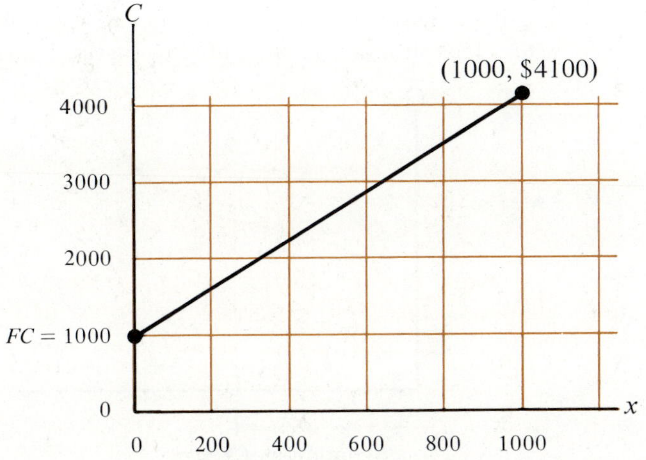

FIGURE 1-18

In the preceding examples we assumed that the variable cost per unit was constant. In many situations this variable cost per unit will become smaller as the value for x is increased. This is precisely the case that an individual faces in terms of utility costs. Consider the following electric rate schedule.

First 5,000 kwh, \$0.20 per 100 kwh

Above 5,000 kwh, \$0.10 per 100 kwh

Graphically, the cost schedule would appear as follows:

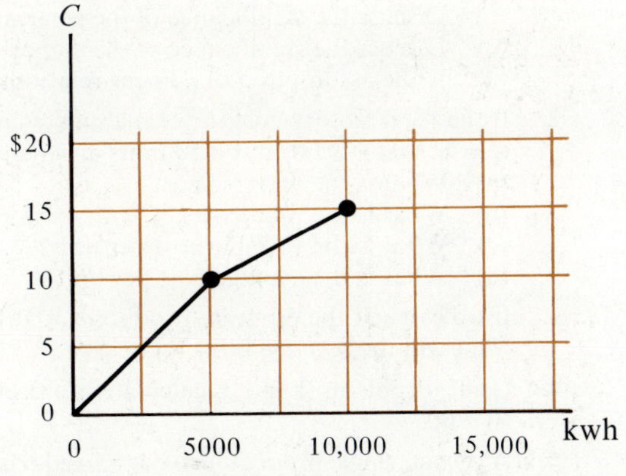

FIGURE 1-19

In many production situations, the variable cost per unit will decrease as a certain level of production is approached. This phenomenon is referred to as *economies of scale*. Beyond that level of production, the opposite happens. For example, the variable cost per unit begins to increase and one incurs what is described as *diseconomies of scale*. This situation is illustrated in Figure 1-20.

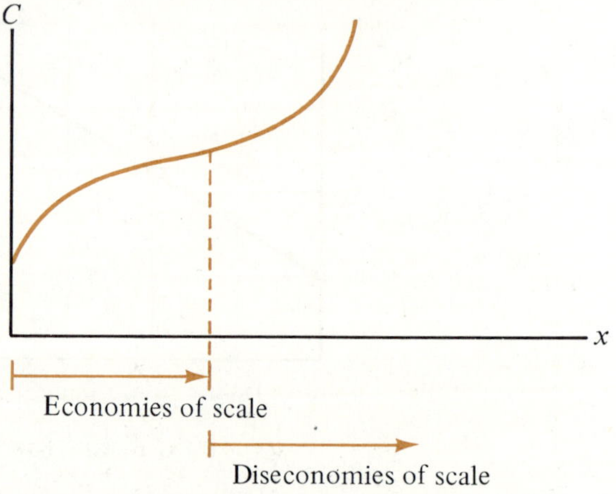

FIGURE 1-20

It should be apparent from studying Figure 1-20 that cost can be modeled by a linear function for a certain range of production.

PROBLEMS

1. Given that the cost-volume relationship is linear, the production cost for 100 units is \$2550, and the production cost for 300 units is \$2650.
  (a) Find the cost function.
  (b) Discuss the significance of the vertical intercept.
  (c) Discuss the significance of the slope.
  (d) Is the assumption of a linear relationship a good one?

2. If the total factory cost, C, of making x units of a product is given by $C = \$3.00x + \$200$, and if 50 units are made:
  (a) What is the variable cost?
  (b) What is the total cost?
  (c) What is the variable cost per unit?
  (d) What is the average cost per unit?

3. If we can sell the 50 units produced in Problem 2 at a price of \$8.00, what will the revenue be?

4. Using Problems 2 and 3, calculate the profit resulting from the sale of 50 units.

5. A certain single product firm has a fixed cost of \$300, a variable cost per unit of \$2.00 and sells the product for \$3.00 per unit. Find the total cost function and the total revenue function for this firm.

6. If you are a salesman and are paid a salary of $400 a month and a commission of 4% on your dollar sales for the month, how would you express your monthly pay as a linear function? What would your earnings be this month if you sold $5000 worth of merchandise?

7. Given the following data:
   (a) Find the cost function.
   (b) Graph the cost function.

Level of production	0	100	200	300
Total cost	$4000	4800	5600	6400

8. Given that the total cost of producing 500 units is $6500 and that when 700 units were produced the total cost was $8500, find the linear cost function for this data.

9. Should one expect the variable cost per unit to be a constant always?

10. The water rates per quarter for the city of Nice and Green are:

    For the first 5000 cubic feet or less, $0.34 per 100 cubic feet.
    For the next 10,000 cubic feet or less, $0.28 per 100 cubic feet.
    For the next 35,000 cubic feet or less, $0.22 per 100 cubic feet.
    For the next 250,000 cubic feet or less, $0.13 per 100 cubic feet.
    For use of over 300,000 cubic feet, the rate for the entire quantity used is $0.13 per 100 cubic feet.

    Express this rate structure graphically using the vertical axis for total cost.

11. Discuss the implications of the rate structure in Problem 10 for users near the 300,000 cubic foot level.

12. The electric rates for the city of Light and Cool are:

    First 6,000 kwh, 0.12 per 100 kwh
    Next 12,000 kwh, 0.10 per 100 kwh
    Above 18,000 kwh, 0.08 per 100 kwh

    Minimum billing per period is $3.00. Express this rate structure graphically using the vertical axis for total cost.

13. Assume that Table 1-4 gives the minimum commission schedule on the New York Stock Exchange for single round lot orders. Subject to the provision that the minimum commission on single round lot orders not exceed $65.00
    (a) Express the commission structure graphically using the vertical axis for commission.
    (b) At what dollar sale is there no further increase in commission, i.e., at what point does the graph become horizontal?

TABLE 1-4 SINGLE ROUND LOT ORDERS

Money Involved in the Order	Minimum Commission
$100–$799.99	2.0% + $6.40
800–2499.99	1.3% + 12.00
2500 and over	0.9% + 22.00

14. Assume that Table 1-5 gives the minimum commission schedule on the New York Stock Exchange for multiple round lot orders.

TABLE 1-5 MULTIPLE ROUND LOT ORDERS

Money Involved in the Order	Minimum Commission
\$100–\$2,499.99	1.3% + \$12.00
2,500–19,999.99	0.9% + 22.00
20,000–29,999.99	0.6% + 82.00
30,000–500,000	0.4% + 142.00

plus

First to tenth round lot, \$6.00 per round lot
Eleventh round lot and over, \$4.00 per round lot

An investor plans to buy 200 shares of stock at \$100.00 per share. He can purchase these shares either (1) all on one day; or (2) 100 shares per day for two days. What would he save (if anything) in commission by buying two round lots of 100 shares each instead of buying a multiple round lot of 200 shares? Consult Problem 13 for needed information. Assume no change in the stock price over the two days.

15. Repeat Problem 14 with the price per share at \$20.00.

16. A college contracts with a time-sharing company for computer time at a cost of \$10.00 per hour. The minimum cost per month is \$100.00, i.e., the first ten hours are considered individually applied to the minimum. Graph this cost function with the vertical axis as cost.

17. Repeat Problem 16 for a maximum cost per month of \$400.00.

18. A supplier agrees to furnish a certain item at the following prices:

\$10.00 per unit for any order of 100 units or less
9.00 per unit for any order from 101 to 200 units
7.50 per unit for any order in excess of 200 units

(a) Graph this function with cost plotted on the vertical axis.
(b) Is it cheaper to buy 105 units than 98 units?

19. Shopping for your Thanksgiving turkey you find the following prices:

Under 10 pounds, \$0.90 per pound
10 to 20 pounds, 0.80 per pound
Over 20 pounds, 0.60 per pound

(a) Graph your cost function with the vertical axis as cost.
(b) Is it true that you can buy a 24-pound turkey for the same price as an 18-pound turkey?

1-6 BREAKEVEN ANALYSIS

In Sections 4 and 5 of this chapter we looked at revenue and cost functions separately; in this section we will look at these functions together. In particular, we will look at the point or points where these two functions intersect. This point is referred to as the breakeven point. When the functions are illustrated graphically the chart, it is called a breakeven chart.

A breakeven chart is a simple mathematical model based on the distinction between fixed and variable costs. This distinction first became of practical importance during the Industrial Revolution and was recognized by a number of nineteenth-century economists. The breakeven chart was developed independently by the American engineers Hess, Knoeppel, Rautenstrauch, and Williams, and their associates, and by the Scottish accountant Sir John Mann.*

Example 1. Given the following data, construct the cost and revenue functions and find the breakeven point. Illustrate graphically.

Fixed cost = \$1500

Variable cost per unit = \$0.75

Selling price per unit = \$1.50

Solution: Let x denote the number of units. Then Total Cost, C, equals $cx+F$.

$$C = (0.75)x + 1500$$

$$R = px \qquad \text{where } R \text{ is the revenue}$$

$$R = 1.50x$$

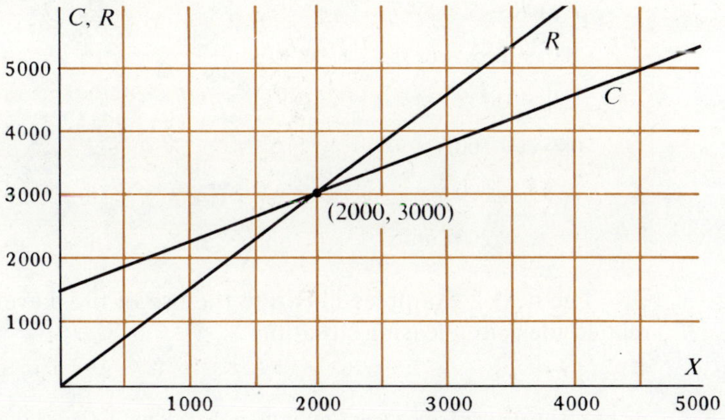

FIGURE 1-21

* R. H. Parker, *Management Accounting: An Historical perspective*, p. 72, Augustus M. Kelley Publishers, New York, 1969.

To find the breakeven point, we set R equal to C and solve for x.

$$R = C$$
$$1.5x = 0.75x + 1500$$
$$0.75x = 1500$$
$$x = 2000$$

To find the other coordinate of the breakeven point, substitute $x = 2000$ into $R = 1.5x$. This gives $R = 1.5(2000) = \$3000$.

As the reader no doubt expects, the *profit* is obtained by subtracting total cost from total revenue.

$$\text{Profit} = \text{Total Revenue} - \text{Total Cost}$$
$$P = R - C$$

It should be clear that $P = R - C = 0$ at a breakeven point. Since the businessman is interested in making a profit and not just breaking even, the desired profit would normally be set equal to something other than zero.

Example 2. Using the functions of Example 1, find the number of units that must be sold in order to make a profit of \$500.

Solution:

$$P = R - C$$
$$500 = 1.5x - (0.75x + 1500)$$
$$500 = 0.75x - 1500$$
$$0.75x = 500 + 1500$$
$$0.75x = 2000$$
$$x = 2667 \text{ units}$$

The next 2 examples illustrate the use of breakeven analysis and related ideas in a leasing situation.

Example 3. The management of a certain firm is faced with the choice of leasing a machine from manufacturer A at an annual cost of \$3000 plus \$0.02 per unit produced on this machine or leasing from manufacturer B at an annual cost of \$5250 plus \$0.015 per unit produced. Calculate the number of units that must be sold to break even in each case. Each unit can be sold for \$0.05.

Solution:

Manufacturer A	*Manufacturer B*
$R = C$	$R = C$
$0.05x = 3000 + 0.02x$	$0.05x = 5250 + 0.015x$
$0.03x = 3000$	$0.035x = 5250$
$x = 100{,}000$	$x = 150{,}000$

If the anticipated sales are 150,000 units, for example, the firm should lease from manufacturer A and make a profit. We note that leasing the machine from manufacturer B would result in a breakeven situation.

Example 4. What decision should be made if anticipated sales are **(a)** 300,000 units, and **(b)** 700,000?

Solution: In both cases we are past the breakeven level. Since revenue is the same for both machines the decision will be based on the total cost involved in leasing from either manufacturer A or manufacturer B.

If the anticipated sales are 300,000 units,

Manufacturer A	*Manufacturer B*
$C = 3000 + 300{,}000(0.02)$	$C = 5250 + 300{,}000(0.015)$
$= 3000 + 6000$	$= 5250 + 4500$
$C = 9000$	$C = 9750$

Hence, the machine should be leased from manufacturer A.

If the anticipated sales are 700,000 units,

Manufacturer A	*Manufacturer B*
$C = 17{,}000$	$C = 15{,}750$

Hence, the machine should be leased from manufacturer B.

In the breakeven situation of Example 1 we assumed that both the cost and revenue functions were linear. A somewhat more realistic situation is presented in Example 5, where the revenue function is a quadratic.

Example 5. Illustrate the following cost and revenue functions graphically, and find the breakeven points:

$$C = 1x + 40, \qquad R = -0.2x^2 + 10x$$

Solution:

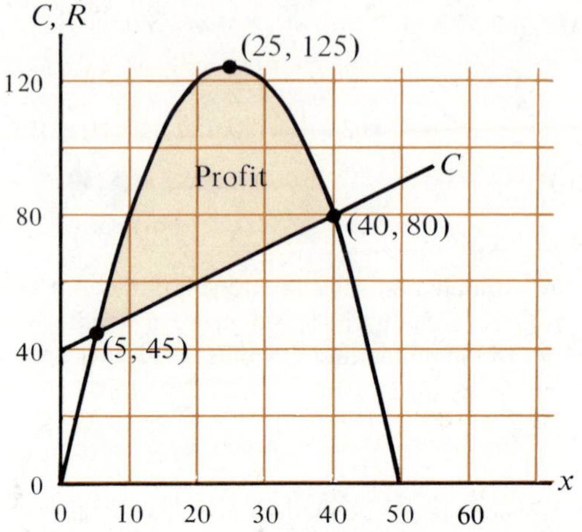

FIGURE 1-22

$$P = R - C$$

$$P = -0.2x^2 + 10x - (1x+40)$$

$$P = -0.2x^2 + 9x - 40$$

Using the quadratic formula to find those values for x that cause $P = 0$, we have:

$$x = \frac{-9 \pm \sqrt{(9)^2 - 4(-0.2)(-40)}}{2(-0.2)}$$

$$x = \frac{-9 \pm \sqrt{49}}{-0.4} = 5, 40$$

PROBLEMS

1. Graph $y = 2x+3$ and $y = 4x$ on the same coordinate system and find their point of intersection.
2. Given the cost and revenue functions, $C = 1000 + 7x$, $R = 9x$.
   (a) Graph these straight lines on a breakeven chart, and identify the breakeven point.
   (b) Find the breakeven point algebraically.
3. Determine the production level necessary to give a profit of \$500 for the cost and revenue functions in Problem 2.
4. Plot the following data on a coordinate system. Take both C and R as the vertical axis.

x	R	C
0	0	2000
100	200	2150

5. Find the equations of the linear cost and revenue functions implied by the data in Problem 4.

6. Find the breakeven point for Problem 5.

7. Determine a policy that your company would follow in renting a car from a concern that offers the following options:

Plan 1: Eighty dollars a month and \$0.08 per mile.
Plan 2: A flat \$0.12 per mile with a minimum of \$40 per month.

Note that neither plan includes the purchase of gasoline. Given that your company will be responsible for this cost (\$0.05 per mile), what would be the point at which you would be indifferent to the choice of plan?

8. Suppose that you need to have 50 copies of a one-page table for your next business meeting. At the present time you have two options for reproducing this table.

Option 1: Use a ZROZ-24 at a cost of \$0.01 per page.
Option 2: Prepare a spirit master at a cost of \$0.30 for the master and a cost of \$0.001 per sheet of duplicator paper.

(a) Which option would you choose for this task?
(b) In general, what sort of policy would you set up concerning the use of these two options?
(c) How would your policy be affected if the ZROZ-24 rate changed to \$0.0425 for the first four copies and \$0.005 for every copy after the initial four?

9. The Good Product Company earns an after-tax profit of \$2400 on sales of \$88,000. The average tax rate of the company is 40%. The only product of the company sells for \$20.00, of which \$15.00 is variable cost. The firm has a capital structure as follows:

Debentures @ 8%	\$62,500
Common Stock	85,000
Retained Earnings	6,600
	154,100

(a) Fill in the following short-form income statement:

Sales	\$88,000
Variable Costs	______
Gross Profit	______
Interest Expense	______
Other Fixed Expenses	______
Earnings before Taxes	______
Taxes	______
Net Income	2,400

(b) What is the breakeven point?

10. There are two alternative methods of acquiring a computer: rental from the manufacturer or outright purchase.

The rental cost can be broken down into three components:

1. Monthly rental payments.
2. Overtime payments for use in excess of a stated amount. Assume these payments to be 10% of the cumulative rental cost.

3. Pre-tax investment credit, which is equal to $\frac{1}{3}(2)(0.07)(P)$ and is to be subtracted from the cumulative rental cost,

where C_r = cumulative cost of renting
N = time in months
R = monthly rental
P = purchase cost

(a) Express C_r as a function of N, R, and P.
(b) Given that $P = \$150{,}000$ and $R = \$3000$, find C_r for $N = 12$, 24, 36, 48, and 60 months.

Excluding a consideration of possible residual value, the cumulative purchase cost is made up of the following components:

1. The purchase cost.
2. Maintenance.
3. Insurance.
4. Cost of money for the purchase. Assume this to be $(N/12)(P)(S/2)$.
5. Investment tax credit, which is equal to $2(0.07)(P)$ and is to be subtracted from the cumulative purchase cost,

where C_p = cumulative cost of purchasing
M = monthly maintenance
S = an acceptable simple interest rate
I = monthly insurance cost
P = purchase cost
N = time in months
R = monthly rental

(c) Express C_p as a function of M, S, I, P, and N.
(d) Given that $P = \$150{,}000$, $R = \$3000$, $M = 10\%$ of the rental cost, $S = 0.08$, and $I = 1\%$ of the rental cost, find C_p for $N = 12$, 24, 36, 48, and 60 months.
(e) Plot the results obtained in (b) and (d) with n on the horizontal axis.
(f) It appears that whether one should buy or lease a computer is determined by how long one is planning on keeping that particular machine. How long would you have to intend to keep your machine for C_p to be less than C_r?

11. **(a)** Graph the profit function found in Example 5, i.e., $P = -0.2x^2 + 9x - 40$.
(b) For what value of x is P at a maximum? (I.e., find vertex of the parabola.)
(c) Is this the same x value at which the revenue function of Example 5 function reached a maximum?

12. Given the following cost and revenue functions:

$$C = x + 2.5$$

$$R = -2x^2 + 7x$$

(a) Graph the functions on a breakeven chart, and identify the breakeven points.
(b) Find and graph the profit function.

Select the best answer for Problems 13–15, which relate to applications of breakeven analysis.*

13. Given the following notations, what is the breakeven sales level in units?

SP = selling price per unit
FC = total fixed cost
VC = variable cost per unit

(a) $\dfrac{SP}{FC \div VC}$

(b) $\dfrac{FC}{VC \div SP}$

(c) $\dfrac{VC}{SP - FC}$

(d) $\dfrac{FC}{SP - VC}$

14. A company increased the selling price for its product from \$1.00 to \$1.10 a unit when total fixed costs increased from \$400,000 to \$480,000 and variable cost per unit remained unchanged. How would these changes affect the breakeven point?
(a) The breakeven point in units would be increased.
(b) The breakeven point in units would be decreased.
(c) The breakeven point in units would remain unchanged.
(d) The effect cannot be determined from the information given.

15. The contribution margin per unit is the difference between the selling price and the variable cost per unit, and the contribution margin ratio is the ratio of the contribution margin to the selling price per unit. If the selling price and the variable cost per unit both increase 10% and fixed costs do not change, what is the effect on the contribution margin ratio?
(a) Contribution margin per unit and the contribution margin ratio both remain unchanged.
(b) Contribution margin per unit and the contribution margin ratio both increase.
(c) Contribution margin per unit increases and the contribution margin ratio remains unchanged.
(d) Contribution margin per unit increases and the contribution margin ratio decreases.

1-7 FUNCTIONS THAT GRAPH AS A HYPERBOLA

In this section we illustrate the use of an equation of a hyperbola to model the relationship between *unit cost* and the *number of units*

* Problems 13–15 are from the Uniform CPA Examinations, Copyright © (1975) by the American Institute of Certified Public Accountants, Inc., reprinted with permission.

produced. A convenient form of the equation for this purpose is

$$y = \frac{a}{x-b} + c$$

where a, b, and c are constants. For this application, let $y =$ unit cost, $c =$ variable cost per unit, $a =$ fixed cost, and $x =$ the number of units produced. Under these conditions *unit cost equals a share of the fixed costs plus the variable cost per unit.* Note that the share of the fixed costs is dependent on the number of units produced. An interesting feature of the model

$$y = \frac{a}{x-b} + c$$

is that y will approach but never reach the limiting value c as x takes on larger and larger values. In terms of *limits,* which are covered later in the text, it would be correct to say that y *approaches* or tends to c as x *approaches* infinity (becomes arbitrarily large). One branch of a typical graph of this model is shown in Figure 1-23.

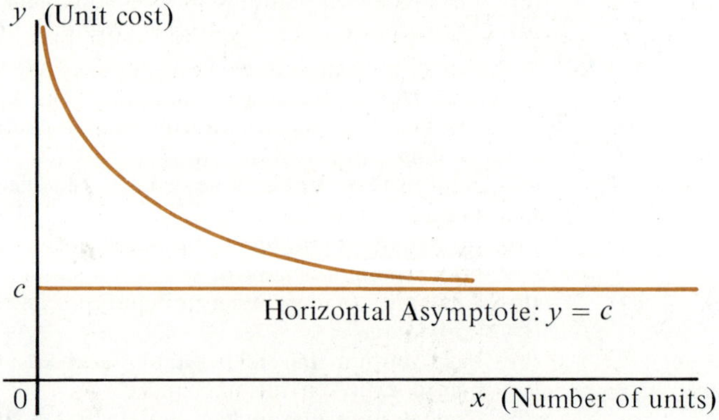

FIGURE 1-23

In Figure 1-23, the horizontal line whose equation is $y = c$ is an example of a horizontal asymptote. Furthermore, in this case, the vertical axis is a vertical asymptote. In general, for the model under discussion here, an equation for the vertical asymptote is $x = b$. Some mathematical examples of an equation of a hyperbola are given next.

Example 1. Graph and discuss the function $y = 2/(x-3)$.

Solution:

x	3	3.01	4	5	7	2.99	2	1	0	-1
y	—	200	2	1	0.5	-200	-2	-1	$-2/3$	-0.5

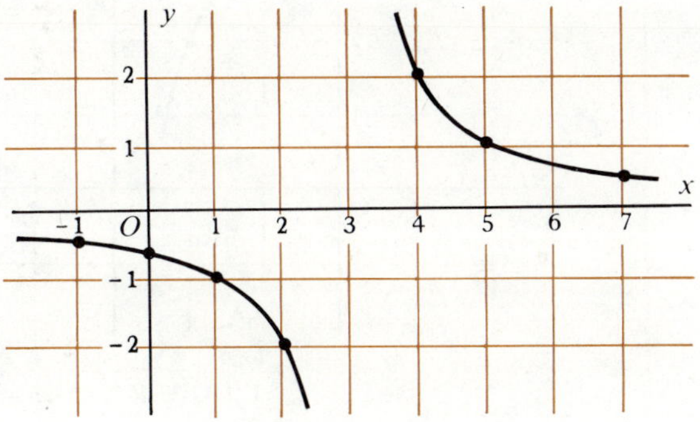

FIGURE 1-24

Discussion: $a = 2$, $b = 3$, and $c = 0$. The vertical line $x = 3$ has a special relationship to the graph of this function. It is called a *vertical asymptote.* It is possible to find points on the graph that are arbitrarily close to this vertical asymptote by choosing x close, but not equal, to 3. Note that y is undefined for $x = 3$. The horizontal line $y = 0$ has a similar relationship to the graph. It is called a *horizontal asymptote.*

Example 2. Express $(x-3)(y-4) = 2$ in the form $y = a/(x-b)+c$.

Solution:

$$y - 4 = \frac{2}{x-3}$$

$$y = \frac{2}{x-3} + 4$$

Example 3. Graph and discuss the function of Example 2.

Solution:

x	-1	1	2	3	4	5	7
y	3.5	3	2	—	6	5	4.5

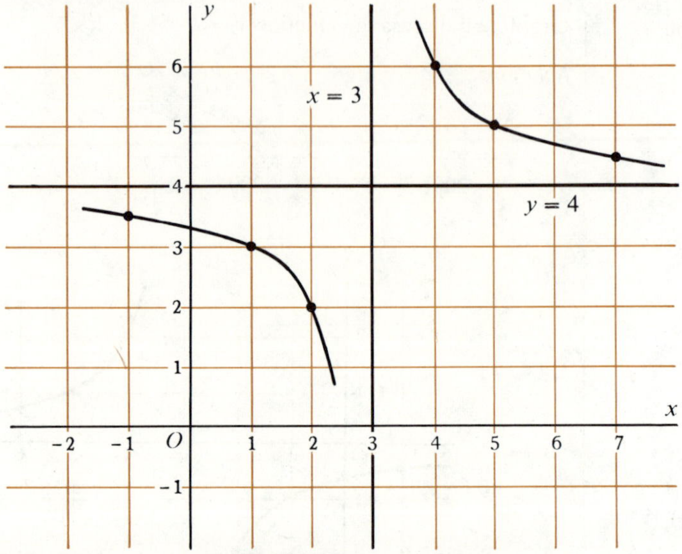

FIGURE 1-25

Discussion: $a = 2$, $b = 3$, and $c = 4$. The line $x = 3$ is a vertical asymptote. The line $y = 4$ is a horizontal asymptote. At $x = 3$, y is undefined.

In Section 1-4 we represented demand by a linear function with a negative slope. One can also model demand by a hyperbola. Consider the following demand function

$$y = \frac{200}{x}$$

where y represents price and x represents the number of units demanded. The demand function would graph as follows:

x	y
10	20
20	10
50	4
100	2

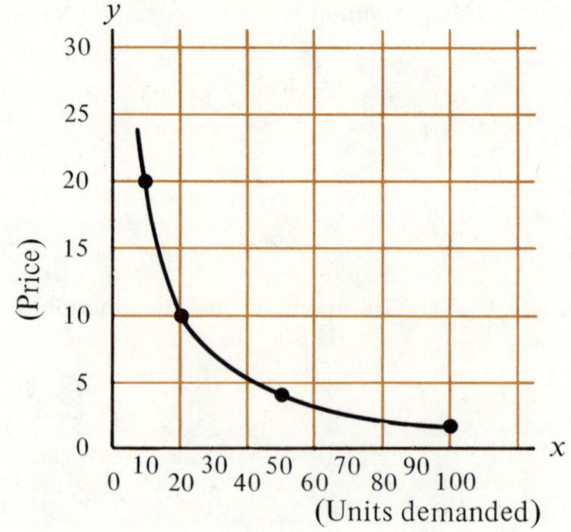

FIGURE 1-26

PROBLEMS

Graph the following functions of the form

$$y = \frac{a}{x-b} + c$$

1. $y = 2/x$

2. $y = 2/(x+3)$

3. $y = 2/x + 4$

4. $y = -2/x + 4$

5. Assume you want to know how much fixed cost is associated with each unit produced. Assume that fixed costs are \$200, and let y be the fixed cost per unit produced and x the number of units produced.
  (a) What is the mathematical relationship between units produced and the fixed cost per unit?
  (b) Graph the relationship.

6. Assume you are planning on renting a 24-foot motor home for a trip to the South during the spring vacation. Rental of the motor home will be \$400. Find the cost per individual if you go alone or take 1, 2, 3, 4, or 5 friends along. Graph your results.

7. Given that the fixed cost of operating a plant is \$100,000 and that the variable cost per unit is \$10.
  (a) What is the mathematical relationship between unit cost y, and the number of units produced x, assuming that fixed costs are to be prorated over the actual units produced?
  (b) Graph the relationship for $x = 1000, 2000, \ldots, 10{,}000, 100{,}000$.
  (c) What does the product xy represent?
  (d) What is the function or mathematical relationship for total cost?
  (e) What is the economic significance of the horizontal asymptote: $y = 10$?

8. Given the following demand function $y = 5{,}000/x$ where x represents selling price and y units demanded and sold.
  (a) How many units will be sold if the selling price is \$10?
  (b) How many units will be sold if the selling price is \$5?
  (c) What does the product $x \times y$ represent?

1-8 SUMMARY

Three types of functions were discussed in this chapter. The three functions graph as a straight line, a parabola, and a hyperbola, respectively. We have also seen that each of these types of functions can be used to model one or more important economic relationships.

Various properties and typical graphs of these functions are included in the following summary. Their applications to business and economics are also reviewed.

At the end of this section, several applied problems are included. These are used by permission from the American Institute of Certified Public Accountants, Inc. and the Institute of Management Accounting.

LINEAR FUNCTION

1. Equation can be written in the form $y = mx + b$, where m = slope and $b = y$ intercept.
2. Graphs as a straight line.

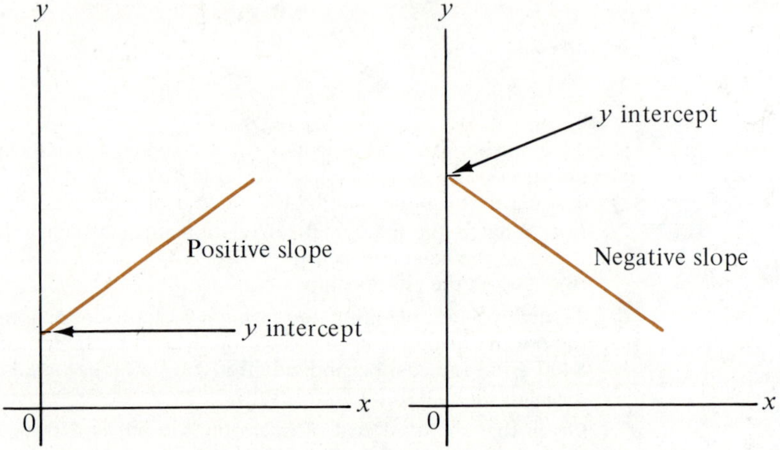

FIGURE 1-27

3. Can be used to model linear cost and linear demand.

   x = quantity, p = price, and C = total cost

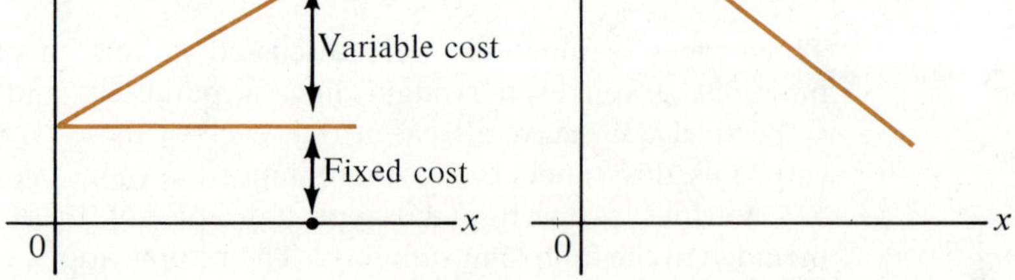

FIGURE 1-28

QUADRATIC FUNCTION

1. Equation can be written in the form $y = ax^2 + bx + c$, where $c = y$ intercept.
2. Graphs as a parabola opening upward or downward.

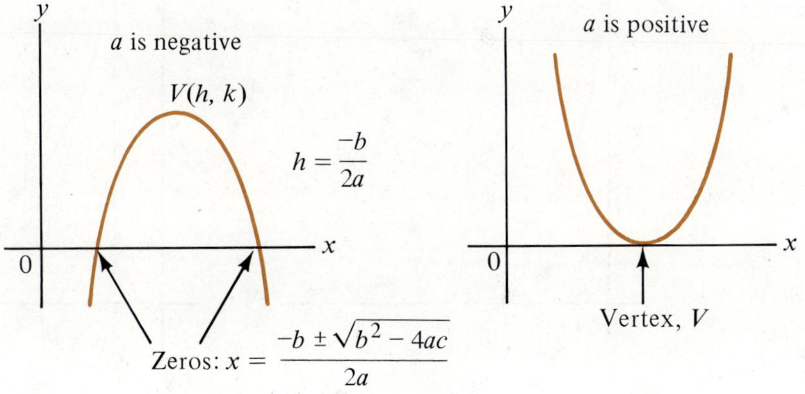

FIGURE 1-29

3. Can be used to model total revenue, when demand is linear.

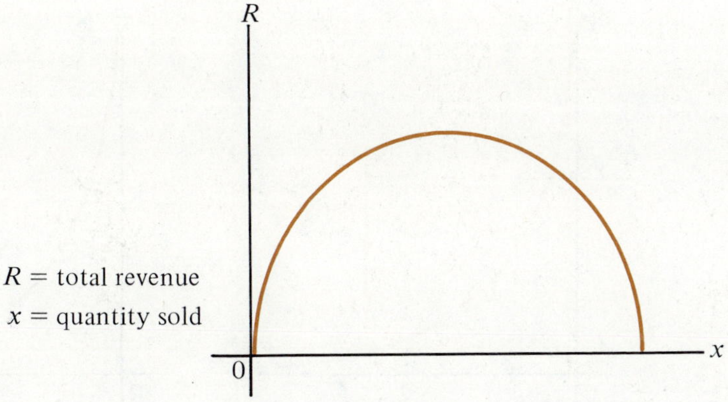

FIGURE 1-30

FUNCTIONS THAT GRAPH AS A HYPERBOLA

1. Equation can be written in the form $y = a/(x-b)+c$, where the line $x = b$ is a vertical asymptote and the line $y = c$ is a horizontal asymptote.
2. Graphs as a hyperbola.

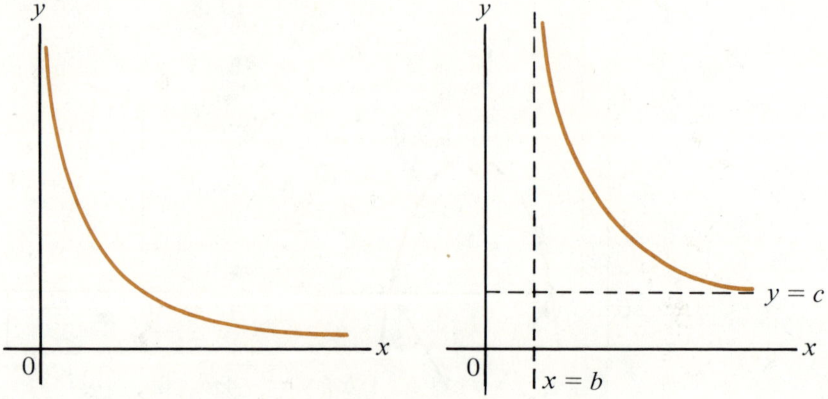

FIGURE 1-31

3. Can be used to model *unit cost* and *demand.*

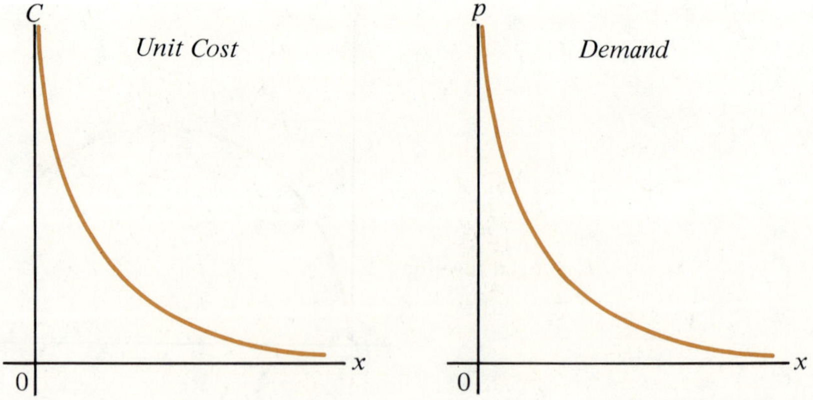

FIGURE 1-32

BREAKEVEN ANALYSIS

Case 1. Linear Cost Function and Linear Revenue Function

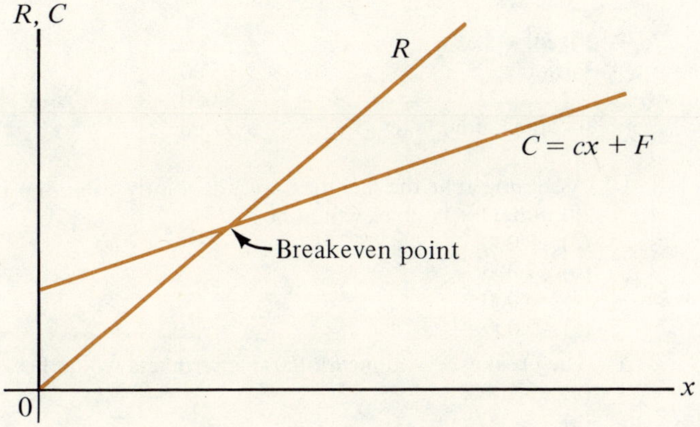

FIGURE 1-33

Case 2. Linear Cost Function and Quadratic Revenue Function

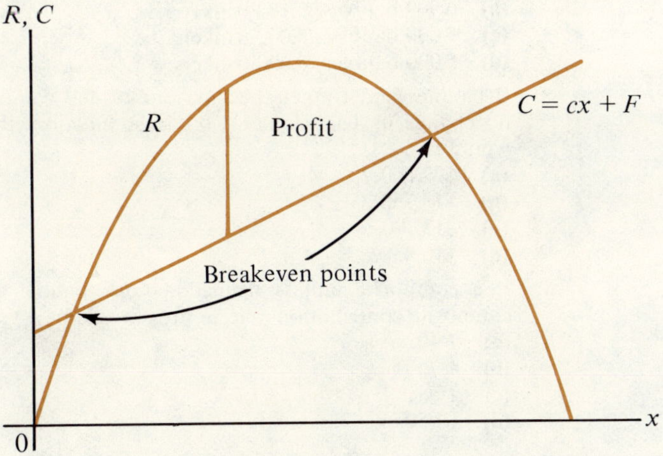

FIGURE 1-34

PROBLEMS

The Dooley Co. manufactures two products, baubles and trinkets. The following are projections for the coming year.

	Baubles		Trinkets		
	Units	*Amount*	*Units*	*Amount*	*Totals*
Sales	10,000	$10,000	7,500	$10,000	$20,000
Costs:					
Fixed		2,000		5,600	7,600
Variable		6,000		3,000	9,000
		8,000		8,600	16,600
Income before taxes		$ 2,000		$ 1,400	$ 3,400

1.* Assuming that the facilities are not jointly used, the breakeven output (in units) for baubles would be
(a) 8,000.
(b) 7,000.
(c) 6,000.
(d) 5,000.

2. The breakeven volume (dollars) for trinkets would be
(a) $8,000.
(b) $7,000.
(c) $6,000.
(d) $5,000.

3. Assuming that consumers purchase composite units of four baubles and three trinkets, the composite unit contribution margin would be
(a) $4.40.
(b) $4.00.
(c) $1.33.
(d) $1.10.

4. If consumers purchase composite units of four baubles and three trinkets, the breakeven output for the two products would be
(a) 6,909 baubles; 6,909 trinkets.
(b) 6,909 baubles; 5,182 trinkets.
(c) 5,000 baubles; 8,000 trinkets.
(d) 5,000 baubles; 6,000 trinkets.

5. If baubles and trinkets become one-to-one complements and there is no change in The Dooley Co.'s cost function, the breakeven volume would be
(a) $22,500.
(b) $15,750.
(c) $13,300.
(d) $10,858.

6. If a composite unit is defined as one bauble and one trinket, the composite contribution margin ratio would be
(a) 7/10.
(b) 4/7.
(c) 2/5.
(d) 19/50.

7. The Vernom Corporation, which produces and sells to wholesalers a highly successful line of summer lotions and insect repellents, has decided to diversify in order to stabilize sales throughout the year. A natural area for the company to consider is the production of winter lotions and creams to prevent dry and chapped skin.

After considerable research, a winter products line has been developed. However, because of the conservative nature of the company management, Vernom's president has decided to introduce only one of the new products for this coming winter. If the product is a success, further expansion in future years will be initiated.

The product selected (called Chap-off) is a lip balm that will be sold in a lipstick-type tube. The product will be sold to wholesalers in boxes of 24 tubes for $8.00 per box. Because of available capacity, no additional fixed charges will be incurred to produce the product. However, a $100,000 fixed charge will be absorbed by the product to allocate a fair share of the company's present fixed costs to the new product.

Using the estimated sales and production of 100,000 boxes of Chap-off as the standard volume, the accounting department has developed the following costs:

Direct Labor	$2.00/box
Direct Materials	$3.00/box
Total Overhead	$1.50/box
Total	$6.50/box

Vernom has approached a cosmetics manufacturer to discuss the possibility of purchasing the tubes for Chap-off. The purchase price of the empty tubes from the cosmetics manufacturer would be $0.90 per 24 tubes. If the Vernom Corporation accepts the purchase proposal, it is estimated that direct labor and variable overhead costs would be reduced by 10 percent and direct material costs would be reduced by 20 percent.

Required:

(a) Should the Vernom Corporation make or buy the tubes? Show calculations to support your answer.

(b) What would be the minimum purchase price acceptable to the Vernom Corporation for the tubes? Support your answer with an appropriate explanation.

(c) Instead of sales of 100,000 boxes, revised estimates show sales volume at 125,000 boxes. At this new volume additional equipment, at an annual rental of $10,000, must be acquired to manufacture the tubes. However, this incremental cost would be the only additional fixed cost required even if sales increased to 300,000 boxes. (The 300,000 level is the goal for the third year of production.) Under these circumstances should the Vernom Corporation make or buy the tubes? Show calculations to support your answer.

(d) The company has the option of making and buying at the same time. What would be your answer to part C if this alternative were considered? Show calculations to support your answer.

(e) What non-quantifiable factors should the Vernom Corporation consider in determining whether it should make or buy the lipstick tubes?

REFERENCES

BOOKS

Allen, R. G. D. *Mathematical Analysis for Economists.* New York: St. Martin's Press, Inc., 1967.

Allendoerfer, Carl B., and Oakley, Cletus O. *Principles of Mathematics.* New York: McGraw-Hill Book Co., Inc., 1969.

Bardell, Ross H., and Spitzbart, Abraham. *College Algebra.* Reading, Mass.: Addison-Wesley, Inc., 1966.

Beatty, William E. *Mathematical Relationships in Business and Economics.* Braintree, Mass.: D. H. Mark Publishing Co., 1970.

Bowen, Earl K. *Mathematics with Applications in Management and Economics.* Homewood, Ill.: Richard D. Irwin, Inc., 1967.

Daus, Paul H., and Whyburn, William M. *Algebra with Applications to Business and Economics.* Reading, Mass.: Addison-Wesley, Inc., 1961.

Hamburg, Morris. *Statistical Analysis for Decision Making.* New York: Harcourt Brace Jovanovich, Inc., 1970.

Kerlinger, Fred N. *Foundations of Behavioral Research.* New York: Holt, Rinehart and Winston, Inc., 1973.

May, W. Graham. *Foundations in Modern Mathematics.* Waltham, Mass.: Blaisdell Publishing Co. Inc., 1967.

Parker, R. H. *Management Accounting: An Historical Perspective*, New York: Augustus M. Kelley Publishers, 1969.

Rivett, Patrick. *Principles of Model Building.* New York: John Wiley & Sons, Inc., 1972.

Sampson, Roy J., and Calmus, Thomas W. *Economic Concepts Applications Analysis.* Boston, Mass.: Houghton Mifflin Co., 1974.

Springer, Clifford H., Herlihy, R. E., and Beggs, R. I. *Basic Mathematics.* Homewood, Illinois: Richard D. Irwin, Inc., 1965.

ARTICLES

Brandon, Dick H. "Computer Acquisition Method Analysis," *Datamation:* Sept. 1972, pp. 76–79.

Hartl, Robert J. "The Linear Total Revenue Curve in Cost-Volume-Profit Analysis," *Management Accounting:* March 1975, pp. 49–52.

Problem 7 adapted from the CMA examination, with permission.

TWO

Introduction to vectors and matrices

Chapters 2 and 3 deal mainly with matrices. A *matrix* is a doubly subscripted set of objects, normally arranged in *rows* and *columns*. The mathematical operations of addition and multiplication can be defined in a meaningful way for certain classes of matrices. These definitions and associated theorems result in a rich and useful mathematical structure.

The following examples of matrix applications are developed in this text:

coefficient matrix	(*systems of linear algebraic equations*)
parts explosion matrix	(*inventory analysis*)
technological matrix	(*Leontief input-output model*)
payoff matrix	(*financial decision making*)
transition matrix	(*Markov chains*)

2-1 DEFINITION, EQUALITY, AND ADDITION OF VECTORS

In this section and in the next section we will introduce the concept of a vector. First we must define what is meant by a vector, then we will describe certain algebraic manipulations that can be carried out with vectors. Finally, we will point out how this new mathematical concept can be used to express a system of linear equations as a single vector equation.

DEFINITION 2.1. *A vector is an ordered n-tuple* $(x_1, x_2, \ldots, x_n)$ *of numbers arranged in a row or in a column.*

If the numbers, which are called components, are arranged in a row, the vector is called a *row vector*, and if the components are arranged in a column, the vector is called a *column vector*. The following are examples of row vectors:

$$(3, 6, 1) \qquad (9, 2) \qquad (4, 6, 1, 8) \qquad (x_1, x_2, x_3, x_4)$$

The following are examples of column vectors:

$$\begin{pmatrix} 4 \\ 3 \\ 2 \\ 1 \end{pmatrix} \qquad \begin{pmatrix} 6 \\ 2 \end{pmatrix} \qquad \begin{pmatrix} 1 \\ 9 \\ 8 \end{pmatrix} \qquad \begin{pmatrix} b_1 \\ b_2 \\ b_3 \end{pmatrix}$$

Even though a vector consists of n distinct components, the entire vector should be considered as a single entity. In keeping with this consideration, vectors, in this text, are denoted by single letters set in boldface print. The following examples illustrate this point.

$$\mathbf{a} = (1, 3, 8) \qquad \mathbf{b} = \begin{pmatrix} 2 \\ 6 \\ 1 \end{pmatrix} \qquad \mathbf{c} = (c_1, c_2, c_3, c_4)$$

DEFINITION 2.2. *Two row vectors are said to be equal if they have the same number of components and their corresponding components are equal. Similarly, two column vectors are said to be equal if they have the same number of components and their corresponding components are equal.*

A *row vector* with two or more components is never equal to a *column vector*. Equality of vectors **a** and **b** is denoted by $\mathbf{a} = \mathbf{b}$. This definition is illustrated by the following examples:

1. $(2, 3, 6) = (2, 3, 6)$
2. $(4, 7, 2)$ does not equal $(4, 8, 2)$ since the middle components are not equal.
3. $\begin{pmatrix} 2 \\ 3 \\ 8 \end{pmatrix}$ does not equal $(2, 3, 8)$ since one is a column vector and the other is a row vector.
4. $\begin{pmatrix} 2 \\ 6 \\ 7 \end{pmatrix}$ and $\begin{pmatrix} 2 \\ 6 \\ 7 \\ 8 \end{pmatrix}$ are not equal since the second vector has more components than the first.

DEFINITION 2.3. *The sum of two row vectors or two column vectors (with the same number of components) is formed by adding the corresponding components of the given row vectors or column vectors.*

Addition of a row vector and a column vector is not defined. The sum of vectors **a** and **b** is denoted by **a** + **b**.

The following examples illustrate this definition:

1. $(1, 4, 2) + (4, 2, 1) = (5, 6, 3)$
2. $\begin{pmatrix} 9 \\ 2 \\ 4 \end{pmatrix} + \begin{pmatrix} 8 \\ 5 \\ 2 \end{pmatrix} = \begin{pmatrix} 17 \\ 7 \\ 6 \end{pmatrix}$
3. The sum of $\mathbf{a} = (2, 1, 9)$ and $\mathbf{b} = \begin{pmatrix} 30 \\ 1 \\ 4 \end{pmatrix}$ is not defined since **a** is a row vector and **b** is a column vector.
4. The sum of $\mathbf{c} = (3, 5, 9, 4)$ and $\mathbf{d} = (1, 5, 2)$ is not defined since **c** and **d** do not have the same number of components.

PROBLEMS

1. Test the following pairs of vectors for equality. If the pairs are not equal, give the reason for the inequality.

(a) $(2, 3, 4); (2, 3, 4)$ **(c)** $(2, 1, 2); (2, 1, 2, 5)$

(b) $(1, 3, 5); \begin{pmatrix} 1 \\ 3 \end{pmatrix}$ **(d)** $(1, 2, 3); (3, 2, 1)$

2. Add each of the pairs of vectors for which the sum is defined. If the sum is not defined, give the reason.

(a) $(2, 3, 7); (1, 4, 2)$ **(c)** $(4, 3, 7); (1, 0, 2, 4)$

(b) $(1, 4, 1); \begin{pmatrix} 2 \\ 3 \\ 5 \end{pmatrix}$ **(d)** $(a, b, c); (0, 0, 0)$

3. Let **x** be a row vector with 5 components and let **y** be a row vector with 6 components.

(a) Write out **x** and **y** using subscript notation.
(b) Is it possible for **x** to equal **y**? Why?
(c) Is it possible to add **x** and **y** together? Why?

4. Vectors with only two components can be represented geometrically as follows. Take the first component of the given vector as the x coordinate of a point in the xy plane and take the second component of the vector as the y coordinate of the point. Plot the point that corresponds to each of the following vectors.

(a) $\mathbf{a} = (2, 3)$ **(c)** $\mathbf{c} = \begin{pmatrix} -1 \\ -2 \end{pmatrix}$

(b) $\mathbf{b} = (-3, 6)$ **(d)** $\mathbf{d} = \begin{pmatrix} 3 \\ -5 \end{pmatrix}$

5. Calculate and illustrate geometrically the sum of the following pairs of vectors:

(a) (2, 3); (1, 5)

(b) (−1, −2); (3, 0)

(c) $\begin{pmatrix} 1 \\ 2 \end{pmatrix}; \begin{pmatrix} -1 \\ 1 \end{pmatrix}$

(d) $\begin{pmatrix} -1 \\ 4 \end{pmatrix}; \begin{pmatrix} 3 \\ 1 \end{pmatrix}$

6. Given that you have a column vector called **x**, which has five components, write out this vector in component form.
7. Give an example of a row vector that has four components and a column vector that has five components.
8. Add the following vectors:

(a) $\begin{pmatrix} x_1 \\ x_2 \\ x_3 \end{pmatrix} + \begin{pmatrix} y_1 \\ y_2 \\ y_3 \end{pmatrix}$

(b) $(a, b, c) + (3, 2, 1)$

9. In order to manufacture a certain product, three raw materials are needed. Assume the product is cement, which requires the following:

2 units of sand	d_1
1 unit of Portland cement	d_2
1 unit of water	d_3

Express these ingredients as a row vector.

10. The demand in units for three types of goods is as follows:

$$g_1 = 100{,}000, \ g_2 = 150{,}000, \ g_3 = 125{,}000$$

Express the demand as a column vector.

11. The total output of goods from three industries in millions of dollars is as follows:

$$i_1 = 1200, \ i_2 = 1000, \ i_3 = 800$$

Express the production data as a column vector.

12. There are three other suppliers of a certain product besides your own company. Each supplier has the following market share in percents:

$$s_1 = 30, \ s_2 = 15, \ s_3 = 25, \ s_4 = 30$$

Express the above market shares as a column vector.

2-2 MULTIPLICATION OF A VECTOR BY A SCALAR AND BY A VECTOR

DEFINITION 2.4. *Let* **a** *be a row or column vector and let k be a real number. Then the product k***a** *is defined to be the vector whose components are given by k times the corresponding components of* **a**.

This type of multiplication is called *multiplication by a scalar*, where the number k is called a *scalar*.

The following examples illustrate Definition 2.4.

1. $\frac{1}{2}(4, 8, 2) = (2, 4, 1)$

2. $0.2\begin{pmatrix} 2 \\ 6 \\ 5 \end{pmatrix} = \begin{pmatrix} 0.4 \\ 1.2 \\ 1 \end{pmatrix}$

3. $4(2, 1, 3, 5) = (8, 4, 12, 20)$

DEFINITION 2.5. *Let* **a** *be a row vector and* **b** *be a column vector, both with n components. Then the product of* **a** *times* **b** *is defined to be the scalar* $a_1 b_1 + a_2 b_2 + \cdots + a_n b_n$. *This product is denoted by* **ab**. *This product can also be written*

$$\mathbf{ab} = \sum_{k=1}^{n} a_k b_k$$

It is called the inner product.

The following examples illustrate the above definition.

1. $(1, 4, 2)\begin{pmatrix} 3 \\ 4 \\ 2 \end{pmatrix} = (1)(3) + (4)(4) + (2)(2) = 23$

2. $\begin{pmatrix} 3 \\ 4 \\ 2 \end{pmatrix}$ times $(1, 4, 2)$ does not exist according to the above definition.

Remark: Definition 2.5 only admits multiplication of a column vector by a row vector and always leads to a scalar quantity as an answer. Later in the chapter the concept of a matrix will be introduced and a definition of matrix multiplication will be given at that time. In the more general matrix setting, a row vector with n components can be multiplied by a column vector with m components, but the answer will be a matrix instead of a scalar.

3. $(2, 8, 6, 5)\begin{pmatrix} 7 \\ 8 \end{pmatrix}$ is not defined because the row vector has four components and the column vector has only two components.

Next we give an example to illustrate how these definitions can be used to express *a system of linear equations* as a *single vector equation.*

Example 1. Write the system

$$2x + 3y + 5z = 4$$
$$3x + 5y + 6z = 2$$
$$x + 2y + 5z = 3$$

as a single vector equation.

Solution: Form the four coloumn vectors

$$\mathbf{a} = \begin{pmatrix} 2 \\ 3 \\ 1 \end{pmatrix}, \quad \mathbf{b} = \begin{pmatrix} 3 \\ 5 \\ 2 \end{pmatrix}, \quad \mathbf{c} = \begin{pmatrix} 5 \\ 6 \\ 5 \end{pmatrix}, \quad \mathbf{d} = \begin{pmatrix} 4 \\ 2 \\ 3 \end{pmatrix}$$

Then the given system can be written in the form

$$x\begin{pmatrix} 2 \\ 3 \\ 1 \end{pmatrix} + y\begin{pmatrix} 3 \\ 5 \\ 2 \end{pmatrix} + z\begin{pmatrix} 5 \\ 6 \\ 5 \end{pmatrix} = \begin{pmatrix} 4 \\ 2 \\ 3 \end{pmatrix}$$

where x, y, and z are treated as unknown scalars. The system can be written even more simply as $x\mathbf{a} + y\mathbf{b} + z\mathbf{c} = \mathbf{d}$.

This can be checked as follows:

$$x\begin{pmatrix} 2 \\ 3 \\ 1 \end{pmatrix} + y\begin{pmatrix} 3 \\ 5 \\ 2 \end{pmatrix} + z\begin{pmatrix} 5 \\ 6 \\ 5 \end{pmatrix} = \begin{pmatrix} 4 \\ 2 \\ 3 \end{pmatrix}$$

$$\begin{pmatrix} 2x \\ 3x \\ 1x \end{pmatrix} + \begin{pmatrix} 3y \\ 5y \\ 2y \end{pmatrix} + \begin{pmatrix} 5z \\ 6z \\ 5z \end{pmatrix} = \begin{pmatrix} 4 \\ 2 \\ 3 \end{pmatrix} \qquad \textit{by Definition 2.4}$$

$$\begin{pmatrix} 2x + 3y + 5z \\ 3x + 5y + 6z \\ 1x + 2y + 5z \end{pmatrix} = \begin{pmatrix} 4 \\ 2 \\ 3 \end{pmatrix} \qquad \textit{by Definition 2.3}$$

$$\begin{aligned} 2x + 3y + 5z &= 4 \\ 3x + 5y + 6z &= 2 \qquad \textit{by Definition 2.2} \\ x + 2y + 5z &= 3 \end{aligned}$$

PROBLEMS

1. Perform the following multiplications:

(a) $0.5\begin{pmatrix} 4 \\ 6 \\ 2 \end{pmatrix}$

(b) $3(4, 8, 7)$

(c) $x\begin{pmatrix} 1 \\ 4 \\ 2 \end{pmatrix}$

(d) $5\begin{pmatrix} 4 \\ 2 \\ 1 \end{pmatrix}$

2. Perform the following multiplications:

(a) $(2, 3)\begin{pmatrix} 7 \\ 1 \end{pmatrix}$

(c) $(7, 8, 2, 1)\begin{pmatrix} 4 \\ 3 \\ 2 \\ 1 \end{pmatrix}$

(b) $(2, 8, 4)\begin{pmatrix} 3 \\ 6 \\ 1 \end{pmatrix}$ **(d)** $(9, 1, 2, 8)\begin{pmatrix} 4 \\ 3 \\ 6 \\ 7 \end{pmatrix}$

3. Perform the following if they are defined. If they are not, explain why.

(a) $(6, 7, 2)\begin{pmatrix} 8 \\ 4 \end{pmatrix}$ **(c)** $(2, 7, 6)\begin{pmatrix} 8 \\ 4 \\ 2 \end{pmatrix}$

(b) $x\begin{pmatrix} 5 \\ 2 \end{pmatrix}$ **(d)** $\begin{pmatrix} 8 \\ 3 \\ 2 \end{pmatrix}\begin{pmatrix} 6 \\ 7 \\ 5 \end{pmatrix}$

4. Why are the following not defined?

(a) $(3, 6, 2)\begin{pmatrix} 4 \\ 8 \end{pmatrix}$ **(c)** $\begin{pmatrix} 2 \\ 1 \end{pmatrix}\begin{pmatrix} 6 \\ 4 \\ 7 \end{pmatrix}$

(b) $(2, 4, 1)(8, 3, 6)$ **(d)** $\begin{pmatrix} 2 \\ 2 \\ 1 \end{pmatrix}(4, 8, 3, 6)$

5. Show that multiplying $3\begin{pmatrix} 6 \\ 4 \\ 2 \end{pmatrix}$ gives the same result as

$$\begin{pmatrix} 6 \\ 4 \\ 2 \end{pmatrix} + \begin{pmatrix} 6 \\ 4 \\ 2 \end{pmatrix} + \begin{pmatrix} 6 \\ 4 \\ 2 \end{pmatrix}$$

6. Write the following system as a vector equation:

$$4x + 6y + 3z = 2$$
$$5x + 4y + z = 1$$
$$5x + 6y = 4$$

7. Perform the scalar multiplications and write as a system of equations.

$$w\begin{pmatrix} 4 \\ 9 \\ 3 \end{pmatrix} + x\begin{pmatrix} 2 \\ 1 \\ 2 \end{pmatrix} + y\begin{pmatrix} 3 \\ 2 \\ 4 \end{pmatrix} = \begin{pmatrix} 4 \\ 9 \\ 3 \end{pmatrix}$$

8. A certain single product firm has variable costs per unit:

Labor	\$0.50
Material	0.30
Variable overhead	0.20

(a) Express these variable costs as a column vector.
(b) Express the variable costs of producing 200 units as a column vector.
(c) Multiply the column vector in (b) by a row vector to determine the total variable cost of producing 200 units of this product.

9. A company buys 3 machines at a cost of \$20,000, \$15,000, and \$13,000, respectively. The salvage value after 4 years is 10% of the original cost.
(a) Express machine costs as a row vector.
(b) Use the row vector in (a) to find the machine costs that are to be depreciated (i.e., the depreciable base).

Problems 10 through 17 are designed to acquaint the reader with the concept of n-space. The set of all n-tuples of real numbers is called n-space, which we denote by the symbol R^n. One usually speaks of a particular n-tuple in n-space as a *vector* or *point* in R^n.

10. Give three numerical examples of a point in 5-space, R^5.

11. Let k be a real number and let $\mathbf{a}$ be a point in R^n. Show that $k\mathbf{a}$ is also a point in R^n.

12. Let $\mathbf{a}$ and $\mathbf{b}$ be points in R^n. Show that $\mathbf{a}+\mathbf{b}$ is also a point in R^n.

13. Let $\mathbf{a}$ and $\mathbf{b}$ be points in R^n, and define $-\mathbf{a}$ to be $-1\mathbf{a}$ and $\mathbf{a}-\mathbf{b}$ to be $\mathbf{a}+(-\mathbf{b})$. Calculate $3\mathbf{a}$, $-\mathbf{b}$, $\mathbf{a}+\mathbf{b}$, and $\mathbf{a}-\mathbf{b}$ for the following pairs of points:
(a) $\mathbf{a} = (2, -3)$, $\mathbf{b} = (-1, 4)$
(b) $\mathbf{a} = (2, 1, -6)$, $\mathbf{b} = (1, 3, 4)$
(c) $\mathbf{a} = (-1, 2, -3, 4, -3)$, $\mathbf{b} = (2, 4, 3, 6, 1)$

14. The *distance* between two points $\mathbf{a}$ and $\mathbf{b}$ in R^n is defined by

$$d(\mathbf{a}, \mathbf{b}) = \sqrt{(a_1-b_1)^2 + (a_2-b_2)^2 + \cdots + (a_n-b_n)^2}$$

where $d(\mathbf{a}, \mathbf{b})$ denotes the distance between $\mathbf{a}$ and $\mathbf{b}$.
(a) Compute the distance between the pairs of points in Problem 13.
(b) Write the expression for $d(\mathbf{a}, \mathbf{b})$ using summation notation.

15. Let $\mathbf{a}$, $\mathbf{b}$, and $\mathbf{c}$ be points in R^n and let k and 1 be real numbers. Show that
(a) $(\mathbf{a}+\mathbf{b})+\mathbf{c} = \mathbf{a}+(\mathbf{b}+\mathbf{c})$
(b) $\mathbf{a}+0 = 0+\mathbf{a} = \mathbf{a}$, where $0 = (0, 0, \ldots, 0)$
(c) $\mathbf{a}+(-\mathbf{a}) = 0$, where $-\mathbf{a} = (-a_1, -a_2, \ldots, -a_n)$
(d) $\mathbf{a}+\mathbf{b} = \mathbf{b}+\mathbf{a}$
(e) $k(\mathbf{a}+\mathbf{b}) = k\mathbf{a}+k\mathbf{b}$
(f) $(k+1)\mathbf{a} = k\mathbf{a}+1\mathbf{a}$
(g) $(k1)\mathbf{a} = k(1\mathbf{a})$
(h) $1\mathbf{a} = \mathbf{a}$
These eight properties are used in the definition of an important mathematical structure called a *vector space*.

2-3 DEFINITION, EQUALITY, AND ADDITION OF MATRICES

In this section we will introduce the concept of a matrix and some of the standard notation that is used in the study of matrices.

DEFINITION 2.6. *An m by n matrix is a rectangular array of numbers consisting of m rows and n columns. The expression m by n or $m \times n$ is called the order of the matrix. The numbers a_{ij} are called elements of the matrix.*

$$\begin{pmatrix} a_{11} & a_{12} & \cdots & a_{1n} \\ a_{21} & a_{22} & \cdots & a_{2n} \\ \vdots & \vdots & \ddots & \vdots \\ a_{m1} & a_{m2} & \cdots & a_{mn} \end{pmatrix}$$

As with vectors, the reader should adopt the point of view that even though an m by n matrix contains m times n elements, the entire matrix should be considered as a single entity. In keeping with this point of view, matrices are denoted by single capital letters. Definition 2.6 is illustrated as:

1. $\begin{pmatrix} 2 & 3 \\ 4 & 2 \\ 1 & 7 \end{pmatrix}$ *is a 3 by 2 matrix*

2. $\begin{pmatrix} 3 & -2 & 4 & 7 \\ 5 & 6 & -8 & 2 \end{pmatrix}$ *is a 2 by 4 matrix*

When a matrix is denoted by a capital letter, as A, it is customary to write $A = (a_{ij})$, where i stands for the row and j for the column in which the element a_{ij} is found. This double subscript notation is illustrated in the next two examples.

1. $A = \begin{pmatrix} a_{11} & a_{12} & a_{13} \\ a_{21} & a_{22} & a_{23} \end{pmatrix}$ *could be written as* (a_{ij}), *where* $i = 1, 2$ *and* $j = 1, 2, 3$

2. If $B = \begin{pmatrix} 2 & 3 & 5 \\ 1 & 0 & 4 \end{pmatrix}$, *then* $b_{11} = 2$, $b_{12} = 3$, $b_{21} = 1$, *and so on*

Next we define equality and addition for matrices.

DEFINITION 2.7. *Two matrices A and B are said to be equal, written $A = B$, if the following two conditions are met:*

a. *A and B have the same number of rows and columns (i.e., A and B are the same order).*
b. *Each element of A equals the corresponding element of B (i.e., $a_{ij} = b_{ij}$ for each i and j involved).*

This definition is illustrated below.

1. $\begin{pmatrix} 2 & 1 & 3 \\ 4 & 2 & 1 \end{pmatrix} = \begin{pmatrix} 2 & 1 & 3 \\ 4 & 2 & 1 \end{pmatrix}$ *since both matrices are the same order and their corresponding elements are equal*

2. $\begin{pmatrix} 2 & 3 \\ 4 & 7 \end{pmatrix}$ and $\begin{pmatrix} 2 & 3 & 1 \\ 4 & 7 & 2 \end{pmatrix}$ *are not equal since they are not the same order*

3. $\begin{pmatrix} 3 & 1 & 7 \\ 4 & 2 & 4 \end{pmatrix}$ and $\begin{pmatrix} 3 & 1 & 7 \\ 4 & 2 & 8 \end{pmatrix}$ *are not equal since the elements in the second row and third column are not equal*

Example 1. Given the matrix equation

$$\begin{pmatrix} x+y & 4 \\ x-y & 6 \end{pmatrix} = \begin{pmatrix} 3 & 4 \\ 5 & 6 \end{pmatrix}$$

solve for x and y.

Solution:

$$\left.\begin{aligned} x + y &= 3 \\ x - y &= 5 \end{aligned}\right\} \quad \textit{by the definition of equality of matrices}$$

First eliminate y by adding the two equations together.

$$\begin{aligned} x + y &= 3 \\ x - y &= 5 \\ \hline 2x + 0 &= 8 \\ x &= 4 \end{aligned}$$

Substituting $x = 4$ into $x + y = 3$ gives $4 + y = 3$ or $y = -1$. Hence, the solution is $x = 4$ and $y = -1$.

Example 2. Given that $a_{11} = 2$, $a_{12} = 3a_{11}$, $a_{13} = 5a_{12}$, $a_{21} = a_{13}$, $a_{22} = a_{12}$, and $a_{23} = 8$, write out the matrix A.

Solution:

$$a_{12} = (3)(2) = 6$$

$$a_{13} = (5)(6) = 30$$

$$a_{21} = a_{13} = 30$$

$$a_{22} = a_{12} = 6$$

$$a_{23} = 8$$

Recall that a_{ij} is in the ith row and jth column of A; therefore, we have

$$A = \begin{pmatrix} 2 & 6 & 30 \\ 30 & 6 & 8 \end{pmatrix}$$

DEFINITION 2.8. *Let A and B be two matrices with the same number of rows and of columns. Then the sum of A and B, written $A + B$, is the matrix obtained by adding corresponding elements from A and B. Symbolically, $A + B = C$, if $c_{ij} = a_{ij} + b_{ij}$ for each i and j involved.*

1. $\begin{pmatrix} 2 & 3 & 4 \\ 1 & 7 & 2 \end{pmatrix} + \begin{pmatrix} 3 & 4 & 1 \\ 2 & 8 & 4 \end{pmatrix} = \begin{pmatrix} 5 & 7 & 5 \\ 3 & 15 & 6 \end{pmatrix}$

2. The sum of $\begin{pmatrix} 2 & 3 & 4 \\ 1 & 7 & 2 \end{pmatrix}$ and $\begin{pmatrix} 2 & 1 \\ 8 & 4 \end{pmatrix}$ is not defined since the two matrices are not the same order.

Example 3. Given the matrix equation

$$\begin{pmatrix} x^2 & 2 \\ 1 & 4 \end{pmatrix} + \begin{pmatrix} 3x & 2 \\ 2 & 5 \end{pmatrix} = \begin{pmatrix} -2 & 4 \\ 3 & 9 \end{pmatrix}$$

solve for x.

Solution: From the definition of equality and addition for matrices, we have the quadratic equation $x^2 + 3x = -2$. This can be written as

$$x^2 + 3x + 2 = 0$$

$$(x+2)(x+1) = 0$$

By equating each factor to zero, we have $x+2 = 0$ or $x+1 = 0$. This gives

$$x = -2, \qquad x = -1$$

Before proceeding to the next section, we must point out that a *vector* is a special case of a *matrix*. For example, a row vector with 5 components is a 1 by 5 matrix, and a column vector with 3 components is a 3 by 1 matrix.

1. The row vector (2, 3, 4, 5) can be viewed as a 1 by 4 matrix.
2. The column vector $\begin{pmatrix} 2 \\ 4 \\ 7 \\ 9 \end{pmatrix}$ can be viewed as a 4 by 1 matrix.

PROBLEMS

1. Test the following pairs of matrices for equality. If they are not equal, give the reason.

(a) $\begin{pmatrix} 4 & 2 & 1 \\ 3 & 6 & 2 \end{pmatrix}, \begin{pmatrix} 4 & 2 & 2 \\ 3 & 6 & 2 \end{pmatrix}$

(b) $\begin{pmatrix} 4 & 3 \\ 2 & 6 \\ 1 & 2 \end{pmatrix}, \begin{pmatrix} 4 & 2 & 1 \\ 3 & 6 & 2 \end{pmatrix}$

2. Give an example of a pair of 3 by 4 matrices that are equal.

3. Perform the indicated matrix additions.

(a) $\begin{pmatrix} 1 & 2 \\ 3 & 4 \end{pmatrix} + \begin{pmatrix} 8 & 7 \\ 6 & 2 \end{pmatrix}$

(b) $\begin{pmatrix} 4 & 6 & 8 \\ 9 & 3 & 2 \end{pmatrix} + \begin{pmatrix} 2 & -1 & 3 \\ -4 & 2 & 5 \end{pmatrix}$

(c) $\begin{pmatrix} 8 & 9 & 10 \\ 6 & 3 & 4 \\ 5 & 2 & 1 \end{pmatrix} + \begin{pmatrix} 8 & 3 & 2 \\ 4 & 7 & 1 \\ 2 & 1 & 4 \end{pmatrix}$

4. Which of the following pairs of matrices can be added? If addition is not defined, give the reason.

(a) $\begin{pmatrix} 2 & 3 & 1 \\ 4 & 5 & 7 \end{pmatrix}, \begin{pmatrix} 1 & 4 \\ 2 & -2 \\ 1 & -3 \end{pmatrix}$

(b) $\begin{pmatrix} 1 & 4 \\ 2 & 3 \\ 3 & 2 \end{pmatrix}, \begin{pmatrix} -2 & 5 \\ 4 & 6 \\ 8 & 7 \end{pmatrix}$

(c) $\begin{pmatrix} 2 \\ 1 \\ 3 \end{pmatrix}, (4, 5, 7)$

5. Given

$$\begin{pmatrix} 2 & 3 & 4 \\ 1 & 2 & 3 \end{pmatrix} + \begin{pmatrix} a & b & 2 \\ 3 & 1 & 4 \end{pmatrix} = \begin{pmatrix} 5 & 8 & 6 \\ 4 & 3 & 7 \end{pmatrix}$$

Find a and b.

6. The entries a_{i3} $(i = 1, 2, 3, 4)$ make up the ________________ (row) or (column) of the matrix A.

7. Given that you have a 6×5 matrix called B. Write the elements (i.e., $b_{11}, \ldots$) in matrix form.

8. A company produces potatoes in three different agricultural regions. These potatoes are then shipped to four warehouses from which they are distributed to retailers. Any region can ship to any warehouse. The available shipping routes are shown in Figure 2-1. The shipping costs for the various routes are shown in Table 2-1.

TABLE 2-1

From	To	Cost per bushel (in cents)
R_1	W_1	5
R_1	W_2	7
R_1	W_3	9
R_1	W_4	16
R_2	W_1	10
R_2	W_2	5
R_2	W_3	11
R_2	W_4	10
R_3	W_1	8
R_3	W_2	7
R_3	W_3	4
R_3	W_4	8

Express the cost schedule in tabular form using regions as row headings and warehouses as column headings.

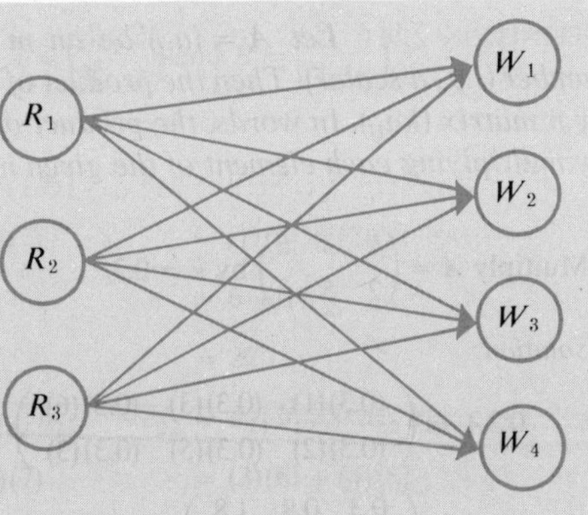

FIGURE 2-1

9. **(a)** Express the cost information in Problem 8 as a 3 by 4 matrix. Call this matrix C.
(b) Give the numerical value of c_{11}.
(c) What route is associated with the element c_{21}? c_{12}?

10. **(a)** Construct a 4 by 2 variable cost matrix from the cost information of Problem 9 in Section 1-6. Use the same order of labeling for rows and columns. Call this matrix V.
(b) Describe v_{11} and v_{32} in words.
(c) Describe $\sum_{i=1}^{4} v_{i1}$ in words.

11. Assume that one can either buy or lease a machine and that demand will be either high or low. Set up a table that illustrates the possible situations that can occur, given the following:

If demand is low and you purchase the machine, a net profit of \$10,000 will result. If demand is low and you lease the machine, a net profit of \$12,000 will result. If demand is high and you purchase the machine, a net profit of \$20,000 will result. Otherwise the net profit is \$5000.

12. Express the payoffs in Problem 11 as a payoff matrix.

13. Given the matrix equation

$$\begin{pmatrix} 2x+y & 4 \\ 3 & x+3y \end{pmatrix} = \begin{pmatrix} 1 & 4 \\ 3 & 2 \end{pmatrix}$$

Solve for x and y.

2-4 MULTIPLICATION OF A MATRIX BY A SCALAR AND BY A MATRIX

To distinguish them from vectors or matrices, real numbers are commonly called *scalars* when dealing with real numbers and vectors or real numbers and matrices. Next we define multiplication of a matrix of any order by a scalar.

5. Multiply the following if multiplication is defined.

(a) $\begin{pmatrix} 1 & 2 \\ 3 & 4 \end{pmatrix}\begin{pmatrix} 8 & -7 \\ -6 & 2 \end{pmatrix}$ (d) $\begin{pmatrix} 6 & 8 & 3 & 1 \\ 4 & 9 & 2 & 3 \\ 7 & 1 & 4 & 2 \end{pmatrix}\begin{pmatrix} 7 & 2 & 3 \\ 8 & 4 & 1 \\ 6 & 5 & 2 \\ 2 & 9 & 8 \end{pmatrix}$

(b) $\begin{pmatrix} 4 & 6 & 8 \\ 9 & 3 & 2 \end{pmatrix}\begin{pmatrix} 6 & 7 \\ 2 & 1 \\ 3 & 2 \end{pmatrix}$ (e) $\begin{pmatrix} 2 & 1 \\ 4 & 3 \end{pmatrix}\begin{pmatrix} 2 & 9 & 8 & 6 & 4 \\ 3 & 2 & 2 & 1 & 4 \end{pmatrix}$

(c) $\begin{pmatrix} 4 & 9 & 6 \\ 8 & 7 & 2 \end{pmatrix}\begin{pmatrix} 4 & 8 & 2 \\ 9 & 1 & 4 \end{pmatrix}$ (f) $\begin{pmatrix} 3 \\ 2 \\ 4 \end{pmatrix}(1, 6, 7)$

6. Consider pairs of matrices with the following orders. Determine which pairs of matrices can be multiplied and indicate the order of the product if multiplication is defined.

(a) $3 \times 2,\ 2 \times 4$
(b) $1 \times 3,\ 1 \times 3$
(c) $2 \times 2,\ 2 \times 3$
(d) $3 \times 3,\ 3 \times 3$
(e) $3 \times 1,\ 1 \times 3$
(f) $6 \times 8,\ 8 \times 7$

7. Let $A = \begin{pmatrix} a_{11} & a_{12} & \cdots & a_{18} \\ a_{21} & a_{22} & \cdots & a_{28} \end{pmatrix}$, $B = \begin{pmatrix} b_{11} & b_{12} \\ b_{21} & b_{22} \\ \vdots & \vdots \\ b_{81} & b_{82} \end{pmatrix}$, and $C = AB$.

Use summation notation to express the four elements of C.

8. A company employs three different types of workers: skilled, semi-skilled, and laborers. Of the skilled workers, 10 have been with the company less than a year, 20 from 1 to 4 years, 50 from 5 to 19 years, and 60 for 20 years or more. Of the semi-skilled employees, 50 have been with the company less than a year, 150 from 1 to 4 years, 250 from 5 to 19 years, and 200 for 20 years or more. Of the laborers, 90 have been with the company less than 1 year, 30 from 1 to 4 years, 50 from 5 to 19 years, and 40 for 20 years or more.

(a) Express the above in tabular form with type of worker as the row heading.
(b) Express the above as a matrix W with type of worker as the row heading.
(c) Using subscript notation, how would you make reference to the number of skilled workers who have been with the company for over 20 years?
(d) Using summation notation, how would you express the total number of laborers?

9. Given Table 2-2:

TABLE 2-2

Length of Time, t, with Company in Years	Vacation Days
$t < 1$	5
$1 \leqslant t < 5$	10
$5 \leqslant t < 20$	15
$20 \leqslant t$	20

(a) Express the number of vacation days as a column vector.
(b) Using the matrix developed in Problem 8 and the column vector from (a), calculate, by type of worker, the number of days of work that will be lost because of vacation.

10. Given Table 2-3:

TABLE 2-3

Type of Worker	Average Pay per Day
skilled	\$40.00
semi-skilled	30.00
laborer	25.00

(a) Express the data as a row vector.
(b) Using matrix multiplication and the result of Problem 9b, find the cost of providing vacations to employees this year.

11. Dancy, Inc. is going to begin producing a new chemical cleaner. It will be produced by combining alcohol, peroxide, and enzyme. Each quart of the new cleaner will require $\frac{1}{2}$ quart of alcohol, one quart of peroxide, and $\frac{1}{3}$ quart of enzyme. The costs per quart are 40¢ for alcohol, 60¢ for peroxide, and 20¢ for enzyme. The matrix operation to determine the cost of producing one quart of cleaner is

(a) $(\frac{1}{2}, 1, \frac{1}{3})\begin{pmatrix} 0.40 \\ 0.60 \\ 0.20 \end{pmatrix}$

(b) $\begin{pmatrix} \frac{1}{2} \\ 1 \\ \frac{1}{3} \end{pmatrix}\begin{pmatrix} 0.40 \\ 0.60 \\ 0.20 \end{pmatrix}$

(c) $(\frac{1}{2}, 1, \frac{1}{3})\,(0.40, 0.60, 0.20)$

(d) $\begin{pmatrix} 0.40 \\ 0.60 \\ 0.20 \end{pmatrix}(\frac{1}{2}, 1, \frac{1}{3})$

(e) None of the above.

2-5 SYSTEMS OF EQUATIONS AS SINGLE MATRIX EQUATIONS

The definitions of equality and multiplication of matrices make it possible to write a system of linear equations in n unknowns as a single matrix equation. The next two examples illustrate this point.

Example 1. Write the matrix equation

$$\begin{pmatrix} 2 & 3 & 1 \\ 4 & 2 & 3 \\ 1 & 1 & 6 \end{pmatrix}\begin{pmatrix} x \\ y \\ z \end{pmatrix} = \begin{pmatrix} 2 \\ 1 \\ 3 \end{pmatrix}$$

as a system of equations.

Solution: If we perform the indicated multiplication we arrive at the vector equation

$$\begin{pmatrix} 2x+3y+\ z \\ 4x+2y+3z \\ x+\ y+6z \end{pmatrix} = \begin{pmatrix} 2 \\ 1 \\ 3 \end{pmatrix}$$

This gives the system of equations

$$\begin{aligned} 2x + 3y + \ z &= 2 \\ 4x + 2y + 3z &= 1 \\ x + \ y + 6z &= 3 \end{aligned}$$

by definition of equality of vectors.

Example 2. Write the system of equations

$$\begin{aligned} x + 2y + 3z &= 1 \\ x + 3y + 5z &= 2 \\ 2x + 5y + 9z &= 3 \end{aligned}$$

as a single matrix equation.

Solution: Form the three matrices

$$\begin{pmatrix} 1 & 2 & 3 \\ 1 & 3 & 5 \\ 2 & 5 & 9 \end{pmatrix}, \quad \begin{pmatrix} x \\ y \\ z \end{pmatrix}, \quad \text{and} \quad \begin{pmatrix} 1 \\ 2 \\ 3 \end{pmatrix}$$

This system can be written as the single matrix equation

$$\begin{pmatrix} 1 & 2 & 3 \\ 1 & 3 & 5 \\ 2 & 5 & 9 \end{pmatrix}\begin{pmatrix} x \\ y \\ z \end{pmatrix} = \begin{pmatrix} 1 \\ 2 \\ 3 \end{pmatrix}$$

as can be verified by the method of Example 1.

In general, a system of equations of the form I,

$$\begin{aligned} a_{11}x_1 + a_{12}x_2 + \cdots + a_{1n}x_n &= b_1 \\ a_{21}x_1 + a_{22}x_2 + \cdots + a_{2n}x_n &= b_2 \\ \vdots \qquad \vdots \qquad \ddots \qquad \vdots \quad &\quad \vdots \\ a_{m1}x_1 + a_{m2}x_2 + \cdots + a_{mn}x_n &= b_m \end{aligned} \qquad (I)$$

where $x_1, x_2, \ldots$ are used instead of $x, y, \ldots$, can be written as a single matrix equation of form *II*.

$$\begin{pmatrix} a_{11} & a_{12} & \cdots & a_{1n} \\ a_{21} & a_{22} & \cdots & a_{2n} \\ \vdots & \vdots & \ddots & \vdots \\ a_{m1} & a_{m2} & \cdots & a_{mn} \end{pmatrix} \begin{pmatrix} x_1 \\ x_2 \\ \vdots \\ x_n \end{pmatrix} = \begin{pmatrix} b_1 \\ b_2 \\ \vdots \\ b_m \end{pmatrix} \qquad (II)$$

The matrix equation *II* can be abbreviated further to read

$$A\mathbf{x} = \mathbf{b}, \text{ where } A = (a_{ij}),\ \mathbf{x} = \begin{pmatrix} x_1 \\ x_2 \\ \vdots \\ x_n \end{pmatrix}, \text{ and } \mathbf{b} = \begin{pmatrix} b_1 \\ b_2 \\ \vdots \\ b_m \end{pmatrix}$$

and A is called the matrix of coefficients.

PROBLEMS

1. Write the following system in matrix form.

$$\begin{aligned} x - 2y + z &= 9 \\ 2x + y - 2z &= -1 \\ x + y + 3z &= 2 \end{aligned}$$

2. Given the following system of equations, write them as a matrix equation.

$$\begin{aligned} x + 2y + 3z &= 1 \\ x + 3y + 5z &= 2 \\ 2x + 5y + 9z &= 3 \end{aligned}$$

3. Write the following system of equations as a single matrix equation.

$$\begin{aligned} 4x + 6y + 3z &= 2 \\ 5x + 4y + z &= 1 \\ 5x + 6y &= 4 \end{aligned}$$

4. Write the following matrix equation as a system of equations.

$$\begin{pmatrix} 4 & 2 & 6 \\ 9 & 1 & 0 \\ 3 & 2 & 4 \end{pmatrix} \begin{pmatrix} x_1 \\ x_2 \\ x_3 \end{pmatrix} = \begin{pmatrix} 4 \\ 9 \\ 3 \end{pmatrix}$$

5. Write the following system as a matrix equation.

$$\begin{aligned} x_1 + 2x_2 + 3x_3 &= 2 \\ 2x_1 + 3x_2 + 4x_3 &= -5 \\ 2x_1 - x_2 &= 3 \end{aligned}$$

6. Write the following vector equation as a matrix equation.

$$x\begin{pmatrix}2\\1\\3\end{pmatrix}+y\begin{pmatrix}4\\1\\2\end{pmatrix}+z\begin{pmatrix}5\\8\\4\end{pmatrix}=\begin{pmatrix}2\\1\\3\end{pmatrix}$$

7. Substitute the following values, $\sum_{i=1}^{n} x_i = 1500$, $\sum_{i=1}^{n} y_i = 400$, $\sum_{i=1}^{n} x_i^2 = 550{,}000$, $\sum_{i=1}^{n} x_i y_i = 133{,}000$, and $n = 5$, in the normal equations of Problem 8 with the third equation and b_2 terms deleted.

8. Verify that $b_0 = 41$ and $b_1 = 0.13$ is the solution to the system in Problem 7.

Problems 9 and 10 are designed to give computational experience in fitting a function of the form $y = b_0 + b_1 x_1 + b_2 x_2$ to data. In this case y is a *linear* function of the two independent variables x_1 and x_2. It can be shown that the resulting normal equations for this three-variable model are:

$$\begin{aligned}\sum y &= b_0 n + b_1 \sum x_1 + b_2 \sum x_2\\ \sum x_1 y &= b_0 \sum x_1 + b_1 \sum x_1{}^2 + b_2 \sum x_1 x_2\\ \sum x_2 y &= b_0 \sum x_2 + b_1 \sum x_1 x_2 + b_2 \sum x_2{}^2\end{aligned}$$

(*normal equations for the three-variable case*)

The summations are all from 1 to n, where n is the number of observations of y.

9. Set up the normal equations for the preceding three-variable case in matrix form.

10. **(a)** Set up the normal equations for fitting the function $\hat{y} = b_0 + b_1 x_1 + b_2 x_2$ to the data from Table 2-4.

TABLE 2-4

x_1	x_2	y
1	1	8.9
1	1	9.1
−1	1	2.8
−1	1	3.2
−1	−1	−5.3
−1	−1	−4.7
1	−1	0.9
1	−1	1.1

(b) Determine b_0, b_1, and b_2.

2-6 BUSINESS APPLICATIONS OF VECTOR AND MATRIX MULTIPLICATION

We will begin this section with an example that involves multiplication of a *vector* by a *scalar*.

Example 1. The material needed to build one wall of a building is known to be five pounds of nails, twenty two-by-fours, four two-by-sixes, and five sheets of plywood. This information can be expressed as the row vector $\mathbf{m} = (5, 20, 4, 5)$. How much material will be needed to construct four walls of this type?

Solution: The amount of material needed can be found by performing the multiplication $4\mathbf{m} = (20, 80, 16, 20)$. Note that the first component of $4\mathbf{m}$ equals 20; this means that 20 pounds of nails are needed to make the four walls.

The next example involves multiplication of a *column vector* by a *row vector* to determine a total material cost.

Example 2. In addition to the information given in Example 1, it is known that nails cost \$0.20/pound, two-by-fours cost \$0.60 each, two-by-sixes cost \$3.00 each, and plywood costs \$6.00/sheet. This information can be expressed by the column vector $\mathbf{c} = \begin{pmatrix} 0.20 \\ 0.60 \\ 3.00 \\ 6.00 \end{pmatrix}$. Find the total cost of materials.

Solution: The scalar quantity $\mathbf{mc} = (5, 20, 4, 5)\begin{pmatrix} 0.20 \\ 0.60 \\ 3.00 \\ 6.00 \end{pmatrix} = \$55.00$ is the cost of the material needed to produce one wall.

The next example involves multiplication of a *matrix* by a *vector*.

Example 3. Consider a manufacturer who can produce two different products, P_1 and P_2. One unit of P_1 requires 3 parts of type a and 2 parts of type b. Also, one unit of P_2 requires 2 parts of type a, 6 parts of type b, and 2 parts of type c. This information can be displayed in matrix form as follows.

		Parts a	b	c
Products	1	3	2	0
	2	2	6	2

Note: One unit of product 1 requires 3 parts of type a, 2 parts of type b, and no parts of type c, and so forth.

Call the matrix $\begin{pmatrix} 3 & 2 & 0 \\ 2 & 6 & 2 \end{pmatrix}$ the parts matrix P. Let us suppose that the company has been requested to fill an order for 300 units of P_1 and 150 units of P_2. Adopting the convention of *representing orders by row vectors*, we have $\mathbf{o} = (300, 150)$. Find the number of parts of each type required to fill the order.

Solution: Multiply the parts matrix P by the order vector $\mathbf{o}$. This gives $(300, 150)\begin{pmatrix} 3 & 2 & 0 \\ 2 & 6 & 2 \end{pmatrix} = (1200, 1500, 300)$. Interpreting this answer, we see that 1200 parts of type a, 1500 parts of type b, and 300 parts of type c are needed to fill the order.

The following type of problem can be handled conveniently by using matrix multiplication.

A manufacturer can produce four products, A, B, C, and D. Each of these products requires sub-assemblies I, II, III, IV, and V. Each sub-assembly requires parts a, b, c, d, e, f, and g. Given an order for four products, determine the number of each part that is required in order to fill the order. To solve such a problem, form the sub-assembly matrix S.

		Sub-Assembly				
		I	*II*	*III*	*IV*	*V*
	A					
Product	*B*					
	C					
	D					

Next form the parts matrix P.

		Parts						
		a	*b*	*c*	*d*	*e*	*f*	*g*
	I							
	II							
Sub-Assembly	*III*							
	IV							
	V							

The matrix obtained by multiplying the *sub-assembly matrix* times the *parts matrix* gives a matrix SP, which expresses the number of parts needed for each product. Finally, the solution is obtained by multiplying the order size expressed as a row vector $\mathbf{o}$ times the matrix, SP. This is an example of a *parts explosion* problem.

Notice that in setting up these matrices, we placed the highest order of production (i.e., a product is of higher order than a sub-assembly and a sub-assembly is of higher order than a part) as *row headings*. This insures that the matrix multiplications will always be possible and that we are always adding like units. In the previous example, the use of this convention leads to the multiplication of a (4×5) matrix times a (5×7) matrix. Without this convention the second factor could have been a (7×5) matrix in which case multiplication would be undefined. The product $\mathbf{o}\,(SP)$ is a (1×7) vector and gives the amount of each part required to fill the order.

Example 4. Given the sub-assembly matrix:

$$S = \begin{array}{c} \\ A \\ B \\ C \\ D \end{array} \begin{array}{c} \textit{Sub-Assembly} \\ \begin{array}{ccccc} \text{I} & \text{II} & \text{III} & \text{IV} & \text{V} \end{array} \\ \begin{pmatrix} 4 & 2 & 1 & 6 & 7 \\ 3 & 1 & 2 & 4 & 4 \\ 2 & 3 & 5 & 2 & 1 \\ 1 & 2 & 4 & 1 & 2 \end{pmatrix} \end{array}$$

and the parts matrix:

$$P = \begin{array}{c} \\ \text{I} \\ \text{II} \\ \text{III} \\ \text{IV} \\ \text{V} \end{array} \begin{array}{c} \textit{Part} \\ \begin{array}{ccccccc} a & b & c & d & e & f & g \end{array} \\ \begin{pmatrix} 2 & 3 & 5 & 7 & 8 & 2 & 1 \\ 1 & 4 & 3 & 6 & 2 & 1 & 5 \\ 5 & 2 & 4 & 6 & 1 & 8 & 4 \\ 1 & 2 & 2 & 1 & 3 & 4 & 6 \\ 8 & 1 & 6 & 7 & 2 & 4 & 2 \end{pmatrix} \end{array}$$

and the order size vector $\mathbf{o} = (20, 10, 10, 5)$. Find the total number of each kind of part needed.

Solution: The solution is simply $\mathbf{o}(SP)$. The element in the upper left-hand corner of SP is given by $4(2)+2(1)+(1)(5)+(6)(1)+(7)(8) = 8+2+5+6+56 = 77$.

Continuing in this manner, we have

$$SP = \begin{array}{c} \\ A \\ B \\ C \\ D \end{array} \begin{array}{c} \textit{Part} \\ \begin{array}{ccccccc} a & b & c & d & e & f & g \end{array} \\ \begin{pmatrix} 77 & 41 & 84 & 101 & 69 & 70 & 68 \\ 53 & 29 & 58 & 71 & 48 & 55 & 48 \\ 42 & 33 & 49 & 71 & 35 & 59 & 51 \\ 41 & 23 & 41 & 58 & 23 & 48 & 37 \end{pmatrix} \end{array}$$

To produce a unit of product A, one needs 77 of part a, 41 of part b, and so forth. If we multiply $\mathbf{o}$ times SP, we get the number of each part required to fill the order.

$$\mathbf{o}(SP) = (20, 10, 10, 5)(SP)$$

$$\mathbf{o}(SP) = (2695, 1555, 2955, 3730, 2325, 2780, 2535)$$

The first component of this vector is given by

$$= 20(77) + 10(53) + 10(42) + 5(41)$$

$$= 1540 + 530 + 420 + 205$$

$$= 2695$$

We now know that 2695 units of part a, 1555 units of part b, and so forth are required in order to fill the order.

Let us carry this one step further. If we know the cost of each part, we can compute the total material cost of the order.

The product $\mathbf{o}(SP)$ is a (1×7) vector that gives the amount of each part required to fill the order. If we write the cost per part as a column vector, we have

$$\mathbf{c} = \begin{pmatrix} \$1 \\ 0.5 \\ 2 \\ 1 \\ 3 \\ 2 \\ 1 \end{pmatrix}$$

where $1.00 is the cost of part *a*
$0.50 is the cost of part *b*

The result gives a Total Cost equal to $[\mathbf{o}(SP)]\mathbf{c}$, where

$$[\mathbf{o}(SP)]\,\mathbf{c} = (2695, 1555, 2955, 3730, 2325, 2780, 2535)\begin{pmatrix} \$1 \\ 0.5 \\ 2 \\ 1 \\ 3 \\ 2 \\ 1 \end{pmatrix}$$

$$= \$28{,}182.50$$

Therefore, the total material cost of the order is $28,182.50.

PROBLEMS

1. Given that a person purchases the following items: 1 stepladder, 4 gallons of red paint, 1 gallon of white paint, 2 paint brushes, and 1 five-quart pail, express these quantities as a row vector.
2. Given that the items in Problem 1 cost $24.00, $5.00, $4.50, $4.00, and $0.25, respectively, express these costs as a column vector.
3. Obtain the total cost of the purchase using Problems 1 and 2 and vector multiplication.
4. Consider a manufacturer who can produce YIPS, YAPS, and ZAPS. Assume that in order to produce one YIP, 2 units of *A* and 6 units of *B* are required. To produce one YAP, 3 units of *A* and 1 unit of *B* are required. One ZAP requires 1 unit of *A* and 2 units of *B*. Construct a parts matrix. Use parts as the column headings.
5. If you received an order for 6 YIPS, 2 YAPS, and 7 ZAPS, would you use a row or column vector to represent the order, assuming that the formula to compute the total number of parts required is determined by multiplying the order vector times the parts matrix?

6. Multiply the order vector of Problem 5 times the parts matrix of Problem 4. Determine the total number of units required.

7. Given that the cost of one unit of A is \$2.00 and the cost of one unit of B is \$3.00, express the cost as a vector.

8. Obtain the total cost of the order using the results of Problems 6 and 7.

9. We can produce two products P_1 and P_2. To produce a P_1 requires four of sub-assembly I, two of sub-assembly II, and three of sub-assembly III. To produce a P_2 requires two of sub-assembly I and one of sub-assembly III. Form the sub-assembly matrix. [*Note:* Place the highest order of production as row headings.]

10. Each sub-assembly in Problem 9 requires the following parts: Sub-assembly I requires two units of part number 061 and three units of part number 072; sub-assembly II requires one of part number 061 and two of part number 072; sub-assembly III requires four units of part number 061 and five units of part number 072. Form the parts matrix. [*Note:* Place the highest order of production as row headings.]

11. Using Problems 9 and 10, determine how many of each part is required to produce each product.

12. Given an order for sixteen P_1's and ten P_2's, write this order as a row vector and find out how many of each part are required in inventory to fill the order. Use the result of Problem 11.

13. If the cost per unit of part number 061 is \$0.50 and the cost per unit of part number 072 is \$1.00, compute the total material cost of the order. Use the result of Problem 12.

14. Given the following information:

$$\text{Product}\ \begin{matrix} & \begin{matrix} \text{I} & \text{II} \end{matrix} \\ \begin{matrix} A \\ B \end{matrix} & \begin{pmatrix} 5 & 4 \\ 6 & 1 \end{pmatrix} \end{matrix} \qquad \text{Sub-Assembly}\ \begin{matrix} & \begin{matrix} a & b & c \end{matrix} \\ \begin{matrix} \text{I} \\ \text{II} \end{matrix} & \begin{pmatrix} 5 & 4 & 2 \\ 3 & 1 & 0 \end{pmatrix} \end{matrix}$$

(Left matrix columns: *Sub-Assembly*; right matrix columns: *Part*.)

- **(a)** What is the significance of the figures in these matrices?
- **(b)** Given the order vector (100, 200), how many parts are required to fill the order?
- **(c)** Given that the cost of part a is \$0.25, part b is \$0.30, and part c is \$0.10, express cost as a column vector.
- **(d)** What is the material cost of the order?

15. If at present we have 10,000 units of part a on hand, 7500 units of part b, and 4000 units of part c, can we fill the order in Problem 14? If not, what part, or parts, are needed?

16. Let the following variable cost information be given for a firm making tables, chairs, and desks.

	Tables (x_1)	Chairs (x_2)	Desks (x_3)
Material	\$14	\$11	\$10
Labor	5	4	8
Variable overhead	3	1	2

- **(a)** Express the information in matrix form. Call this matrix A.
- **(b)** Given an order for 5 tables, 20 chairs, and 100 desks, use the matrix A in (a) to compute the total variable cost.

17. Find the total variable cost function for the firm described in Problem 16.

18. A firm purchases two machines at a cost of $10,000 and $20,000, respectively. Both machines have a lifetime of four years and have zero salvage values at the end of the four years. The two machines are depreciated using the "straight line method" (i.e., equal amounts each year). Express the costs as a row vector and the percentages as a column vector. By matrix multiplication, find the depreciation schedule.

THREE

Algebra of square matrices

3-1 INTRODUCTION TO THE ALGEBRA OF SQUARE MATRICES

There are many interesting similarities, but some important differences as well, between the algebra of square matrices and elementary algebra. Our objective in this chapter is to compare these two algebras. Elementary algebra deals with the real number system. The real number system is discussed briefly in the preparation chapter, in connection with the *number line*. The real number system can also be viewed as a mathematical system that satisfies certain axioms. A partial list of the axioms for the real number system is given next. One additional axiom, called a "completeness" axiom, is needed to fully describe the real number system. Any completeness axiom involves some concepts that are not treated in this text, so we have not included one here.

The axioms that are included are sometimes called the *field axioms*. The rational numbers satisfy the field axioms.

FIELD AXIOMS

A-1	Every pair of real numbers a and b has a unique sum $a+b$ and a unique product ab which are real numbers.	*Law of closure*
A-2	$a+b=b+a$, $ab=ba$ for all real numbers a and b.	*Commutative laws*
A-3	$a+(b+c)=(a+b)+c$, $a(bc)=(ab)c$ for all real numbers a, b, and c.	*Associative laws*
A-4	$a(b+c)=ab+ac$ for all real numbers a, b, and c.	*Distributive law*

A-5	There exists a unique real number 0, called the *additive identity* with the property that $a+0=0+a=a$ for every real number a. The additive identity is called zero.	*Additive identity*
A-6	For every real number a there exists a unique real number $-a$, called the *additive inverse* of a, with the property that $a+(-a)=0$.	*Additive inverse*
A-7	There exists a unique real number 1, called the *multiplicative identity*, with the property that $a \cdot 1 = 1 \cdot a = a$ for every real number a. The multiplicative identity is called one.	*Multiplicative identity*
A-8	For every nonzero real number a there exists a unique real number a^{-1}, called the *multiplicative inverse* of a, with the property that $a^{-1}a = aa^{-1} = 1$.	*Multiplicative inverse*

The eight properties provide a partial description of the real number system. Many of the familiar rules of elementary algebra can be derived from these eight properties, or axioms. The axioms are illustrated with examples, but no attempt is made to include proofs of the rules of algebra derived from these axioms. We have included this list of axioms to provide a basis for comparison of the algebra of square matrices with the algebra of real numbers.

Example 1. Use the distributive law to rewrite $2x+2y$.

Solution: This is a simple example of factoring. Therefore, by A-4,

$$2x + 2y = 2(x+y)$$

Example 2. Verify the associative law for the numerical example

$$2 + (3+7) = (2+3) + 7$$

Solution:

$$2 + (3+7) = 2 + 10 = 12 \qquad (2+3) + 7 = 5 + 7 = 12$$

Example 3. Find the multiplicative inverse of 3.

Solution: Recall that $3 \cdot \frac{1}{3} = 1$. Therefore, by A-8,

$$3^{-1} = \tfrac{1}{3}$$

Example 4. Show that $(a+b)^2 = a^2+2ab+b^2$.

Solution: $(a+b)^2 = (a+b)(a+b)$.

Let $c = a+b$. Then

$$\begin{aligned}
(a+b)(a+b) &= c(a+b) \\
&= ca + cb && \text{A-4} \\
&= ac + bc && \text{A-2} \\
&= a(a+b) + b(a+b) \\
&= a^2 + ab + ba + b^2 && \text{A-4} \\
&= a^2 + ab + ab + b^2 && \text{A-2} \\
&= a^2 + ab(1+1) + b^2 && \text{A-4} \\
&= a^2 + ab(2) + b^2 \\
&= a^2 + 2ab + b^2 && \text{A-2}
\end{aligned}$$

The familiar operations of subtraction and division are defined as follows: $a-b = a+(-b)$, where $-b$ is the additive inverse of b; $a \div b = ab^{-1}$, $b \neq 0$, where b^{-1} is the multiplicative inverse of b.

The next example illustrates the fact that division is not associative.

Example 5. Compare $12 \div (6 \div 2)$ with $(12 \div 6) \div 2$.

Solution:

$$12 \div (6 \div 2) = 12 \div 3 = 4$$

$$(12 \div 6) \div 2 = 2 \div 2 = 1$$

Hence, $12 \div (6 \div 2) \neq (12 \div 6) \div 2$, and division is not associative.

We point out that in the more common notation for division, the result of Example 5 takes the form $\frac{12}{6/2} = \frac{12}{3} = 4$, and $\frac{12/6}{2} = \frac{2}{2} = 1$ as before. This means that we must be precise and distinguish between $\frac{a}{b/c}$ and $\frac{a/b}{c}$ because division is not an associative operation.

Example 6. Solve the equation $3x+4 = 7$ for x.

Solution:

$$3x + 4 = 7$$

$$3x + 4 + (-4) = 7 + (-4)$$

$$3x + 0 = 3 \qquad \text{A-6}$$

$3x = 3$	*A-5*
$3^{-1}(3x) = 3^{-1}(3)$	
$1x = 1$	*A-8*
$x = 1$	*A-7*

Note that multiplying both sides of an equation by a^{-1}, where $a \neq 0$, is the same as dividing both sides of the equation by a.

Now that we have recalled some basic facts from elementary algebra, we are ready to resume our study of matrices. As the title of this chapter suggests, the emphasis here is on square (n by n) matrices.

In Chapter 2 we discussed the addition and multiplication of matrices. The reader will recall that two matrices cannot be *added* unless they are both the *same order*. In the case of *multiplication*, the left factor must have the *same number of columns* as the right factor has *rows* in order for the product to be defined. In this chapter we will restrict our attention to the class of n by n matrices (i.e., square matrices). For a fixed value of n, both the sum and the product of any pair of n by n matrices are defined.

We restrict our attention for the moment to the class of 3 by 3 matrices as a typical class of n by n matrices. Consider the 3 by 3 matrix

$$Z = \begin{pmatrix} 0 & 0 & 0 \\ 0 & 0 & 0 \\ 0 & 0 & 0 \end{pmatrix}$$

If we compute the sum $A + Z$, where $A = (a_{ij})$, $i, j = 1, 2, 3$, we have

$$\begin{pmatrix} a_{11} & a_{12} & a_{13} \\ a_{21} & a_{22} & a_{23} \\ a_{31} & a_{32} & a_{33} \end{pmatrix} + \begin{pmatrix} 0 & 0 & 0 \\ 0 & 0 & 0 \\ 0 & 0 & 0 \end{pmatrix} = \begin{pmatrix} a_{11} & a_{12} & a_{13} \\ a_{21} & a_{22} & a_{23} \\ a_{31} & a_{32} & a_{33} \end{pmatrix} \quad \textit{by definition of addition of matrices}$$

Briefly, $A + Z = A$. In other words, the matrix Z behaves in the class of 3 by 3 matrices like the additive identity 0 of the real number system.

Assisted by the fact that there is a 3 by 3 matrix which behaves like a *zero*, it seems natural to seek a matrix which behaves like a *one* in the sense that $AI = IA = A$. We might test the 3 by 3 matrix with all ones as its elements for this property. A quick check will reveal that the matrix just described does not have the desired property. However, if we retain three of the ones and consider the matrix

$$I = \begin{pmatrix} 1 & 0 & 0 \\ 0 & 1 & 0 \\ 0 & 0 & 1 \end{pmatrix}$$

we find that $AI = A$.

Example 7. Show that $AI = A$, where $A = (a_{ij})$, $i, j = 1, 2, 3$, and

$$I = \begin{pmatrix} 1 & 0 & 0 \\ 0 & 1 & 0 \\ 0 & 0 & 1 \end{pmatrix}$$

Solution:

$$AI = \begin{pmatrix} a_{11} & a_{12} & a_{13} \\ a_{21} & a_{22} & a_{23} \\ a_{31} & a_{32} & a_{33} \end{pmatrix}\begin{pmatrix} 1 & 0 & 0 \\ 0 & 1 & 0 \\ 0 & 0 & 1 \end{pmatrix}$$

$$= \begin{pmatrix} a_{11}+0+0 & 0+a_{12}+0 & 0+0+a_{13} \\ a_{21}+0+0 & 0+a_{22}+0 & 0+0+a_{23} \\ a_{31}+0+0 & 0+a_{32}+0 & 0+0+a_{33} \end{pmatrix} \quad \textit{by definition of matrix multiplication}$$

$$= \begin{pmatrix} a_{11} & a_{12} & a_{13} \\ a_{21} & a_{22} & a_{23} \\ a_{31} & a_{32} & a_{33} \end{pmatrix}$$

$$= A$$

DEFINITION 3.1. *The elements $a_{11}, a_{22}, a_{33}, \ldots, a_{nn}$ of the n by n matrix A constitute what is called the principal diagonal of A.*

DEFINITION 3.2. *The n by n matrix Z, all of whose elements equal zero, is called the additive identity for the class of n by n matrices.*

DEFINITION 3.3. *The n by n matrix I, with ones on the principal diagonal and zeros elsewhere, is called the multiplicative identity for the class of n by n matrices.*

Example 8. Write the additive and multiplicative identities for the class of 4 by 4 matrices.

Solution:

$$Z = \begin{pmatrix} 0 & 0 & 0 & 0 \\ 0 & 0 & 0 & 0 \\ 0 & 0 & 0 & 0 \\ 0 & 0 & 0 & 0 \end{pmatrix} \quad \text{and} \quad I = \begin{pmatrix} 1 & 0 & 0 & 0 \\ 0 & 1 & 0 & 0 \\ 0 & 0 & 1 & 0 \\ 0 & 0 & 0 & 1 \end{pmatrix}$$

The multiplicative identity I for a given value of n is often called simply the *identity matrix* or just the *identity*.

PROBLEMS

Perform the indicated operations in Problems 1–10.

1. $\begin{pmatrix} 2 & -4 \\ 1 & 7 \end{pmatrix} + \begin{pmatrix} 0 & 0 \\ 0 & 0 \end{pmatrix}$

2. $\begin{pmatrix} 1 & 0 \\ 0 & 1 \end{pmatrix}\begin{pmatrix} 4 & -3 \\ -6 & 5 \end{pmatrix}$

3. $\begin{pmatrix} 1 & 2 \\ 3 & 4 \end{pmatrix}\begin{pmatrix} -2 & 1 \\ \frac{3}{2} & -\frac{1}{2} \end{pmatrix}$

4. $\begin{pmatrix} 4 & -2 & -5 \\ 6 & 1 & 3 \\ 2 & 4 & 7 \end{pmatrix}\begin{pmatrix} 1 & 0 & 0 \\ 0 & 1 & 0 \\ 0 & 0 & 1 \end{pmatrix}$

5. $\begin{pmatrix} 1 & 0 & 0 \\ 0 & 1 & 0 \\ 0 & 0 & 0 \end{pmatrix} - \begin{pmatrix} 4 & -3 & 2 \\ 1 & 6 & 5 \\ 4 & 3 & 8 \end{pmatrix}$

6. $\begin{pmatrix} 1 & 0 & 0 \\ 0 & 1 & 0 \\ 0 & 0 & 1 \end{pmatrix}\begin{pmatrix} 1 & 0 & 0 \\ 0 & 1 & 0 \\ 0 & 0 & 1 \end{pmatrix}$

7. $\begin{pmatrix} 0 & 0 \\ 0 & 0 \end{pmatrix}\begin{pmatrix} 1 & 0 \\ 0 & 1 \end{pmatrix}$

8. $\begin{pmatrix} 4 & 3 \\ -2 & -6 \end{pmatrix} - \begin{pmatrix} 4 & 3 \\ -2 & -6 \end{pmatrix}$

9. $\begin{pmatrix} 6 & 8 \\ -3 & -5 \end{pmatrix} + \begin{pmatrix} -6 & -8 \\ 3 & 5 \end{pmatrix}$

10. $\left[\begin{pmatrix} 1 & 3 \\ 2 & 4 \end{pmatrix} + \begin{pmatrix} 3 & 5 \\ 6 & 7 \end{pmatrix}\right] + \begin{pmatrix} 4 & 8 \\ 4 & 9 \end{pmatrix}$

11. What is the appropriate additive identity for the following matrices?

(a) $\begin{pmatrix} 2 & 4 & 1 & 6 \\ 8 & 3 & 2 & 4 \\ 5 & 9 & 6 & 2 \\ 1 & 8 & 3 & 4 \end{pmatrix}$ **(b)** $\begin{pmatrix} 2 & 1 \\ 3 & 8 \end{pmatrix}$

12. What is the appropriate multiplicative identity for the matrices in Problem 11?

13. Show that $AI = IA$ for the class of 3 by 3 matrices.

14. Show that $ZA = Z$ for the class of 3 by 3 matrices. The trace of a square matrix A is defined to be the sum of the elements in the principal diagonal and is denoted by tr A.

15. Observe that tr $A = \sum_{i=1}^{n} a_{ii}$.

16. Compute the trace of A $\begin{pmatrix} 2 & 1 & 3 \\ 4 & 1 & 2 \\ 1 & 8 & 3 \end{pmatrix}$.

17. Show that $\operatorname{tr}(kA) = k \operatorname{tr} A$, where k is a real number.

18. Show that $\operatorname{tr}(A+B) = \operatorname{tr} A + \operatorname{tr} B$.

A solid waste manager is in charge of the collection system described as follows. The system consists of three trucks (with crews), three garages, three collection zones, and one land fill site. The garages are built at various places throughout the collection region other than the land fill site. Problems 19 through 22 relate to this system.

19. Let L denote the land fill site, g_i $(i = 1,2,3)$ denote the garages, and z_i $(i = 1,2,3)$ denote the collection zones. Draw a schematic diagram for the system using the suggested notation.

20. If

$$a_{ij} = \text{distance from } g_i \text{ to } z_i$$

$$x_j = \text{distance from } z_j \text{ to } L$$

$$b_i = \text{distance from } L \text{ to } g_i$$

$$y_j = \text{distance through } z_j$$

$$d_{ij} = \text{total distance traveled from garage } i\text{, collecting in zone } j\text{, traveling to the land fill site, and back to garage } i.$$

express d_{ij} in terms of the variables defined.

21. Use the result of Problem 20 to determine the total distance matrix, $D = (d_{ij})$, for the waste collecting system in Figure 3-1.

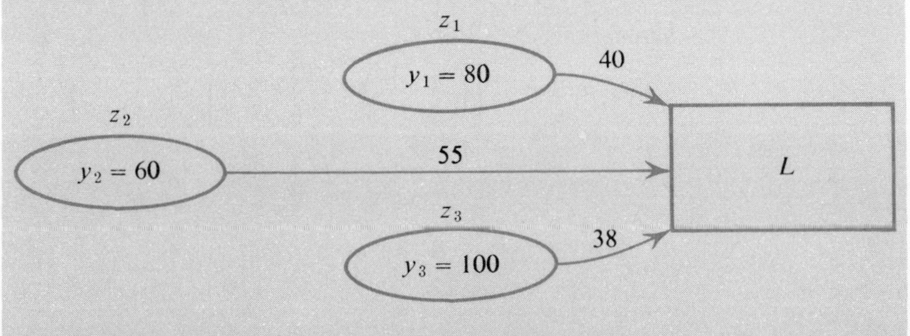

FIGURE 3-1

$$\mathbf{b} = (b_1, b_2, b_3) = (110, 170, 38) \qquad A = (a_{ij}) = \begin{pmatrix} 12 & 18 & 14 \\ 6 & 10 & 12 \\ 45 & 12 & 25 \end{pmatrix}$$

22. An assignment problem.

(a) Use the total distance matrix D from Problem 21 with garages as row headings and zones as column headings to determine the sum of the distances of all three trucks if the manager assigns zone 1 to garage 1, zone 2 to garage 2, and zone 3 to garage 3.

(b) How many different ways can the manager assign zones to garages?

(c) What is the best assignment in the sense of minimum combined distance?

3-2 SOME ALGEBRAIC LAWS FOR SQUARE MATRICES

In addition to possessing both an additive and a multiplicative identity, square matrices also satisfy the following important *theorems* where A, B, and C are n by n matrices.

1. $A+B=B+A$ — Commutative Law for Addition
2. $(A+B)+C=A+(B+C)$ — Associative Law for Addition
3. $(AB)C=A(BC)$ — Associative Law for Multiplication
4. $A(B+C)=AB+AC$ — Left-hand Distributive Law
5. $(A+B)C=AC+BC$ — Right-hand Distributive Law

Remark: Even though we are restricting our attention to square matrices in this chapter, these properties hold for rectangular (non-square) matrices as well, provided that the orders are such that the operations are defined.

Example 1. Let $A=\begin{pmatrix}1 & 3\\2 & 4\end{pmatrix}$, $B=\begin{pmatrix}3 & 5\\6 & 7\end{pmatrix}$, and $C=\begin{pmatrix}4 & 8\\4 & 9\end{pmatrix}$. Verify the associative law for addition (i.e., calculate $(A+B)+C$ and $A+(B+C)$ and compare the resulting matrices for equality).

Solution:

$$(A+B)+C=\left[\begin{pmatrix}1 & 3\\2 & 4\end{pmatrix}+\begin{pmatrix}3 & 5\\6 & 7\end{pmatrix}\right]+\begin{pmatrix}4 & 8\\4 & 9\end{pmatrix}$$

$$=\begin{pmatrix}4 & 8\\8 & 11\end{pmatrix}+\begin{pmatrix}4 & 8\\4 & 9\end{pmatrix}$$

$$=\begin{pmatrix}8 & 16\\12 & 20\end{pmatrix}$$

$$A+(B+C)=\begin{pmatrix}1 & 3\\2 & 4\end{pmatrix}+\left[\begin{pmatrix}3 & 5\\6 & 7\end{pmatrix}+\begin{pmatrix}4 & 8\\4 & 9\end{pmatrix}\right]$$

$$=\begin{pmatrix}1 & 3\\2 & 4\end{pmatrix}+\begin{pmatrix}7 & 13\\10 & 16\end{pmatrix}$$

$$=\begin{pmatrix}8 & 16\\12 & 20\end{pmatrix}$$

Therefore, $(A+B)+C=A+(B+C)$.

Example 2. Use the matrices A, B, and C given in Example 1 to verify the left-hand distributive law for square matrices.

Solution: Show that $A(B+C) = AB + AC$.

$$A(B+C) = \begin{pmatrix} 1 & 3 \\ 2 & 4 \end{pmatrix}\left[\begin{pmatrix} 3 & 5 \\ 6 & 7 \end{pmatrix} + \begin{pmatrix} 4 & 8 \\ 4 & 9 \end{pmatrix}\right]$$

$$= \begin{pmatrix} 1 & 3 \\ 2 & 4 \end{pmatrix}\begin{pmatrix} 7 & 13 \\ 10 & 16 \end{pmatrix}$$

$$= \begin{pmatrix} 37 & 61 \\ 54 & 90 \end{pmatrix}$$

$$AB + AC = \begin{pmatrix} 1 & 3 \\ 2 & 4 \end{pmatrix}\begin{pmatrix} 3 & 5 \\ 6 & 7 \end{pmatrix} + \begin{pmatrix} 1 & 3 \\ 2 & 4 \end{pmatrix}\begin{pmatrix} 4 & 8 \\ 4 & 9 \end{pmatrix}$$

$$= \begin{pmatrix} 21 & 26 \\ 30 & 38 \end{pmatrix} + \begin{pmatrix} 16 & 35 \\ 24 & 52 \end{pmatrix}$$

$$\begin{pmatrix} 37 & 61 \\ 54 & 90 \end{pmatrix}$$

Therefore, $A(B+C) = AB + AC$, as expected.

PROBLEMS

Use the matrices $A = \begin{pmatrix} 2 & 4 \\ 6 & 3 \end{pmatrix}$, $B = \begin{pmatrix} 8 & 7 \\ 2 & 1 \end{pmatrix}$ and $C = \begin{pmatrix} 4 & 9 \\ 2 & 3 \end{pmatrix}$ in Problems 1 through 8.

1. Verify the associative law $(A+B)+C = A+(B+C)$.
2. Verify the associative law $(AB)C = A(BC)$.
3. Verify the left-hand distributive law $A(B + C) = AB + AC$.
4. Verify the right-hand distributive law $(A + B)C = AC + BC$.
5. Verify the commutative law, $A + B = B + A$
6. Does $AB = BA$?
7. Does $BC = CB$?
8. Does $AC = CA$?

Use the matrices $D = \begin{pmatrix} 1 & 2 & 4 \\ -2 & 1 & -6 \end{pmatrix}$, $E = \begin{pmatrix} 2 \\ -1 \\ 3 \end{pmatrix}$, and $F = \begin{pmatrix} 5 \\ 2 \\ -4 \end{pmatrix}$ in Problems 9 through 12.

9. **(a)** Is $(D+E)+F$ defined? Explain.
   (b) Is $D+(E+F)$ defined? Explain.
   (c) Is $(D+E)+F = D+(E+F)$ a meaningful statement in this example? Explain.
10. **(a)** Is $(DE)F$ defined? Explain.
    (b) Is $D(EF)$ defined? Explain.
    (c) Is $(DE)F = D(EF)$ a meaningful statement in this example? Explain.

11. **(a)** Is $D(E+F)$ defined?
(b) Is $DE+DF$ defined?
(c) Verify that the left-hand distributive law $D(E + F) = DE + DF$.

12. **(a)** Is $(D+E)F$ defined? Explain.
(b) Is $DF+EF$ defined? Explain.
(c) Is $(D+E)F = DF+EF$ a meaningful statement in this example? Explain.

13. Let $A = \begin{pmatrix} a_{11} & a_{12} \\ a_{21} & a_{22} \end{pmatrix}$ and $B = \begin{pmatrix} b_{11} & b_{12} \\ b_{21} & b_{22} \end{pmatrix}$. Show that $A+B = B+A$. [*Hint:* Use Axiom A-2 from Section 1-1.]

14. Compute $\begin{pmatrix} 2 & 1 & 3 \\ 3 & 1 & 2 \\ 1 & 1 & 4 \end{pmatrix}\begin{pmatrix} 4 & 1 & 2 \\ 1 & 3 & 2 \\ 2 & 1 & 1 \end{pmatrix}$.

15. Compute $\begin{pmatrix} 4 & 1 & 2 \\ 1 & 3 & 2 \\ 2 & 1 & 1 \end{pmatrix}\begin{pmatrix} 2 & 1 & 3 \\ 3 & 1 & 2 \\ 1 & 1 & 4 \end{pmatrix}$.

16. Are the results in Problems 14 and 15 equal? Can you see any implications of this situation for the existence of a commutative law for multiplication of 3 by 3 matrices?

17. Show that $A+B = B+A$, where A and B are both m by n matrices with real elements. [*Hint:* Let $X = A+B$ and $Y = B+A$. Then show that $X = Y$ by using appropriate matrix definitions and properties of real numbers.]

18. Show that $(A+B)+C = A+(B+C)$, where A, B, and C are all m by n matrices with real elements.

19. Prove the distributive law $A(B+C) = AB+AC$ for the class of 2 by 2 matrices.

3-3 FAILURE OF THE COMMUTATIVE LAW FOR MATRIX MULTIPLICATION

Conspicuously absent from the list of theorems in the previous section is the commutative law for multiplication. To see that this law does not hold in general even for square matrices, we will cite the following examples involving

$$A = \begin{pmatrix} 1 & 2 & 1 \\ 3 & 1 & 2 \\ 4 & 2 & 1 \end{pmatrix} \qquad \text{and} \qquad B = \begin{pmatrix} 2 & 1 & 2 \\ 1 & 2 & 3 \\ 2 & 1 & 3 \end{pmatrix}$$

Example 1. | Compute AB.

Solution:

$$\begin{pmatrix} 1 & 2 & 1 \\ 3 & 1 & 2 \\ 4 & 2 & 1 \end{pmatrix}\begin{pmatrix} 2 & 1 & 2 \\ 1 & 2 & 3 \\ 2 & 1 & 3 \end{pmatrix} = \begin{pmatrix} 6 & 6 & 11 \\ 11 & 7 & 15 \\ 12 & 9 & 17 \end{pmatrix}$$

Example 2. Compute BA.

Solution:

$$\begin{pmatrix} 2 & 1 & 2 \\ 1 & 2 & 3 \\ 2 & 1 & 3 \end{pmatrix}\begin{pmatrix} 1 & 2 & 1 \\ 3 & 1 & 2 \\ 4 & 2 & 1 \end{pmatrix} = \begin{pmatrix} 13 & 9 & 6 \\ 19 & 10 & 8 \\ 17 & 11 & 7 \end{pmatrix}$$

These two examples taken together constitute a counterexample for a commutative law for multiplication of square matrices.

Now, since the *commutative law* for multiplication is included as an axiom for the real number system and since *it does not hold for square matrices*, we would expect to find significant differences between the familiar algebra of the real number system and the algebra of square matrices. The following two examples provide, it seems to us, a somewhat dramatic manifestation of these differences. Both examples involve the matrices $A = \begin{pmatrix} 1 & -2 \\ 3 & 1 \end{pmatrix}$ and $B = \begin{pmatrix} 3 & 2 \\ 4 & 1 \end{pmatrix}$.

Example 3. Compute $(A+B)^2$ where $(A+B)^2$ has the usual meaning $(A+B)(A+B)$.

Solution:

$$\left[\begin{pmatrix} 1 & -2 \\ 3 & 1 \end{pmatrix} + \begin{pmatrix} 3 & 2 \\ 4 & 1 \end{pmatrix}\right]^2 = \begin{pmatrix} 4 & 0 \\ 7 & 2 \end{pmatrix}\begin{pmatrix} 4 & 0 \\ 7 & 2 \end{pmatrix}$$

Therefore, $(A+B)^2 = \begin{pmatrix} 16 & 0 \\ 42 & 4 \end{pmatrix}$.

Example 4. Compute $A^2 + 2AB + B^2$.

Solution:

$$A^2 = \begin{pmatrix} 1 & -2 \\ 3 & 1 \end{pmatrix}\begin{pmatrix} 1 & -2 \\ 3 & 1 \end{pmatrix} \qquad B^2 = \begin{pmatrix} 3 & 2 \\ 4 & 1 \end{pmatrix}\begin{pmatrix} 3 & 2 \\ 4 & 1 \end{pmatrix}$$

$$A^2 = \begin{pmatrix} -5 & -4 \\ 6 & -5 \end{pmatrix} \qquad B^2 = \begin{pmatrix} 17 & 8 \\ 16 & 9 \end{pmatrix}$$

$$AB = \begin{pmatrix} 1 & -2 \\ 3 & 1 \end{pmatrix}\begin{pmatrix} 3 & 2 \\ 4 & 1 \end{pmatrix} \qquad 2AB = AB + AB$$

$$= \begin{pmatrix} -5 & 0 \\ 13 & 7 \end{pmatrix} \qquad\qquad = \begin{pmatrix} -5 & 0 \\ 13 & 7 \end{pmatrix} + \begin{pmatrix} -5 & 0 \\ 13 & 7 \end{pmatrix}$$

$$= \begin{pmatrix} -10 & 0 \\ 26 & 14 \end{pmatrix}$$

Therefore,

$$A^2 + 2AB + B^2 = \begin{pmatrix} -5 & -4 \\ 6 & -5 \end{pmatrix} + \begin{pmatrix} -10 & 0 \\ 26 & 14 \end{pmatrix} + \begin{pmatrix} 17 & 8 \\ 16 & 9 \end{pmatrix}$$

$$= \begin{pmatrix} 2 & 4 \\ 48 & 18 \end{pmatrix}$$

The fact that the results of Examples 3 and 4 are not equal shows that the very familiar identity from high school algebra, $(a+b)^2 = a^2 + 2ab + b^2$, does not hold for square matrices. To see why this formula is not valid for matrices, we expand $(A + B)^2$ as far as possible, using the theorems of Section 3-2.

$$(A+B)^2 = (A+B)(A+B)$$

$$= (A+B)A + (A+B)B \qquad \textit{Left-hand Distributive Law}$$

$$= AA + BA + AB + BB \qquad \textit{Right-hand Distributive Law}$$

$$= A^2 + BA + AB + B^2$$

Therefore, $(A+B)^2 = A^2 + BA + AB + B^2$.

It is now clear why the familiar identity fails. Namely, to obtain $2AB$ from $BA + AB$ would require the use of the *commutative law.* The commutative law *does not hold* for matrices.

PROBLEMS

Use the matrices $A = \begin{pmatrix} 1 & 4 \\ 2 & 6 \end{pmatrix}$ and $B = \begin{pmatrix} 3 & 5 \\ 4 & 2 \end{pmatrix}$ in Problems 1 through 4.

1. Illustrate the *failure* of the commutative law for matrix multiplication.
2. Compute $(A-B)(A-B)$.
3. Compute $A^2 - 2AB + B^2$.
4. Are the results of Problem 2 and Problem 3 equal? Why?
5. Find the value of x for which

$$\begin{pmatrix} 1 & 0 \\ 0 & x \end{pmatrix}\begin{pmatrix} 2 & 1 \\ 1 & 4 \end{pmatrix} = \begin{pmatrix} 2 & 1 \\ 1 & 4 \end{pmatrix}\begin{pmatrix} 1 & 0 \\ 0 & x \end{pmatrix}$$

6. Find all values of x and y for which the matrix $\begin{pmatrix} x & 1 \\ 2 & 3 \end{pmatrix}$ commutes with the matrix $\begin{pmatrix} 2 & 1 \\ 2 & y \end{pmatrix}$.
7. Consider the matrix equation $A\mathbf{x}+\mathbf{a} = B\mathbf{x}+\mathbf{b}$, where A and B are n by n, and $\mathbf{a}$, $\mathbf{b}$, and $\mathbf{x}$ are n-component column vectors. Rewrite $A\mathbf{x}+\mathbf{a} = B\mathbf{x}+\mathbf{b}$ in the form $C\mathbf{x} = \mathbf{c}$, showing and giving a reason for each step.
8. Consider the matrix equation $A\mathbf{x} = \mathbf{b}$, where A is n by n. Show that if there exists an n by n matrix B with the property that $BA = I$, then $\mathbf{x} = B\mathbf{b}$. Give reasons.
9. Consider the matrix equation $XA = B$, where X is n by m, A is m by m, and B is n by m. Show that if there exists an m by m matrix C with the property that $AC = I$, then $X = BC$. Give reasons.

3-4 THE INVERSE OF A SQUARE MATRIX

The expressions $a+(-a) = 0$ and $aa^{-1} = 1$ are included in the list of axioms for the real numbers in Section 3-1. In the list of axioms A-1 to A-8, we find the statements that every real number has an additive inverse and every nonzero real number has a multiplicative inverse. We are now ready to study additive and multiplicative inverses for square matrices.

DEFINITION 3.4. *Let A be an n by n matrix. Then the matrix $-A$ with the property that $A+(-A) = Z$, where Z is the n by n matrix with all elements equal to zero, is called the additive inverse of A.*

DEFINITION 3.5. *Let A be an n by n matrix. Then the matrix A^{-1} with the property that $AA^{-1} = I$, where I is the n by n matrix with ones on the principal diagonal and zeros elsewhere, is called the multiplicative inverse of A.*

As can be seen from the following examples, it is easy to obtain the additive inverse of a matrix.

Example 1. Find the additive inverse of $A = \begin{pmatrix} a & b \\ c & d \end{pmatrix}$.

Solution:

$$\begin{pmatrix} a & b \\ c & d \end{pmatrix} + \begin{pmatrix} x & y \\ z & w \end{pmatrix} = \begin{pmatrix} 0 & 0 \\ 0 & 0 \end{pmatrix}$$

$$a+x = 0,\ b+y = 0,\ c+z = 0, \text{ and } d+w = 0$$

$$x = -a,\ y = -b,\ z = -c, \text{ and } w = -d$$

Therefore, the additive inverse of A is $-A = \begin{pmatrix} -a & -b \\ -c & -d \end{pmatrix}$.

Example 2.

$$\begin{pmatrix} a_{11} & a_{12} & a_{13} \\ a_{21} & a_{22} & a_{23} \\ a_{31} & a_{32} & a_{33} \end{pmatrix} + \begin{pmatrix} -a_{11} & -a_{12} & -a_{13} \\ -a_{21} & -a_{22} & -a_{23} \\ -a_{31} & -a_{32} & -a_{33} \end{pmatrix} = \begin{pmatrix} 0 & 0 & 0 \\ 0 & 0 & 0 \\ 0 & 0 & 0 \end{pmatrix}$$

We see from these examples that the additive inverse for the n by n matrix $A = (a_{ij})$ can be obtained by simply forming the new matrix $-A = (-a_{ij})$.

When one speaks of *the inverse of a matrix* in practice, he is referring to *the multiplicative inverse.* The multiplicative inverse is somewhat troublesome for two reasons.

1. The given matrix may not possess a multiplicative inverse, and
2. It is much more difficult to determine A^{-1} than $-A$ even if it does exist.

For the moment, we will restrict our study to 2 by 2 matrices.

For example, let $A = \begin{pmatrix} 1 & 2 \\ 3 & 4 \end{pmatrix}$. To find A^{-1} we must determine four numbers x, y, z, and w such that

$$\begin{matrix} A & A^{-1} & = & I \end{matrix}$$

$$\begin{pmatrix} 1 & 2 \\ 3 & 4 \end{pmatrix}\begin{pmatrix} x & y \\ z & w \end{pmatrix} = \begin{pmatrix} 1 & 0 \\ 0 & 1 \end{pmatrix}$$

$$\begin{pmatrix} x+2z & y+2w \\ 3x+4z & 3y+4w \end{pmatrix} = \begin{pmatrix} 1 & 0 \\ 0 & 1 \end{pmatrix}$$

Equating corresponding elements, we have

$$\begin{aligned} x + 2z &= 1 \\ 3x + 4z &= 0 \end{aligned} \quad \text{and} \quad \begin{aligned} y + 2w &= 0 \\ 3y + 4w &= 1 \end{aligned}$$

If you solve these two systems of linear equations by the method of elimination or substitution, you will obtain:

$$x = -2,\ y = 1,\ z = \tfrac{3}{2},\ w = -\tfrac{1}{2}$$

Therefore

$$\begin{pmatrix} 1 & 2 \\ 3 & 4 \end{pmatrix}^{-1} = \begin{pmatrix} -2 & 1 \\ \frac{3}{2} & -\frac{1}{2} \end{pmatrix}$$

In the next example, we use the same approach to derive a formula for A^{-1} if

$$A = \begin{pmatrix} a & b \\ c & d \end{pmatrix}.$$

The desired result is

$$\begin{pmatrix} a & b \\ c & d \end{pmatrix}^{-1} = \frac{1}{ad-bc}\begin{pmatrix} d & -b \\ -c & a \end{pmatrix}$$

if $ad - bc \neq 0$. If $ad - bc = 0$, then

$$\begin{pmatrix} a & b \\ c & d \end{pmatrix}$$

has no inverse.

Example 3. Let $A = \begin{pmatrix} a & b \\ c & d \end{pmatrix}$. Find A^{-1}.

Solution:

$$\begin{pmatrix} a & b \\ c & d \end{pmatrix}\begin{pmatrix} x & y \\ z & w \end{pmatrix} = \begin{pmatrix} 1 & 0 \\ 0 & 1 \end{pmatrix}$$

(I) $ax + bz = 1$ (III) $ay + bw = 0$

(II) $cx + dz = 0$ (IV) $cy + dw = 1$

To solve for x and z, we solve I and II simultaneously by the method of substitution.

Equation II can be rewritten in the form

$$z = -\frac{c}{d}x$$

If we substitute this into I, we have

$$ax + b\left(-\frac{c}{d}x\right) = 1$$

$$x\left(a - \frac{bc}{d}\right) = 1$$

$$x\left(\frac{ad-bc}{d}\right) = 1$$

$$x(ad - bc) = d$$

$$x = \frac{d}{ad-bc} \qquad (ad - bc \neq 0)$$

Substituting this value of x into

$$z = (-c/d)x$$

gives

$$z = \left(-\frac{c}{d}\right)\frac{d}{ad - bc}$$

$$z = \frac{-c}{ad - bc}$$

In a similar manner one obtains from equations III and IV the solution

$$y = \frac{-b}{ad - bc} \quad \text{and} \quad w = \frac{a}{ad - bc}$$

Therefore,

$$A^{-1} = \begin{pmatrix} \dfrac{d}{ad - bc} & \dfrac{-b}{ad - bc} \\[2ex] \dfrac{-c}{ad - bc} & \dfrac{a}{ad - bc} \end{pmatrix}$$

$$= \frac{1}{ad - bc}\begin{pmatrix} d & -b \\ -c & a \end{pmatrix}$$

Example 4. Find A^{-1} for $A = \begin{pmatrix} 2 & 3 \\ 4 & 7 \end{pmatrix}$.

Solution:

$$a = 2,\ b = 3,\ c = 4, \quad \text{and} \quad d = 7$$

$$\begin{aligned} ad - bc &= (2)(7) - (4)(3) \\ &= 2 \end{aligned}$$

Therefore,

$$A^{-1} = \begin{pmatrix} \frac{7}{2} & -\frac{3}{2} \\ -2 & 1 \end{pmatrix}$$

Checking this, we have

$$\begin{aligned} AA^{-1} &= \begin{pmatrix} 2 & 3 \\ 4 & 7 \end{pmatrix}\begin{pmatrix} \frac{7}{2} & -\frac{3}{2} \\ -2 & 1 \end{pmatrix} \\ &= \begin{pmatrix} 1 & 0 \\ 0 & 1 \end{pmatrix} \\ &= I \end{aligned}$$

Example 5. The matrix $A = \begin{pmatrix} 3 & 6 \\ 2 & 4 \end{pmatrix}$ has no inverse since the expression $ad - bc$ equals zero. This means that the determination of the elements of A^{-1} by the method of Example 1 would involve *division by zero*, which *is undefined*; hence A^{-1} does not exist.

The result obtained in Example 3 essentially completes the study of the multiplicative inverse for the class of 2 by 2 matrices. A similar result for matrices larger than 2 by 2 would be both difficult to obtain and to use. A practical algorithm for finding the inverse of a matrix larger than 2 by 2 will be given in Section 3-6. A second method of finding the inverse as well as a test to determine if the inverse exists will be taken up in Section 4-5. The next section illustrates an application of the inverse of a matrix once it has been found.

PROBLEMS

1. Compute the additive inverse for the following matrices:

(a) $\begin{pmatrix} 2 & 1 \\ 5 & 6 \end{pmatrix}$ **(b)** $\begin{pmatrix} 3 & 2 & 8 \\ 4 & 5 & 2 \\ 1 & 7 & 6 \end{pmatrix}$

2. Compute the multiplicative inverse, A^{-1}, for the matrix $A = \begin{pmatrix} 5 & 2 \\ 6 & 1 \end{pmatrix}$.

3. Check your result in Problem 2 by multiplication.

4. Compute A^{-1} if $A = \begin{pmatrix} 3 & 4 \\ 5 & 2 \end{pmatrix}$ and check your result by multiplication.

5. Determine the matrices that do not have a multiplicative inverse.

(a) $\begin{pmatrix} 2 & 4 \\ 4 & 8 \end{pmatrix}$ **(c)** $\begin{pmatrix} 1 & 3 \\ 3 & 9 \end{pmatrix}$

(b) $\begin{pmatrix} 3 & 6 \\ 6 & 18 \end{pmatrix}$ **(d)** $\begin{pmatrix} 1 & 2 \\ 2 & 4 \end{pmatrix}$

6. Do the matrices that do not have an inverse in Problem 5 have any common characteristic that you can notice?

7. Verify by multiplication that $\begin{pmatrix} 1 & 0 & -2 \\ 0 & \frac{1}{3} & 0 \\ 0 & 0 & 1 \end{pmatrix}$ is the inverse of $\begin{pmatrix} 1 & 0 & 2 \\ 0 & 3 & 0 \\ 0 & 0 & 1 \end{pmatrix}$.

8. Solve equations III and IV in Example 3 of this section for y and w.

9. Use the definition of A^{-1} to find the inverse of

$$A = \begin{pmatrix} a & 0 & 0 \\ 0 & b & 0 \\ 0 & 0 & c \end{pmatrix} \qquad \text{where } a, b, \text{ and } c \neq 0$$

10. Use the result of Problem 9 to find

$$\begin{pmatrix} 2 & 0 & 0 \\ 0 & 3 & 0 \\ 0 & 0 & -4 \end{pmatrix}^{-1}$$

Remark: Finding A^{-1} is usually not this easy.

3-5 THE APPLICATION OF THE INVERSE OF A MATRIX TO SOLVE SYSTEMS OF EQUATIONS

The reader will recall that in Section 2-5 we discussed the possibility of writing a system of linear equations in n unknowns as a single matrix equation. For the case where $m = n$, such matrix equations have the form $A\mathbf{x} = \mathbf{b}$, where A is the n by n matrix of coefficients of the given system, $\mathbf{x}$ is the n by 1 matrix of unknowns, and $\mathbf{b}$ is the n by 1 matrix of right-hand sides. We now have enough algebra for square matrices to do the following examples.

Example 1. Let A be an n by n matrix which possesses an inverse. Solve the matrix equation $A\mathbf{x} = \mathbf{b}$, where $\mathbf{x}$ and $\mathbf{b}$ are both n by 1 matrices (i.e., column vectors).

Solution:

$$A\mathbf{x} = \mathbf{b}$$

$$A^{-1}A\mathbf{x} = A^{-1}\mathbf{b} \qquad \textit{multiplication by } A^{-1}\textit{, which is known to exist}$$

$$I\mathbf{x} = A^{-1}\mathbf{b} \qquad \textit{definition of } A^{-1}$$

$$\mathbf{x} = A^{-1}\mathbf{b} \qquad \textit{definition of the multiplicative identity}$$

Example 2. Solve the matrix equation

$$\begin{pmatrix} 2 & 3 \\ 4 & 7 \end{pmatrix}\begin{pmatrix} x \\ y \end{pmatrix} = \begin{pmatrix} 1 \\ 2 \end{pmatrix}$$

Solution: We know from Example 4 of the previous section that the inverse of the coefficient matrix $A = \begin{pmatrix} 2 & 3 \\ 4 & 7 \end{pmatrix}$ is $A^{-1} = \begin{pmatrix} \frac{7}{2} & -\frac{3}{2} \\ -2 & 1 \end{pmatrix}$.

$$\begin{pmatrix} \frac{7}{2} & -\frac{3}{2} \\ -2 & 1 \end{pmatrix}\begin{pmatrix} 2 & 3 \\ 4 & 7 \end{pmatrix}\begin{pmatrix} x \\ y \end{pmatrix} = \begin{pmatrix} \frac{7}{2} & -\frac{3}{2} \\ -2 & 1 \end{pmatrix}\begin{pmatrix} 1 \\ 2 \end{pmatrix} \qquad \textit{multiplying both sides by } A^{-1}$$

$$\begin{pmatrix} 1 & 0 \\ 0 & 1 \end{pmatrix}\begin{pmatrix} x \\ y \end{pmatrix} = \begin{pmatrix} \frac{1}{2} \\ 0 \end{pmatrix}$$

$$\begin{pmatrix} x \\ y \end{pmatrix} = \begin{pmatrix} \frac{1}{2} \\ 0 \end{pmatrix} \qquad x = \tfrac{1}{2} \textit{ and } y = 0 \textit{ is the solution}$$

Example 3. Solve the system of equations

$$2x + 3y = 1$$

$$4x + 7y = 2$$

by the method of elimination.

Solution:

$$\begin{aligned} 4x + 6y &= 2 \\ 4x + 7y &= 2 \quad \textit{subtracting the first equation from the second} \\ \hline y &= 0 \end{aligned}$$

$$4x + 7(0) = 2$$

$$x = \tfrac{1}{2} \quad \textit{as found in Example 2}$$

A study of these examples reveals that the inverse of a matrix can be used to solve systems of equations. In fact, if we look carefully at the solution of Example 1, we see that the n by 1 matrix of unknowns is equal to the n by n inverse of the coefficient matrix times the n by 1 matrix of right-hand sides. Once this multiplication has been performed, the solution of the system can be written down immediately by definition of equality of vectors.

Example 4. Given that the inverse of $\begin{pmatrix} 1 & 2 & 3 \\ 1 & 3 & 5 \\ 2 & 5 & 9 \end{pmatrix}$ is $\begin{pmatrix} 2 & -3 & 1 \\ 1 & 3 & -2 \\ -1 & -1 & 1 \end{pmatrix}$, solve the following system of linear equations.

$$\begin{aligned} x + 2y + 3z &= 1 \\ x + 3y + 5z &= 2 \\ 2x + 5y + 9z &= 3 \end{aligned}$$

Solution: If we write this system as a matrix equation, we have

$$\begin{pmatrix} 1 & 2 & 3 \\ 1 & 3 & 5 \\ 2 & 5 & 9 \end{pmatrix}\begin{pmatrix} x \\ y \\ z \end{pmatrix} = \begin{pmatrix} 1 \\ 2 \\ 3 \end{pmatrix}$$

If we multiply both sides by the inverse we get

$$\begin{pmatrix} 2 & -3 & 1 \\ 1 & 3 & -2 \\ -1 & -1 & 1 \end{pmatrix}\begin{pmatrix} 1 & 2 & 3 \\ 1 & 3 & 5 \\ 2 & 5 & 9 \end{pmatrix}\begin{pmatrix} x \\ y \\ z \end{pmatrix} = \begin{pmatrix} 2 & -3 & 1 \\ 1 & 3 & -2 \\ -1 & -1 & 1 \end{pmatrix}\begin{pmatrix} 1 \\ 2 \\ 3 \end{pmatrix}$$

$$\begin{pmatrix} 1 & 0 & 0 \\ 0 & 1 & 0 \\ 0 & 0 & 1 \end{pmatrix}\begin{pmatrix} x \\ y \\ z \end{pmatrix} = \begin{pmatrix} -1 \\ 1 \\ 0 \end{pmatrix} \quad \textit{by definition of } A^{-1}$$

$$\begin{pmatrix} x \\ y \\ z \end{pmatrix} = \begin{pmatrix} -1 \\ 1 \\ 0 \end{pmatrix} \quad \textit{by definition of the multiplicative identity}$$

Therefore, the solution is $x = -1$, $y = 1$, $z = 0$, by definition of equality of vectors.

If we check this answer by direct substitution, we get

$$-1+2+0=1$$

$$-1+3+0=2$$

$$-2+5+0=3$$

which proves that this is indeed the solution of the system.

Example 5. Solve the new system obtained from the system given in Example 4 by changing the constant column from $\begin{pmatrix}1\\2\\3\end{pmatrix}$ to $\begin{pmatrix}2\\3\\4\end{pmatrix}$.

Solution: Using the work done for Example 4, we can go directly to

$$\begin{pmatrix}x\\y\\z\end{pmatrix}=\begin{pmatrix}2&-3&1\\1&3&-2\\-1&-1&1\end{pmatrix}\begin{pmatrix}2\\3\\4\end{pmatrix}$$

$$\begin{pmatrix}x\\y\\z\end{pmatrix}=\begin{pmatrix}-1\\3\\-1\end{pmatrix}$$

$$x=-1,\ y=3,\ z=-1$$

PROBLEMS

1. Given the system of equations

$$2x+3y=4$$

$$x-2y=-5$$

(a) Write the system as a matrix equation.

(b) Verify that $\begin{pmatrix}\frac{2}{7}&\frac{3}{7}\\\frac{1}{7}&-\frac{2}{7}\end{pmatrix}$ is the inverse of the matrix of coefficients.

(c) Use the inverse to solve for x and y.

Perform the following operations for the systems of equations given in Problems 2 through 4.

(a) Write the system as a matrix equation.
(b) Find the inverse of the matrix of coefficients.
(c) Solve the system for x and y.

2. $2x-y=3$

$3x-y=4$

3. $x+y=4$

$3x-y=2$

4. $2x + 3y = -4$

$4x + 6y = -5$

5. Given the following system:

$$3x_1 + 2x_2 + 2x_3 = 4$$
$$x_2 = 5$$
$$2x_1 + 2x_2 + 2x_3 = 6$$

(a) Write the system as a matrix equation.

(b) Verify that $\begin{pmatrix} 1 & 0 & -1 \\ 0 & 1 & 0 \\ -1 & -1 & \frac{3}{2} \end{pmatrix}$ is the inverse of the matrix of coefficients.

(c) Use the inverse to solve for x_1, x_2, and x_3.

6. Write the following system of linear equations in matrix form.

$$\begin{aligned} x_1 &= a_{11}x_1 + a_{12}x_2 + \cdots + a_{1n}x_n + d_1 \\ x_2 &= a_{21}x_1 + a_{22}x_2 + \cdots + a_{22}x_n + d_2 \\ &\vdots \\ x_n &= a_{n1}x_1 + a_{n2}x_2 + \cdots + a_{nn}x_n + d_n \end{aligned}$$

7. Express $\mathbf{x} = A\mathbf{x} + \mathbf{d}$ in the form $(I - A)\mathbf{x} = \mathbf{d}$.

8. Express the solution of $(I - A)\mathbf{x} = \mathbf{d}$ in terms of the inverse of $I - A$ and the column vector $\mathbf{d}$.

3-6 AN ALGORITHM FOR FINDING THE INVERSE OF A MATRIX

In the last section we saw that the inverse of a matrix can be used to solve systems of linear equations once the inverse of the coefficient matrix is known. In this section we will describe a method of finding the inverse of a matrix. The process of finding the multiplicative inverse of a matrix is sometimes called inverting the matrix.

Let A be an n by n matrix. Form an n by $2n$ matrix by writing A and the n by n identity matrix side by side with A on the left. This matrix will be called the *composite matrix*. Then perform row operations (to be described) on this composite matrix until the first n columns are transformed into the identity matrix. If this can be done, then the other n columns (i.e., the right half) of the composite matrix will be the inverse of A.

The permissible row operations are:

1. A row can be multiplied by any nonzero real number.
2. A constant times any row can be added to or subtracted from any other row.
3. Any pair of rows can be interchanged.

The procedure described is illustrated in the following examples. A justification of this algorithm is given after the examples.

Example 1. Use the procedure described above to find the inverse of

$$A = \begin{pmatrix} 2 & 1 \\ 1 & 1 \end{pmatrix}$$

Solution:

a. $\left(\begin{array}{cc|cc} 2 & 1 & 1 & 0 \\ 1 & 1 & 0 & 1 \end{array}\right)$

b. $\left(\begin{array}{cc|cc} 1 & 0 & 1 & -1 \\ 1 & 1 & 0 & 1 \end{array}\right)$ *by subtracting row 2 from row 1 of* (*a*)

c. $\left(\begin{array}{cc|cc} 1 & 0 & 1 & -1 \\ 0 & 1 & -1 & 2 \end{array}\right)$ *by subtracting row 1 from row 2 of* (*b*)

Since the matrix $\begin{pmatrix} 2 & 1 \\ 1 & 1 \end{pmatrix}$ in the left half of (a) has been transformed into the identity $\begin{pmatrix} 1 & 0 \\ 0 & 1 \end{pmatrix}$ in (c), it follows that A^{-1} can be read off from the right half of (c). This gives $A^{-1} = \begin{pmatrix} 1 & -1 \\ -1 & 2 \end{pmatrix}$.

Example 2. Find the inverse of $A = \begin{pmatrix} 3 & 2 & 2 \\ 0 & 1 & 0 \\ 2 & 2 & 2 \end{pmatrix}$

Solution:

a. $\left(\begin{array}{ccc|ccc} 3 & 2 & 2 & 1 & 0 & 0 \\ 0 & 1 & 0 & 0 & 1 & 0 \\ 2 & 2 & 2 & 0 & 0 & 1 \end{array}\right)$

b. $\left(\begin{array}{ccc|ccc} 1 & 0 & 0 & 1 & 0 & -1 \\ 0 & 1 & 0 & 0 & 1 & 0 \\ 2 & 2 & 2 & 0 & 0 & 1 \end{array}\right)$ *subtracting row 3 from row 1 of* (*a*)

c. $\left(\begin{array}{ccc|ccc} 1 & 0 & 0 & 1 & 0 & -1 \\ 0 & 1 & 0 & 0 & 1 & 0 \\ 0 & 2 & 2 & -2 & 0 & 3 \end{array}\right)$ *subtracting two times row 1 from row 3 of* (*b*)

d. $\left(\begin{array}{ccc|ccc} 1 & 0 & 0 & 1 & 0 & -1 \\ 0 & 1 & 0 & 0 & 1 & 0 \\ 0 & 0 & 2 & -2 & -2 & 3 \end{array}\right)$ *subtracting two times row 2 from row 3 of* (*c*)

e. $\left(\begin{array}{ccc|ccc} 1 & 0 & 0 & 1 & 0 & -1 \\ 0 & 1 & 0 & 0 & 1 & 0 \\ 0 & 0 & 1 & -1 & -1 & \frac{3}{2} \end{array}\right)$ *dividing row 3 of (d) by 2, i.e., multiplying by $\frac{1}{2}$*

Then

$$A^{-1} = \begin{pmatrix} 1 & 0 & -1 \\ 0 & 1 & 0 \\ -1 & -1 & \frac{3}{2} \end{pmatrix}$$

Example 3. Find the inverse of $A = \begin{pmatrix} 1 & 2 & 3 \\ 1 & 3 & 5 \\ 2 & 5 & 9 \end{pmatrix}$.

Solution:

$$\left(\begin{array}{ccc|ccc} 1 & 2 & 3 & 1 & 0 & 0 \\ 1 & 3 & 5 & 0 & 1 & 0 \\ 2 & 5 & 9 & 0 & 0 & 1 \end{array}\right)$$

We subtract row one from row two, and subtract twice row one from row three. This gives

a. $\left(\begin{array}{ccc|ccc} 1 & 2 & 3 & 1 & 0 & 0 \\ 0 & 1 & 2 & -1 & 1 & 0 \\ 0 & 1 & 3 & -2 & 0 & 1 \end{array}\right)$

If we subtract row two from row three of (a), we get

b. $\left(\begin{array}{ccc|ccc} 1 & 2 & 3 & 1 & 0 & 0 \\ 0 & 1 & 2 & -1 & 1 & 0 \\ 0 & 0 & 1 & -1 & -1 & 1 \end{array}\right)$

Subtract twice row two from row one of (b) to get $a_{12} = 0$.

c. $\left(\begin{array}{ccc|ccc} 1 & 0 & -1 & 3 & -2 & 0 \\ 0 & 1 & 2 & -1 & 1 & 0 \\ 0 & 0 & 1 & -1 & -1 & 1 \end{array}\right)$

Adding row three to row one of (c) to get $a_{13} = 0$ gives

d. $\left(\begin{array}{ccc|ccc} 1 & 0 & 0 & 2 & -3 & 1 \\ 0 & 1 & 2 & -1 & 1 & 0 \\ 0 & 0 & 1 & -1 & -1 & 1 \end{array}\right)$

Subtract twice row three from row one of (d) to get $a_{23} = 0$.

e. $\left(\begin{array}{ccc|ccc} 1 & 0 & 0 & 2 & -3 & 1 \\ 0 & 1 & 0 & 1 & 3 & -2 \\ 0 & 0 & 1 & -1 & -1 & 1 \end{array}\right)$

Therefore,

$$A^{-1} = \begin{pmatrix} 2 & -3 & 1 \\ 1 & 3 & -2 \\ -1 & -1 & 1 \end{pmatrix}$$

Remarks: If the given matrix A has no inverse, it will be impossible to complete the transformation of A into the identity matrix using only permissible row operations. An alternative test for the existence of the inverse of a matrix is given in Section 4-5 of the next chapter.

The following explanation for the algorithm described was suggested by Dr. Allan E. Andersen.

Let A be a given n by n matrix. Form the matrix equation $A\mathbf{x} = I\mathbf{y}$, where $\mathbf{x}$ and $\mathbf{y}$ are both n by 1 column vectors. This matrix equation is equivalent to the system of linear equations (1).

$$\begin{aligned} a_{11}x_1 + a_{12}x_2 + \cdots + a_{1n}x_n &= y_1 + 0y_2 + \cdots + 0y_n \\ a_{21}x_1 + a_{22}x_2 + \cdots + a_{2n}x_n &= 0y_1 + y_2 + \cdots + 0y_n \\ \vdots \quad\quad \vdots \quad \ddots \quad \vdots \quad &\quad\; \vdots \quad\quad \vdots \quad \ddots \quad \vdots \\ a_{n1}x_1 + a_{n2}x_2 + \cdots + a_{nn}x_n &= 0y_1 + 0y_2 + \cdots + y_n \end{aligned} \tag{1}$$

$$A\mathbf{x} = I\mathbf{y}$$

Assume system (1) has a solution and that it has been solved by the method of elimination to give (2).

$$\begin{aligned} x_1 + 0x_2 + \cdots + 0x_n &= m_{11}y_1 + m_{12}y_2 + \cdots + m_{1n}y_n \\ 0x_1 + x_2 + \cdots + 0x_n &= m_{21}y_1 + m_{22}y_2 + \cdots + m_{2n}y_n \\ \vdots \quad\quad \vdots \quad \ddots \quad \vdots \quad &\quad\; \vdots \quad\quad \vdots \quad \ddots \quad \vdots \\ 0x_1 + 0x_2 + \cdots + x_n &= m_{n1}y_1 + m_{n2}y_2 + \cdots + m_{nn}y_n \end{aligned} \tag{2}$$

$$I\mathbf{x} = M\mathbf{y}$$

A procedure for obtaining (2) from (1) would involve a repeated application of the three rules listed:

1. Any pair of equations can be interchanged.
2. An equation can be multiplied by any nonzero real number.
3. A constant times any equation can be added to or subtracted from any other equation.

The system (1) is equivalent to the composite matrix (3).

$$\left(\begin{array}{cccc|cccc} a_{11} & a_{12} & \cdots & a_{1n} & 1 & 0 & \cdots & 0 \\ a_{21} & a_{22} & \cdots & a_{2n} & 0 & 1 & \cdots & 0 \\ \vdots & \vdots & \ddots & \vdots & \vdots & \vdots & \ddots & \vdots \\ a_{n1} & a_{n2} & \cdots & a_{nn} & 0 & 0 & \cdots & 1 \end{array}\right) \tag{3}$$

$$(A \mid I)$$

Also, system (2) is equivalent to the composite matrix (4).

$$\left(\begin{array}{cccc|cccc} 1 & 0 & \cdots & 0 & m_{11} & m_{12} & \cdots & m_{1n} \\ 0 & 1 & \cdots & 0 & m_{21} & m_{22} & \cdots & m_{2n} \\ \vdots & \vdots & \ddots & \vdots & \vdots & \vdots & \ddots & \vdots \\ 0 & 0 & \cdots & 1 & m_{n1} & m_{n2} & \cdots & m_{nn} \end{array}\right) \qquad \begin{array}{c} (4) \\ (I\,|\,M) \end{array}$$

It is important to note that the permissible *row operations* listed at the beginning of this section can be obtained from the three rules given above by simply replacing the word "equation" with the word "row." It follows that the use of row operations to go from (3) to (4) is equivalent to solving system (1) for x in terms of y. This statement plus the fact that $M = A^{-1}$ should convince the reader that the algorithm is valid.

To see that $M = A^{-1}$, we rewrite the equations $A\mathbf{x} = I\mathbf{y}$ and $I\mathbf{x} = M\mathbf{y}$ in the forms $A\mathbf{x} = \mathbf{y}$ and $\mathbf{x} = M\mathbf{y}$, respectively. Combining these two equations we have

$$\mathbf{y} = A\mathbf{x}$$

$$\mathbf{y} = A(M\mathbf{y})$$

$$\mathbf{y} = (AM)\mathbf{y} \qquad \text{by use of the associative law for matrix multiplication}$$

Since the matrix AM times y equals y for any column vector y, it follows that $AM = I$ and, therefore, $M = A^{-1}$.

PROBLEMS

1. Set up the composite matrix for the following matrices:

(a) $\begin{pmatrix} 2 & 0 \\ 1 & 1 \end{pmatrix}$ **(b)** $\begin{pmatrix} 1 & 0 \\ 2 & 2 \end{pmatrix}$

2. Using row operations, find the inverse of the matrices given in Problem 1.

3. Set up the composite matrix and find the inverse for the following matrices:

(a) $\begin{pmatrix} 5 & 2 \\ 6 & 1 \end{pmatrix}$ **(b)** $\begin{pmatrix} 3 & 4 \\ 5 & 2 \end{pmatrix}$

4. Set up the composite matrix for the following matrices:

(a) $\begin{pmatrix} 2 & 3 \\ 1 & -2 \end{pmatrix}$ **(b)** $\begin{pmatrix} 1 & 3 & -2 \\ 0 & 1 & 4 \\ 0 & 0 & 2 \end{pmatrix}$

5. Use row operations to find the inverse of the matrices given in Problem 4.

6. Use row operations to find the inverse of $\begin{pmatrix} 3 & 0 & 1 \\ 4 & 2 & 1 \\ 3 & 0 & 1 \end{pmatrix}$, if it exists.

7. Use row operations to find the inverse of $\begin{pmatrix} 4 & 2 & 0 \\ 3 & 0 & 0 \\ 6 & 1 & 1 \end{pmatrix}$. Check your result by multiplication.

8. Write

$$\begin{aligned} y_1 &+ ax_1 + bx_2 \\ y_2 &= cx_1 + dx_2 \end{aligned} \qquad \text{where } ad - bc \neq 0 \tag{1}$$

in the form

$$\begin{aligned} x_1 &= a'y_1 + b'y_2 \\ x_2 &= c'y_1 + d'y_2 \end{aligned} \qquad (\textit{i.e., solve for } x\textit{'s in terms of } y\textit{'s}) \tag{2}$$

9. Show that (2) in Problem 8 has the form $\begin{pmatrix} x_1 \\ x_2 \end{pmatrix} = \begin{pmatrix} a & b \\ c & d \end{pmatrix}^{-1} \begin{pmatrix} y_1 \\ y_2 \end{pmatrix}$.

10. Find the inverse of $A = \begin{pmatrix} 3 & 2 & 2 \\ 0 & 1 & 0 \\ 2 & 2 & 2 \end{pmatrix}$ as follows:

(a) Rewrite the system

$$\begin{aligned} y_1 &= 3x_1 + 2x_2 + 2x_3 \\ y_2 &= x_2 \\ y_3 &= 2x_1 + 2x_2 + 2x_3 \end{aligned}$$

in the form

$$\begin{aligned} x_1 &= b_{11}y_1 + b_{12}y_2 + b_{13}y_3 \\ x_2 &= b_{21}y_1 + b_{22}y_2 + b_{23}y_3 \\ x_3 &= b_{31}y_1 + b_{32}y_2 + b_{33}y_3 \end{aligned} \qquad (\textit{i.e., solve for } x\textit{'s in terms of } y\textit{'s})$$

(b) Observe that $B = (b_{ij}) = A^{-1}$, hence this procedure provides a method of inverting a matrix.

Remark: This matrix is inverted in Example 2.

11. Use the procedure outlined in Problem 10 to find the inverse of

$$A = \begin{pmatrix} 1 & 2 & 3 \\ 1 & 3 & 5 \\ 2 & 5 & 9 \end{pmatrix} \text{ of Example 3.}$$

3-7 INPUT-OUTPUT ANALYSIS

Input-output analysis is a concept invented by Wassily Leontief about 40 years ago. He has the distinction not only of having single-handedly invented this field, but also of having trained a good number of its practitioners. The value of this concept is now being realized with the development of the high-speed computer. Originally the purpose of the input-output model was to try to relate one industry's activities to broad economic aggregates, such as gross national product or personal consumption expenditures. It is now being used by companies rather than government to forecast market trends. A comprehension of the concepts of identity matrix, matrix inversion, and matrix multiplication is a requirement for an understanding of this section.

In input-output analysis the basic tool is a table which, as we have seen in previous sections, can also be called a matrix. On the left side of the table is a list of industries, one row for each industry. The rows show the sales, called *outputs*, that each industry makes to the others. Across the top of the table, the original industries appear again, one to each column. The columns show the purchases, called *inputs*, that each industry makes from the others. These industries compose what we will call the processing sector. There are two additional columns. The first represents final demand, which includes items such as purchases by households and government, exports, and accumulation to inventory. The last column represents total output.

Example 1. Interpret the following table which assumes an economy of two industries.

		Purchases (inputs) in millions of $		Final Demand	Total Output
		Industry 1	Industry 2		
Sales (outputs) in millions of $	Industry 1	280	1420	1100	2800
	Industry 2	840	710	2000	3550

Solution: Reading across row 1 we see that Industry 1 had a total output (sales) of $2800 million. This is accounted for by intraindustry sales of $280 million, sales of $1420 million to Industry 2, and sales of $1100 million to households, etc. Reading down column 1, we see that there were $280 million of intraindustry purchases (inputs) and purchases of $840 million from Industry 2.

After an input-output table has been constructed for a given year, a table of input, or *technological coefficients*, can be developed from it. This is done by dividing the total output of a given industry into each of the inputs of the supplying industries. Each result is a fraction expressed as a decimal. A technological coefficient is the amount of input required from each industry to produce one dollar's worth of the output of a given industry. These figures are more important than dollar volume in examining the structure of an industry's market, because they can be applied to any level of sales.

Example 2. Formulate a table of technological coefficients using the data presented in Example 1.

Solution: Dividing all the entries in each industry's column by the total output for that industry results in the following table:

	Industry 1	Industry 2
Industry 1	0.1	0.4
Industry 2	0.3	0.2

This table can be interpreted as follows. Each dollar's worth of production in Industry 1 will require

Intraindustry purchase	\$0.10
Purchase from Industry 2	0.30

Each dollar's worth of production in Industry 2 will require

Purchase from Industry 1	\$0.40
Intraindustry purchase	0.20

To supply an increase of \$1 in final demand in just one industry, we must

1. Increase total output in this industry by more than \$1.
2. Increase total output in other industries even though they experience no increase in final demand.

These increases result from the existence of intraindustry requirements.

The determination of the new output levels required in all industries to meet a change in final demand is the main problem of input-output analysis. A method of solution based on matrix concepts follows.

Example 3. Express the table of technological coefficients as a matrix called A, and total output as a column vector called $\mathbf{x}$. Use the data in Examples 1 and 2.

Solution:

$$A = \begin{pmatrix} a_{11} & a_{12} \\ a_{21} & a_{22} \end{pmatrix} = \begin{pmatrix} 0.1 & 0.4 \\ 0.3 & 0.2 \end{pmatrix}$$

$$\mathbf{x} = \begin{pmatrix} x_1 \\ x_2 \end{pmatrix} = \begin{pmatrix} 2800 \\ 3550 \end{pmatrix}$$

Using the matrix of technological coefficients and the total output vector we can then express interindustry requirements as

$$\text{Interindustry requirements} = A\mathbf{x}$$

As we mentioned before, we must also satisfy final demand. If we let final demand for the industries be represented by the column vector **d**, we can then express total output as

$$\text{Total output} = \text{Interindustry requirements} + \text{Final demand}$$

$$\mathbf{x} = A\mathbf{x} + \mathbf{d}$$

The question is: What must total output be to satisfy the interindustry requirements and final demand? We can answer this question by solving the matrix equation above for **x**.

$$\mathbf{x} = A\mathbf{x} + \mathbf{d}$$

$$I\mathbf{x} = A\mathbf{x} + \mathbf{d} \qquad \textit{i.e., } \mathbf{x} = I\mathbf{x}$$

$$I\mathbf{x} - A\mathbf{x} = \mathbf{d}$$

$$(I - A)\mathbf{x} = \mathbf{d} \qquad \textit{Right-hand Distributive Law}$$

$$(I - A)^{-1}(I - A)\mathbf{x} = (I - A)^{-1}\mathbf{d}$$

$$I\mathbf{x} = (I - A)^{-1}\mathbf{d}$$

$$\mathbf{x} = (I - A)^{-1}\mathbf{d}$$

Example 4. Using the result obtained above, solve for **x** given $A = \begin{pmatrix} 0.1 & 0.4 \\ 0.3 & 0.2 \end{pmatrix}$ and $\mathbf{d} = \begin{pmatrix} 1100 \\ 2000 \end{pmatrix}$.

Solution:

$$(I - A)^{-1} = \begin{pmatrix} \frac{4}{3} & \frac{2}{3} \\ \frac{1}{2} & \frac{3}{2} \end{pmatrix}$$

$$\mathbf{x} = (I - A)^{-1}\mathbf{d}$$

$$\mathbf{x} = \begin{pmatrix} \frac{4}{3} & \frac{2}{3} \\ \frac{1}{2} & \frac{3}{2} \end{pmatrix}\begin{pmatrix} 1100 \\ 2000 \end{pmatrix} = \begin{pmatrix} 2800 \\ 3550 \end{pmatrix}$$

The reader should check to see that this value for final demand is in fact the same as the one in the table in Example 1, since A was developed from this table.

Given any expected level of final demand, one can forecast required output by computing $(I - A)^{-1}$. Note that this inverse must only be calculated once.

Example 5. If a decrease from 1100 to 1000 occurs in final demand for Industry 1, and the final demand for Industry 2 remains constant, what would be the required total output for both industries?

Solution:

$$\mathbf{x} = (I - A)^{-1}\mathbf{d}$$

$$\mathbf{x} = \begin{pmatrix} \frac{4}{3} & \frac{2}{3} \\ \frac{1}{2} & \frac{3}{2} \end{pmatrix} \begin{pmatrix} 1000 \\ 2000 \end{pmatrix} = \begin{pmatrix} 2667 \\ 3500 \end{pmatrix}$$

Examining these results, we see that a \$100 million decrease in final demand for Industry 1 results in the following decreases in total output:

Industry 1	\$133 million
Industry 2	50 million

This is caused by the decrease in interindustry requirements as well as the decrease in final demand. Hence, if you were in charge of production for Industry 2, you would decrease total output by \$50 million if you anticipated that your final demand would remain constant as Industry 1's final demand decreased from \$1100 to \$1000 million.

We have looked at a simple situation. If there were 286 industries, we would have a little trouble inverting a 286 by 286 matrix. However, it should be clear that input-output analysis is an important tool in forecasting. For large-scale problems we would use a computer.

PROBLEMS

1. Verify that $\begin{pmatrix} 0.9 & -0.4 \\ -0.3 & 0.8 \end{pmatrix}^{-1} = \begin{pmatrix} \frac{4}{3} & \frac{2}{3} \\ \frac{1}{2} & \frac{1}{2} \end{pmatrix}$.
2. What would $\mathbf{x}$ have been in Example 5 if $\mathbf{d}$ were equal to $\begin{pmatrix} 1200 \\ 2000 \end{pmatrix}$?
3. Assuming $\mathbf{d} = \begin{pmatrix} 1100 \\ 2000 \end{pmatrix}$ as in Example 4, but that $A = \begin{pmatrix} 0.2 & 0.4 \\ 0.3 & 0.2 \end{pmatrix}$ or $a_{11} = 0.2$ instead of 0.1, find $(I - A)^{-1}$.
4. Solve for $\mathbf{x}$ using the information in Problem 3.

Given the following in millions:

	Industry 1	Industry 2	Industry 3	Final Demand	Total Output
Industry 1	20	50	40	90	200
Industry 2	10	5	20	65	100
Industry 3	10	15	5	70	100

5. Construct the matrix of technological coefficients A.

6. Interpret a_{21}.

7. Find $(I - A)$.

8. Find **x** if

$$(I - A)^{-1} = \begin{pmatrix} 1.18788 & 0.728387 & 0.653506 \\ 0.07828 & 1.13683 & 0.272294 \\ 0.07488 & 0.217835 & 1.13002 \end{pmatrix}$$

and

$$\mathbf{d} = \begin{pmatrix} 90 \\ 65 \\ 70 \end{pmatrix}$$

9. Using the inverse in Problem 8, find **x** if

$$\mathbf{d} = \begin{pmatrix} 100 \\ 65 \\ 70 \end{pmatrix}$$

10. Given $A = \begin{pmatrix} 0.2 & 0.4 \\ 0.3 & 0.1 \end{pmatrix}$.

(a) Find $I - A$.
(b) Find $(I - A)^{-1}$.

11. Using the result of Problem 10, find **x** if $\mathbf{d} = \begin{pmatrix} 300 \\ 600 \end{pmatrix}$.

12. Using the result of Problem 10, find **x** if $\mathbf{d} = \begin{pmatrix} 330 \\ 600 \end{pmatrix}$.

13. Interpret the results of Problems 11 and 12.

14. It follows from $\mathbf{x} = A\mathbf{x} + \mathbf{d}$ that

$$b_{11}\Delta x_1 + b_{12}\Delta x_2 = \Delta d_1$$

$$b_{21}\Delta x_1 + b_{22}\Delta x_2 = \Delta d_2, \qquad \textit{where } B = I - A$$

Let $\Delta d_1 = 1$ and $\Delta d_2 = 0$.

(a) Solve for Δx_1, Δx_2 in terms of the elements of B.
(b) Use the results of (a) to solve for Δx_1 and Δx_2 given the technological coefficient matrix $A = \begin{pmatrix} 0.1 & 0.4 \\ 0.3 & 0.2 \end{pmatrix}$.

3-8 SUMMARY AND SOME PROBLEMS THAT INVOLVE MATRIX ALGEBRA

In Chapters 2 and 3 we looked at the conditions for equality, addition, subtraction, and multiplication of vectors and matrices. We also defined the operations of addition, subtraction and multiplication of vectors and matrices. Some interesting applications of matrices were presented, the most notable being the so-called parts explosion problem and input-output analysis. Also of special interest is how one can use the inverse of a matrix to solve systems of equations. This procedure can be summarized as follows:

1. Rewrite the system of equations as a matrix equation.
2. Find the inverse of the matrix of coefficients.
3. Multiply each side of the matrix equation by the inverse.
4. The solution is the product of the inverse times the right-hand side of the matrix equation.

Some basic properties of square matrices can be summarized as follows:

PROPERTIES OF SQUARE MATRICES

M-1 Every pair of $n \times n$ matrices A and B has a unique *sum*, $A + B$, and a unique *product*, AB, which are also $n \times n$ matrices.

M-2 $A + B = B + A$ for all $n \times n$ matrices A and B.
Note: $AB \neq BA$, generally.

M-3 $A + (B + C) = (A + B) + C$, $A(BC) = (AB)C$ for all $n \times n$ matrices A, B, and C.

M-4 (a) $A(B + C) = AB + AC$ ⎫ for all $n \times n$ matrices A,
(b) $(A + B)C = AC + BC$ ⎭ B, and C.
Note: In contrast to the real numbers, the right-hand distributive law (b) for square matrices cannot be derived from the left-hand distributive law (a) since this would involve the use of the commutative law of multiplication, which fails to hold for square matrices.

M-5 There exists a unique $n \times n$ matrix, Z, called the *additive identity*, with the property that $A + Z = Z + A = A$, for every $n \times n$ matrix A. This matrix is called the *zero* $n \times n$ matrix.

M-6 For every $n \times n$ matrix, A, there exists an $n \times n$ matrix, $-A$, called the *additive inverse* of A with the property that $A+(-A) = Z$.

M-7 There exists a unique $n \times n$ matrix, I, called the *multiplicative identity*, with the property that $AI = IA = A$ for every $n \times n$ matrix A.

M-8 For every $n \times n$ matrix, A, whose determinant is not zero (see Chapter 4), there exists a unique $n \times n$ matrix, A^{-1}, called the *multiplicative inverse* of A, with the property that $A^{-1}A = AA^{-1} = I$.

It is interesting to compare this list of properties with the field axioms given at the beginning of Chapter 3.

PROBLEMS*

A manufacturer's plant has two service departments (designated as S_1 and S_2) and three production departments (designated as P_1, P_2, and P_3) and wishes to allocate all factory overhead to production departments. A primary distribution of overhead to all departments has already been made and is indicated. The company makes the secondary distribution of overhead from service departments to production departments on a reciprocal basis, recognizing the fact that services of one service department are utilized by another. Data regarding costs and allocation percentages are as follows:

SERVICE DEPARTMENT OVERHEAD COST ALLOCATION

Service Department	Percentages to Be Allocated to Departments S_1	S_2	P_1	P_2	P_3
S_1	0%	10%	20%	40%	30%
S_2	20	0	50	10	20
	Primary Overhead to Be Allocated				
	\$98,000	\$117,600	\$1,400,000	\$2,100,000	\$640,000

Matrix algebra is to be used in the secondary allocation process. The amount of overhead to be allocated to the service departments you express in two simultaneous equations as:

$$S_1 = 98{,}000 + .20S_2 \quad \text{or} \quad S_1 - .20S_2 = \$\ 98{,}000$$

$$S_2 = 117{,}600 + .10S_1 \quad \text{or} \quad S_2 - .10S_1 = \$117{,}600$$

* Problems 1–7 from the Uniform CPA Examinations, Copyright © (1970) by the American Institute of Certified Public Accountants, Inc., are reprinted with permission.

1. The system of simultaneous equations may be stated in matrix form as

(a) $$\overset{A}{\begin{pmatrix} 1 & -.20 \\ -.10 & 1 \end{pmatrix}} \overset{S}{\begin{pmatrix} S_1 \\ S_2 \end{pmatrix}} = \overset{b}{\begin{pmatrix} \$\ 98{,}000 \\ \$117{,}600 \end{pmatrix}}.$$

(b) $$\overset{A}{\begin{pmatrix} 1 & \$\ 98{,}000 & 1 \\ -.20 & \$117{,}600 & -.10 \end{pmatrix}} \overset{S}{\begin{pmatrix} S_1 \\ S_2 \end{pmatrix}} = \overset{b}{\begin{pmatrix} \$\ 98{,}000 \\ \$117{,}600 \end{pmatrix}}.$$

(c) $$\overset{A}{\begin{pmatrix} 1 & S_1 & 1 \\ -.20 & S_2 & -.10 \end{pmatrix}} \overset{S}{\begin{pmatrix} S_1 \\ S_2 \end{pmatrix}} = \overset{b}{\begin{pmatrix} \$\ 98{,}000 \\ \$117{,}600 \end{pmatrix}}.$$

(d) $$\overset{A}{\begin{pmatrix} 1 & 1 & S_1 \\ -.20 & -.10 & S_2 \end{pmatrix}} \overset{S}{\begin{pmatrix} S_1 \\ S_2 \end{pmatrix}} = \overset{b}{\begin{pmatrix} \$\ 98{,}000 \\ \$117{,}600 \end{pmatrix}}.$$

2. For the correct matrix A in item 1, there exists a unique inverse matrix A^{-1}. Multiplication of the matrix A^{-1} by the matrix A will produce
   (a) The matrix A
   (b) Another inverse matrix.
   (c) The correct solution to the system.
   (d) An identity matrix.

3. Without prejudice to your previous answers, assume that the correct matrix form in item 1 was

$$\overset{A}{\begin{pmatrix} 1 & -.20 \\ -.10 & 1 \end{pmatrix}} \overset{S}{\begin{pmatrix} S_1 \\ S_2 \end{pmatrix}} = \overset{b}{\begin{pmatrix} \$\ 98{,}000 \\ \$117{,}600 \end{pmatrix}}$$

Then the correct inverse matrix A^{-1} is

(a) $$\begin{pmatrix} \frac{1}{.98} & \frac{.20}{.98} \\ \frac{.10}{.98} & \frac{1}{.98} \end{pmatrix}$$
(c) $$\begin{pmatrix} \frac{1}{.30} & \frac{.20}{.30} \\ \frac{.10}{.30} & \frac{1}{.30} \end{pmatrix}$$

(b) $$\begin{pmatrix} \frac{1}{.98} & \frac{1}{.98} \\ \frac{.20}{.98} & \frac{.10}{.98} \end{pmatrix}$$
(d) $$\begin{pmatrix} \frac{1}{.98} & -\frac{1}{.98} \\ -\frac{.20}{.98} & \frac{.10}{.98} \end{pmatrix}$$

4. The total amount of overhead allocated to department S_1 after receiving the allocation from department S_2 is
   (a) \$141,779 **(c)** \$121,520
   (b) \$124,000 **(d)** \$117,600

5. The total amount of overhead allocated to department S_2 after receiving the allocation from department S_1 is
   (a) \$392,000 **(c)** \$130,000
   (b) \$220,000 **(d)** \$127,400

6. Without prejudice to your previous answers, assume that the answer to item 4 is $100,000 and to item 5 is $150,000; then the total amount of overhead allocated to production department P_1 would be
(a) $1,508,104 **(c)** $1,489,800
(b) $1,495,000 **(d)** $108,104

7. A company that makes humidifiers has just received an order for 100 commercial models and 200 residential models. Both of these humidifiers have three subassemblies I, II, and III as follows:

	I	II	III
Commercial	6	4	2
Residential	1	3	5

Three types of parts #013, #014, #043 costing $2.00, $3.00, and $0.10 respectively are used to make the subassemblies. The allocation of parts to subassemblies is shown in the following table:

	#013	#014	#043
I	6	1	10
II	5	2	40
III	4	1	100

Determine the cost of the parts to fill this order, using matrix algebra.

REFERENCES

Books

Andree, Richard V. *Selections from Modern Abstract Algebra.* New York: Henry Holt and Co., 1958.

Campbell, Hugh G. *Matrices with Applications.* New York: Appleton-Century-Crofts, 1968.

Corcoran, A. Wayne. *Mathematical Applications in Accounting.* New York: Harcourt Brace Jovanovich, 1968.

Curtis, Charles W. *Linear Algebra, An Introductory Approach.* Boston: Allyn and Bacon, Inc., 1963.

Draper, N. R., and Smith, H. *Applied Regression Analysis.* New York: John Wiley & Sons, Inc., 1966.

Kemeny, John G., Schleifer, A., Snell, J. R., and Thompson, G. L. *Finite Mathematics with Business Applications.* Englewood Cliffs, N.J.: Prentice-Hall, Inc., 1964.

Leontief, Wassily. *Input-Output Economics.* New York: Oxford University Press, Inc., 1966.

Levin, Richard I., and Kirkpatrick, C. A. *Quantitative Approaches to Management.* New York: McGraw-Hill Book Co., Inc., 1965.

Lial, Margaret L., and Miller, Charles David. *Finite Mathematics: With Applications in Business, Biology, and Behavioral Sciences.* Glenview, Illinois: Scott, Foresman and Company, 1977.

Miernyk, William H. *The Elements of Input-Output Analysis.* New York: Random House, Inc., 1965.

Parker, William Vann, and Eaves, James Clifton. *Matrices.* New York: The Ronald Press Co., 1960.

Shank, John K. *Matrix Methods In Accounting.* Reading, Mass.: Addison-Wesley Publishing Co., 1972.

Springer, Clifford H., Herlihy, R. E., and Beggs, R. I. *Basic Mathematics.* Homewood, Ill.: Richard D. Irwin, Inc., 1965.

ARTICLES

"Costing with Matrix Algebra." *Management Accounting:* April 1976.

"Criteria for Entering an Industry." *Sales Management:* Oct. 15, 1969.

"Marketers Meet 'Input-Output'." *Business Week:* Oct. 29, 1966.

"Planners Put Big Picture on a Grid." *Business Week:* Sept. 23, 1967.

"They Call It 'Instant Research'." *Business Week:* Jan. 25, 1969.

"How Input-Output Helped Map the Marketing Strategy of Glass Containers Corp." *Industrial Marketing:* January 1970.

"Input-Output Analysis For Financial Decision-Making." *Management Accounting:* September 1976.

FOUR

Determinants and solutions of systems of linear equations

4-1 DEFINITION OF THE DETERMINANT OF AN n by n MATRIX

In this chapter we will introduce the concept of the determinant of a square matrix. Determinants can be used to determine if a given square matrix has a multiplicative inverse, or if certain systems of linear equations have solutions. Determinants can also be used to solve systems of n linear equations in n unknowns.

In Section 3-4 of Chapter 3 we derived a formula for finding the inverse of a 2 by 2 matrix. It was found that if

$$A = \begin{pmatrix} a & b \\ c & d \end{pmatrix}$$

then

$$A^{-1} = \begin{pmatrix} \dfrac{d}{ad-bc} & \dfrac{-b}{ad-bc} \\[2ex] \dfrac{-c}{ad-bc} & \dfrac{a}{ad-bc} \end{pmatrix}$$

The expression $ad-bc$, which appears as the denominator of each of the elements of A^{-1}, is called the determinant of A. We will study this expression further before giving a definition of the determinant of an n by n matrix.

Writing the expression $ad-bc$ in the form $ad+(-bc)$, we see that it is a sum of signed products. Furthermore, each product contains exactly one element from each row and one element from each column. In the first product, ad, "a" is from the first row and the first column while "d" is from the second row and the second column. In the second product, $-bc$, "b" is from the first row and the second column. We shall soon see that the determinant of an n by n matrix is defined in terms of these ideas.

Next we introduce the term *transposition* as used in the definition of determinant given below. Imagine a set of positive integers listed in some definite order, let us say 5, 2, 1, 3, 4. We use the term transposition to describe an *interchange* in the order of any single *adjacent pair* of integers in such a list. One example of a transposition of the given list of integers is 5, 2, 3, 1, 4, and another example is 2, 5, 1, 3, 4.

Example 1. Use transpositions to rearrange the integers 5, 2, 1, 3, 4 into the *natural order* 1, 2, 3, 4, 5 in two different ways.

Solution:

5, 2, 1, 3, 4	5, 2, 1, 3, 4
5, 1, 2, 3, 4	5, 2, 1, 4, 3
1, 5, 2, 3, 4	2, 5, 1, 4, 3
1, 2, 5, 3, 4	2, 5, 1, 3, 4
1, 2, 3, 5, 4	2, 1, 5, 3, 4
1, 2, 3, 4, 5	1, 2, 5, 3, 4
	1, 2, 3, 5, 4
	1, 2, 3, 4, 5

In the first case *five* transpositions are used and in the second case *seven* transpositions are used. In both cases we needed an *odd* number of steps to restore the set of integers 5, 2, 1, 3, 4 to the natural order. This situation did not happen by chance. Mathematicians have proved that given a list of integers the number of steps needed to restore the natural order $1, 2, 3, \ldots, n$, using only transpositions, remains either *even* or *odd*, regardless of the number of steps actually taken to do this. The "evenness" or "oddness" is invariant, relative to transpositions.

DEFINITION 4.1., PART A. *Let A be an n by n matrix. The determinant of A denoted by $|A|$ is defined to be the sum of all possible signed n-fold products that can be formed by choosing exactly one element from each row and exactly one element from each column of A.*
PART B. *Let each product be arranged so that the first subscripts of the factors are in the natural order, $1, 2, 3, \ldots, n$. The number of transpositions required to restore the second subscripts of a particular product to the natural order, $1, 2, 3, \ldots, n$, determines the sign of that product. If the number of transpositions required is odd, the product receives a*

minus sign, and if the number of transpositions is even or zero, the product receives a plus sign.

Example 2. Let $A = \begin{pmatrix} a_{11} & a_{12} \\ a_{21} & a_{22} \end{pmatrix}$. Find the determinant of A.

Solution: There are two possible products, $a_{11}a_{22}$ and $a_{12}a_{21}$.

$a_{11}a_{22}$ receives a plus sign since the second subscripts are in the natural order. $a_{12}a_{21}$ receives a *minus sign* since it requires *one* transposition to *restore* the *second subscripts* to the natural order.

Therefore, the determinant of the matrix $\begin{pmatrix} a_{11} & a_{12} \\ a_{21} & a_{22} \end{pmatrix}$ is $a_{11}a_{22}+(-a_{12}a_{21}) = a_{11}a_{22}-a_{12}a_{21}$.

Example 3. Given the 3 by 3 matrix $A = \begin{pmatrix} a_{11} & a_{12} & a_{13} \\ a_{21} & a_{22} & a_{23} \\ a_{31} & a_{32} & a_{33} \end{pmatrix}$, find the determinant of A by Definition 4.1.

Solution: The possible products are $a_{11}a_{22}a_{33}$, $a_{12}a_{23}a_{31}$, $a_{13}a_{21}a_{32}$, $a_{13}a_{22}a_{31}$, $a_{11}a_{23}a_{32}$, and $a_{12}a_{21}a_{33}$.

Analyzing the product $a_{13}a_{21}a_{32}$, we have the order 312 for the *second subscripts.* $312 \to 132 \to 123$. *Two* transpositions are required to *restore* the natural order. Therefore, the product $a_{12}a_{21}a_{32}$ receives a *plus sign.* Similarly, for the product $a_{13}a_{22}a_{31}$, we have $321 \to 312 \to 132 \to 123$ or *three* necessary transpositions. Therefore, this product receives a *minus sign.*

Further analysis on the part of the reader will reveal that the first three products receive plus signs and the second three products receive minus signs. Hence, the determinant of A is

$$|A| = a_{11}a_{22}a_{33} + a_{12}a_{23}a_{31} + a_{13}a_{21}a_{32}$$
$$- a_{13}a_{22}a_{31} - a_{11}a_{23}a_{32} - a_{12}a_{21}a_{33}$$

Example 4. Compute $\begin{vmatrix} 2 & 3 & 4 \\ 1 & 7 & 6 \\ 4 & 2 & 3 \end{vmatrix}$.

Solution: Using the result of Example 3, we have

$$\begin{vmatrix} 2 & 3 & 4 \\ 1 & 7 & 6 \\ 4 & 2 & 3 \end{vmatrix} = (2)(7)(3) + (3)(6)(4) + (4)(1)(2) - (4)(7)(4)$$
$$- (2)(6)(2) - (3)(1)(3)$$
$$= 42 + 72 + 8 - 112 - 24 - 9$$
$$= 122 - 145$$
$$= -23$$

Beginning students sometimes confuse the terms matrix and determinant. A *matrix* is simply a rectangular array of numbers. A *determinant*, which is only defined for square matrices, is a real number associated with a given square matrix that can be calculated by Definition 4.1.

For example, $A = \begin{pmatrix} 2 & 3 & 4 \\ 1 & 7 & 6 \\ 4 & 2 & 3 \end{pmatrix}$ is a 3 by 3 matrix. On the other hand, $|A| = \begin{vmatrix} 2 & 3 & 4 \\ 1 & 7 & 6 \\ 4 & 2 & 3 \end{vmatrix} = -23$ is the determinant of this matrix.

PROBLEMS

1. Find all the possible products that can be formed by choosing exactly one element from each row and exactly one element from each column of the following matrices:

   (a) $\begin{pmatrix} 4 & 2 \\ 1 & 3 \end{pmatrix}$ **(c)** $\begin{pmatrix} 2 & 3 & 4 \\ 5 & 6 & 2 \\ 1 & 7 & 4 \end{pmatrix}$

   (b) $\begin{pmatrix} c_{11} & c_{12} \\ c_{21} & c_{22} \end{pmatrix}$ **(d)** $\begin{pmatrix} x & y \\ 2 & 8 \end{pmatrix}$

2. Find all the possible products that can be formed by choosing exactly one element from each row and exactly one element from each column of the following matrices:

   (a) $\begin{pmatrix} x_{11} & x_{12} \\ x_{21} & x_{22} \end{pmatrix}$ **(b)** $\begin{pmatrix} x & 3 & 9 \\ 8 & 4 & 2 \\ 6 & 8 & 1 \end{pmatrix}$

3. How many transpositions are required to rearrange the following lists of integers into the natural order?

   (a) 2, 3, 1 **(c)** 4, 5, 2, 3, 1
   (b) 3, 2, 1, 4 **(d)** 5, 7, 8, 10, 9

4. Using only transpositions, rearrange the integers 4, 3, 2, 1 into the natural order 3 different ways.

5. Using only transpositions, rearrange the integers 1, 3, 5, 2, 4 into the natural order 2 different ways.

6. Was the number of transpositions in Problem 5 even or odd?

7. Given the following products:

   (a) $a_{33} \cdot a_{22} \cdot a_{11}$ **(b)** $a_{31} \cdot a_{23} \cdot a_{12}$

   Arrange each product so that the first subscripts of the factors are in the natural order.

8. Given the following products whose first subscripts are in the natural order, how many transpositions are required to get the second subscripts in the natural order? (Show all steps.)

(a) $b_{12} \cdot b_{21} \cdot b_{33}$ **(b)** $b_{13} \cdot b_{22} \cdot b_{31}$

9. What sign should be attached to the products in Problem 8?

10. $a_{13} a_{21} \, x \, a_{42}$ is one of twenty-four possible signed products of a 4 by 4 determinant.

(a) Solve for x.

(b) Choose the correct sign for this particular product.

11. Given the following products:

(a) $a_{32} \cdot a_{21} \cdot a_{13}$ **(c)** $a_{32} \cdot a_{23} \cdot a_{11}$

(b) $a_{31} \cdot a_{22} \cdot a_{13}$ **(d)** $a_{33} \cdot a_{21} \cdot a_{12}$

After arranging the products so that the first subscripts of the factors are in the natural order, restore the second subscript to the natural order by transpositions and attach the correct sign to the product.

12. Given the 3 by 3 matrix

$$B = \begin{pmatrix} b_{11} & b_{12} & b_{13} \\ b_{21} & b_{22} & b_{23} \\ b_{31} & b_{32} & b_{33} \end{pmatrix}$$

find the determinant of B by Definition 4.1.

13. Given that the matrix $A = \begin{pmatrix} a_{11} & a_{12} \\ a_{21} & a_{22} \end{pmatrix}$ has as its determinant $|A| = a_{11}a_{22} - a_{12}a_{21}$, compute

(a) $\begin{vmatrix} 6 & 3 \\ 2 & 1 \end{vmatrix}$ **(b)** $\begin{vmatrix} 7 & 8 \\ 16 & 4 \end{vmatrix}$

14. Given that

$$|A| = a_{11}a_{22}a_{33} + a_{12}a_{23}a_{31} + a_{13}a_{21}a_{32}$$
$$- a_{13}a_{22}a_{31} - a_{11}a_{23}a_{32} - a_{12}a_{21}a_{33},$$

compute

(a) $\begin{vmatrix} 2 & 4 & 8 \\ 1 & 6 & 2 \\ 3 & 0 & 4 \end{vmatrix}$ **(b)** $\begin{vmatrix} 6 & 0 & 0 \\ 2 & 1 & 3 \\ 4 & 2 & 1 \end{vmatrix}$

4-2 EVALUATION OF DETERMINANTS OF 2 BY 2 AND 3 BY 3 MATRICES

The following scheme can be used to evaluate determinants of 2 by 2 and 3 by 3 matrices.

Case 1. 2 by 2 matrices

$$\begin{vmatrix} a_{11} & a_{12} \\ a_{21} & a_{22} \end{vmatrix} = a_{11}a_{22} - a_{12}a_{21} \qquad \textit{by Definition 4.1}$$

Now draw 2 arrows in the matrix $\begin{pmatrix} a_{11} & a_{12} \\ a_{21} & a_{22} \end{pmatrix}$ as follows:

$$\begin{vmatrix} a_{11} & a_{12} \\ a_{21} & a_{22} \end{vmatrix}$$

Interpret the arrow pointing *down and to the right* as an instruction to *multiply and use a plus sign.* Interpret the arrow pointing *upward and to the right* as an instruction to *multiply and use a minus sign.* This scheme gives

$$\begin{vmatrix} a_{11} & a_{12} \\ a_{21} & a_{22} \end{vmatrix} = a_{11}a_{22} - a_{12}a_{21}$$

Example 1. Evaluate $\begin{vmatrix} 2 & 3 \\ 1 & 5 \end{vmatrix}$ by the preceding scheme.

Solution:

$$\begin{aligned} \begin{vmatrix} 2 & 3 \\ 1 & 5 \end{vmatrix} &= +(2)(5) - (1)(3) \\ &= 10 - 3 \\ &= 7 \end{aligned}$$

Example 2. Evaluate the determinant of $\begin{pmatrix} 3 & -2 \\ -3 & 5 \end{pmatrix}$ using this scheme.

Solution: Determinant of $\begin{pmatrix} 3 & -2 \\ -3 & 5 \end{pmatrix}$

$$\begin{aligned} &= \begin{vmatrix} 3 & -2 \\ -3 & 5 \end{vmatrix} \\ &= +(3)(5) - (-3)(-2) \\ &= 15 - 6 \\ &= 9 \end{aligned}$$

Case 2. 3 by 3 matrices

It was shown in the previous section using Definition 4.1 that

$$\begin{vmatrix} a_{11} & a_{12} & a_{13} \\ a_{21} & a_{22} & a_{23} \\ a_{31} & a_{32} & a_{33} \end{vmatrix} = \begin{aligned} &a_{11}a_{22}a_{33} + a_{12}a_{23}a_{31} + a_{13}a_{32}a_{21} \\ &- a_{31}a_{22}a_{13} - a_{32}a_{23}a_{11} - a_{33}a_{12}a_{21} \end{aligned}$$

The correct products with the appropriate sign are indicated by the following scheme where the arrows that *start down and to the right* indicate a *plus sign* and the arrows *starting up and to the right indicate a minus sign.*

$$\begin{vmatrix} a_{11} & a_{12} & a_{13} \\ a_{21} & a_{22} & a_{23} \\ a_{31} & a_{32} & a_{33} \end{vmatrix}$$

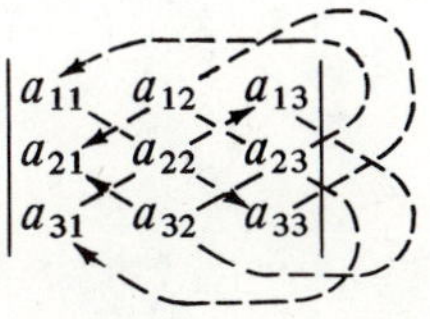

$$= a_{11}a_{22}a_{33} + a_{12}a_{23}a_{31} + a_{13}a_{32}a_{21} - a_{31}a_{22}a_{13} - a_{32}a_{23}a_{11} - a_{33}a_{12}a_{21}$$

Example 3. Use the above scheme to evaluate the determinant in Example 4 of Section 4-1.

Solution:

$$\begin{vmatrix} 2 & 3 & 4 \\ 1 & 7 & 6 \\ 4 & 2 & 3 \end{vmatrix}$$

$$= (2)(7)(3) + (3)(6)(4) + (4)(2)(1) - (4)(7)(4) - (2)(6)(2) - (3)(3)(1)$$

$$= 42 + 72 + 8 - 112 - 24 - 9$$

$$= 122 - 145 = -23,$$

as found previously

Example 4. Evaluate the determinant of $\begin{pmatrix} 2 & 3 & 4 \\ 1 & -2 & 1 \\ 4 & 6 & 8 \end{pmatrix}$.

Solution:

$$\begin{vmatrix} 2 & 3 & 4 \\ 1 & -2 & 1 \\ 4 & 6 & 8 \end{vmatrix}$$

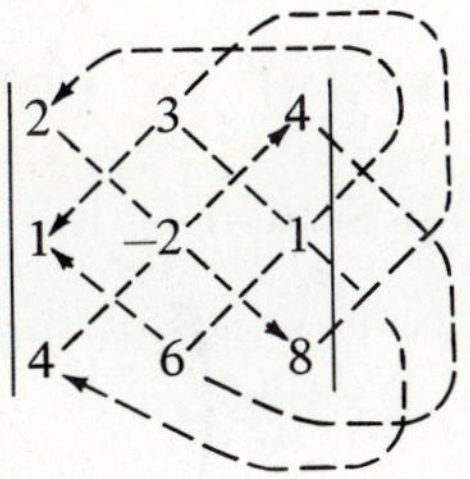

$$= -32 + 12 + 24 + 32 - 12 - 24$$

$$= 0$$

The scheme given in Example 4 *does not work* for determinants of matrices with order greater than 3 by 3!

An alternative method of evaluation of determinants of any order is given in the next section.

PROBLEMS

Use the scheme of drawing arrows to evaluate the following determinants:

1. $\begin{vmatrix} 3 & 2 \\ 6 & 1 \end{vmatrix}$

2. $\begin{vmatrix} 4 & 9 \\ 8 & 0 \end{vmatrix}$

3. $\begin{vmatrix} 4 & 2 & 6 \\ 1 & 8 & 4 \\ 3 & 2 & 1 \end{vmatrix}$

4. $\begin{vmatrix} 5 & 7 & 2 \\ 1 & 8 & 7 \\ 3 & 4 & 4 \end{vmatrix}$

5. $\begin{vmatrix} b_{11} & b_{12} & b_{13} \\ b_{21} & b_{22} & b_{23} \\ b_{31} & b_{32} & b_{33} \end{vmatrix}$

6. Compare the result of Problem 5 with the result of Problem 12, Section 4-1.

7. Solve for x in the equation

$$\begin{vmatrix} x & 3 \\ 2 & 2 \end{vmatrix} = 0$$

8. Show that $\begin{vmatrix} x & y & z \\ 2 & 1 & 3 \\ 3x & 3y & 3z \end{vmatrix} = 0$ for any choice of x, y, and z.

4-3 THE METHOD OF COFACTORS

In Section 4-1 it was found that the determinant of the matrix

$$A = \begin{pmatrix} a_{11} & a_{12} & a_{13} \\ a_{21} & a_{22} & a_{23} \\ a_{31} & a_{32} & a_{33} \end{pmatrix}$$

is given by the expression

$$\begin{aligned} |A| = {} & a_{11}a_{22}a_{33} + a_{12}a_{23}a_{31} + a_{13}a_{21}a_{32} \\ & - a_{13}a_{22}a_{31} - a_{11}a_{23}a_{32} - a_{12}a_{21}a_{33} \end{aligned} \tag{1}$$

One of several possible rearrangements of this expression is as follows:

$$|A| = a_{11}(a_{22}a_{33} - a_{23}a_{32}) - a_{12}(a_{21}a_{33} - a_{23}a_{31}) + a_{13}(a_{21}a_{32} - a_{22}a_{31}) \tag{2}$$

A careful look at the three "differences" inside the parentheses above reveals that they are simply the values of certain 2 by 2 determinants. Therefore, we can write $|A|$ in terms of other determinants as

$$|A| = a_{11}\begin{vmatrix} a_{22} & a_{23} \\ a_{32} & a_{33} \end{vmatrix} - a_{12}\begin{vmatrix} a_{21} & a_{23} \\ a_{31} & a_{33} \end{vmatrix} + a_{13}\begin{vmatrix} a_{21} & a_{22} \\ a_{31} & a_{32} \end{vmatrix} \tag{3}$$

This is a specific case of the very important theorem for determinants that asserts that an n by n determinant can be expressed in terms of the elements of any row or column times certain $n-1$ by $n-1$ determinants. Before stating this theorem, we need to define two more terms.

DEFINITION 4.2. *Let a_{ij} be the element in the ith row and jth column of the n by n matrix A. Then the minor of a_{ij} is the $n-1$ by $n-1$ determinant formed by deleting the ith row and jth column of A.*

DEFINITION 4.3. *Let a_{ij} be an element of the matrix A. The cofactor of a_{ij} = the minor of a_{ij} if $i+j$ is even. If $i+j$ is odd, then the cofactor of a_{ij} = the negative of the minor of a_{ij}. The cofactor of a_{ij} is denoted by A_{ij}.*

Example 1. Let $A = \begin{pmatrix} 2 & 3 & 1 \\ 4 & 7 & 2 \\ 3 & 1 & 2 \end{pmatrix}$. Find the minors of a_{12} and a_{33}.

Solution:

$$\begin{matrix} 2 & 3 & 1 \\ 4 & 7 & 2 \\ 3 & 1 & 2 \end{matrix} \qquad \text{minor of } a_{12} \text{ is } \begin{vmatrix} 4 & 2 \\ 3 & 2 \end{vmatrix} = 8 - 6 = 2$$

$$\begin{matrix} 2 & 3 & 1 \\ 4 & 7 & 2 \\ 3 & 1 & 2 \end{matrix} \qquad \text{minor of } a_{33} \text{ is } \begin{vmatrix} 2 & 3 \\ 4 & 7 \end{vmatrix} = 14 - 12 = 2$$

Example 2. Using the determinant and the results of Example 1. find the cofactors a_{12} and a_{33}.

Solution:

$$\text{Cofactor of } a_{12} = -\begin{vmatrix} 4 & 2 \\ 3 & 2 \end{vmatrix} = -2$$

The minus sign is used since $1+2=3$ is odd.

$$\text{Cofactor of } a_{33} = \begin{vmatrix} 2 & 3 \\ 4 & 7 \end{vmatrix}$$

$$= 2 \qquad \textit{since } 3+3=6 \textit{ is even}$$

THEOREM 4.1. *Let A be an n by n matrix. Then the determinant of A can be found by computing the sum of the elements in any row or column times their corresponding cofactors.*

The proof of this theorem for the general case is beyond the scope of this text. However, our result (3), which was obtained from Definition 4.1, is in agreement with Theorem 4.1.

$$A_{11} = \text{Cofactor of } a_{11} = \begin{vmatrix} a_{22} & a_{23} \\ a_{32} & a_{33} \end{vmatrix}$$

$$A_{12} = \text{Cofactor of } a_{12} = -\begin{vmatrix} a_{21} & a_{23} \\ a_{31} & a_{33} \end{vmatrix}$$

$$A_{13} = \text{Cofactor of } a_{13} = \begin{vmatrix} a_{21} & a_{22} \\ a_{31} & a_{32} \end{vmatrix}$$

Hence, by Theorem 4.1,

$$\begin{vmatrix} a_{11} & a_{12} & a_{13} \\ a_{21} & a_{22} & a_{23} \\ a_{31} & a_{32} & a_{33} \end{vmatrix} = a_{11} \text{ cofactor of } a_{11} + a_{12} \text{ cofactor of } a_{12} + a_{13} \text{ cofactor of } a_{13}$$

$$= a_{11}\begin{vmatrix} a_{22} & a_{23} \\ a_{32} & a_{33} \end{vmatrix} - a_{12}\begin{vmatrix} a_{21} & a_{23} \\ a_{31} & a_{33} \end{vmatrix} + a_{13}\begin{vmatrix} a_{21} & a_{22} \\ a_{31} & a_{32} \end{vmatrix} \tag{3}$$

Example 3. Evaluate the determinant $\begin{vmatrix} 2 & 1 & 3 \\ 1 & 2 & 4 \\ 2 & 1 & 5 \end{vmatrix}$ by the method of cofactors in two different ways.

Solution: By using row two, we have

$$\begin{vmatrix} 2 & 1 & 3 \\ 1 & 2 & 4 \\ 2 & 1 & 5 \end{vmatrix} = -1\begin{vmatrix} 1 & 3 \\ 1 & 5 \end{vmatrix} + 2\begin{vmatrix} 2 & 3 \\ 2 & 5 \end{vmatrix} - 4\begin{vmatrix} 2 & 1 \\ 2 & 1 \end{vmatrix}$$

$$= -1(5-3) + 2(10-6) - 4(2-2)$$
$$= -1(2) + 2(4) - 4(0)$$
$$= -2 + 8 - 0$$
$$= 6$$

By using column one, we have

$$\begin{vmatrix} 2 & 1 & 3 \\ 1 & 2 & 4 \\ 2 & 1 & 5 \end{vmatrix} = 2\begin{vmatrix} 2 & 4 \\ 1 & 5 \end{vmatrix} - 1\begin{vmatrix} 1 & 3 \\ 1 & 5 \end{vmatrix} + 2\begin{vmatrix} 1 & 3 \\ 2 & 4 \end{vmatrix}$$
$$= 2(10-4) - 1(5-3) + 2(4-6)$$
$$= 2(6) - 1(2) + 2(-2)$$
$$= 12 - 2 - 4$$
$$= 6$$

Next we illustrate a fact about determinants that will be needed in Section 4-5.

THEOREM 4.2. *If the cofactors of the elements of any row or column of an n by n determinant are multiplied by the corresponding elements in some other row or column and then added together, the result is zero.*

Example 4. (Partial proof of Theorem 4.2 for 3 by 3 case.) Compute the quantity $a_{11}A_{21}+a_{12}A_{22}+a_{13}A_{23}$ for the matrix $A = \begin{pmatrix} a_{11} & a_{12} & a_{13} \\ a_{21} & a_{22} & a_{23} \\ a_{31} & a_{32} & a_{33} \end{pmatrix}$.

Solution:

$$A_{21} = \text{Cofactor of } a_{21} = -\begin{vmatrix} a_{12} & a_{13} \\ a_{32} & a_{33} \end{vmatrix}$$

$$A_{22} = \text{Cofactor of } a_{22} = \begin{vmatrix} a_{11} & a_{13} \\ a_{31} & a_{33} \end{vmatrix}$$

$$A_{23} = \text{Cofactor of } a_{23} = -\begin{vmatrix} a_{11} & a_{12} \\ a_{31} & a_{32} \end{vmatrix}.$$

The desired quantity is given by

$$-a_{11}(a_{12}a_{33}-a_{32}a_{13}) + a_{12}(a_{11}a_{33}-a_{31}a_{13}) - a_{13}(a_{11}a_{32}-a_{31}a_{12})$$
$$= -a_{11}a_{12}a_{33} + a_{11}a_{32}a_{13} + a_{12}a_{11}a_{33} - a_{12}a_{31}a_{13}$$
$$- a_{11}a_{13}a_{32} + a_{13}a_{31}a_{12}$$

$$= (a_{12}a_{11}a_{33} - a_{11}a_{12}a_{33}) + (a_{11}a_{32}a_{13} - a_{11}a_{13}a_{32})$$
$$+ (a_{13}a_{31}a_{12} - a_{12}a_{31}a_{13})$$
$$= 0 + 0 + 0$$
$$= 0$$

PROBLEMS

1. Find the minor of the elements b_{31} and b_{23} given that

$$B = \begin{pmatrix} b_{11} & b_{12} & b_{13} \\ b_{21} & b_{22} & b_{23} \\ b_{31} & b_{32} & b_{33} \end{pmatrix}$$

2. Find the minor of a_{31}, a_{32}, a_{33} if

$$A = \begin{pmatrix} 2 & 6 & 1 \\ 3 & 4 & 2 \\ 8 & 1 & 6 \end{pmatrix}$$

3. Find the cofactor of a_{31}, a_{32}, a_{33} in Problem 2.
4. Use the results of Example 3 to evaluate the determinant of A in Problem 2.
5. Given the matrix $\begin{pmatrix} 4 & 1 & 0 \\ 3 & 6 & 2 \\ 2 & 0 & 1 \end{pmatrix}$, evaluate the determinant by computing the sum of the elements in column three times their corresponding cofactors.
6. Evaluate the determinant of the matrix in Example 5 by computing the sum of the elements in column two times their corresponding cofactors.
7. Evaluate the following determinants by the method of cofactors:

(a) $\begin{vmatrix} 2 & 1 & 4 \\ 3 & 6 & 1 \\ 2 & 1 & 2 \end{vmatrix}$ **(b)** $\begin{vmatrix} 4 & 1 & 3 \\ 3 & 0 & 0 \\ 2 & 1 & 6 \end{vmatrix}$

8. Evaluate the following determinants by the method of cofactors:

(a) $\begin{vmatrix} 3 & 1 & 2 \\ 5 & 1 & 3 \\ 2 & 4 & 6 \end{vmatrix}$ **(b)** $\begin{vmatrix} 4 & 1 & 3 \\ 3 & 0 & 1 \\ 4 & 0 & 2 \end{vmatrix}$

9. Illustrate Theorem 4.2 by summing the cofactors of column three times the corresponding elements of column two for the following:

$$\begin{vmatrix} 6 & 2 & 1 \\ 4 & 0 & 2 \\ 3 & 1 & 4 \end{vmatrix}$$

10. Illustrate Theorem 4.2 for the following:

(a) $\begin{vmatrix} 3 & 2 & 1 \\ 6 & 8 & 4 \\ 2 & 1 & 3 \end{vmatrix}$ **(b)** $\begin{vmatrix} 6 & 2 & 1 \\ 3 & 1 & 1 \\ 2 & 4 & 5 \end{vmatrix}$

11. Evaluate the determinant of the following matrix:

$$\begin{pmatrix} 6 & 8 & 4 & 2 \\ 1 & 3 & 2 & 0 \\ 0 & 1 & 4 & 2 \\ 0 & 1 & 2 & 3 \end{pmatrix}$$

i.e., continue the following partial solution:

$$6\begin{vmatrix} 3 & 2 & 0 \\ 1 & 4 & 2 \\ 1 & 2 & 3 \end{vmatrix} + 8(-)\begin{vmatrix} 1 & 2 & 0 \\ 0 & 4 & 2 \\ 0 & 2 & 3 \end{vmatrix} + 4\begin{vmatrix} 1 & 3 & 0 \\ 0 & 1 & 2 \\ 0 & 1 & 3 \end{vmatrix} + 2(-)\begin{vmatrix} 1 & 3 & 2 \\ 0 & 1 & 4 \\ 0 & 1 & 2 \end{vmatrix}$$

$$6\begin{vmatrix} 3 & 2 & 0 \\ 1 & 4 & 2 \\ 1 & 2 & 3 \end{vmatrix} - 8\begin{vmatrix} 1 & 2 & 0 \\ 0 & 4 & 2 \\ 0 & 2 & 3 \end{vmatrix} + 4\begin{vmatrix} 1 & 3 & 0 \\ 0 & 1 & 2 \\ 0 & 1 & 3 \end{vmatrix} - 2\begin{vmatrix} 1 & 3 & 2 \\ 0 & 1 & 4 \\ 0 & 1 & 2 \end{vmatrix}$$

12. Evaluate the following determinants:

(a) $\begin{vmatrix} 2 & 1 & 0 & 3 \\ 4 & 2 & 1 & 6 \\ 2 & 3 & 2 & 1 \\ 4 & 5 & 0 & 2 \end{vmatrix}$ **(b)** $\begin{vmatrix} 3 & 2 & 1 & 1 \\ 0 & 2 & 1 & 0 \\ 4 & 6 & 3 & 1 \\ 2 & 5 & 2 & 2 \end{vmatrix}$

4-4 CRAMER'S RULE

Determinants can be used to solve any system of linear equations which has a unique solution. The procedure for doing this involves the use of ratios of determinants and it is called Cramer's rule. Cramer's rule is illustrated in the next example.

Example 1. Given the linear system

$$\begin{aligned} 2x + 3y + 4z &= 4 \\ x + 2y + 5z &= 0 \\ 4x + y + z &= 5, \end{aligned}$$

solve for y using Cramer's rule.

Solution:

(Please note use of column of right-hand sides.)

$$y = \frac{\begin{vmatrix} 2 & 4 & 4 \\ 1 & 0 & 5 \\ 4 & 5 & 1 \end{vmatrix}}{\begin{vmatrix} 2 & 3 & 4 \\ 1 & 2 & 5 \\ 4 & 1 & 1 \end{vmatrix}}$$

$$y = \frac{0+80+20-0-50-4}{4+60+4-32-10-3}$$

$$y = \frac{46}{23} = 2$$

Cramer's rule is explained next in general terms for the special case of two linear equations in two unknowns.

In Section 4-1 it was pointed out that the expression $ad - bc$ occurs in a natural way when we attempt to find the inverse of the square matrix $A = \begin{pmatrix} a & b \\ c & d \end{pmatrix}$. The reader will recall that $ad-bc = \begin{vmatrix} a & b \\ c & d \end{vmatrix}$.

Next we show that this same quantity also occurs when the system of equations

$$\begin{aligned} ax + by &= k_1 \\ cx + dy &= k_2 \end{aligned}$$

is solved by the method of elimination.

Multiply the first equation by d and the second equation by b to get

$$\begin{aligned} adx + bdy &= k_1 d \\ bcx + bdy &= k_2 b \end{aligned}$$

Subtract the second equation from the first to get

$$\begin{aligned} adx - bcx &= k_1 d - k_2 b \\ (ad-bc)x &= k_1 d - k_2 b \end{aligned}$$

For $ad-bc \neq 0$, we have $x = \dfrac{k_1 d - k_2 b}{ad-bc}$. Similarly, multiplying the first equation by c and the second equation by a and subtracting gives $y = \dfrac{k_2 a - k_1 c}{ad-bc}$.

Note that both denominators are given by $ad-bc$, i.e., $\begin{vmatrix} a & b \\ c & d \end{vmatrix}$. In fact, the numerators can also be expressed as the determinants $\begin{vmatrix} k_1 & b \\ k_2 & d \end{vmatrix}$ and $\begin{vmatrix} a & k_1 \\ c & k_2 \end{vmatrix}$. *Note:* The k column has been substituted into the two determinants.

In summary, the system

$$\begin{aligned} ax + by &= k_1 \\ cx + dy &= k_2 \end{aligned} \qquad ad - bc \neq 0$$

has the solution

$$x = \frac{\begin{vmatrix} k_1 & b \\ k_2 & d \end{vmatrix}}{\begin{vmatrix} a & b \\ c & d \end{vmatrix}}, \quad y = \frac{\begin{vmatrix} a & k_1 \\ c & k_2 \end{vmatrix}}{\begin{vmatrix} a & b \\ c & d \end{vmatrix}}$$

This result is a special case of a theorem called *Cramer's rule.*

THEOREM 4.3. *The system of n equations in n unknowns*

$$\begin{aligned} a_{11}x_1 + a_{12}x_2 + \cdots + a_{1n}x_n &= c_1 \\ a_{21}x_1 + a_{22}x_2 + \cdots + a_{2n}x_n &= c_2 \\ \vdots \quad\quad \vdots \quad\quad \ddots \quad\quad \vdots \quad &\quad \vdots \\ a_{n1}x_1 + a_{n2}x_2 + \cdots + a_{nn}x_n &= c_n \end{aligned}$$

where

$$D_n = \begin{vmatrix} a_{11} & a_{12} & \cdots & a_{1n} \\ a_{21} & a_{22} & \cdots & a_{2n} \\ \vdots & \vdots & \ddots & \vdots \\ a_{n1} & a_{n2} & \cdots & a_{nn} \end{vmatrix} \neq 0$$

has as its solution

$$x_1 = \frac{\begin{vmatrix} c_1 & a_{12} & a_{13} & \cdots & a_{1n} \\ c_2 & a_{22} & a_{23} & \cdots & a_{2n} \\ \vdots & \vdots & \vdots & \ddots & \vdots \\ c_n & a_{n2} & a_{n3} & \cdots & a_{nn} \end{vmatrix}}{D_n},$$

$$x_2 = \frac{\begin{vmatrix} a_{11} & c_1 & a_{13} & \cdots & a_{1n} \\ a_{21} & c_2 & a_{23} & \cdots & a_{2n} \\ \vdots & \vdots & \vdots & \ddots & \vdots \\ a_{n1} & c_n & a_{n3} & \cdots & a_{nn} \end{vmatrix}}{D_n}, \ \ldots, \ x_n = \frac{\begin{vmatrix} a_{11} & a_{12} & \cdots & c_1 \\ a_{21} & a_{22} & \cdots & c_2 \\ \vdots & \vdots & \ddots & \vdots \\ a_{n1} & a_{n2} & \cdots & c_n \end{vmatrix}}{D_n}$$

Example 2. Use Cramer's rule to solve the linear system

$$\begin{aligned} x + 2y + 3z &= 1 \\ x + 3y + 5z &= 2 \\ 2x + 5y + 9z &= 3 \end{aligned}$$

Solution:

$$D_3 = \begin{vmatrix} 1 & 2 & 3 \\ 1 & 3 & 5 \\ 2 & 5 & 9 \end{vmatrix} = 1(27-25) - 1(18-15) + 2(10-9) = 1$$

By using the method of cofactors and column one, we have

$$x = \frac{\begin{vmatrix} 1 & 2 & 3 \\ 2 & 3 & 5 \\ 3 & 5 & 9 \end{vmatrix}}{D_3} = \frac{1(27-25)-2(18-15)+3(10-9)}{1} = -1$$

$$y = \frac{\begin{vmatrix} 1 & 1 & 3 \\ 1 & 2 & 5 \\ 2 & 3 & 9 \end{vmatrix}}{D_3} = \frac{1}{1}$$

$$z = \frac{\begin{vmatrix} 1 & 2 & 1 \\ 1 & 3 & 2 \\ 2 & 5 & 3 \end{vmatrix}}{D_3} = \frac{0}{1} = 0$$

Hence, the solution is $x = -1$, $y = 1$, $z = 0$.

Example 3. Evaluate the determinant of the coefficient matrix for the system

$$\begin{pmatrix} 1 & 0 & 1 \\ 2 & 1 & 0 \\ 3 & 0 & 1 \end{pmatrix}\begin{pmatrix} x \\ y \\ z \end{pmatrix} = \begin{pmatrix} 1 \\ 3 \\ 5 \end{pmatrix}$$

Solution: The coefficient matrix is

$$A = \begin{pmatrix} 1 & 0 & 1 \\ 2 & 1 & 0 \\ 3 & 0 & 1 \end{pmatrix} \quad \text{and} \quad |A| = 1 + 0 + 0 - 3 + 0 + 0 = -2$$

using the scheme of Section 4-2

PROBLEMS

1. **(a)** By Cramer's rule, solve

$$\begin{aligned} x + y &= 1 \\ x - y &= 4 \end{aligned}$$

(b) Verify your answer by the method of elimination.

2. Given the following matrix equation:

$$\begin{pmatrix} 1 & 2 & 3 \\ 0 & 1 & 2 \\ 2 & 1 & 4 \end{pmatrix}\begin{pmatrix} x \\ y \\ z \end{pmatrix} = \begin{pmatrix} 4 \\ -2 \\ 6 \end{pmatrix}$$

(a) What is the determinant of the matrix of coefficients?
(b) Solve for y using Cramer's rule.

3. Given the system

$$\begin{aligned} 2x + 3y + 4z &= 4 \\ x + 2y + 5z &= 0 \\ 4x + y + z &= 5 \end{aligned}$$

Solve for x by Cramer's rule.

4. Given the system

$$\begin{aligned} 2x + 2z &= 2 \\ 5x + 4y + z &= 3 \\ 2x + 6z &= 4 \end{aligned}$$

(a) What is the determinant of the matrix of coefficients?
(b) Solve for x, y, and z using Cramer's rule.

5. Given the system

$$\begin{aligned} a + b + c &= -2 \\ 8a + 4b + 2c &= 6 \\ -a + b - c &= 0 \end{aligned}$$

Solve for a, b, and c using Cramer's rule.

Problems 6 through 9 are based on Leontief's input-output analysis concepts.

6. Use Cramer's rule to solve Problem 14 of Section 3-7.

7. Recall the system

$$\begin{aligned} b_{11}\Delta x_1 + b_{12}\Delta x_2 &= \Delta d_1 \\ b_{21}\Delta x_1 + b_{22}\Delta x_2 &= \Delta d_2 \end{aligned}$$

from Problem 14 of Section 3-7. Use Cramer's rule to solve for Δx_1 and Δx_2 in terms of Δd_1, Δd_2, and the elements of B, where $B = I - A$.

8. Using the results of Problem 7, determine Δx_1 and Δx_2 if $\Delta d_1 = 2$, $\Delta d_2 = 3$, and $B = \begin{pmatrix} 0.9 & -0.4 \\ -0.3 & 0.8 \end{pmatrix}$.

9. Use the results of Problem 7 to obtain the required output for both industries in Examples 4 and 5 of Section 3-7.

4-5 THE USE OF DETERMINANTS FOR FINDING THE INVERSE OF A MATRIX

In Section 3-4 we solved a system of 4 equations in 4 unknowns and found that if

$$A = \begin{pmatrix} a & b \\ c & d \end{pmatrix}$$

then

$$A^{-1} = \begin{pmatrix} \dfrac{d}{ad-bc} & \dfrac{-b}{ad-bc} \\[2ex] \dfrac{-c}{ad-bc} & \dfrac{a}{ad-bc} \end{pmatrix}$$

Noting that $ad - bc = |A|$ and using the definition of multiplication of a matrix by a scalar, we have

$$A^{-1} = \frac{1}{|A|}\begin{pmatrix} d & -b \\ -c & a \end{pmatrix}$$

If A is written in subscript notation

$$A = \begin{pmatrix} a_{11} & a_{12} \\ a_{21} & a_{22} \end{pmatrix}$$

then

$$A^{-1} = \frac{1}{|A|}\begin{pmatrix} a_{22} & -a_{12} \\ -a_{21} & a_{11} \end{pmatrix} \tag{1}$$

Example 1. Use the formula to find the inverse of $A = \begin{pmatrix} 2 & 3 \\ 2 & 4 \end{pmatrix}$.

Solution:

$$a_{11} = 2,\ a_{12} = 3,\ a_{21} = 2,\ a_{22} = 4$$

$$\begin{vmatrix} 2 & 3 \\ 2 & 4 \end{vmatrix} = 8 - 6 = 2$$

$$A^{-1} = \frac{1}{2}\begin{pmatrix} a_{22} & -a_{12} \\ -a_{21} & a_{11} \end{pmatrix}$$

$$= \frac{1}{2}\begin{pmatrix} 4 & -3 \\ -2 & 2 \end{pmatrix}$$

$$A^{-1} = \begin{pmatrix} 2 & -\frac{3}{2} \\ -1 & 1 \end{pmatrix}$$

Before we continue with a discussion of the inverse of a square matrix, we need the following matrix concept.

DEFINITION 4.4. *Let A be an m by n matrix. Then the n by m matrix A^t, which is obtained by interchanging the rows and columns of A, is called the transpose of A.*

The next two examples illustrate the definition of the transpose of a matrix.

1. If $A = \begin{pmatrix} 1 & 3 & 5 \\ 4 & 5 & 3 \end{pmatrix}$, then $A^t = \begin{pmatrix} 1 & 4 \\ 3 & 5 \\ 5 & 3 \end{pmatrix}$.

2. If $B = \begin{pmatrix} 3 & 8 \\ 6 & 2 \end{pmatrix}$, then $B^t = \begin{pmatrix} 3 & 6 \\ 8 & 2 \end{pmatrix}$.

Example 2. Given the matrix $A = \begin{pmatrix} a_{11} & a_{12} \\ a_{21} & a_{22} \end{pmatrix}$, form a new matrix B by replacing each element of A by its cofactor.

Solution:

$$A_{11} = a_{22}$$

$$A_{21} = -a_{12}$$

$$A_{12} = -a_{21}$$

$$A_{22} = a_{11}$$

Therefore, the desired matrix is

$$B = \begin{pmatrix} a_{22} & -a_{21} \\ -a_{12} & a_{11} \end{pmatrix}$$

Example 3. Find the transpose of the matrix B of Example 2.

Solution:

$$B = \begin{pmatrix} a_{22} & -a_{21} \\ -a_{12} & a_{11} \end{pmatrix}$$

$$B^t = \begin{pmatrix} a_{22} & -a_{12} \\ -a_{21} & a_{11} \end{pmatrix}$$

If we multiply the matrix B^t of Example 3 by the scalar quantity $1/|A|$ and compare the resulting matrix with (1), we see that this is A^{-1}, the multiplicative inverse of A. Of course, $|A|$ must not be zero. This would be written symbolically as

$$A^{-1} = \frac{1}{|A|} B^t$$

if

$$|A| \neq 0$$

Next we give an extension of this method. This extension can be used to find the inverse of an n by n matrix.

PROCEDURE

Step 1: Calculate the determinant of the given matrix. If this determinant is not zero, go on to step 2.

Step 2: Form a new matrix B by replacing each element in the given matrix by its cofactor. B will be called the *matrix of cofactors.*

Step 3: Take the transpose of the matrix of cofactors.

Step 4: The inverse of the given matrix is found by multiplying the matrix in step 3 by the scalar $1/|A|$.

Before commenting on the validity and the usefulness of the procedure, we will illustrate the procedure by a numerical example.

Example 4. Use the above procedure to find the inverse of the matrix

$$A = \begin{pmatrix} 1 & 2 & 3 \\ 1 & 3 & 5 \\ 2 & 5 & 9 \end{pmatrix}$$

Solution:

Step 1.

$$\begin{vmatrix} 1 & 2 & 3 \\ 1 & 3 & 5 \\ 2 & 5 & 9 \end{vmatrix} = 1 \qquad \textit{from Example 1 in the previous section}$$

Step 2. The cofactor of 9 is

$$\begin{vmatrix} 1 & 2 \\ 1 & 3 \end{vmatrix} = 3 - 2 = 1 \qquad (\textit{sample calculation})$$

The matrix of cofactors is

$$B = \begin{pmatrix} 2 & 1 & -1 \\ -3 & 3 & -1 \\ 1 & -2 & 1 \end{pmatrix}$$

Step 3.

$$B^t = \begin{pmatrix} 2 & -3 & 1 \\ 1 & 3 & -2 \\ -1 & -1 & 1 \end{pmatrix} \qquad \textit{interchanging rows and columns}$$

Step 4.

$$A^{-1} = \frac{1}{|A|} B^t$$

$$\begin{pmatrix} 1 & 2 & 3 \\ 1 & 3 & 5 \\ 2 & 5 & 9 \end{pmatrix}^{-1} = \frac{1}{1}\begin{pmatrix} 2 & -3 & 1 \\ 1 & 3 & -2 \\ -1 & -1 & 1 \end{pmatrix}$$

$$A^{-1} = \begin{pmatrix} 2 & -3 & 1 \\ 1 & 3 & -2 \\ -1 & -1 & 1 \end{pmatrix}$$

Note: This matrix was also inverted by row operations in Section 3-6.

Remarks:

1. If the determinant of the given matrix turns out to be zero, then the matrix does not have a multiplicative inverse. This test for the existence of the inverse of a matrix can also be used in conjunction with the algorithm of Section 3-6.

2. We must point out that even though the four-step procedure can be used to find the inverse of a matrix of order greater than 3 by 3, the procedure is very long and tedious for matrices of order 4 by 4 and greater. Our use of this method in later sections will be limited to 3 by 3 matrices.

Finally, we prove the validity of the procedure for the special case of a 3 by 3 matrix. Let

$$A = \begin{pmatrix} a_{11} & a_{12} & a_{13} \\ a_{21} & a_{22} & a_{23} \\ a_{31} & a_{32} & a_{33} \end{pmatrix}$$

If we denote the cofactor of a_{ij} by A_{ij}, we have

$$B = \begin{pmatrix} A_{11} & A_{12} & A_{13} \\ A_{21} & A_{22} & A_{23} \\ A_{31} & A_{32} & A_{33} \end{pmatrix} \qquad \textit{the matrix of cofactors}$$

$$B^t = \begin{pmatrix} A_{11} & A_{21} & A_{31} \\ A_{12} & A_{22} & A_{32} \\ A_{13} & A_{23} & A_{33} \end{pmatrix}$$

$$\frac{1}{|A|}\begin{pmatrix} A_{11} & A_{21} & A_{31} \\ A_{12} & A_{22} & A_{32} \\ A_{13} & A_{23} & A_{33} \end{pmatrix} \qquad \textit{is the alleged inverse by step 4}$$

To check this we must show that $A\left(\dfrac{1}{|A|}B^t\right) = \begin{pmatrix} 1 & 0 & 0 \\ 0 & 1 & 0 \\ 0 & 0 & 1 \end{pmatrix}.$

Now,

$$\begin{pmatrix} a_{11} & a_{12} & a_{13} \\ a_{21} & a_{22} & a_{23} \\ a_{31} & a_{32} & a_{33} \end{pmatrix} \frac{1}{|A|} \begin{pmatrix} A_{11} & A_{21} & A_{31} \\ A_{12} & A_{22} & A_{32} \\ A_{13} & A_{23} & A_{33} \end{pmatrix}$$

$$= \frac{1}{|A|}\begin{pmatrix} a_{11}A_{11}+a_{12}A_{12}+a_{13}A_{13} & a_{11}A_{21}+a_{12}A_{22}+a_{13}A_{31} & a_{11}A_{31}+a_{12}A_{32}+a_{13}A_{33} \\ a_{21}A_{11}+a_{22}A_{12}+a_{23}A_{13} & a_{21}A_{21}+a_{22}A_{22}+a_{23}A_{23} & a_{21}A_{31}+a_{22}A_{32}+a_{23}A_{33} \\ a_{31}A_{31}+a_{32}A_{12}+a_{33}A_{13} & a_{31}A_{21}+a_{32}A_{22}+a_{33}A_{23} & a_{31}A_{31}+a_{32}A_{32}+a_{33}A_{33} \end{pmatrix}$$

$$= \frac{1}{|A|}\begin{pmatrix} |A| & 0 & 0 \\ 0 & |A| & 0 \\ 0 & 0 & |A| \end{pmatrix} \qquad \text{by Theorems 4.1 and 4.2}$$

$$= \begin{pmatrix} 1 & 0 & 0 \\ 0 & 1 & 0 \\ 0 & 0 & 1 \end{pmatrix} \qquad \text{as shown}$$

PROBLEMS

1. Use the formula $A^{-1} = \frac{1}{|A|}\begin{pmatrix} a_{22} & -a_{12} \\ -a_{21} & a_{11} \end{pmatrix}$ to find the inverse of the following matrices:

(a) $\begin{pmatrix} 6 & 4 \\ 3 & 1 \end{pmatrix}$ **(b)** $\begin{pmatrix} 9 & 2 \\ 4 & 3 \end{pmatrix}$

2. Find the transpose of the following matrices:

(a) (3, 2, 6) **(c)** $\begin{pmatrix} 2 & 1 \\ 5 & 6 \end{pmatrix}$

(b) $\begin{pmatrix} 4 & 2 & 1 \\ 8 & 3 & 4 \end{pmatrix}$ **(d)** $\begin{pmatrix} a_{11} & a_{12} \\ a_{21} & a_{22} \end{pmatrix}$

3. Find the transpose of the following matrices:

(a) $\begin{pmatrix} 6 & 5 & 4 \\ 2 & 3 & 1 \\ 6 & 8 & 9 \end{pmatrix}$ **(c)** $\begin{pmatrix} 2 \\ 6 \\ 1 \end{pmatrix}$

(b) $\begin{pmatrix} c_{11} & c_{12} \\ c_{21} & c_{22} \\ c_{31} & c_{32} \end{pmatrix}$ **(d)** $\begin{pmatrix} 2 & 3 \\ 4 & 5 \\ 6 & 7 \end{pmatrix}$

4. Given the matrix $\begin{pmatrix} 4 & 3 \\ 6 & 9 \end{pmatrix}$

(a) Find the matrix of cofactors.
(b) Find the transpose of the matrix of cofactors.

5. Given that $B = \begin{pmatrix} 4 & 2 \\ 8 & 6 \end{pmatrix}$

(a) Find the determinant of B.
(b) Find the matrix of cofactors.
(c) Find the transpose of the matrix of cofactors.
(d) Find B^{-1}.

6. Show that the result in Example 5, part (d) is indeed the inverse of B (i.e., $B \cdot B^{-1} = I$).

7. Calculate the inverse of each of the following matrices using the 4-step procedure outlined in this section.

(a) $\begin{pmatrix} 3 & 0 & 1 \\ 4 & 2 & 1 \\ 3 & 0 & 1 \end{pmatrix}$ **(b)** $\begin{pmatrix} 4 & 2 & 0 \\ 3 & 0 & 0 \\ 6 & 1 & 1 \end{pmatrix}$

8. Does the following system of equations have a unique solution?

$$x + 2y = 3$$

$$2x + 4y = 6$$

[*Hint:* Is the determinant of the matrix of coefficients equal to zero?]

9. Calculate the inverse of C and show that this is indeed the inverse, i.e.,

$$C\left[\frac{1}{|C|}B^t\right] = I, \qquad C = \begin{pmatrix} c_{11} & c_{12} & c_{13} \\ c_{21} & c_{22} & c_{23} \\ c_{31} & c_{32} & c_{33} \end{pmatrix}$$

10. **(a)** Show that $(A^t)^t = A$.
(b) Show that $(A+B)^t = A^t + B^t$.

11. Recalling the definition of the *trace* of a matrix as the sum of elements on its principal diagonal, write out the trace of AB^t, where A and B are n by n matrices.

12. The following result is given by Campbell*:

The total revenue of a transportation industry maintaining regular service between n stations is given by the trace of AB^t, where $a_{ij} =$ number of passengers transported from station i to station j, and $b_{ij} =$ cost of a ticket from station i to station j. See Problem 11 for the definition of the trace of a matrix.

Draw a schematic diagram and establish the result for the special case $n = 3$.

13. Use the result in Problem 12 to compute the average daily revenue for the Fast Service Transit Company given the following data for cities x, y, and z.

AVERAGE PASSENGER USE

		To		
		x	y	z
	x	0	200	105
From	y	160	0	125
	z	80	110	0

TICKET PRICES IN DOLLARS

		To		
		x	y	z
	x	0	15	50
From	y	15	0	30
	z	50	30	0

* Hugh G. Campbell, *Matrices with Applications.* New York: Appleton-Century-Crofts, 1968, p. 55.

14. Does $B^t = B$ in Problem 13? Could we use AB instead of AB^t in this case to compute the total revenue?

4-6 SUMMARY OF METHODS FOR SOLVING SYSTEMS OF EQUATIONS

In this section we will solve a system of linear equations by three different methods. In the first example, the system is solved by the method of elimination. In the second example, the system is solved by Cramer's rule. Finally, in the last example, the system is solved by first finding the inverse of the coefficient matrix by the algorithm of Section 3-6, and then multiplying both sides of the matrix equation by this inverse.

Example 1. Solve the following linear system by the method of elimination.

$$e_1:\quad 2x + 6y + 4z = -12$$

$$e_2:\quad 6x + 6y + 4z = 8$$

$$e_3:\quad 3x + 2y + 4z = 13$$

Solution: By subtracting e_2 from e_1, we have

$$\begin{array}{r} 2x + 6y + 4z = -12 \\ 6x + 6y + 4z = 8 \\ \hline -4x + 0 + 0 = -20 \end{array}$$

Therefore, $e_4: x = 5$.

Subtract e_3 from e_2.

$$\begin{array}{r} 6x + 6y + 4z = 8 \\ 3x + 2y + 4z = 13 \\ \hline \end{array}$$

$$e_5:\quad 3x + 4y \qquad = -5$$

By substituting e_4 into e_5, we get

$$3(5) + 4(y) = -5$$

$$4y = -20$$

$$y = -5$$

If we substitute $x = 5$ and $y = -5$ into e_1, we have

$$2(5) + 6(-5) + 4z = -12$$

$$4z = -12 - 10 + 30$$

$$4z = 8$$

$$z = 2$$

The solution of the given system is $x = 5$, $y = -5$, and $z = 2$.

Example 2. Use Cramer's rule to solve the system in Example 1.

Solution:

$$D_3 = \begin{vmatrix} 2 & 6 & 4 \\ 6 & 6 & 4 \\ 3 & 2 & 4 \end{vmatrix}$$

$$= 2(24-8) - 6(24-12) + 4(12-18)$$

$$= 2(16) - 6(12) + 4(-6)$$

$$= 32 - 72 - 24 \qquad \textit{by the method of cofactors and the first row}$$

$$= -64$$

$$x = \frac{\begin{vmatrix} -12 & 6 & 4 \\ 8 & 6 & 4 \\ 13 & 2 & 4 \end{vmatrix}}{D_3}, \quad y = \frac{\begin{vmatrix} 2 & -12 & 4 \\ 6 & 8 & 4 \\ 3 & 13 & 4 \end{vmatrix}}{D_3}, \quad \text{and}$$

$$z = \frac{\begin{vmatrix} 2 & 6 & -12 \\ 6 & 6 & 8 \\ 3 & 2 & 13 \end{vmatrix}}{D_3}, \quad \text{where} \quad D_3 = -64$$

By the method of cofactors and column three we have

$$x = \frac{4(16-78)-4(-24-78)+4(-72-48)}{-64}$$

$$x = \frac{4(-62+102-120)}{-64} = \frac{4(-80)}{-64} = 5$$

Use the scheme of Section 4-2, which works for 2 by 2 and 3 by 3 determinants only, and

$$y = \frac{64-144+312-96-104+288}{-64}$$

$$y = \frac{320}{-64} = -5$$

Finally, a similar calculation gives $z = -128/-64 = 2$, and the solution is $x = 5$, $y = -5$, and $z = 2$, the same answers found in Example 1.

Example 3. Solve the system given in Example 1 by finding the inverse of the coefficient matrix and multiplying by this inverse.

Solution: The coefficient matrix of the system is $A = \begin{pmatrix} 2 & 6 & 4 \\ 6 & 6 & 4 \\ 3 & 2 & 4 \end{pmatrix}$. The given system can be written as the matrix equation

$$\begin{pmatrix} 2 & 6 & 4 \\ 6 & 6 & 4 \\ 3 & 2 & 4 \end{pmatrix}\begin{pmatrix} x \\ y \\ z \end{pmatrix} = \begin{pmatrix} -12 \\ 8 \\ 13 \end{pmatrix}$$

Use the algorithm from Section 3-6.

$$(A\,|\,I) = \left(\begin{array}{ccc|ccc} 2 & 6 & 4 & 1 & 0 & 0 \\ 6 & 6 & 4 & 0 & 1 & 0 \\ 3 & 2 & 4 & 0 & 0 & 1 \end{array}\right) \tag{a}$$

Subtract row 2 from row 1 of (a) to get $b_{12} = 0$, $b_{13} = 0$; this gives

$$(B\,|\,X_1) = \left(\begin{array}{ccc|ccc} -4 & 0 & 0 & 1 & -1 & 0 \\ 6 & 6 & 4 & 0 & 1 & 0 \\ 3 & 2 & 4 & 0 & 0 & 1 \end{array}\right) \tag{b}$$

Subtract row 3 from row 2 of (b) to get $c_{23} = 0$; this gives

$$(C\,|\,X_2) = \left(\begin{array}{ccc|ccc} -4 & 0 & 0 & 1 & -1 & 0 \\ 3 & 4 & 0 & 0 & 1 & -1 \\ 3 & 2 & 4 & 0 & 0 & 1 \end{array}\right) \tag{c}$$

Subtract row 2 from row 3 of (c) to get $d_{31} = 0$.

$$(D\,|\,X_3) = \left(\begin{array}{ccc|ccc} -4 & 0 & 0 & 1 & -1 & 0 \\ 3 & 4 & 0 & 0 & 1 & -1 \\ 0 & -2 & 4 & 0 & -1 & 2 \end{array}\right) \tag{d}$$

Multiply row 1 by $\frac{3}{4}$ and then add the result to row 2 of (d) to get $e_{21} = 0$.

$$(E\,|\,X_4) = \left(\begin{array}{ccc|ccc} -4 & 0 & 0 & 1 & -1 & 0 \\ 0 & 4 & 0 & \frac{3}{4} & \frac{1}{4} & -1 \\ 0 & -2 & 4 & 0 & -1 & 2 \end{array}\right) \tag{e}$$

Multiply row 2 by $-\frac{1}{2}$ and then subtract row 2 from row 3 of (e) to get $f_{32} = 0$. This gives

$$(F\,|\,X_5) = \left(\begin{array}{ccc|ccc} -4 & 0 & 0 & 1 & -1 & 0 \\ 0 & -2 & 0 & -\frac{3}{8} & -\frac{1}{8} & \frac{1}{2} \\ 0 & 0 & 4 & \frac{3}{8} & -\frac{7}{8} & \frac{3}{2} \end{array}\right) \tag{f}$$

Multiply row 1 by $-\frac{1}{4}$, row 2 by $-\frac{1}{2}$, row 3 by $\frac{1}{4}$. We now have

$$(I\,|\,A^{-1}) = \left(\begin{array}{ccc|ccc} 1 & 0 & 0 & -\frac{1}{4} & \frac{1}{4} & 0 \\ 0 & 1 & 0 & \frac{3}{16} & \frac{1}{16} & -\frac{1}{4} \\ 0 & 0 & 1 & \frac{3}{32} & -\frac{7}{32} & \frac{3}{8} \end{array}\right) \tag{g}$$

Therefore,

$$A^{-1} = \begin{pmatrix} -\frac{1}{4} & \frac{1}{4} & 0 \\ \frac{3}{16} & \frac{1}{16} & -\frac{1}{4} \\ \frac{3}{32} & -\frac{7}{32} & \frac{3}{8} \end{pmatrix}$$

The solution of the given system can now be obtained by multiplying both sides of the matrix equation

$$\begin{pmatrix} 2 & 6 & 4 \\ 6 & 6 & 4 \\ 3 & 2 & 4 \end{pmatrix} \begin{pmatrix} x \\ y \\ z \end{pmatrix} = \begin{pmatrix} -12 \\ 8 \\ 13 \end{pmatrix}$$

by A^{-1}.

This gives

$$\begin{pmatrix} 1 & 0 & 0 \\ 0 & 1 & 0 \\ 0 & 0 & 1 \end{pmatrix} \begin{pmatrix} x \\ y \\ z \end{pmatrix} = \begin{pmatrix} -\frac{1}{4} & \frac{1}{4} & 0 \\ \frac{3}{16} & \frac{1}{16} & -\frac{1}{4} \\ \frac{3}{32} & -\frac{7}{32} & \frac{3}{8} \end{pmatrix} \begin{pmatrix} -12 \\ 8 \\ 13 \end{pmatrix}$$

$$\begin{pmatrix} x \\ y \\ z \end{pmatrix} = \begin{pmatrix} 3+2+0 \\ -\frac{36}{16}+\frac{8}{16}-\frac{13}{4} \\ -\frac{36}{32}-\frac{56}{32}+\frac{39}{8} \end{pmatrix} = \begin{pmatrix} 5 \\ -5 \\ 2 \end{pmatrix}$$

Therefore, $x = 5$, $y = -5$, and $z = 2$, as found in Examples 1 and 2.

PROBLEMS

Solve the following by the method of elimination, by the use of the inverse of the matrix of coefficients, and by Cramer's rule.

1. $x + y = 2$
$x - y = 5$

2. $3x - 2y = 4$
$-x - 2y = 6$

3. $x + y + z = 1$
$x - y = 4$
$x + y - 2z = 4$

4. $2x + 2z = 2$
$5x + 4y + z = 3$
$2x + 6z = 4$

5. $2x + 3y + 4z = 4$
$x + 2y + 5z = 0$
$4x + y + z = 5$

6. $$\begin{pmatrix} 1 & 2 & 3 \\ 0 & 1 & 2 \\ 2 & 1 & 4 \end{pmatrix} \begin{pmatrix} x \\ y \\ z \end{pmatrix} = \begin{pmatrix} 4 \\ -2 \\ 6 \end{pmatrix}$$

7. $x_1 + x_2 + x_3 = -2$
$8x_1 + 4x_2 + 2x_3 = 6$
$-x_1 + x_2 - x_3 = 0$

REFERENCES

Books

Andree, Richard V. *Selections from Modern Abstract Algebra.* New York: Henry Holt and Co., 1958

Campbell, Hugh G. *Matrices with Applications.* New York: Appleton-Century-Crofts, 1968.

Corcoran, A. Wayne. *Mathematical Applications in Accounting.* New York: Harcourt Brace Jovanovich, 1968.

Curtis, Charles W. *Linear Algebra, An Introductory Approach.* Boston: Allyn and Bacon, Inc., 1963.

Draper, N. R., and Smith, H. *Applied Regression Analysis.* New York: John Wiley & Sons, Inc., 1966.

Kemeny, John G., Schleifer, A., Snell, J. R., and Thompson, G. L. *Finite Mathematics with Business Applications.* Englewood Cliffs, N.J.: Prentice-Hall, Inc., 1964.

Leontief, Wassily. *Input-Output Economics.* New York: Oxford University Press, Inc., 1966.

Levin, Richard I., and Kirkpatrick, C. A. *Quantitative Approaches to Management.* New York: McGraw-Hill Book Co., Inc., 1965.

Lial, Margaret L., and Miller, Charles David. *Finite Mathematics: With Applications in Business, Biology, and Behavioral Sciences.* Glenview, Illinois: Scott, Foresman and Company, 1977.

Miernyk, William H. *The Elements of Input-Output Analysis.* New York: Random House, Inc., 1965.

Parker, William Vann, and Eaves, James Clifton. *Matrices.* New York: The Ronald Press Co., 1960.

Shank, John K. *Matrix Methods In Accounting.* Reading, Mass.: Addison-Wesley Publishing Co., 1972.

Springer, Clifford H., Herlihy, R. E., and Beggs, R. I. *Basic Mathematics.* Homewood, Ill.: Richard D. Irwin, Inc., 1965.

Articles

"Criteria for Entering an Industry." *Sales Management:* Oct. 15, 1969.

"Marketers Meet 'Input-Output'." *Business Week:* Oct. 29, 1966.

"Planners Put Big Picture on a Grid." *Business Week:* Sept. 23, 1967.

"They Call It 'Instant Research'." *Business Week:* Jan. 25, 1969.

"How Input-Output Helped Map the Marketing Strategy of Glass Containers Corp." *Industrial Marketing:* January 1970.

FIVE

Inequalities and linear programming

Linear programming (to be described in Section 5-3) is the main topic of this chapter. Linear programming involves the use of linear functions and linear inequalities. Inequalities are treated first in Sections 5-1 and 5-2 to provide background material for the study of linear programming (LP).

Linear programming problems that involve only two variables can be solved either graphically or numerically. The graphical method, which provides much insight into the nature of linear programming, is limited to the two-variable case. The larger, more realistic LP problems must be done numerically, preferably on a computer. The graphical method presented in Section 5-4 draws heavily from Section 5-2, which deals with systems of linear inequalities. The remainder of the chapter deals with a numerical algorithm called the simplex method, which can be used to solve linear programming problems.

It should also be noted that linear programming is an optimization technique. That is, once the linear constraints and linear objective function have been formalized (i.e., descriptive model) one goes on and uses either the graphical method or the simplex method to arrive at an optimum solution.

5-1 INTRODUCTION TO INEQUALITIES

As noted in the introduction to this chapter, linear programming problems involve the use of inequalities. The inequality $3x+5y \leqslant 8$

is an example of a typical linear inequality encountered in linear programming problems. Before studying inequalities of this type, which involves variables x and y, we must discuss the following important relationships between pairs of real numbers. All four of these relationships are called inequalities. An understanding of inequalities is basic to the study of many topics, both in mathematics and applied mathematics, such as linear programming.

$a < b$	is read	a is *less than* b
$a > b$	is read	a is *greater than* b
$a \leqslant b$	is read	a is *less than* or *equal to* b
$a \geqslant b$		a is *greater than* or *equal to* b

If we represent the real numbers by the points on a straight line, then the relationship $a < b$ means that the point corresponding to a lies to the left of the point corresponding to b.

Example 1. Let $a = 3$, $b = 7$, $c = -4$, $d = 0$, and $e = -6$. Represent these numbers as points on a line and discuss the relative position of these points.

Solution: Denote the point corresponding to the real number a by A, and so forth.

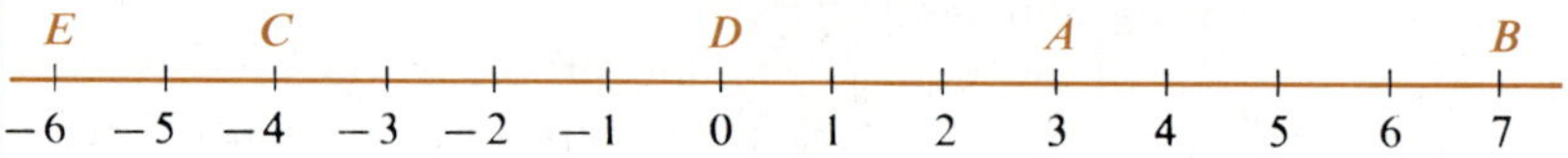

FIGURE 5-1

A is to the left of B, $\therefore\ a < b$.

B is to the right of A, $\therefore\ b > a$.

E is to the left of C, $\therefore\ e < c$, that is, $-6 < -4$.

A is to the right of C, $\therefore\ a > c$, that is, $3 > -4$.

D is to the left of A, $\therefore\ d < a$, or $0 < 3$.

E is to the left of D, $\therefore\ e < d$, or $-6 < 0$.

DEFINITION 5.1. *We say that $a < b$, a is less than b, if and only if there is a positive number h with the property that $a + h = b$. The notation $b > a$ has the same meaning as $a < b$.*

Example 2. Find the numerical value of h for $3 < 8$ and $-6 < -4$.

Solution:

$$3 + 5 = 8, \therefore h = 5$$

$$-6 + 2 = -4, \therefore h = 2$$

The following rules for working with inequalities can be proven from this definition. The letters a, b, c, and d all denote real numbers.

RULES FOR INEQUALITIES

Rule 1. If $a < b$, then $a+c < b+c$.
Rule 2. If $a < b$ and $b < c$, then $a < c$. (*transitivity*)
Rule 3. If $a < b$ and $c < d$, then $a+c < b+d$.
Rule 4. If $a < b$ and $0 < c$ (c is positive), then $ca < cb$.
Rule 5. If $a < b$ and $c < 0$ (c is negative), then $ca > cb$.

Proof of Rule 2. Assume $a < b$ and $b < c$. Then, by Definition 5.1, there are positive numbers h_1 and h_2 that give $a+h_1 = b$ and $b+h_2 = c$. By adding these two equations together, we have

$$a + h_1 + b + h_2 = b + c$$

$$a + h_1 + h_2 = c$$

Since h_1 and h_2 are positive, h_1+h_2 is also positive. Let

$$h_3 = h_1 + h_2$$

Then $a+h_3 = c$. Therefore,

$$a < c \qquad \textit{by Definition 5.1}$$

Proof of Rule 4. Assume $a < b$ and $0 < c$. There exists h such that $a+h - b$, and h is positive. If we multiply both sides by c, we have

$$c(a+h_1) = cb$$

$$ca + ch_1 = cb$$

If we define $h_2 = ch_1$, h_2 is positive since the product of two positive numbers is positive.

$$ca + h_2 = cb$$

$$ca < cb \qquad \textit{by Definition 5.1}$$

DEFINITION 5.2. *The notation $a \leqslant b$ means $a < b$ or $a = b$. Similarly, $a \geqslant b$ means $a > b$ or $a = b$.*

Example 3. Illustrate Definition 5.2 numerically.

Solution:

$3 \leqslant 4$ since $3 < 4$

$4 \leqslant 4$ since $4 = 4$

$-5 \geqslant -7$ since $-5 > -7$

$-5 \geqslant -5$ since $-5 = -5$

Example 4. Solve the inequality $2x+5 \leqslant 13$ for x. Illustrate the solution graphically.

Solution:

$2x + 5 \leqslant 13$

$2x + 5 + (-5) \leqslant 13 + (-5)$ *by Rule 1*

$2x + 0 \leqslant 8$ *by A-6 of Chapter Three*

$2x \leqslant 8$ *by A-5 of Chapter Three*

$\frac{1}{2}(2x) \leqslant \frac{1}{2}(8)$ *by Rule 4*

$x \leqslant 4$

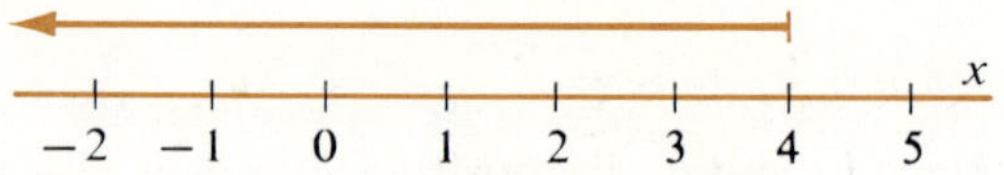

FIGURE 5-2

Inequalities are frequently used to describe sets of real numbers. An important example of this is an *interval.* The following *interval notation* is standard in the mathematical literature.

1. $(a,b) = \{x : a < x < b\}$, *called an open interval*
2. $[a,b] = \{x : a \leqslant x \leqslant b\}$, *called a closed interval*
3. $(a,b] = \{x : a < x \leqslant b\}$,
4. $(a,\infty) = \{x : a < x\}$, *where the symbol* ∞ *is read infinity.*

These four cases are sufficient to indicate the meaning of (, [, and ∞ when used in this connection.

Finally, the solution of Example 4 shown graphically in Figure 5-2 can be written in the form $(-\infty, 4]$, using interval notation.

PROBLEMS

1. Let $a = 2$, $b = -3$, $c = 6$, $d = -5$, and $e = 0$. Represent these numbers as points on a line and discuss the relative position of the following points:

(a) *A* and *C*

(b) *D* and *B*

(c) A and B
(d) D and E
(e) C and E

2. Refer to Definition 5.1. Find the numerical value of h.
(a) $6 < 7$
(b) $-10 < -4$
(c) $-3 < 0$
(d) $10 > 7$ [*Hint:* Translate to less than symbol first.]

3. Use Definition 5.1 to prove the following: If $a < b$, then $a+c < b+c$.

4. Use $a = 2$, $b = 4$, $c = 3$, $d = 6$ to illustrate Rule 3.

5. Use Definition 5.1 to prove the following: If $a < b$ and $c < 0$ (c is negative), then $ca > cb$. [*Hint:* Modify the proof of Rule 4.]

6. Solve the following inequalities for x:
(a) $3 - 2x \leqslant 9$
(b) $2x + 3 \leqslant 5$
(c) $2x + 7 \geqslant 13$

7. Graphically illustrate the solutions to Problem 6.

8. Solve the following inequalities for x:
(a) $4x + 3 \leqslant x - 2$
(b) $3(x-2) < 4$
(c) $4 - 5x \geqslant 14$

9. Graph the following function:

$$f(x) = \begin{cases} 2x & \text{if } 0 \leqslant x \leqslant 1 \\ x+1 & \text{if } 1 \leqslant x \leqslant 2 \end{cases}$$

10. Express the function in Figure 5-3 in the following form:

$$f(x) = \begin{cases} ax & \text{if } 0 \leqslant x \leqslant 2 \\ bx + c & \text{if } 2 \leqslant x \leqslant 4 \end{cases}$$

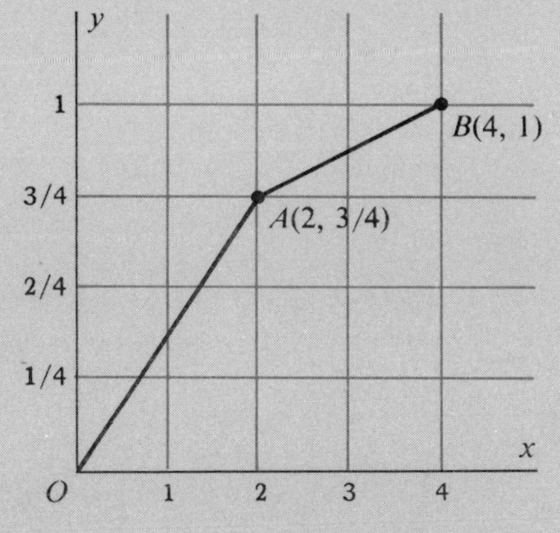

FIGURE 5-3

11. Express the water rate structure for the city of Nice and Green as a function, given the following: For the first 5000 cubic feet or less,

$0.34 per 100 cubic feet. For the next 10,000 cubic feet or less, $0.28 per 100 cubic feet. For the next 35,000 cubic feet or less, $0.22 per 100 cubic feet. For the next 250,000 cubic feet or less, $0.13 per 100 cubic feet. For use of over 300,000 cubic feet, the rate for the entire quantity used is $0.13 per 100 cubic feet.

12. Express the electric rate structure for the city of Light and Cool as a function.

First:	6,000 kwh, $0.12 per 100 kwh
Next:	12,000 kwh, 0.10 per 100 kwh
Above:	18,000 kwh, 0.08 per 100 kwh

Minimum billing per period is $3.00.

13. Express the commission schedule for the New York Stock Exchange as a function.

Money Involved in the Order	Minimum Commission
$100–$799.99	2.0% + $6.40
800–2499.99	1.3% + 12.00
2500 and over	0.9% + 22.00

Subject to the provision that the minimum commission on single round lot orders not exceed $65.00.

14. Express the following cost structure as a function.
(a) $10.00 per unit for any order of 100 units or less
(b) 9.00 per unit for any order from 101 to 200 units
(c) 7.50 per unit for any order in excess of 200 units

15. Assume you are an office equipment manager interested in purchasing some optical mark document readers. In order to accelerate the popularity of his machine, a supplier has made you the offer in Table 5-1.

TABLE 5-1

Quantity (in units)	Price (per unit)	Discount
1	$2140	
2–10	1926	10%
11–20	1712	20%
21–30	1498	30%
31 or more	1284	40%

(a) Express this cost structure graphically using the vertical axis for cost.
(b) Express this cost structure functionally.

16. When analyzing a company's balance sheet, two measures of liquidity are working capital and current ratio. The current ratio (CR) is defined as current assets (CA) divided by current liabilities (CL). Working capital (WC) is defined as current assets minus current liabilities. Expressed as formulas we have:

$$CR = \frac{CA}{CL}, \qquad WC = CA - CL$$

At year end many companies take cash (a current asset) to pay off short term debt (a current liability). This has no effect on working capital but in many cases will increase the current ratio. This is frequently referred to as "window dressing."

Mathematically describe the circumstances that are necessary for the current ratio to increase when a company takes cash to pay a short term debt.

5-2 SYSTEMS OF LINEAR INEQUALITIES

Inequalities involving two variables are considered in this section. The section begins with a treatment of inequalities of the form $y \leqslant mx+b$. This is an example of a linear inequality in two variables. The reader may recall studying linear equations of the form $y = mx+b$ in Section 1-2. Indeed, it is assumed at this point that the reader knows how to graph linear functions quickly and accurately.

We are interested in determining and describing the set of all points in the xy plane whose coordinates satisfy a given inequality involving the variables x and y. The coordinates of a point are said to satisfy an inequality if their substitution into the given inequality results in a true statement. The set of points whose coordinates satisfy a given inequality is called the *solution set* or the *solution space* of the inequality. The term *solution space* is used in this text.

Example 1. Determine the points in the xy plane whose coordinates satisfy the inequality $y \leqslant 2x-1$.

Solution:

x	-2	-1	0	1	2	3
y	$y \leqslant -5$	$y \leqslant -3$	$y \leqslant -1$	$y \leqslant 1$	$y \leqslant 3$	$y \leqslant 5$

Sample Calculation: Let $x = -2$.

$$y \leqslant 2(-2)-1$$

$$y \leqslant -4-1$$

$$y \leqslant -5$$

This table suggests that we draw the graph of $y = 2x-1$ and illustrate graphically the above information.

It should be clear from the graph that the inequality $y \leqslant 2x - 1$ is satisfied by the coordinates of any point that lies on the graph of $y = 2x - 1$ or below it. The solution space of $y \leqslant 2x - 1$ is indicated in Figure 5-4 by the line and the three arrows. Note that the arrows point into the solution space.

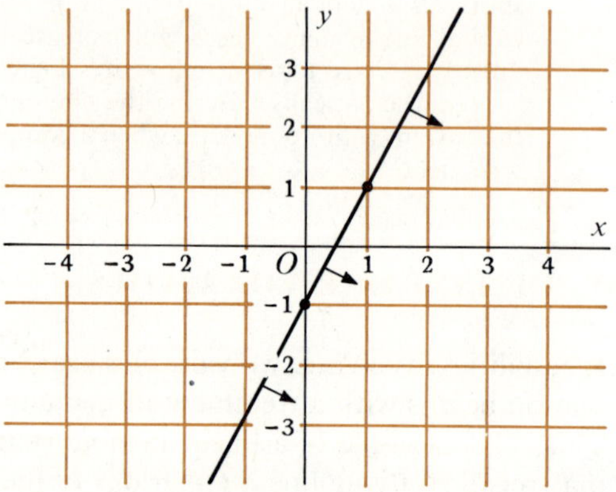

FIGURE 5-4

Example 2. Find the points whose coordinates satisfy the inequality $2x + 3y \leqslant 12$.

Solution:

$2x + 3y \leqslant 12$

$3y \leqslant 12 - 2x$ *by Rule 1*

$y \leqslant 4 - \frac{2}{3}x$ *by Rule 4*

The solution consists of the points on or below the line $y = 4 - \frac{2}{3}x$.

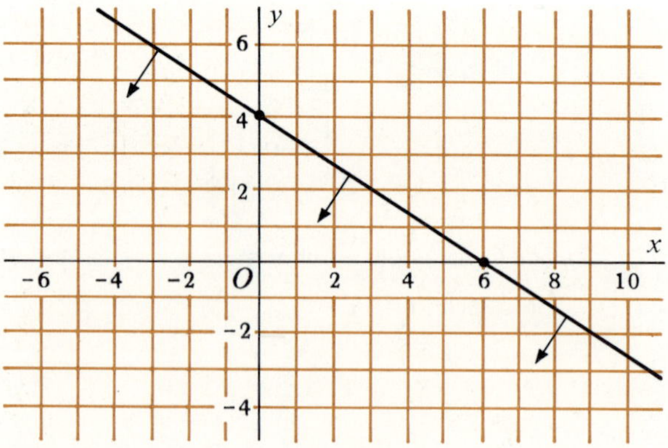

FIGURE 5-5

Systems of linear inequalities are introduced next.

DEFINITION 5.3.

$$
\begin{array}{ccccccccc}
a_{11}x_1 & + & a_{12}x_2 & + & \cdots & + & a_{1n}x_n & \leqslant & b_1 \\
a_{21}x_1 & + & a_{22}x_2 & + & \cdots & + & a_{2n}x_n & \leqslant & b_2 \\
\vdots & & \vdots & & \ddots & & \vdots & & \vdots \\
a_{m1}x_1 & + & a_{m2}x_2 & + & \cdots & + & a_{mn}x_n & \leqslant & b_m
\end{array} \tag{1}
$$

is called a system of m linear inequalities in n unknowns.

DEFINITION 5.4 *We use the term solution space of an inequality or of a system of inequalities to mean the set of points whose coordinates satisfy the given inequality or system of inequalities.*

An examination of the graphs for Examples 1 and 2 illustrates the fact that the solution space for a linear inequality of the form $Ax+By < C$ consists of all the points on one side of the line $Ax+By = C$.

This observation suggests the following method for determining the solution space of a linear inequality.

PROCEDURE FOR DETERMINING INEQUALITY SOLUTION SPACE

Step 1: Draw the graph of the straight line $Ax+By = C$.

Step 2: Test one point that is not on this line to see if its coordinates satisfy the inequality $Ax+By < C$. If the coordinates of this point satisfy the inequality, then all points on the same side of the line $Ax+By = C$ are in the solution space. If the coordinates of the point fail to satisfy the given inequality, then the solution space consists of the points on the opposite side of the line, in addition to the points on the line.

Example 3. Find the solution space for the inequality $2x+3y \leqslant 6$.

Solution:

$2x + 3y = 6$

Step 1

x	0	3
y	2	0

see graph in Figure 5-6

Step 2

$$2(0) + 3(0) = 0, \qquad \textit{using the convenient point } (0,0)$$

$$0 < 6$$

Therefore, the origin is in the solution space and following the procedure outlined above we obtain the solution space indicated in Figure 5-6.

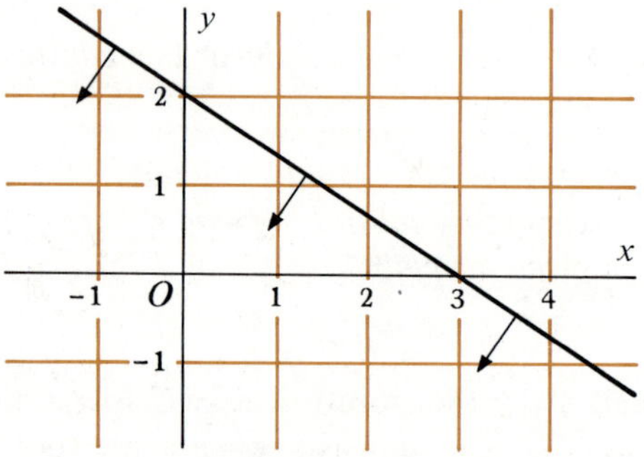

FIGURE 5-6

The solution space for a system of inequalities (1) can be found as follows.

PROCEDURE FOR DETERMINING THE SOLUTION SPACE FOR A SYSTEM OF LINEAR INEQUALITIES

Step 1: Indicate the *solution space* of each inequality in the system by the graph of the boundary along with arrows pointing into the solution space. This results in m individual solution spaces.

Step 2: Shade the *intersection* ($\cap$) of the m sets of points found in Step 1. These are the points that satisfy all m of the inequalities simultaneously. The shaded region is the solution space for the given system.

Example 4. Find the solution space for the system:

$$2x + 3y \leqslant 6$$

$$x \geqslant 0$$

Solution:

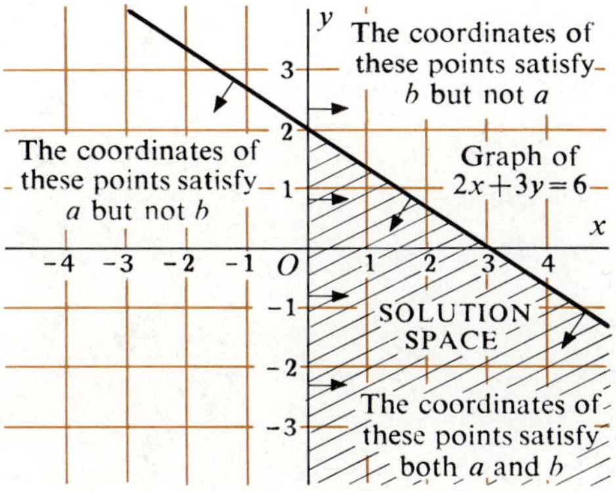

FIGURE 5-7

Example 5. Find the solution space for the system:

$$2x + 3y \geqslant 6$$

$$y - x \geqslant 1$$

$$x \geqslant 0$$

Solution:

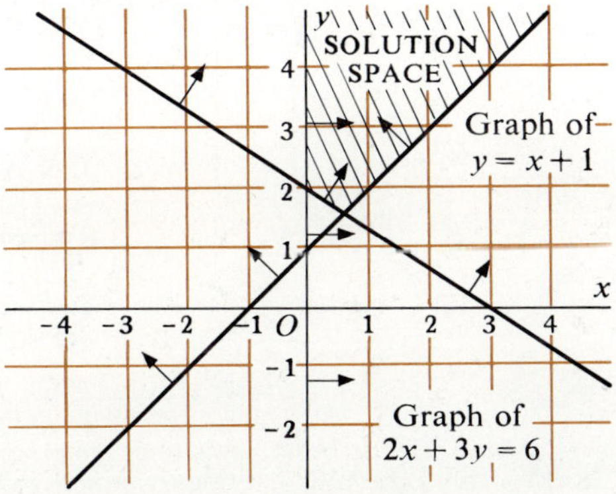

FIGURE 5-8

Example 6. Find the solution space for the system:

$$2x + y \leqslant 3$$

$$3x + y \leqslant 4$$

$$x \geqslant 0$$

$$y \geqslant 0$$

Solution: If we solve the equations $2x+y=3$ and $3x+y=4$ simultaneously, we have

$$(y=4-3x)-(y=3-2x)$$

After subtraction we have

$$0=1-x$$

Therefore,

$$x=1$$

and

$$y=4-3$$

$$y=1$$

The coordinates of the point B are $x=1$ and $y=1$.

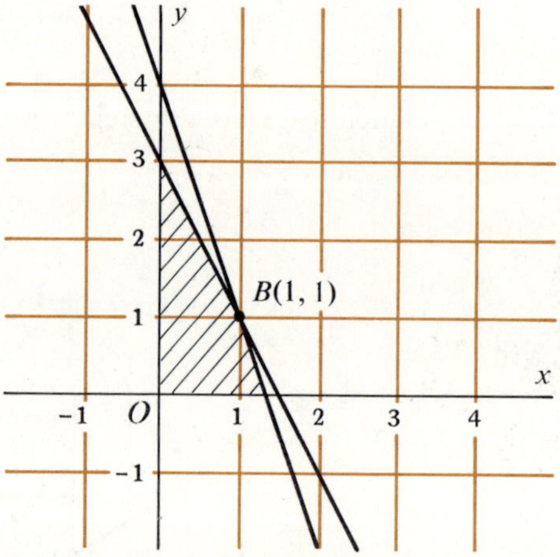

FIGURE 5-9

PROBLEMS

In Problems 1 through 4, shade the points in the xy plane whose coordinates satisfy the given inequalities.

1. $y \leqslant 4$

2. $y \leqslant 4-2x$

3. $x+y \geqslant 4$

4. $4y \geqslant 6x+24$

Find the solution space for the systems of inequalities given in Problems 5 through 8.

5. $x \geqslant 0$

$y \geqslant 0$

6. $2x + y \leqslant 8$
$x + 2y \leqslant 8$

7. $3x + 4y \geqslant 0$
$x + 3y \leqslant 6$
$3x + 2y \leqslant 6$

8. $x + y \leqslant 4$
$x + 2y \leqslant 6$
$x \geqslant 0$
$y \geqslant 0$

9. Find the coordinates of each of the corners of the solution space in Problems 5 through 8.

10. You are about to buy a kayak and have the following desires and constraints:

You would like to take a companion with you on your trip.

You are going to be flown to your departure point and cannot carry a kayak that is longer than 16 feet aboard the aircraft.

Because you will be making frequent portages, you would like the weight to be less than 80 pounds.

The trip will take a month, requiring you to bring quite a few supplies. It is therefore desirable to have a load capacity of between 700 and 750 pounds.

Your budget allows \$350 for the purchase of the kayak. Express these specifications mathematically using the following variables:

z = weight $\qquad$ q = load capacity

x = number of seats $\qquad$ w = dollar cost

y = length

11. Express the following company objectives symbolically:
(a) Sales of \$1,000,000 or more
(b) Cost of goods sold not to exceed \$650,000
(c) Net income to exceed \$100,000
(d) Rate of return of at least 11%

where S = sales
CGS = cost of goods sold
NI = net income
RR = rate of return

12. You are planning to buy a corporate aircraft and want to meet the following specifications:

Altitude	(ALT)	in excess of 18,000 feet
Airspeed	(AS)	in excess of 330 m.p.h.
Number of passengers	(NP)	minimum of 6
Useful load	(UL)	minimum of 2 tons
Landing run	(LR)	within 3000 feet
Willing to pay	(S)	up to \$800,000

Express the above symbolically using inequalities.

13. In accepting employment with a company you have the following objectives:

Starting salary of at least \$10,000
Two-week vacation, minimum
Location within 20 miles of your present home

Express the above mathematically using the following variable names:

D = distance from home

S = starting salary

V = weeks of vacation

5-3 INTRODUCTION TO LINEAR PROGRAMMING

"Linear programming is a nonstatistical mathematical technique whereby the maximization or minimization of a linear expression of variables, called the objective function, is determined in the presence of known or assumed restrictions, called constraints. In essence, it is a procedure for solving the problems in which there are more variables than simultaneous equations in which the variables are expressed".* This excerpt by Dr. Robert B. Sweeney, CPA, is an excellent brief description of the nature of linear programming. Five of his points are discussed in the following paragraphs.

First, linear programming is a nonstatistical mathematical technique. No probability or statistics are needed to study linear programming per se. The mathematics involved in linear programming is relatively easy to understand and to manipulate in contrast to calculus, for example. Linear equations and inequations (inequalities) form the mathematical skeleton around which linear programming (LP) is built.

A linear function called the *objective function* is to be maximized or minimized in some sense, i.e., *optimized*. Most real-world problems have many possible solutions. The purpose of optimization is to choose from among many possible solutions the "best" possible solution (decision). Some examples of "best" are highest profit, lowest cost, largest sales, lowest production time, etc.

The optimization of the objective function takes place in the presence of known or assumed restrictions. The technical term *constraints* is used to describe the restrictions present in a linear programming problem. The constraints are expressed mathematically as inequalities. In a practical real-world situation, the constraints are generated by the presence of limited resources or commodities such as capital, manpower, and raw materials. Mathematically, inequations can be converted to equations by the introduction of slack variables.

* Robert B. Sweeney "Business Use of Linear Programming," *Management Accounting*, September 1965, pp. 39–47.

Finally, as Dr. Sweeney points out, linear programming involves problems in which there are more variables than simultaneous linear equations. These systems have an infinite number of solutions. Any solution (if it exists) that makes the objective function optimal is said to be a solution of the linear programming problem in question.

The solution of a linear programming problem represents an *allocation* of the *scarce resources* involved in such a way that the objective function is optimized. It also provides the basis for a decision by someone in management with decision making responsibility.

Example 1. A certain manufacturer produces two products called ZIPS and ZAPS. The production of ZIPS and ZAPS requires the use of two machines. The production of a ZIP requires 3 minutes on machine I and 1 minute on machine II. The production of one ZAP requires 3 minutes on machine I and 5 minutes on machine II. The total time available on machine I is 2400 minutes, and the total time available on machine II is 1600 minutes. The company realizes a profit of $4.00 for each ZIP that is produced and a $4.50 profit for each ZAP that is produced. Determine the number of ZIPS and ZAPS that should be produced to give the largest possible profit.

Solution: (*set-up only*) To translate this word problem into a mathematical model, we let x denote the number of ZIPS and y denote the number of ZAPS. Clearly, the variables x and y must take on nonnegative values only. This involves two restrictions or constraints on the variables x and y. Using the symbols for inequalities, we have

$$x \geqslant 0$$

$$y \geqslant 0$$

Since only 2400 minutes are available on machine I, we have $3x+3y \leqslant 2400$. For machine II, we obtain the inequality $x+5y \leqslant 1600$.

From the profit contribution of $4.00 for each ZIP and $4.50 for each ZAP, we can express the total profit as a linear function of the variables x and y as follows: $P = 4x+4.5y$, where P denotes total profit in dollars.

Finally, the mathematical model is:

Maximize the *objective function* $P = 4x+4.5y$ subject to the *constraints:*

$x \geqslant 0$ (*x and y denote amounts of products*)

$y \geqslant 0$

$3x + 3y \leqslant 2400$ (*machine I*)

$x + 5y \leqslant 1600$ (*machine II*)

Example 2. In addition to the data given in Example 1, assume that each ZIP and each ZAP requires 4 minutes on a third machine that has 1600 minutes of available time. Set up the mathematical model for this new situation.

Solution: (*set-up only*) The new mathematical model is:

Maximize the *objective function* $P = 4x + 4.5y$ subject to the *constraints:*

$x \geqslant 0$

$y \geqslant 0$

$3x + 3y \leqslant 2400$

$x + 5y \leqslant 1600$

$4x + 4y \leqslant 1600$ (*machine III*)

PROBLEMS

Set up the mathematical model corresponding to the realistic situations described below. Assume that the goal is to maximize profit.

1. A manufacturer can produce two products, *A* and *B*. To produce each product requires processing in two departments, Department I and Department II. Department I has 36 hours available. Department II has 40 hours available. To manufacture a unit of Product *A* requires 4 hours in Department I and 2 hours in Department II. Product *B* requires 1 hour in Department I and 4 hours in Department II. The contribution to profit is \$6 per unit of Product *A* and \$5 per unit of Product *B*.

2. Assume that the products in Problem 1 also require processing in a third department. This department has 20 hours available. Product *A* requires 2 hours and Product *B* requires 1 hour in Department III. Redo Problem 1 with these new conditions.

3. A manufacturer produces two products, Product *A* and Product *B*. To produce each product requires processing in two departments, Department I and Department II. Department I has 200 hours available. Department II has 150 hours available. To manufacture a unit of Product *A* requires 5 hours in Department I and 5 hours in Department II. Product *B* requires 4 hours in Department I and 2 hours in Department II. The contribution to profit per unit is \$2 for Product *A* and \$1 for Product *B*.

4. The products in Problem 3 require additional processing in a third department. This department has 350 hours available. Product *A* requires 10 hours and Product *B* requires 7 hours in Department III. Redo Problem 3 with these new conditions.

5. A company markets two products, Alpha and Gamma. The marginal contributions per gallon are \$5 for Alpha and \$4 for Gamma. Both products consist of two ingredients, D and K. Alpha contains 80% D and 20% K, while the proportions of the same ingredients in Gamma are 40% and 60% respectively. The current inventory is 16,000 gallons of D and 6,000 gallons of K. The only company producing D and K is on strike and will neither deliver nor produce them in the foreseeable future. The company wishes to know the numbers of gallons of Alpha and Gamma that it should produce with its present stock of raw materials in order to maximize its total revenue.

A. The objective function for this problem could be expressed as
(a) $f\max = 0X_1 + 0X_2 + 5X_3 + 5X_4$
(b) $f\min = 5X_1 + 4X_2 + 0X_3 + 0X_4$
(c) $f\max = 5X_1 + 4X_2 + 0X_3 + 0X_4$
(d) $f\max = X_1 + X_2 + 5X_3 + 4X_4$
(e) $f\max = 4X_1 + 5X_2 + X_3 + X_4$

B. The constraint imposed by the quantity of D on hand could be expressed as
(a) $X_1 + X_2 \geqslant 16{,}000$
(b) $X_1 + X_2 \leqslant 16{,}000$
(c) $.4X_1 + .6X_2 \leqslant 16{,}000$
(d) $.8X_1 + .4X_2 \geqslant 16{,}000$
(e) $.8X_1 + .4X_2 \leqslant 16{,}000$

C. The constraint imposed by the quantity of K on hand could be expressed as
(a) $X_1 + X_2 \geqslant 6{,}000$
(b) $X_1 + X_2 \leqslant 6{,}000$
(c) $.8X_1 + .2X_2 \leqslant 6{,}000$
(d) $.8X_1 + .2X_2 \geqslant 6{,}000$
(e) $.2X_1 + .6X_2 \leqslant 6{,}000$

D. To maximize total revenue the company should produce and market
(a) 106,000 gallons of Alpha only.
(b) 90,000 gallons of Alpha and 16,000 gallons of Gamma.
(c) 16,000 gallons of Alpha and 90,000 gallons of Gamma.
(d) 18,000 gallons of Alpha and 4,000 gallons of Gamma.
(e) 4,000 gallons of Alpha and 18,000 gallons of Gamma.

E. Assuming that the marginal contributions per gallon are $7 for Alpha and $9 for Gamma, the company should produce and market
(a) 106,000 gallons of Alpha only.
(b) 90,000 gallons of Alpha and 16,000 gallons of Gamma.
(c) 16,000 gallons of Alpha and 90,000 gallons of Gamma.
(d) 18,000 gallons of Alpha and 4,000 gallons of Gamma.
(e) 4,000 gallons of Alpha and 18,000 gallons of Gamma.

6. The Ball Company manufactures three types of lamps which are labeled A, B, and C. Each lamp is processed in two departments—I and II. Total available man-hours per day for departments I and II are 400 and 600, respectively. No additional labor is available. Time requirements and profit per unit for each lamp type as follows:

	A	B	C
Man-hours required in department I	2	3	1
Man-hours required in department II	4	2	3
Profit per unit (Sales price less all variable costs)	$5	$4	$3

The Company has assigned you as the accounting member of its profit planning committee to determine the numbers of types of A, B, and C lamps that it should produce in order to maximize its total profit from the sale of lamps. The following questions relate to a linear programming model that your group has developed.

A. The coefficients of the objective function would be
(a) 4, 2, 3
(b) 2, 3, 1
(c) 5, 4, 3
(d) 400, 600

B. The constraints in the model would be
(a) 2, 3, 1
(b) 5, 4, 3
(c) 4, 2, 3
(d) 400, 600

C. The constraint imposed by the available man-hours in department I could be expressed as
(a) $4X_1 + 2X_2 + 3X_3 \leqslant 400$
(b) $4X_1 + 2X_2 + 3X_3 \geqslant 400$
(c) $2X_1 + 3X_2 + 1X_3 \leqslant 400$
(d) $2X_1 + 3X_2 + 1X_3 \geqslant 400$

D. The most types of lamps that would be included in the optimal solution would be
(a) 2
(b) 1
(c) 3
(d) 0

7. The Golden Hawk Manufacturing Company wants to maximize the profits on products A, B, and C. The contribution margin for each product follows:

Product	*Contribution Margin*
A	\$2
B	\$5
C	\$4

The production requirements and departmental capacities, by departments, are as follows:

Department	*Production Requirements By Product (Hours)*		
	A	B	C
Assembling	2	3	2
Painting	1	2	2
Finishing	2	3	1

Department	*Departmental Capacity (Total Hours)*
Assembling	30,000
Painting	38,000
Finishing	28,000

A. What is the profit-maximization formula for the Golden Hawk Company?
(a) \$2A + \$5B + \$4C = X (where X = profit).
(b) 5A + 8B + 5C $\leqslant$ 96,000
(c) \$2A + \$5B + \$4C $\leqslant$ X (where X = profit).
(d) \$2A + \$5B + \$4C = 96,000

B. What is the constraint for the Painting Department of the Golden Hawk Company?
(a) 1A + 2B + 2C $\geqslant$ 38,000
(b) \$2A + \$5B + \$4C $\geqslant$ 38,000
(c) 1A + 2B + 2C $\leqslant$ 38,000
(d) 2A + 3B + 2C $\leqslant$ 30,000

5-4 THE GRAPHICAL METHOD OF LINEAR PROGRAMMING

We will now present the graphical method for solving linear programming problems that involve only two independent variables. Although the graphical method is not practical for the solution of linear programming problems that involve more than two independent variables, it does provide some insight into the more involved problems.

The graphical method involves three basic steps. These steps are explained in this section in connection with the mathematical model that was obtained in Example 1 of the last section (i.e., maximize $P = 4x+4.5y$ subject to $x \geqslant 0$, $y \geqslant 0$, $3x+3y \leqslant 2400$, and $x+5y \leqslant 1600$).

Step 1: Determine the solution space for the system of inequalities involved in the given problem. Applying the techniques of Section 5-2 to the above system of inequalities leads to the solution space shown in Figure 5-10.

Step 2: Determine the value of the objective function at the *upper left-hand corner* of the solution space in Step 1. For the present problem, this is the point (0, 320) and the value of the objective function is given by

$$P = 4(0) + 4.5(320)$$
$$= 0 + 1440$$
$$P = 1440$$

Next, graph the equation of the straight line that results when this value is substituted into the objective function. For this problem we have the equation $4x+4.5y = 1440$. The graph of this equation is shown as Figure 5-11.

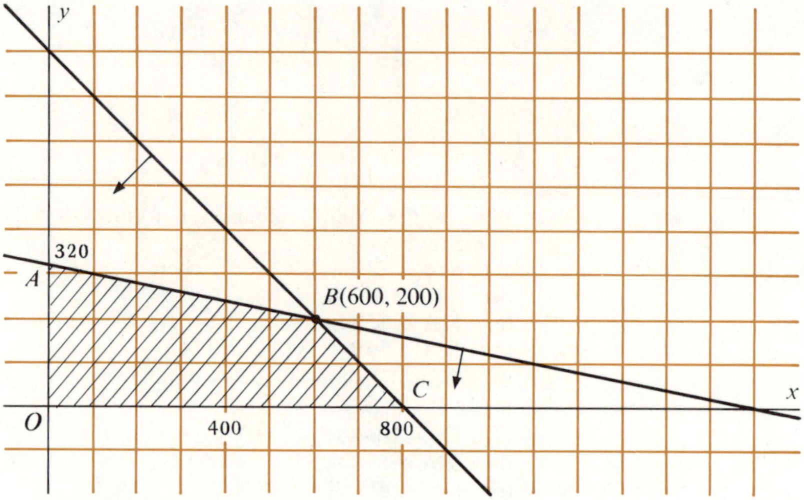

FIGURE 5-10

L_1: $4x + 4.5y = 1440$

x	360	0
y	0	320

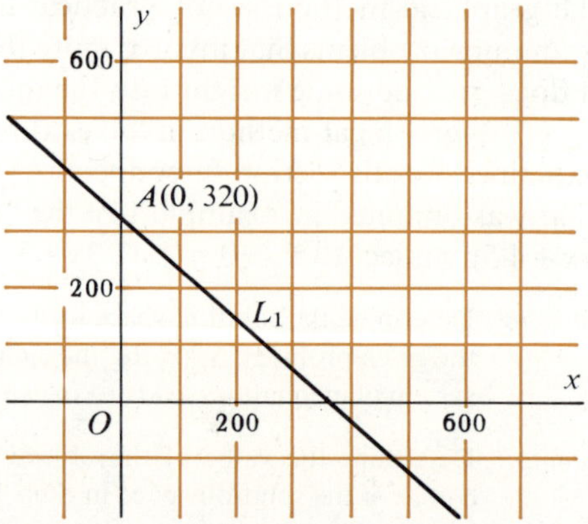

FIGURE 5-11

Step 3: Determine the point or points of the solution space through which a line L_2 parallel to L_1 of Step 2 can be drawn with the restriction that L_2 be *as far from the origin* as possible. (*Note:* L_2 may coincide with L_1.) The location of such a point or points and the corresponding value of the objective function constitute the solution of the given linear programming problem. The line L_2 for the present problem is shown in Figure 5-12.

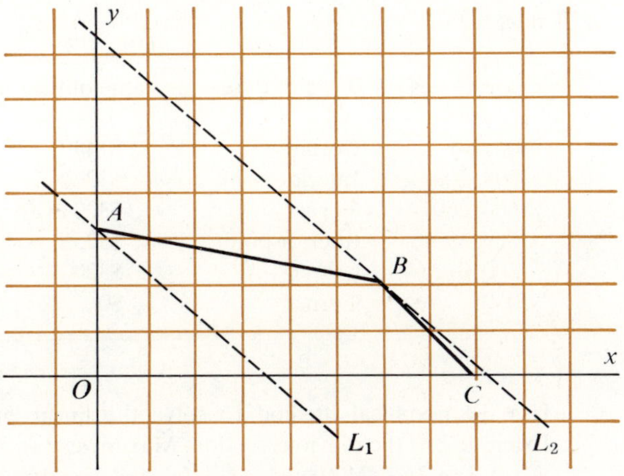

FIGURE 5-12

It is clear from the figure above that point B (600, 200) is the desired point.

$$\begin{aligned} P_{\max} &= 4(600) + 4.5(200) \\ &= 2400 + 900 \\ &= 3300 \end{aligned}$$

EXPLANATION OF THE GRAPHICAL METHOD

The three step procedure that we have outlined is based on the fact that the maximum value of the objective function occurs at a corner of the *solution space* of the system of inequalities given by the constraints. It should be clear that the maximum value *never* occurs at an *interior point* of the solution space since the value of the objective function could then be increased by moving either horizontally to the right-hand boundary or vertically to the upper boundary. This means that the solution of the given problem is to be sought somewhere on the boundary of the solution space. It can be shown further that the maximum value of the objective function occurs at a corner of this boundary. It can happen that more than one corner gives the same maximum value of the objective function. In this case, the maximum value occurs at any point on the line segment joining the two points just described.

This point is illustrated by the following table. This table is based on the problem solved in this section.

TABLE 5-2

Coordinates	Type of Point	Value of the Objective Function
(0, 320)	Corner	\$1440
(590, 200)	Interior point	\$3260
(600, 200)	Corner	\$330 (Maximum Value)
(600, 190)	Interior point	\$3255
(800, 0)	Corner	\$3200
(0, 0)	Corner	\$0

Example 1. Use the graphical method to solve the linear programming problem of Example 2 of the previous section. Maximize $P = 4x+4.5y$ subject to $x \geqslant 0$, $y \geqslant 0$, $3x+3y \leqslant 2400$, $x+5y \leqslant 1600$, and $4x+4y \leqslant 1600$.

Solution: Lines L_1 and L_2 and the solution space are shown in Figure 5-13.

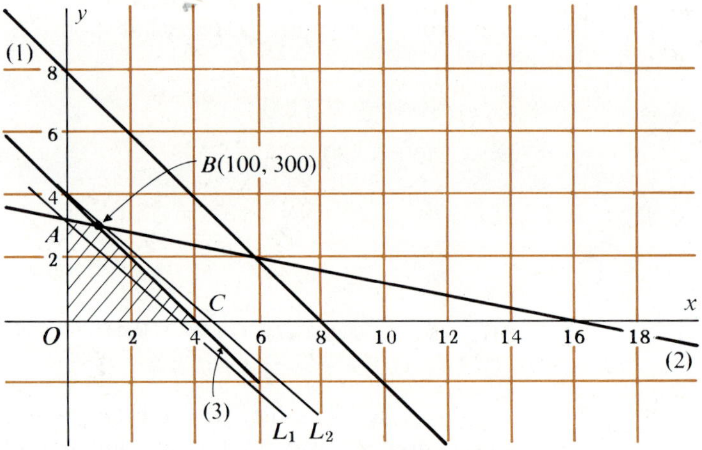

FIGURE 5-13

The point can be found algebraically by solving the system:

$$\begin{aligned} x + 5y &= 1600 \\ 4x + 4y &= 1600 \\ 4x + 20y &= 6400 \qquad y = 300 \\ 4x + 4y &= 1600 \qquad x = 1600 - 1500 = 100 \\ \hline 16y &= 4800 \end{aligned}$$

P_{max}, which occurs at (100, 300), is given by

$$\begin{aligned} P_{max} &= 4(100) + 4.5(300) \\ &= 400 + 1350 \\ P_{max} &= \$1750 \end{aligned}$$

Example 2. Check the profit at the corners O, A, and C and compare them with the profit at B.

Solution: At the origin, $P = 4(0) + 4.5(0) = 0$.
At A, $P = 4(0) + 4.5(320) = \$1440$.
At C, $P = 4(400) + 4.5(0) = \$1600$.

Corner	Profit
O	0
A	\$1440
B	\$1750 (P_{max})
C	\$1600

SUMMARY

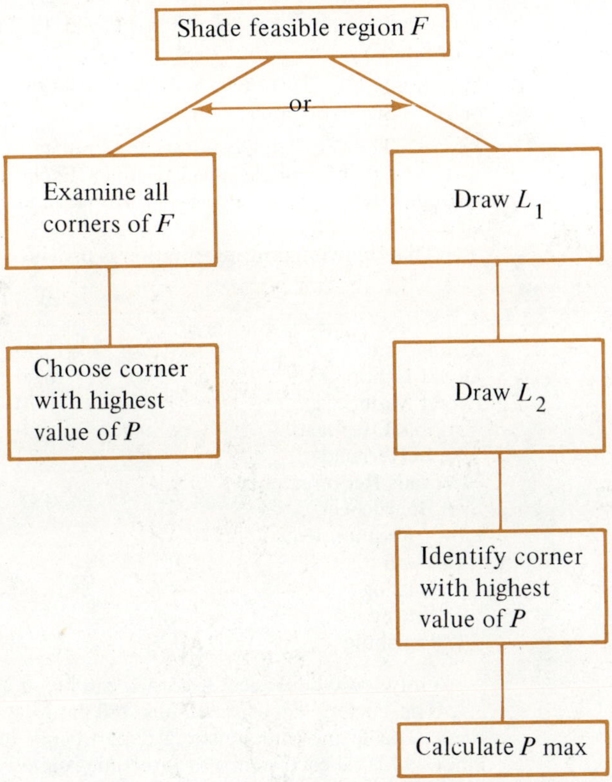

FIGURE 5-13a

PROBLEMS

Problems 1 through 4 are the mathematical models for the first four problems of Section 5-3. Use the graphical method to find P_{max} for each of these problems.

1. Maximize $P = 6x+5y$ subject to $x \geqslant 0$, $y \geqslant 0$, $4x+y \leqslant 36$, and $2x+4y \leqslant 40$.

2. Maximize $P = 6x+5y$ subject to $x \geqslant 0$, $y \geqslant 0$, $4x+y \leqslant 36$, $2x+4y \leqslant 40$, and $2x+y \leqslant 20$.

3. Maximize $P = 2x+y$ subject to $x \geqslant 0$, $y \geqslant 0$, $5x+4y \leqslant 200$, $5x+2y \leqslant 150$.

4. Maximize $P = 2x+y$ subject to $x \geqslant 0$, $y \geqslant 0$, $5x+4y \leqslant 200$, $5x+2y \leqslant 150$, and $10x+7y \leqslant 350$.

5. Solve Problem 1 by finding the coordinates of each corner of the solution space and comparing the value of P at these points. Your result should be the same as in Problem 1.

6. Solve Problem 4 by checking the value of P at each corner of the solution space.

7. Maximize $P = 3x+y$ subject to $x \geqslant 0$, $y \geqslant 1$, and $x+y \leqslant 8$.

8. Maximize $P = x+2y$ subject to $x \geqslant 0$, $y \geqslant 1$, and $x+y \leqslant 8$.

9. Draw the graphs of $2x+3y = P$ for $P = 6$, 12, 18, and 24 all in the same coordinate system.

10. Which line, $L_1 : 2x + 3y = 6$ or $L_2 : 2x + 3y = 24$, is "farther from the origin"? See Problem 9.

11. *Part A* The Witchell Corporation manufactures and sells three grades, A, B, and C, of a single wood product. Each grade must be processed through three phases—cutting, fitting, and finishing—before it is sold.

The following unit information is provided:

	A	B	C
Selling Price	\$10.00	\$15.00	\$20.00
Direct Labor	5.00	6.00	9.00
Direct Materials	.70	.70	1.00
Variable Overhead	1.00	1.20	1.80
Fixed Overhead	.60	.72	1.08
Materials Requirements in Board Feet	7	7	10
Labor Requirements in Hours			
Cutting	3/6	3/6	4/6
Fitting	1/6	1/6	2/6
Finishing	1/6	2/6	3/6

Only 5,000 board feet per week can be obtained.

The cutting department has 180 hours of labor available each week. The fitting and finishing departments have 120 hours each of labor available each week. No overtime is allowed.

Adapted from the CMA examination.

Contract commitments require the company to make 50 units of A per week. In addition, company policy is to produce at least 50 additional units of A and 50 units of B and 50 units of C each week to actively remain in each of the three markets. Because of competition only 130 units of C can be sold each week.

Required: Formulate and label the linear objective function and the constraint functions necessary to maximize the contribution margin.

Part B The graph provided presents the constraint functions for a chair manufacturing company whose production problem can be solved by linear programming. The company earns $8.00 for each kitchen chair sold and $5.00 for each office chair sold.

Required:

(a) What is the profit maximizing production schedule?
(b) How did you select this production schedule?

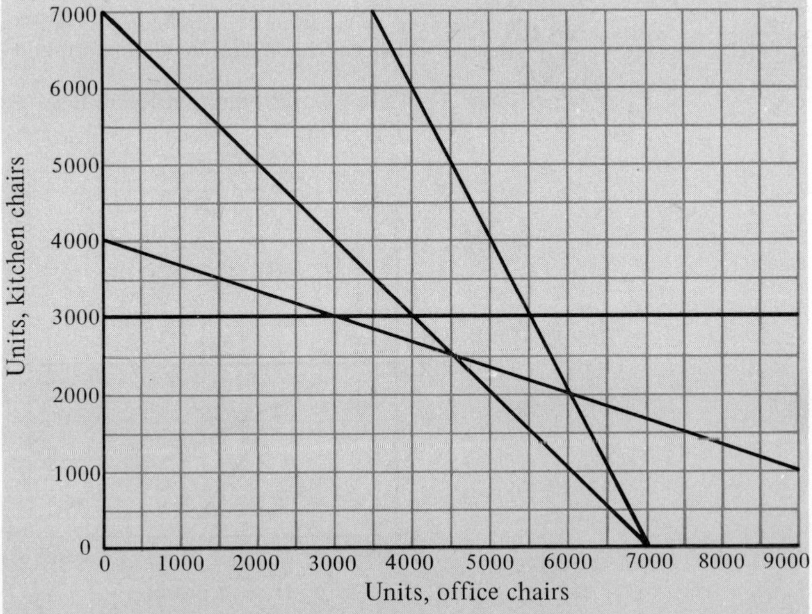

FIGURE 5-13b

12. The cost accountant of the Stangren Corporation, your client, wants your opinion of a technique suggested to him by a young accounting graduate he employed as a cost analyst. The following information was furnished you for the Corporation's two products, trinkets and gadgets:

Problem 12 from the Uniform CPA Examinations, Copyright © (1969) by the American Institute of Certified Public Accountants, Inc., is reprinted with permission.

1. *Exhibit A*

	Daily Capacities in Units			
	Cutting Department	*Finishing Department*	*Sales Price Per Unit*	*Variable Cost Per Unit*
Trinkets	400	240	$50	$30
or				
Gadgets	200	320	$70	$40

2. The daily capacities of each department represent the maximum production for *either* trinkets *or* gadgets. However, any combination of trinkets and gadgets can be produced as long as the maximum capacity of the department is not exceeded. For example, two trinkets can be produced in the Cutting Department for each gadget not produced and three trinkets can be produced in the Finishing Department for every four gadgets not produced.
3. Material shortages prohibit the production of more than 180 gadgets per day.
4. The following is a graphic expression of simultaneous linear equations developed from the production information.

Graph of Production Relationships

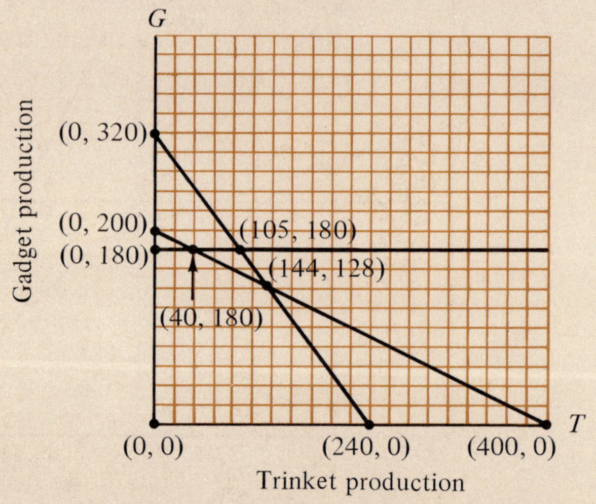

Required:

(a) For what kinds of decisions are contribution margin data (revenue in excess of variable cost) useful?

(b) Comparing the information in Exhibit A with the graph in Exhibit B, identify and list of the graphic location (coordinates) of the
1. Cutting Department's capacity.
2. Production limitation for gadgets because of the materials shortage.
3. Area of feasible (possible) production combinations.

(c) 1. Compute the contribution margin per unit for trinkets and gadgets.
2. Compute the total contribution margin of each of the points of intersections of lines bounding the feasible (possible) production area.
3. Identify the best production alternative.

5-5 THE SIMPLEX METHOD

The simplex method is a numerical algorithm for solving linear programming problems. This method involves a sequence of tables the last of which indicates the solution of the given problem. Our introduction to the simplex method is built around the problem that was solved graphically in the previous section, namely, maximize $P = 4x + 4.5y$ subject to $x \geqslant 0$, $y \geqslant 0$, $3x + 3y \leqslant 2400$, and $x + 5y \leqslant 1600$.

First we introduce two new variables, w and s, called *slack variables.* These variables enable us to rewrite the inequalities $3x + 3y \leqslant 2400$ and $x + 5y \leqslant 1600$ as a pair of linear equations in x, y, w, and s.

$$
\begin{aligned}
3x + 3y + 1w + 0s &= 2400 \\
x + 5y + 0w + 1s &= 1600 \qquad (1)
\end{aligned}
$$

Next we interpret variables w and s. The reader will recall that x and y represent amounts of *real* products, in this case ZIPS and ZAPS. In contrast, the slack variables w and s can be thought of as representing amounts of two *imaginary* products. The coefficients of w and s in Equation (1) indicates that the imaginary product corresponding to w requires one minute on machine I and no time on machine II. Similarly, the other imaginary product requires no time on machine I and one minute on machine II. The fact that the company cannot realize a profit by producing an imaginary product is expressed mathematically by introducing the new objective function

$$P = 0w + 0s + 4x + 4.5y \qquad (2)$$

To provide a smoother lead-in to Table 1 of the simplex method, we write Equation (1) in matrix form as follows:

$$\begin{pmatrix} 1 & 0 & 3 & 3 \\ 0 & 1 & 1 & 5 \end{pmatrix} \begin{pmatrix} w \\ s \\ x \\ y \end{pmatrix} = \begin{pmatrix} 2400 \\ 1600 \end{pmatrix} \qquad (3)$$

Using a standard simplex method format, we rewrite the above information in tabular form as follows:

TABLE 1 (Partial)

	Solution Variables		Variables Not in the Solution		
	w	s	x	y	Constant
w	1	0	3	3	2400
s	0	1	1	5	1600
C	0	0	4	4.5	0

The reader should note the similarity of the w and s rows of this table to the matrix equation, Equation 3, and the similarity of the C row to the objective function, Equation 2. The entries in the C row represent the contribution to profit per unit for each of the variables.

Variables which are used as row hearings are called solution variables. In Table 1 (partial), the solution variables are w and s. The remaining variables, in this case x and y, are said to be *not in the solution.*

If no ZIPS and ZAPS are to be produced, then $x = 0$ and $y = 0$. If we substitute these values into Equation 3, we have

$$\begin{pmatrix} 1 & 0 & 3 & 3 \\ 0 & 1 & 1 & 5 \end{pmatrix} \begin{pmatrix} w \\ s \\ 0 \\ 0 \end{pmatrix} = \begin{pmatrix} 2400 \\ 1600 \end{pmatrix}$$

$$\begin{pmatrix} w \\ s \end{pmatrix} = \begin{pmatrix} 2400 \\ 1600 \end{pmatrix}$$

This gives $w = 2400$ and $s = 1600$. Note that these values can also be read directly from the constant column of Table 1 (partial). From the new objective function, Equation 2, we have the following profit calculation:

$$P = 0(2400) + 0(1600) + 4(0) + 4.5(0)$$

$$P = 0$$

This value can also be obtained by multiplying the row vector $(0, 0)$, which comes from the C row (*using the solution variable entries only*), times the constant column $\begin{pmatrix} 2400 \\ 1600 \end{pmatrix}$ as follows:

$$(0, 0)\begin{pmatrix} 2400 \\ 1600 \end{pmatrix} = 0(2400) + 0(1600) = 0$$

This value is placed in the row labeled C in the constant column.

A complete Table 1 for the simplex method consists of Table 1 (partial), which we have described, plus two additional rows with headings Z and NET, respectively. (i.e., NET $= C - Z$). The row Z represents the decrease in the objective function per unit, and NET denotes the net increase in the objective function per unit. The only entries made in these two rows are for the variables not in the solution.

The introduction of one unit of x requires three units of w and one unit of s.

The resulting decrease in the objective function is given by the following calculation.

$$Z = 3(\text{contribution to profit per unit of } w)$$
$$+ 1(\text{contribution to profit per unit of } s)$$
$$Z = 3(0) + 1(0)$$
$$Z = 0$$

This result is expected since only imaginary products with a price of zero are being produced.

We then enter 0 in the row labeled Z and under the column x. Note that this entry can be obtained by multiplying the C row (using the solution variable entries only) times the column $\begin{pmatrix} 3 \\ 1 \end{pmatrix}$ from the x column.

$$(0, 0)\begin{pmatrix} 3 \\ 1 \end{pmatrix} = 0$$

TABLE 1 (Complete)

	w	s	x	y	Constant
w	1	0	3	3	2400
s	0	1	1	5	1600
C	0	0	\$4.00	\$4.50	0
Z			0	0	
NET			\$4.00	\$4.50	

The decrease in the objective function, which results when a unit of y is brought into the solution, is zero, as in the case for x. This follows from $(0, 0)\begin{pmatrix} 3 \\ 5 \end{pmatrix} = 0$.

The net increase in the objective function is then calculated by subtracting the Z row from the C row. Doing this we get \$4.00 and

\$4.50 for x and y, respectively. This completes the first table, and we move on to the second table. We will continue forming new tables as long as there is at least one positive entry in the NET row. The reader might note that the solution indicated by Table 1 is equivalent to the corner point (0, 0) which yielded no profit when we used the graphical method to arrive at a solution (i.e., to produce imaginary products only and no ZIPS nor ZAPS).

It is clear from Table 1 that for each unit of x brought into the solution the objective function is increased by \$4.00, and for each unit of y brought into the solution the objective function is increased by \$4.50. This means that the next step in our search for production levels that give the maximum profit will be to increase y as much as possible. There are now two questions to be answered:

1. *How much can y be increased with the machine time which is available?*
2. *Which of the solution variables w or z should be replaced by y?*

The production levels indicated by Table 1 are $x = 0$, $y = 0$, $w = 2400$, and $z = 1600$. For each unit of y brought into the solution, 3 units of w must be taken out of the solution. The maximum number of units of y that could be brought into the solution if this were the only constraint present is obtained by dividing 2400 by 3, which gives 800. Similarly, 5 units of z must be taken out of the solution for each unit of y that is included. This leads to a maximum value of y of 320, which is obtained by dividing 1600 by 5. Choosing the minimum of these two numbers so that both constraints will be satisfied, we see that the largest number of units of y which can be brought into the solution is 320. This means that y replaces z and the solution variables are now w and y.

It might be helpful at this point to summarize what we have done by outlining the following steps:

Step 1. Introduce slack variables and rewrite the constraints and objective function and form Table 1 Partial. Note that the initial table involves production of only imaginary products with no profit whatsoever.

Step 2. Find entries for the Z row, i.e., decrease in objective function per unit. Note that these entries are found only for variables that are not in the solution and will be equal to zero in table 1 because they will only require that we stop producing imaginary products which do not contribute to profit.

Step 3. Find entries for the NET row by subtracting the Z row from the C row. The variable with the *largest* positive NET value is brought into the solution. This column is referred to as the *pivot column*.

Step 4. Determine the variable that will depart from the solution by forming a new column called Q by dividing each entry in the constant column by the corresponding entry in the pivot column. The *smallest* quotient indicates which variable will leave the solution.

We now have what we will call Table 1 (Modified).

TABLE 1 (Modified)

	w	s	x	y	Constant	Q
w	1	0	3	3	2400	800
s	0	1	1	5	1000	320 ← OUT
C	0	0	\$4.00	\$4.50	\$0	
Z			0	0		
NET			\$4.00	\$4.50		
				↑ Incoming		

Our objective now is to construct Table 2 and see how much profit can be made by producing ZAPS (y) and an imaginary product (w).

We will first replace the s row with a row labeled y. Entries for the y row are found by dividing each element in the old s row by the entry in the pivot column in that row, i.e., 5.

	w	s	x	y	Constant
new y	0	.2	.2	1	320

The entries for the new w row are found by deducting the entries in the new y row multiplied by 3 (i.e., the pivot element in the old row w) from the old w row.

	w	s	x	y	Constant
old w	1	0	3	3	2400
− 3 new y	−0	−.6	−.6	−3	− 960
new w	1	−.6	2.4	0	1440

The entries in the C row are the same as in Table 1 except for the element in the constant column.

This entry which represents the profit of the current solution, i.e., $y = 320$ and $w = 1440$ is found by multiplying the solution values by their corresponding contributions to profit per unit, i.e., C values.

$$P = 320(\$4.50) + 1440(\$0) = \$1440$$

The above information is summarized in Table 2 (Partial).

TABLE 2 (Partial)

	w	s	x	y	Constant
w	1	−.6	2.4	0	1440
y	0	.2	.2	1	320
C	0	0	\$4.00	\$4.50	\$1440

To obtain the elements in the Z row for those variables not in the solution, i.e., s and x, one totals the product of each element in that column multiplied by the contribution to profit per unit of the solution variable in that row. For example, the Z value for column s is found as follows:

$$-.6(\$0) + .2(\$4.50) = \$.9$$

The z value for column x is found as follows:

$$2.4(\$0) + .2(4.50) = \$.9$$

The reader is reminded that this number represents the decrease in the objective function that would result if one unit of x were produced. The net row is found simply by subtracting the z row from the C row.

TABLE 2

	w	s	x	y	Constant	Q
w	1	−.6	2.4	0	1440	600 ← OUT
y	0	.2	.2	1	320	1600
C	\$0	\$0	\$4.00	\$4.50	\$1440	
Z		\$.9	.9			
NET		\$−.9	\$3.10			
			↑ Incoming			

Examining the NET row of Table 2 we see that for each x brought into the solution we can increase profit by \$3.10. To do this we must see which variable will leave the solution. Examining the column headed Q, we see that w will leave the solution. We will now form Table 3 in the same manner that Table 2 was constructed.

new $x \rightarrow$ elements in old s divided by 2.4

	w	s	x	y	Constant
new $x \rightarrow$	.4166	$-.25$	1	0	600

new $y \rightarrow$ old $y - .2$ times new x

	w	s	x	y	
old y	0	.2	.2	1	320
$-.2$ new x	$-.0833$	$+.05$	$-.2$	-0	-120
new y	$-.0833$	$+.25$	0	1	200

TABLE 3

	w	s	x	y	Constant
x	.4166	$-.25$	1	0	600
y	$-.0833$	.25	0	1	200
C	\$0	\$0	\$4.00	\$4.50	\$3300
Z	\$1.2917	\$.125			
NET	\$−1.2917	\$−.125			

(600)(\$4.00)+(200)(\$4.50) [→ \$3300]

(−.25)(\$4.00)+.25(\$4.50) [→ \$.125]

It is impossible to increase the objective function any further, since both entries in the net row are negative. This means that the optimum solution has been obtained and is $x = 600$ and $y = 200$, with $P_{max} =$ \$3300, as found graphically in Section 5-4. In other words, the company can make the maximum profit by producing 600 ZIPS and 200 ZAPS.

Before we go on to another example, it is interesting and quite important to look at the Z row of the last table. Note that to bring one w into the solution results in a decrease in the objective function of \$1.2917. Checking the original column vectors, we see that to make a unit of w requires one minute on machine I and no time on machine II. By making a unit of w, we are reducing the time available on machine I by one minute. Since this results in a decrease in the objective function of \$1.2917, one minute on machine I must be worth \$1.2917. Similarly, to produce one unit of s requires one minute on machine II and no minutes on machine I. Since the objective function is reduced by \$0.125, if one unit of z is produced, one minute of time on machine II must be worth \$0.125.

Therefore, the total worth of machine time is

$$2400(\$1.2917) + 1600(\$0.125) = \$3300$$

The fact that this is equal to the maximum value of the objective function is no mere coincidence. This aspect is discussed in more extensive treatments of linear programming.

PROBLEMS

For the systems of inequalities in Problems 1 through 4, introduce slack variables and write the resulting system of equations as a single matrix equation.

1. $4x+y \leqslant 36$ and $2x+4y \leqslant 40$.

2. $4x+y \leqslant 36$, $2x+4y \leqslant 40$, and $2x+y \leqslant 20$.

3. $5x+4y \leqslant 200$ and $5x+2y \leqslant 150$.

4. $5x+4y \leqslant 200$, $5x+2y \leqslant 150$, and $10x+7y \leqslant 350$.

5. Set up Table 1 of the simplex method for Problem 1, $P = 6x+5y$.

6. Set up Table 1 of the simplex method for Problem 2, $P = 6x+5y$.

7. Set up Table 1 of the simplex method for Problem 3, $P = 2x+y$.

8. Set up Table 1 of the simplex method for Problem 4, $P = 2x+y$.

9. Use the simplex method to find maximum profit for Problem 5.

10. Use the simplex method to find maximum profit for Problem 6.

11. Use the simplex method to find maximum profit for Problem 7.

12. Use the simplex method to find maximum profit for Problem 8.

13. Beekley, Inc. manufactures widgets, gadgets, and trinkets and has asked for advice in determining the best production mix for its three products. Demand for the Company's products is excellent, and management finds that it is unable to meet potential sales with existing plant capacity.

Each product goes through three operations: milling, grinding and painting. The effective weekly departmental capacities in minutes are: milling—10,000, grinding—14,000 and painting—10,000.

The following data are available on the three products:

	Selling Price per Unit	Variable Cost per Unit	Per Unit Production Time (in Minutes) Milling	Grinding	Painting
Widgets	\$5.25	\$4.45	4	8	4
Gadgets	5.00	3.90	10	4	2
Trinkets	4.50	3.30	4	8	2

A. The objective function for this problem using the simplex method might be expressed

(a) $f\min = 4.45X_1 + 3.90X_2 + 3.30X_3 + 0X_4 + 0X_5 + 0X_6$

(b) $f\max = 5.25X_1 + 5.00X_2 + 4.50X_3 + X_4 + X_5 + X_6$

(c) $f\max = .80X_1 + 1.10X_2 + 1.20X_3 + X_4 + X_5 + X_6$
(d) $f\max = .80X_1 + 1.10X_2 + 1.20X_3 + 0X_4 + 0X_5 + 0X_6$

B. The requirement that total production time in the painting department may not exceed 10,000 minutes per week might be expressed
(a) $4X_1 + 2X_2 + 2X_3 \geqslant 10{,}000$
(b) $4X_1 + 2X_2 + 2X_3 > 10{,}000$
(c) $4X_1 + 2X_2 + 2X_3 \leqslant 10{,}000$
(d) $4X_1 + 2X_2 + 2X_3 < 10{,}000$

C. The variables X_4, X_5, and X_6 included in the answers to item **A** are referred to as
(a) Artificial variables.
(b) Primary variables.
(c) Stochastic variables.
(d) Slack variables.

D. The variables X_1, X_2, and X_3 included in the answers to item **A** are referred to as
(a) Artificial variables.
(b) Solution variables.
(c) Stochastic variables.
(d) Slack variables.

E. A significant advantage of applying the simplex method to certain problems having four or more variables of a single class is that solutions may be arrived at quickly using
(a) Graphic analysis.
(b) Electronic computer routines.
(c) Simple algebraic methods.
(d) Set theory.

5-6 THE SIMPLEX METHOD—A MAXIMIZATION PROBLEM AND A SUMMARY

In this section we will illustrate the simplex method further by solving the linear programming problem given in Example 2 of Section 5-3. The reader will recall that one additional constraint is present in that problem. This constraint resulted from a third machine. This additional constraint causes the optimum solution to involve a slack variable. Some of the data from that example are repeated below.

In order to produce a ZIP, four minutes on machine III are now required in addition to the three minutes on machine I and one minute on machine II. In order to produce a ZAP, four minutes are also required on machine III in addition to three minutes on machine I and five minutes on machine II. The three constraints are:

$3x + 3y \leqslant 2400$ min. (*machine I*)

$1x + 5y \leqslant 1600$ (*machine II*)

$4x + 4y \leqslant 1600$ (*machine III*)

Remember: The coefficients of x read as follows: to produce one unit of x requires 3 minutes on machine I, one minute on machine II, and four minutes on machine III; it does not mean that 3 x's and 3 y's can be made in one minute using machine I.

To change the inequalities to equalities it is necessary to introduce three slack variables, each of which requires one minute on one machine only. This results in the following three equations:

$$3x + 3y + 1w + 0s + 0t = 2400 \qquad (machine\ I)$$

$$1x + 5y + 0w + 1s + 0t = 1600 \qquad (machine\ II)$$

$$4x + 4y + 0w + 0s + 1t = 1600 \qquad (machine\ III)$$

We now have the same problem as before, which is to allocate machine time in order to maximize the objective function,

$$P = 4.00x + 4.50y + 0w + 0s + 0t$$

TABLE 1 (Complete)

	w	s	t	x	y	Constant	Q	
w	1	0	0	3	3	2400	800	
s	0	1	0	1	5	1600	320	OUT
t	0	0	1	4	4	1600	400	
C				\$4.00	\$4.50			
Z				0	0			
NET				\$4.00	\$4.50			
					↑ IN			

As usual, the initial solution involves only slack variables and no profit. The entries in the last row of Table 1 above correspond to the fact that if one unit of x is introduced into the solution, the objective function will be increased by \$4.00; whereas if one unit of y is brought into the solution the objective function will be increased by \$4.50. Since $4.5 > 4.0$, we bring y into the solution.

Next, we must determine how many units of y can be introduced. The largest amount of y that can be brought into the solution is given by the positive minimum of the ratios: 2400/3, 1600/5, and 1600/4, or 320 units of y. Since the minimum ratio $320 = 1600/5$ corresponds to s, we know that z must be taken out of the solution.

At this point we would like to recall the steps involved in forming the first table of the simplex method and also formulate those steps necessary to formulate all subsequent tables.

Formulation of Table 1.

Step 1. Introduce slack variables and rewrite the constraints and objective function and form Table 1 (Partial). Note that the initial table involves production of only imaginary products with no profit whatsoever.

Step 2. Find entries for the Z row, i.e., decrease in objective function per unit. Note that these entries are found only for variables that are not in the solution and will be equal to zero in Table 1 because they will only require that we stop producing imaginary products which do not contribute to profit.

Step 3. Find entries for the NET row by subtracting the Z row from the C row. The variable with the *largest* positive NET value is brought into the solution. This column is referred to as the *pivot column.*

Step 4. Determine the variable that will depart from the solution by forming a new column called Q by dividing each entry in the constant column by the corresponding entry in the pivot column. The *smallest* quotient indicates which variable will leave the solution.

Formation of all subsequent tables.

Step 5. Find entries for the row which represents the incoming variable. This is done by dividing all the entries in the row that is leaving the solution by the entry in the pivot column in that row.

Step 6. Rows for all variables that remain in the solution are found by taking all of the entries for that row in the preceding table and subtracting from them the entries in the row for the incoming variable (i.e., result of Step 5) after they have been multiplied by the element in the pivot column for the particular row we are forming.

Step 7. The entries in the C row are the same as in Table 1. These entries simply represent the contribution to profit per unit. The entry in the C row and constant column is the profit that is realized from the current solution (i.e., values in the constant column) and can be found by multiplying the solution values by their corresponding C values.

Step 8. Entries for the Z row are found only for those variables not in the solution. The entry for any particular column is the sum of the product of each element in that column multiplied by the corresponding C value of the solution variable in that row. Again these entries represent the decrease in the objective function that would result if one of these variables not in the solution were brought into the solution.

Step 9. Entries for the NET row are found by subtracting the Z row from the C row. This difference tells us if it is advantageous to bring into the solution a variable that is not in the solution. It is advantageous if the increase in profit caused by the introduction of this unit is greater than the decrease in profit or in short if the Z value is positive.

Step 10. If there are no positive NET entries we have an optimal solution and have finished the problem. Otherwise one will look for another solution by introducing the variable with the largest positive NET entry. This variable identifies the pivot column. We then find which variable will depart from the solution by forming a new column called Q by dividing each entry in the constant column by the corresponding entry in the pivot column. The smallest quotient indicates which variable will leave the solution. One must go to Step 5 to form the next table.

We will now use Steps 5 through 10 to arrive at a solution to our problem. It will require the formation of two subsequent tables and the reader should trace their development beginning with a reference to Table 1.

TABLE 2

	w	s	t	x	y	Constant	Q
w	1	−.6	0	2.4	0	1440	600
y	0	.2	0	.2	1	320	1600
t	0	−.8	1	3.2	0	320	100 ← OUT
C	0	0	0	$4.00	$4.50	$1440	
Z		$.9		$.9			
NET		$.9		$3.10			
				↑ IN			

TABLE 3

	w	s	t	x	y	Constant
w	1	0	−.75	0	0	1200
y	0	.25	−.0625	0	1	300
x	0	−.25	.3125	1	0	100
C	0	0	0	$4.00	$4.50	
Z		.125	.96875			
NET		−.125	−.96875			

A look at the NET row of Table 3 reveals that it is impossible to further increase the objective function. The optimum solution then is to produce 100 units of x and 300 units of y, giving \$1750 as the maximum obtainable profit (i.e., $P_{\max} = 100(4.00) + 300(4.50) = \$1750$). This result was also obtained graphically in Example 1 of Section 5-4.

The total time used is 1200 minutes on machine I, 1600 minutes on machine II, and 1600 minutes on machine III, as shown by

$$300\begin{pmatrix}3\\5\\4\end{pmatrix} + 100\begin{pmatrix}3\\1\\4\end{pmatrix} = \begin{pmatrix}1200\\1600\\1600\end{pmatrix}$$

The fact that 1200 minutes went unused on machine I is indicated by the presence of the slack variable w in the final solution. This happens because to produce one unit of w requires one minute on machine

I only. It should be noted that the presence of a slack variable in the solution is always an indication of unused capacity.

If we were to decide on expanding capacity, clearly it would not be to our advantage to increase the capacity of machine I.

Note that the entries in the Z row times the capacity of the machine needed to produce 1600 units of s and 1600 units of t results in

$$\$0.125(1600) + \$0.96875(1600) = \$200 + \$1550 = \$1750$$

which is the maximum value of the objective function. Hence, in this case, the value of one minute on machine II is \$0.125 and the value of one minute of time on machine III is \$0.96875.

PROBLEMS

Problems 1 and 2 refer to material in the body of the text in this section.

1. Write the three equations that precede Table 1 in this section as a matrix equation.
2. Trace the development of Tables 2 and 3 in this section.
3. Use the simplex method to solve Problem 2 of Section 5-3, i.e., maximize $P = 6x+5y$ subject to $x \geqslant 0$, $y \geqslant 0$, $4x+y \leqslant 36$, $2x+4y \leqslant 40$, and $2x+y \leqslant 20$. Compare your result with the graphical solution of Problem 2, Section 5-4.
4. Use the simplex method to solve Problem 4 of Section 5-3, i.e., maximize $P = 2x+y$ subject to $x \geqslant 0$, $y \geqslant 0$, $5x+4y \leqslant 260$, $5x+2y \leqslant 150$, and $10x+7y \leqslant 350$. Compare your result with the graphical solution of Problem 4, Section 5-4.
5. A certain firm makes two products, type A and type B. Each product requires 10 man-hours per unit in Department I. Each unit of product A requires 4 man-hours in Department II and each unit of product B requires 8 man-hours in Department II. There are 160 man-hours available in Department I and 80 man-hours available in Department II per week. The firm earns a profit of \$3.00 per unit of product A and a profit of \$4.00 per unit of product B. Assume that it is possible to sell all units produced each week. Find the levels of production that will give the maximum profit.
6. Use the same data given in Problem 5 except that the firm earns a profit of \$6.00 per unit of product A and \$4.00 per unit of product B.
7. Maximize $P = 2x+3y$ subject to $x \geqslant 0$, $y \geqslant 0$, $x+y \geqslant 10$, and $x+y \leqslant 12$.
8. Use the same data given in Problem 7 except $P = 9x+3y$.
9. Review Problem. Select the best answer for each of the following items which relate to applications of linear programming.

   Items to be Answered:

   A. In a linear programming maximization problem for business problem solving, the coefficients of the objective function usually are

(a) Marginal contributions per unit.
(b) Variable costs.
(c) Profit based upon allocations of overhead and all indirect costs.
(d) Usage rates for scarce resources.
(e) None of the above.

B. The constraints in a linear programming problem usually model
(a) Profits.
(b) Restrictions.
(c) Dependent variables.
(d) Goals.
(e) None of the above.

C. If there are four activity variables and two constraints in a linear programming problem, the most products that would be included in the optimal solution would be
(a) 6.
(b) 4.
(c) 2.
(d) 0.
(e) None of the above.

D. Linear programming is used most commonly to determine
(a) That mix of variables which will result in the largest quantity.
(b) The best use of scarce resources.
(c) The most advantageous prices.
(d) The fastest timing.
(e) None of the above.

E. Assume the following data for the two products produced by Wagner Company:

	Product *A*	Product *B*
Raw material requirements (units)		
X	3	4
Y	7	2
Contribution margin per unit	$10	$4

If 300 units of raw material *X* and 400 units of raw material *Y* are available, the set of relationships appropriate for maximization of revenue using linear programming would be

(a) $3A + 4B \geqslant 300$
$7A + 2B \geqslant 400$
$10A + 4B$ MAX

(b) $3A + 7B \geqslant 300$
$4A + 2B \geqslant 400$
$10A + 4B$ MAX

(c) $3A + 7B \leqslant 300$
$4A + 2B \leqslant 400$
$10A + 4B$ MAX

(d) $3A + 4B \leqslant 300$
$7A + 2B \leqslant 400$
$10A + 4B$ MAX

(e) None of the above.

F. A final tableau for a linear programming profit maximization problem is shown below:

	X_1	X_2	X_3	S_1	S_2	
X_1	1	0	4	3	−7	50
X_2	0	1	−2	−6	2	60
C	0	0	5	1	9	1,200

If X_1, X_2, and X_3 represent products, S_1 refers to square feet (in thousands) of warehouse capacity and S_2 refers to labor hours (in hundreds), the number of X_1 that should be produced to maximize profit would be

(a) 60.
(b) 50.
(c) 1.
(d) 0.
(e) None of the above.

G. Assuming the same facts as in item **F**, the contribution to profit of an additional 100 hours of labor would be

(a) 9.
(b) 2.
(c) 1.
(d) −7.
(e) None of the above.

H. Assuming the same facts as in item **F**, an additional 1,000 square feet of warehouse space would

(a) Increase X_1 by 3 units and decrease X_2 by 6 units.
(b) Decrease X_2 by 6 units and increase X_1 by 2 units.
(c) Decrease X_1 by 7 units and increase X_2 by 2 units.
(d) Increase X_1 by 3 units and decrease X_2 by 7 units.
(e) Do none of the above.

I. The following is the final tableau of a linear programming profit maximization problem:

	X_1	X_2	S_1	S_2	
X_1	1	0	−5	3	125
X_2	0	1	1	−1	70
C	0	0	5	7	500

The marginal contribution to profit of 5 for each added resource unit S_1 can be maintained if the added resource units do not exceed

(a) 125.
(b) 100.
(c) 70.
(d) 25.
(e) None of the above.

J. Assume the following per unit raw material and labor requirements for the production of products *A* and *B*.

	Product A	Product B
Pounds of lead	5	7
Hours of labor	3	4

Assuming that 13,400 pounds of lead and 7,800 hours of labor are available, the production of products A and B required to use all of the available lead and labor hours is shown in the following final Gaussian tableau.

$$\left(\begin{array}{cc|cc|c} 1 & 0 & -4 & 7 & 1{,}000 \\ 0 & 1 & 3 & -5 & 1{,}200 \end{array}\right)$$

If the available amounts were increased to 15,000 pounds of lead and 8,800 hours of labor, the matrix operation to perform to determine the production schedule which would fully utilize these resources is

(a) $\begin{pmatrix} 5 & 7 \\ 3 & 4 \end{pmatrix}\begin{pmatrix} 15{,}000 \\ 8{,}800 \end{pmatrix}$

(b) $\begin{pmatrix} 15{,}000 \\ 8{,}800 \end{pmatrix}\begin{pmatrix} -4 & 7 \\ 3 & -5 \end{pmatrix}$

(c) $\begin{pmatrix} -4 & 7 \\ 3 & -5 \end{pmatrix}\begin{pmatrix} 1{,}000 \\ 1{,}200 \end{pmatrix}$

(d) $\begin{pmatrix} -4 & 7 \\ 3 & -5 \end{pmatrix}\begin{pmatrix} 15{,}000 \\ 8{,}800 \end{pmatrix}$

(e) None of the above.

K. The following schedule provides data for product A, which is processed through processes 1 and 2, and product B, which is processed through process 1 only:

	Product A	Product B
Raw material cost per gallon	\$ 4	\$ 9
Process 1 (500 gallon input capacity per hour):		
Processing cost per hour	\$60	\$60
Loss in processing	30%	20%
Process 2 (300 gallon input capacity per hour):		
Processing cost per hour	\$50	
Loss in processing	10%	
Selling price per gallon	\$20	\$40

If the objective is to maximize profit per eight-hour day, the objective function of a profit-maximizing linear programming problem would be

(a) $20A + 40B - 4A - 4B$.

(b) $20A + 40B - 4A - 4B - 60(A+B) - 50A$.

(c) $20(.63A) + 40(.80B) - 4(.63A) - 9(.8B)$
$$-\,60\left(\frac{A+B}{500}\right) - 50\left(\frac{.7A}{300}\right).$$

(d) $20(.63A) + 40(.80B) - 4A - 9B$
$$- 60\left(\frac{A}{500} + \frac{B}{500}\right) - 50\left(\frac{.7A}{300}\right).$$

(e) None of the above.

L. Assuming the same facts as in item **K**, a constraint of the problem would be

(a) $.63A \leqslant 2{,}400$.
(b) $.8A \leqslant 2{,}400$.
(c) $.7A + .8B \leqslant 4{,}000$.
(d) $.9A \leqslant 4{,}000$.
(e) None of the above.

REFERENCES

Books

Boulding, Kenneth E. *Linear Programming and the Theory of the Firm.* New York: The MacMillan Company, 1960.

Charnes, A., Cooper, W. W., and Henderson, A. *An Introduction to Linear Programming.* New York: John Wiley and Sons, Inc., 1953.

Dantzig, George B. *Linear Programming and Extensions.* Princeton, N.J.: Princeton University Press, 1963.

Dorfman, Robert, Samuelson, Paul A., and Solow, Robert M. *Linear Programming and Economic Analysis.* New York: McGraw-Hill Book Co., Inc., 1958.

Ferguson, Robert O., and Sargent, Lauren F. *Linear Programming: Fundamentals and Applications.* New York: McGraw-Hill Book Co., Inc., 1958.

Gass, Saul I. *Linear Programming: Methods and Applications.* New York: McGraw-Hill Book Co., Inc., 1964.

Hanna, Samuel C., and Saber, John C. *Linear Programming and Matrix Algebra.* Babson Park, Mass.: Babson College Press, 1973.

Levin, Richard I., and Lamone, Rudolph P. *Linear Programming for Management Decisions.* Homewood, Illinois: Richard D. Irwin, Inc., 1969.

Springer, Clifford H., Herlihy, R. E., and Beggs, R. I. *Advanced Methods and Models.* Homewood, Illinois: Richard D. Irwin, Inc., 1965.

Stockton, R. Stansbury. *Introduction to Linear Programming.* Boston: Allyn and Bacon, Inc., 1963.

Vajda, Steven. *Readings in Linear Programming.* New York: John Wiley and Sons, Inc., 1958.

Articles

Henderson, Alexander, and Schlaifer, R. "Mathematical Programming Better Information for Better Decision Making." *Harvard Business Review* 32: 73–100.

Mulligan, James E. "Basic Optimization Techniques—A Brief Survey." *The Journal of Industrial Engineering* XVI, 3: 192–197.

Raun, Donald L. "Product-Mix Analysis by Linear Programming." *Management Accounting* XLVII: 3–13.

Sweeney, Robert B. "Business Use of Linear Programming". *Management Accounting* September 1965, pp. 39–47.

SIX

Mathematics of finance

All of you, no doubt, are aware of the fact that if you deposit a sum of money in the bank you will be paid interest by the bank. In effect, the bank is paying the saver for the use of his or her money. Just as the use of a building has a cost, rent, the use of money has a cost, interest. The reader should also be aware of the fact that one dollar received today is worth more than one dollar to be received two years from today. That is, money in effect has a time value. In this chapter we look at four basic time values of money calculations. They are:

1. future value of an amount,
2. future value of an annuity,
3. present value of a future amount,
4. present value of a future annuity.

A common application of the future value of an amount is a single deposit in a savings account; one can determine how much would be in that account n years from now. We extend our discussion of compounding to include instantaneous compounding. An application of the future value of an annuity is a bond sinking fund, where one will make more than a single deposit to an account. In our discussion of the future value of an annuity it is assumed that deposits are made at the end of each period or year. The present value of a future amount is the amount you would pay now for an amount to be received n years or periods from now. An application would be the proceeds one would lend today for a non-interest-bearing note receivable, due at some future date. The present value of a future annuity is the value today of a series of future payments. Applications include the capitalization of lease payments and capital budgeting. Capital budgeting is the making of long-term planning decisions for investments and their financing.

Discounted cash flows are used in making capital budgeting decisions. In the last section we will examine the after tax effect of depreciation on discounted cash flows. In this chapter we assume that the cash flows will be certain when in fact they are subject to varying degrees of uncertainty. We will return to the cash flow model using expected values after the topic of probability is introduced.

The underlying mathematics of compound interest is a geometric progression that is a special kind of sequence. We will begin this chapter with a discussion of sequences and progressions before looking at the calculations involving the time value of money.

6-1 SEQUENCES

In Chapter 1 we introduced the concept of a function. In this section we will discuss a special class of functions called sequences.

DEFINITION 6.1. *A sequence of real numbers is a real-valued function whose domain is the set of positive integers.*

In this text the term *sequence* will be understood to mean a sequence of real numbers.

Example 1. Discuss the following sequence:

x	$f(x)$
1	1
2	1/2
3	1/3
⋮	⋮
n	$1/n$
⋮	⋮

Note that the function values or terms of the sequence are not necessarily positive integers.

Solution: This function is a sequence since its domain is the set of positive integers, $1, 2, 3, \ldots, n, \ldots$. The functional form is given by $f(n) = 1/n$. The range of this function is $1, 1/2, 1/3, \ldots, 1/n, \ldots$.

Example 1 stresses the idea that a sequence is a special case of the more general notion of a function. Ordinarily, however, one does not use the tabular format of Example 1 to display a given sequence. For example, consider a typical sequence:

x	1	2	3	$\cdots$	n	$\cdots$
$f(x)$	$f(1)$	$f(2)$	$f(3)$	$\cdots$	$f(n)$	$\cdots$

If we denote $f(n)$ by a_n, all of the above information is contained in the listing $a_1, a_2, a_3, \ldots, a_n, \ldots$. This conventional notation for sequences will generally be used in this text.

Example 2. Discuss the sequence $0, 1/2, 2/3, \ldots, 1-1/n, \ldots$.

Solution: The nth term of this sequence is $f(n) = a_n = 1-1/n$.

$$f(1) = a_1 = 1 - 1/1 = 1 - 1 = 0$$

$$f(2) = a_2 = 1 - 1/2 = 1/2, \text{ etc.}$$

There are times when one is interested in finding the sum of the first n terms of a sequence. This process, viewed as a function of n, generates a new sequence, called the *sequence of partial sums.*

Example 3. Form the first five terms of the sequence of partial sums for the given sequence $1, 4, 7, 10, 13, 16, \ldots$.

Solution: Let $a_1 = 1$, $a_2 = 4$, etc. Define $s_n = a_1 + a_2 + \cdots + a_n$. Then

$$s_1 = 1$$

$$s_2 = 1 + 4 = 5$$

$$s_3 = 1 + 4 + 7 = 12$$

$$s_4 = 1 + 4 + 7 + 10 = s_3 + 10 = 22$$

$$s_5 = 1 + 4 + 7 + 10 + 13 = s_4 + 13 = 35$$

The sequence of partial sums is $1, 5, 12, 22, 35, \ldots$.

PROBLEMS

1. Write out the first five terms of the following sequences:
   (a) $a_n = 2 + 4/n$
   (b) $a_n = n^2$
   (c) $a_n = \sqrt{n}$
   (d) $a_n = \log n$
   (e) $a_n = \ln n$
   (f) $a_n = 2^n$
   (g) $a_n = e^n$
   (h) $a_n = e^{-n}$
2. Plot the graph of $y = f(x)$, where $x = 1, 2, 3, 4, 5, 6$ and $f(x) = 2 + (x-1)\,3$.
3. Plot the graph of $y = f(x)$, where $x = 1, 2, 3, 4, 5, 6$ and $f(x) = (1.06)^x$.
4. Write the first six terms of the sequence $1, 2, \ldots, n!, \ldots$, where $n!$ denotes n factorial and by definition $n! = n(n-1)(n-2)\cdots 2(1)$.
5. Write the first six terms of $n!/n$, where $n! = n(n-1)(n-2)\cdots 2(1)$.
6. Write the first five terms of the following sequence: $a_1 = 2$ and $a_n = 2 + (n-1)\,3$, $n = 1, 2, 3, \ldots$.

7. Write the first five terms of the following sequence: $a_1 = a$ and $a_n = a+(n-1)d$, $n = 1, 2, 3, \ldots$.

8. Form the first five terms of the sequence of partial sums for:

(a) the sequence in Problem 6

(b) the sequence in Problem 7

9. Show that the sequence with $a_1 = a$, $a_n = a_{n-1}+d$, $n = 2, 3, \ldots$ is the same sequence as the one defined in Problem 7.

10. Show that the nth term of the sequence of partial sums for the sequence $a, a+d, \ldots, a+(n-1)d, \ldots$ is equal to $na+d(1+2+3+\cdots+n-1)$.

11. Write the first five terms of the following sequence: $a_1 = 2$ and $a_n = 3a_{n-1}$, $n = 2, 3, \ldots$. Note that $a_n = 2(3)^{n-1}$.

12. Write the first five terms of the following sequence: $a_1 = a$ and $a_n = ra_{n-1}$, $n = 2, 3, \ldots$. Note that $a_n = ar^{n-1}$.

13. Show that the nth term of the sequence of partial sums for the sequence $a, ar, \ldots, ar^{n-1}, \ldots$ is given by $s_n = a(1+r+r^2+\cdots+r^{n-1})$.

14. Form the first five terms of the sequence of partial sums for the sequence $a, b, c, d, e, \ldots$.

15. Form the first three terms of the sequence of partial sums for the sequence $a_n = (1.06)^{-n}$, $n = 0, 1, 2, \ldots$.

6-2 ARITHMETIC AND GEOMETRIC PROGRESSIONS

In the previous section we introduced the notion of a sequence and pointed out that a sequence is a special case of a function. In this section we single out two special kinds of sequences for more detailed investigation. These two special cases are called arithmetic and geometric progressions.

DEFINITION 6.2. *An arithmetic progression is a sequence of the form* a, $a+d$, $a+2d, \ldots, a+(n-1)d, \ldots$, *where* d *is called the common difference.* For example, 1, 3, 5, 7, ... is an arithmetic progression with $a = 1$ and $d = 2$.

An arithmetic progression has the distinctive property that the kth term can be obtained by adding d, the common difference, to the $k-1$st term.

In the previous section we also introduced the concept of a sequence of partial sums. In the case of an arithmetic progression it is possible to obtain a formula for the nth term of the sequence of partial sums. The desired formula is

$$s_n = \sum_{k=1}^{n} a + (k-1)d = na + \frac{n(n-1)d}{2}$$

Proof:

$$s_n = a + (a+d) + (a+2d) + \cdots + [a+(n-1)d]$$

Reversing the order of the terms, we get

$$s_n = [a+(n-1)d] + \cdots + (a+2d) + (a+d) + a$$

Combining these two expressions for s_n gives $2s_n = n[2a+(n-1)d]$. Therefore,

$$s_n = na + \frac{n(n-1)d}{2} \tag{1}$$

This formula for the sum of the first n terms of an arithmetic progression is given in terms of a, n, and d. The right-hand side can also be expressed in terms of a_1, n, and a_n, where $a_n = a+(n-1)d$ and $a_1 = a$.

$$\begin{aligned} s_n &= \frac{n}{2}[2a+(n-1)d] \\ &= \frac{n}{2}[a+a+(n-1)d] \\ &= \frac{n}{2}(a_1+a_n) \end{aligned} \tag{2}$$

Example 1. Compute the sum of the first ten terms of the arithmetic progression 1, 3, 5, ... in two ways.

Solution: $a = 1$, $n = 10$, $d = 3-1 = 5-3 = 2$.

$$s_{10} = 10a + \frac{10(9)d}{2} \qquad \textit{from (1)}$$

$$s_{10} = 10 + 45(2)$$

$$s_{10} = 100$$

or

$$a_{10} = a + 9d \qquad \textit{from (2)}$$

$$a_{10} = 19$$

$$s_{10} = \frac{10}{2}(1+19)$$

$$s_{10} = 100$$

Next we define the term *geometric progression.*

DEFINITION 6.3. *A geometric progression is a sequence of the form a, ar, ar^2, ar^3, ..., ar^{n-1}, ..., where r is called the common ratio.*

A geometric progression has the distinctive property that the kth term can be obtained by multiplying the $k-1$st term by r, the common ratio. An example of a geometric progression is $1, 2, 4, 8, \ldots, 256$, where $a = 1$ and $r = 2$.

Geometric progressions are used extensively in the remaining sections of this chapter in connection with the mathematics of finance. An understanding of geometric progressions is basic to the study of the implications of investing money at compound interest.

As with arithmetic progressions, it is possible to develop a formula for the sum of the first n terms of a geometric progression. In this case we seek a formula for $s_n = a + ar + ar^2 + \cdots + ar^{n-1}$ (nth partial sum). The correct formula is $s_n = a(r^n - 1)/(r-1)$.

Proof:

$$s_n = a + ar + ar^2 + \cdots + ar^{n-1}$$

$$rs_n = ar + ar^2 + \cdots + ar^n \qquad \textit{multiplying both sides by } r$$

$$rs_n - s_n = ar^n - a$$

$$(r-1)\,s_n = a(r^n - 1)$$

$$\therefore \; s_n = \frac{a(r^n - 1)}{r-1} \tag{3}$$

Example 2. Compute $1 + 2 + 4 + 8 + \cdots + 128 + 256$.

Solution: This is the sum of the first nine terms of a geometric progression with $a = 1$ and $r = 2$. Applying (3) we have

$$s_9 = \frac{1(2^9 - 1)}{2-1}$$

$$s_9 = 511$$

Formula (3) is very important and is used later in this chapter to obtain the formula for the amount of an annuity.

PROBLEMS

1. 4, 7, 10, ..., a_n, ... is an arithmetic progression. Determine the following quantities:
   (a) the common difference, d
   (b) a_4, a_{10}, and a_{20}
   (c) the sum of the first 20 terms
2. $a_1, a_2, a_3, \ldots$ is an arithmetic progression with $a_5 = 40$ and $a_{11} = 80$. Determine the common difference, d. [*Hint:* Use some elementary algebra.]

3. Find the sum of the first 100 positive integers using a result from this section.

4. $2, 6, 18, \ldots, a_n, \ldots$ is a geometric progression. Determine the following quantities.
(a) the common ratio, r
(b) a_2, a_4, and a_6
(c) the sum of the first six terms

5. $a_1, a_2, a_3, \ldots$ is a geometric progression with $a_4 = 30$ and $a_8 = 480$. Determine the common ratio, r.

6. Find the sum of the first eight terms of the geometric progression $1, 2, 4, 8, \ldots$.

7. Graph the following functions on the same coordinate system.
(a) $f(x) = 1 + (x-1)2$, $x = 1, 2, 3, 4, 5$, i.e., the first five terms of the arithmetic progression with $a = 1$ and $d = 2$
(b) $g(x) = 2^{x-1}$, $x = 1, 2, 3, 4, 5$, i.e., the first five terms of the geometric progression with $a = 1$ and $r = 2$
(c) Do the points of the graph in (a) lie along a straight line?

8. Consider the geometric progression with $a_n = 3(2)^{n-1}$.
(a) Write the first six terms of this progression.
(b) Discuss the sequence $b_1, b_2, \ldots, b_n, \ldots$, where $b_n = \log a_n$.
(c) Do the six points of the graph of $f(x) = b_x$, $x = 1, 2, \ldots, 6$, lie along a straight line? If so, what is its slope?

9. Let $x_1, x_2, \ldots, x_n$ be the first n terms of an arithmetic progression. Give a formula for $\bar{x} = (1/n)\sum_{i=1}^{n} x_i$, where $\bar{x}$ is called the mean or arithmetic average of $(x_1, x_2, \ldots, x_n)$.

10. Use the result of Problem 9 to find the mean of $\frac{1}{2}, 1, \frac{3}{2}, 2, \frac{5}{2}, 3, \frac{7}{2}$.

11. Write the first four terms of the geometric progression with
(a) $a = 1$ and $r = 1.08$
(b) $a = 100$ and $r = 1 + i$

12. Write the first four terms of the geometric progression with $a = 1$ and $r = 1/1.08$.

13. Find the fourth partial sum of the geometric progression with $a = 1000$ and $r = 1.06$, i.e., the fourth term in the sequence of partial sums.

14. Show that $\sum_{k=1}^{n} a[1+i)^{k-1}] = a\left(\dfrac{(1+i)^n - 1}{i}\right)$.

15. Use the concept of area to prove $\sum_{k=1}^{n} a + (k-1)d = (n/2)(a_1 + a_n)$, where $a_1 = a$ and $a_n = a + (n-1)d$.

6-3 COMPOUND INTEREST

The concept of compound interest is familiar to most readers. Assume that \$1,000.00 is deposited in a bank and left for one year. The interest rate is 6% per year. Figure 6-1 illustrates the principal on deposit at the end of Year 1. At the end of Year 1 the accumulated amount is \$1060.00. If this principal were allowed to remain deposited for an

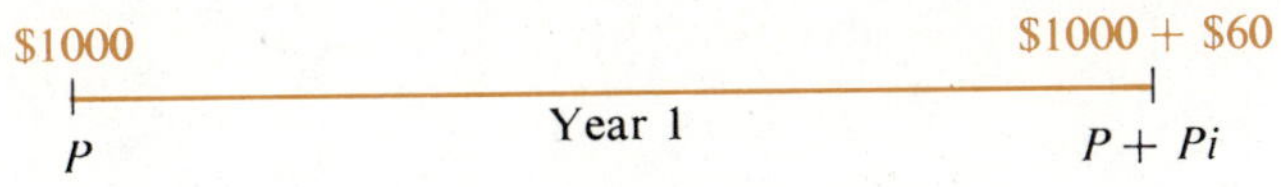

FIGURE 6-1

additional year under the same conditions, we would have Figure 6-2, where V_n denotes the value of the original principal at the end of Year n.

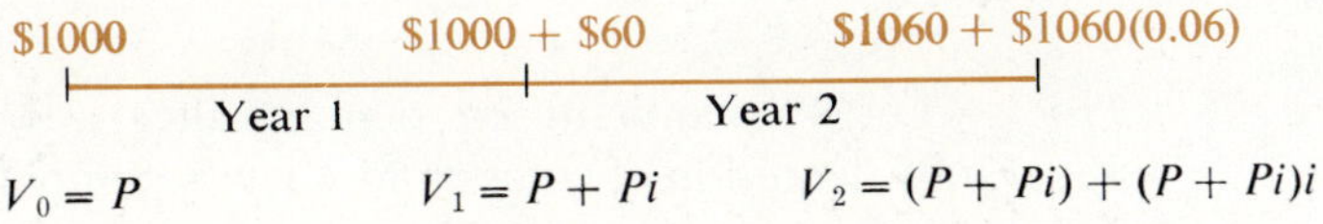

FIGURE 6-2

Note that the interest at the end of Year 2 is paid on the amount on deposit at the beginning of the period, which is equivalent to the amount at the end of the previous period.

This process is called compound interest because the interest is *converted* to principal; that is, in the next period one earns interest on the interest. In this case the number of conversions is two.

Example 1. Show that $(P+Pi)+(P+Pi)i = P(1+i)^2$.

Solution:

$$\begin{aligned} &(P+Pi)+(P+Pi)i \\ &= (P+Pi)(1+i) \qquad \textit{taking out the common factor } P+Pi \\ &= P(1+i)(1+i) \\ &= P(1+i)^2 \end{aligned}$$

We can see from the above example that the value of the original principal at the end of the second year is given by the formula $V_2 = P(1+i)^2$.

Example 2. Use this formula to compute the amount that would result if $1000.00 were deposited at 6% for two years.

Solution:

$$\begin{aligned} V_2 &= P(1+i)^2 \\ &= 1000(1+0.06)^2 \end{aligned}$$

$$= 1000(1.1236)$$

$$= \$1123.60$$

The formula $V_2 = P(1+i)^2$ suggests the following formula for the amount on deposit at the end of n years. The amount, V_n, on deposit after n years at compound interest i is given by

$$V_n = P(1+i)^n$$

Proof: Let V_n = amount on deposit after n years. Then $V_{n+1} = V_n + V_n i$ by definition of compound interest. Simplified,

$$V_{n+1} = V_n(1+i)$$

This means that the progression $P, V_1, V_2, \ldots, V_n$ is a geometric progression. Explicitly, we have

$$P, P(1+i), P(1+i)^2, P(1+i)^3, \ldots, P(1+i)^n$$

$$\therefore\ V_n = P(1+i)^n$$

The geometric progression written above has $n+1$ terms and n conversions. Furthermore, the $n+1$st term gives V_n, the amount on deposit at the end of n years.

Example 3. Apply the formula $V_n = P(1+i)^n$. Find the amount on deposit after four years if $P = \$2000$ and $i = 0.06$.

Solution:

$$V_4 = \$2000(1+0.06)^4$$

$$= \$2000(1.06)^4$$

$$= \$2000(1.2625)$$

$$= \$2525$$

If the number of years in Example 3 were 20, the calculation $(1.06)^{20}$ would be tedious. One could multiply 1.06×1.06 twenty times, use logs, or use a calculator that has an exponential key. The first two options are time-consuming and can be inaccurate. The third option may not be available. In light of these difficulties, we suggest the use of Table A-4 in the Appendix. This table provides the compound interest of \$1 for $n = 1$ through 50 periods.

To use this table for values of P other than \$1 simply multiply the table entry by P, i.e., this table gives values of $(1+i)^n$.

Example 4. Find the amount on deposit if \$1000 is deposited at an interest rate of 8% for 20 years.

Solution:

$$V_{20} = 1000(1+0.08)^{20}$$

$$= 1000(4.6610) \qquad \textit{from Table A-4}$$

$$= \$4661$$

In all of the previous examples the interest was compounded annually. What would happen if interest were compounded, or converted to principal, before the year end, i.e., semi-annually, quarterly, or daily? The following example illustrates what has to be done to handle this type of problem.

Example 5. Calculate the amount on deposit at the end of one year if \$1.00 is deposited at an interest rate of 8% compounded semi-annually.

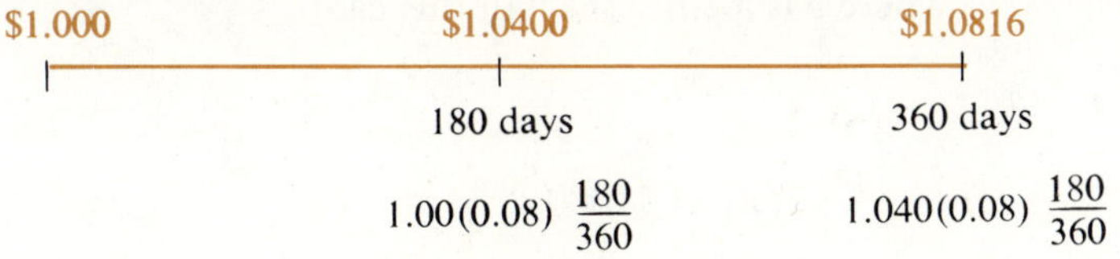

FIGURE 6-3

Note that at the first conversion at the end of 180 days, or one-half of a year,

$$\text{Interest} = 1.0000(0.08)(\tfrac{1}{2})$$

$$= 1.0000(0.04)$$

The amount at the end of the first period is given by

$$\text{Amount} = 1\left(1+\frac{i}{2}\right)^1$$

At the end of the first year (360 days) the amount is \$1.0816. The same result could have been obtained from Table A-4 using $i = 0.04$ and $n = 2$.

In general, to compute the amount on deposit, P, when interest is compounded k times per year at i percent per year for n years, we use Table A-4 and the following formula:

$$V_n = P\left(1+\frac{i}{k}\right)^{kn}$$

Example 6. Calculate the amount on deposit if interest on \$1000 is compounded quarterly for ten years at 8% per year.

Solution:

$$V_{10} = 1000\left(1 + \frac{0.08}{4}\right)^{4(10)}, \qquad k = 4,\ n = 10$$

$$= 1000(1.02)^{40}$$

$$= \$2208$$

It is interesting to note that if we had compounded annually, rather than quarterly, the amount would have been \$2,158.90. It can be shown that if we compounded instantaneously or continuously, the amount would be

$$\text{Amount} = Pe^{in}$$

where e is a constant.* In this case,

$$\text{Amount} = \$1000e^{(0.08)10}$$

$$= \$1000e^{0.8}$$

$$= \$1000(2.23) \qquad \textit{using Table A-3}$$

$$= \$2230$$

In summary, if we had invested \$1000 at 8% at the end of ten years we would have

TABLE 6-1

Compounding	Amount
annually	\$2158.90
quarterly	2208.00
instantaneously	2230.00

* The irrational number e rounded to four decimal places is 2.7183 and is the base for the natural log system.

PROBLEMS

1. Compute the amount that would result if \$3000, deposited at 4%, were compounded annually for five years.
2. Given that one deposits \$2000 at a rate of 6% compounded annually for ten years.
   (a) Find the amount on deposit by using the formula for V_n.
   (b) Find the amount on deposit by using logarithms.

   Note: One would expect the answers to (a) and (b) to be the same. This will not be the case because of truncation errors in the tables.
3. Given that $a = 100$ and $r = (1 + 0.08)$, write the first five terms of the corresponding geometric progression.
4. Given the geometric progression 200, 208, 216.32, 224.97, 233.97, ...
   (a) What is a?
   (b) What is r?
   (c) Interpreting this as a compound interest problem, what is the interest rate?
5. Calculate the amount on deposit if interest on \$2000 is compounded semi-annually for 12 years at 8% per year.
6. Complete Table 6-2 for $P = \$1000$, $n = 1$, and $i = 0.08$.

TABLE 6-2

Compounding	V_n
$k = 1$ (annually)	
$k = 2$ (semi-annually)	
$k = 4$ (quarterly)	

7. Given that $(1.00666)^{12} = 1.0829995$ and $(1.0002222)^{360} = 1.083277$, extend the table of Problem 6 to:
   (a) $k = 12$ (compounding monthly)
   (b) $k = 360$ (compounding daily)
8. Given that $e^{0.08} = 1.083287$, extend the table in Problem 6 assuming instantaneous compounding.
9. What would be the result of Example 5 if one compounded quarterly?
10. Complete the following table. Use compound interest and $P = \$1.00$. Assume annual compounding.

	n	1	2	3	4	5
$i = 0.04$	V_n					
$i = 0.08$	V_n					

11. Plot the data of Problem 10 on a single coordinate system with time as the horizontal axis.
12. A competing bank is paying interest compounded semi-annually. Assume that you are currently paying 8% compounded annually, and this

is the maximum amount that can be paid by law. What rate, compounded semi-annually, would yield the same return as 8% compounded annually?

13. A certain bank is currently paying 4% per year compounded annually on \$2,000,000. A branch manager who graduated from college the previous year suggests to top management that they convert to a semi-annual compounding policy. Playing the role of a member of the higher echelon,

(a) Evaluate the extra interest cost that would result if this suggestion were implemented.

(b) Would you accept or reject the suggestion?

6-4 PRESENT VALUE

As we saw in the previous section, if one were to put \$1.00 in the bank at the beginning of the year, by year end he would have more than \$1.00 as the interest converted to principal. The amount that results, V_n, is determined by the interest rate and the length of time that the money is on deposit.

Example 1. Using the formula $V_n = P(1+i)^n$ with $P = \$1.00$, $n = 1$, and $i = 0.08$ (compounding annually), solve for V_1.

Solution:

$$\begin{aligned} V_1 &= \$1.00(1+0.08)^1 \\ &= \$1.00(1.08) \\ &= \$1.08 \end{aligned}$$

In this section we would like to answer the following question: What is the worth now (i.e., *present value*) of a given amount that is to be received in the future, given a specific discount rate? By *discount rate* we mean a rate at which we discount the worth of a dollar because it is not received at this moment.

Example 2. Given a discount rate of 0.08, what would be the current worth of \$1.08 to be received one year from now?

Solution: Using $V_n = P(1+i)^n$ and solving for P, we have

$$\begin{aligned} P &= \frac{V_n}{(1+i)^n} \\ &= \frac{1.08}{1.08} \\ &= 1.00 \end{aligned}$$

In general, P_n will denote the present value of an amount V to be received n periods from now using a discount rate i. The formula for present value is

$$P_n = \frac{V}{(1+i)^n}$$

Using laws of exponents and writing the right-hand side as a product, we have

$$P_n = V(1+i)^{-n}$$

Example 3. Using a discount rate of 6%, calculate the present value of \$2000 that is to be received two years from now.

Solution:

$$\begin{aligned} P_2 &= \$2000(1+0.06)^{-2} \\ &= \$2000(0.8900) \\ &= \$1780 \end{aligned}$$

Note that in the above example one is faced with evaluating $(1.06)^{-2}$ or $1/(1.06)^2$. This can easily be done longhand. However, if n were equal to 14, one would be forced to use logarithms or some type of calculator. Because these calculations occur often, Table A-5 in the Appendix lists present values of \$1 for $n = 1$ through 50 periods. To use this table for values of V other than \$1.00, simply multiply the table entry by V, i.e., this table gives values of $(1+i)^{-n}$.

Example 4. Using a discount rate of 6%, find the present value of \$10,000 that is to be received five years from now.

Solution:

$$\begin{aligned} P_n &= V(1+i)^{-n} \\ P_5 &= \$10{,}000(1.06)^{-5} \\ &= \$10{,}000(0.7474) \qquad \textit{using Table A-5} \\ &= \$7473 \end{aligned}$$

In all of the above examples the discount rate assumed annual discounting. As in the previous section, we have the formula

$$P_n = V\left(1+\frac{i}{k}\right)^{-kn}$$

where k is the number of times per year that P is discounted.

Example 5. Using a discount rate of 8% discounted quarterly, find the present value of $10,000 that is to be received five years from now.

Solution:

$$P_n = V\left(1 + \frac{i}{k}\right)^{-kn}$$

$$V_5 = \$10{,}000(1.02)^{-4(5)}$$

$$= \$10{,}000(1.02)^{-20}$$

$$= \$10{,}000(0.6730)$$

$$= \$6730$$

In some cases one may know the values of P_n and V and wish to solve for i, such as in the purchase of an item bought now at P_n for the purpose of resale at price V at a time n periods from now.

Example 6. Given that one can purchase now for $15,000 an item he expects to be able to sell for $18,000 four years from now. Solve for i assuming annual discounting.

Solution:

$$P_n = V(1+i)^{-n}$$

$$\$15{,}000 = \$18{,}000(1+i)^{-4}$$

$$\frac{\$15{,}000}{\$18{,}000} = \frac{1}{(1+i)^4}$$

$$(1+i)^4 = \frac{18}{15}$$

$$1 + i = \sqrt[4]{1.2}$$

$$1 + i = 1.0466$$

$$i = 0.0466$$

PROBLEMS

1. Using an annual discount rate of 0.08, find the present value of $100 that is to be received three years from now.
2. Given that one is going to receive $2000 upon return home four years from now. Assume an annual discount rate of 6%.
   (a) Find the present value using the formula for P_n.
   (b) Find the present value using logarithms.

   Note: The results of *a* and *b* will differ because of truncation errors in the tables.

3. Given that $a = 100$, $r = 1/1.06$ in a geometric progression. Write the first five terms of this geometric progression.

4. Given the geometric progression 1.0000, 0.9434, 0.8900, 0.8396, ...
   (a) What is a?
   (b) What is r?
   (c) Interpreting this as a present value problem, what is the discount rate?

5. Find the present value of \$10,000 to be received three years from now discounted semi-annually at 8% per year.

6. Complete Table 6-3 for $V = \$1000$, $n = 1$, and $i = 0.08$.

TABLE 6-3

Discounting	P_n
$k = 1$ (annually)	
$k = 2$ (semi-annually)	
$k = 4$ (quarterly)	

7. Redo Example 6 of this section assuming quarterly discounting. [*Hint:* $\sqrt[16]{1.2} = 1.01146$.]

8. Complete the following table for $V = \$1.00$.

	n	1	2	3	4	5
$i = 0.04$	P_n					
$i = 0.08$	P_n					

9. Plot the data of Problem 8 on a single coordinate system with time as the horizontal axis.

10. One can purchase a building that is expected to resell at \$120,000 five years from now. The cost of the building is \$110,000.
    (a) Solve for i assuming annual discounting.
    (b) Would this be a wise decision?

6-5 ANNUITIES

In Section 6-3 we looked at compound interest and how a *single* amount deposited can grow as the interest becomes principal. In this section we will do something quite similar, except that, rather than a single deposit, we will look at an amount being deposited each term (i.e., year). By definition, an annuity is a series of payments or receipts of a fixed amount for a specified number of years. Note that the payments will be made at the end of the period rather than at the beginning of the period.

Example 1. How much would one have on deposit if he were to make payments of \$1000 a year for four years (i.e., a four-year annuity)? The interest being paid is 6%.

Solution:

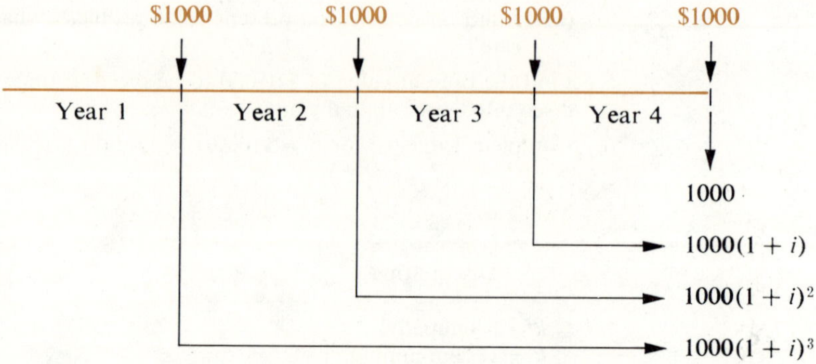

$$\begin{aligned} \text{Sum} &= \$1000 + \$1000(1.06) + \$1000(1.06)^2 + \$1000(1.06)^3 \\ &= 1000 + 1000(1.06) + 1000(1.1236) + 1000(1.1910) \\ &= 1000.00 + 1060.00 + 1123.60 + 1191.00 \\ &= \$4374.60 \end{aligned}$$

The solution was obtained by using Table A-4. We could have made use of the formula developed in Section 6-2 for the first n terms of a geometric progression,

$$s_n = \frac{a(r^n - 1)}{r - 1}$$

where a = first term in the progression

r = common ratio

n = number of terms

Example 2. Use the formula for the nth partial sum of a geometric progression to solve the problem posed in Example 1.

Solution: We have the following progression:

$$1000, 1000(1.06), 1000(1.06)^2, 1000(1.06)^3, \ldots$$

The sum of these first four terms is

$$s_4 = \frac{1000(1.06^4 - 1)}{1.06 - 1}$$

$$= \frac{1000(1.262476 - 1)}{0.06}$$

$$= \frac{1000(0.262476)}{0.06}$$

$$= \$4374.60$$

Using the formula for the nth partial sum of a geometric progression, we would like to develop a formula for the *amount of an annuity* with payments or receipts of R dollars and an interest rate i.

Let $r = 1 + i$. Then we have

$$S_n = \frac{R[(1+i)^n - 1]}{1 + i - 1}$$

$$S_n = R\left[\frac{(1+i)^n - 1}{i}\right]$$

Next, let $s_{\overline{n}|i} = \dfrac{(1+i)^n - 1}{i}$. Then $S_n = Rs_{\overline{n}|i}$. The values of $s_{\overline{n}|i}$ are tabulated in Table A-6.

The quantity $s_{\overline{n}|i}$ is the amount of an annuity of \$1.00, i.e., the special case $R = 1$.*

Example 3. Redo Example 1 using the preceding formula and Table A-6.

Solution: $i = 0.06$, $n = 4$, and $R = 1000$.

$$S_n = Rs_{\overline{n}|i}$$

$$S_4 = 1000s_{\overline{4}|0.06}$$

$$= 1000(4.3746)$$

$$= \$4374.60, \qquad \textit{as expected}$$

Thus far we have been concerned with the worth of a sequence of payments at some time in the future. As a businessman, you may also be interested in the present value of a sequence of receipts to be received at year end for n years.

Example 4. What is the present value of receipts of \$1000 to be received at year end for four years? Assume a discount rate of 6%.

---

* The symbol $s_{\overline{n}|i}$ may be read as "s angle n at the rate i."

Solution:

$$\text{Present value} = 1000(1.06)^{-1} + 1000(1.06)^{-2} + 1000(1.06)^{-3} + 1000(1.06)^{-4}$$
$$= 1000(0.9434) + 1000(0.8900) + 1000(0.8396) + 1000(0.7921)$$
$$= 1000(0.9434 + 0.8900 + 0.8396 + 0.7921)$$
$$= 10000(3.4651)$$
$$= \$3465.10$$

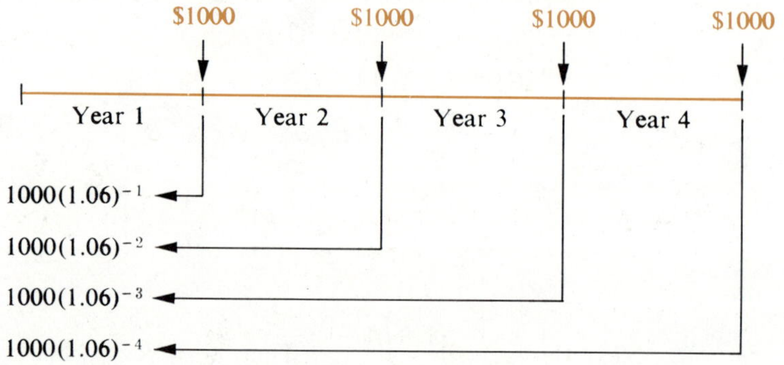

FIGURE 6-5

The present value of an annuity of $1000 to be received over a period of four years is $3465.10.

In the previous problem we saw that to calculate the present value of an annuity we simply multiplied the sum of the first four terms of a geometric progression by $R = 1000$. In general, we will denote these sums by $a_{\overline{n}|i}$, where n is the number of years and i is the appropriate discount rate. We then have the following formula for the *present value of an annuity.*

$$A_n = Ra_{\overline{n}|i}$$

A table of values of $a_{\overline{n}|i}$ is provided in Table A-7. Note that the entry in column 3 of this table for $n = 4$ is the same as the sum of the first four entries in column 3 of Table A-5. There will be a slight discrepancy due to truncation.

Example 5. Use Table A-7 to find the present value of an annuity of $10,000 that is to be received for five years. Use an 8% discount rate.

Solution:

$$A_n = Ra_{\overline{n}|i}$$

$$A_5 = \$10{,}000(3.9927)$$

$$A_5 = \$39{,}927$$

In many applications, the receipts in the sequence, sometimes referred to as the cash flow, are not all equal. One may receive \$2000 for the first three years and \$1000 for the next three years (Figure 6-6).

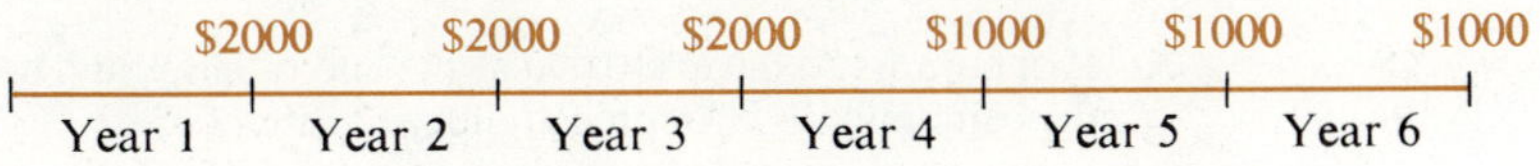

FIGURE 6-6

Clearly, we can use our formula to evaluate the present value of the receipts of the first three years. It is simply

$$A_n = Ra_{\overline{n}|i}$$

$$A_3 = 2000(2.5771) \qquad \textit{assuming 8\% as the discount rate}$$

$$A_3 = \$5154.20$$

Evaluating the flow over the last three years, we have for the present value of the last three receipts

$$\$1000[(1.08)^{-4} + (1.08)^{-5} + (1.08)^{-6}]$$

$$= 1000(0.7350 + 0.6806 + 0.6302)$$

$$= 1000(2.0458)$$

$$= \$2045.80$$

The same result could have been obtained as follows:

$$\$1000[a_{\overline{6}|0.08} - a_{\overline{3}|0.08}]$$

$$= 1000(4.6229 - 2.5771)$$

$$= 1000(2.0458)$$

$$= \$2045.80$$

Combining these two sums, we see that the present value of all six payments, A_6, is given by

$$A_6 = 2000a_{\overline{3}|0.08} + 1000(a_{\overline{6}|0.08} - a_{\overline{3}|0.08})$$

$$= 5154.20 + 2045.80$$

$$= \$7200$$

PROBLEMS

1. How much would one have on deposit after three years if he made annual payments of $10,000? Assume an 8% interest rate.
   (a) Arrive at the solution intuitively.
   (b) Arrive at the solution using the formula for the partial sum of a geometric series.
   (c) Arrive at the solution using the formula for the amount of an annuity.

2. How much would one have on deposit at the end of 15 years given that he deposited $1200 at the end of each year? Use a discount rate of 6%.

3. If one wants to have $10,000 at the end of ten years, how much must he deposit annually? Assume an interest rate of 6%.

4. What is the present value of receipts of $10,000 received at year end for three years? Assume a discount rate of 10%.
   (a) Arrive at a solution intuitively.
   (b) Arrive at a solution using the formula for the present value of an annuity.

5. What is the present value of receipts of $2000 received for 15 years? Assume a discount rate of 12%.

6. Given that one can purchase a machine for $10,000. Assume a discount rate of 12% and a flow of five years. How much must the annual receipts be in order to give a 12% return on the money invested in the machine?

7. Find the present value of an annuity of $10,000 for three years followed by an annuity of $5000 for three additional years. Assume a discount rate of 10%.

8. One can purchase a machine for $28,300. It is anticipated that this will result in a cash flow of $5000 a year for the next 12 years. What is the rate of return on this investment?

Problems 9–12 are based on the following symbols and formulas from the indicated tables:

$$S^n = (1+i)^n \qquad \text{A-4 (compound interest of \$1)}$$

$$V^n = (1+i)^{-n} \qquad \text{A-5 (present value of \$1)}$$

$$s_{\overline{n}|i} = \frac{(1+i)^n - 1}{i} \qquad \text{A-6 (amount of an annuity)}$$

$$a_{\overline{n}|i} = \frac{1-(1+i)^{-n}}{i} \qquad \text{A-7 (present value of an annuity)}$$

9. Show that $V^n = 1/S^n$ (*A-5 in terms of A-4*).

10. Show that $s_{\overline{n}|i} = (S^n - 1)/i$ (*A-6 in terms of A-4*).

11. Show that $a_{\overline{n}|i} = (1 - V^n)/i$ (*A-7 in terms of A-5*).

12. Show that $a_{\overline{n}|i} = (S^n - 1)/iS^n$ (*A-7 in terms of A-6*).

6-6 METHODS OF DEPRECIATION AND AFTER TAX EFFECT ON CASH FLOW

Depreciation is a method of allocating the cost of an asset such as a building or machine over a period of years. Those assets which can be depreciated are called depreciable assets. There are a number of methods to allocate the cost of these assets, some of which are related to the passage of time and others that are related to use or output. Two of the more popular methods that are related to the passage of time are the *straight-line* method and the *sum-of-the-years-digits* method. In applying either of these methods one must be familiar with the following terms: (a) *cost*, (b) *salvage* or *residual value*, and (c) *life of the asset.* The cost includes all expenditures relating to the acquisition and preparation for use of the asset. The salvage value is the amount one expects to realize when the asset is retired. The life of an asset is determined by both physical and functional factors.

The straight-line method of allocation involves the following formula.

$$D = \frac{C - S}{L}, \tag{3}$$

where D = annual depreciation

C = cost of the asset

S = salvage value

L = life of the asset in years

The *book value* of an asset is the cost of that asset minus the accumulated depreciation.

Example 1. Given that an asset costs \$50,000 and will have a salvage value of \$10,000 at the end of 4 years, calculate the annual depreciation using the straight-line method.

Solution:

$$D = \frac{C - S}{L} \tag{3}$$

$$D = \frac{\$50{,}000 - \$10{,}000}{4} = \frac{\$40{,}000}{4} = \$10{,}000$$

Example 2. Using the information given in Example 1, calculate the book value and the accumulated depreciation for the date of purchase and years 1–4.

Solution:

	Date of Purchase	Year 1	Year 2	Year 3	Year 4
Depreciation	$ 0	$10,000	$10,000	$10,000	$10,000
Accumulated Depreciation	0	10,000	20,000	30,000	40,000
Book Value	50,000	40,000	30,000	20,000	10,000

It can be seen from the above example that the accumulated depreciation increases by the constant amount of $10,000 each year. Similarly the book value of the asset decreases by $10,000 each year.

Example 3. Using the data given in Example 2, plot the book value of the asset on the vertical axis versus years on the horizontal axis.

Solution:

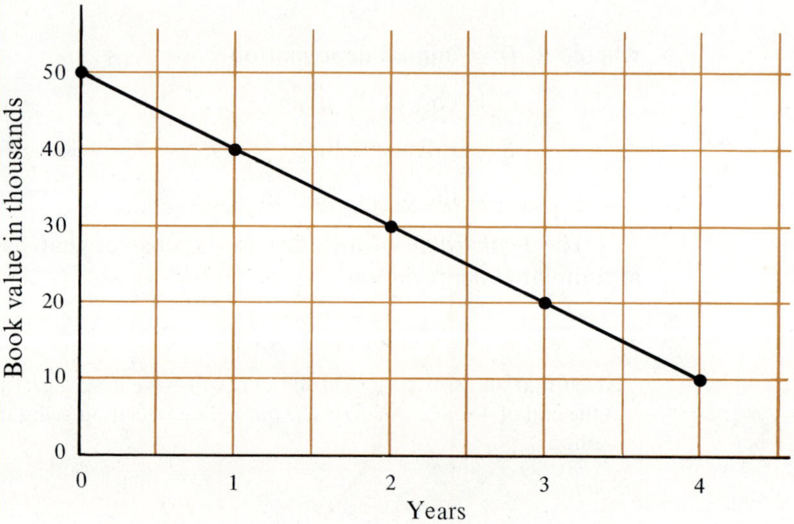

FIGURE 6-7

The graph should make it clear why this method of depreciation is called the straight-line method.

The sum-of-the-years-digits method of allocation is an accelerated method of depreciation. By accelerated we mean that more of the cost (a greater fraction) of the asset is to be depreciated in the earlier years of the life of the asset.

The fraction of the asset to be depreciated each year is determined by putting the digit of the year in reverse order over the sum of the digits of the life periods. This method can be best understood by looking at an example.

Example 4. Determine how much annual depreciation should be taken if an asset cost $15,000, has no salvage value, a life of five years and the sum-of-the-years-digits method of depreciation is utilized.

Solution:

	Digit of Years in Reverse Order	Fraction of Asset to be Depreciated	Annual Depreciation	Accumulated Depreciation
Year 1	5	5/15	$ 5,000	$ 5,000
Year 2	4	4/15	4,000	9,000
Year 3	3	3/15	3,000	12,000
Year 4	2	2/15	2,000	14,000
Year 5	1	1/15	1,000	15,000
	15	1	$15,000	

↑ Sum of the years digits

It can be seen from the above example that the accumulated depreciation increases by a smaller amount each year. Similarly the book value of the asset decreases by a smaller amount each year. Graphically we have the following result:

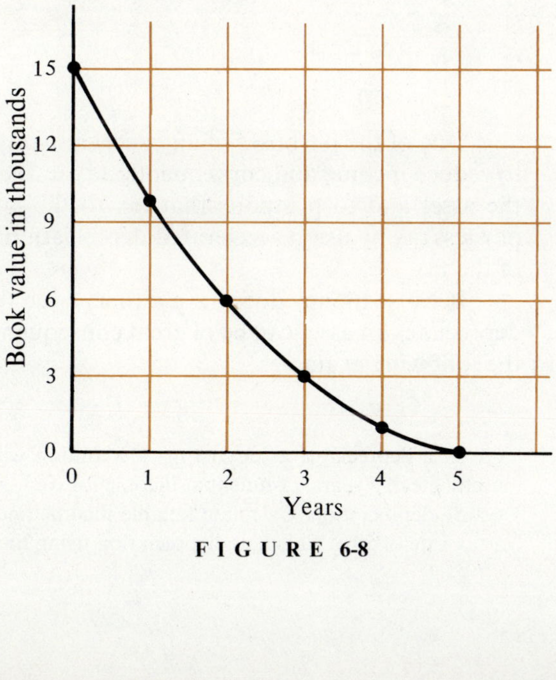

FIGURE 6-8

In the case of an asset with a short life, such as five years, the computation of the sum of the years digits is no problem. However, for an asset with a longer life such as forty years, the calculation of the sum of the years digits involves the addition of forty numbers. A convenient method for computing the sum of the years digits is given next. It is based on the formula for the sum of the first n terms of an arithmetic progression. Arithmetic progressions are treated in Section 6-2.

The sum of the first n terms of an arithmetic progression is given by the following formula.

$$s_n = na + \frac{n(n-1)}{2}d \tag{4}$$

where a is the first term of the progression
d is the common difference
n is the number of terms in the sum

When applied to a sum-of-the-years-digits depreciation problem, both a and d are taken to be 1.

Example 5. Find the sum of the years digits if an asset has a life of 40 years.

Solution: $a = 1,\ d = 1,\ n = 40$

$$s_{40} = (40)(1) + \frac{40(40-1)1}{2}$$

$$s_{40} = 40 + \frac{1560}{2}$$

$$s_{40} = 40 + 780$$

$$s_{40} = 820$$

One of the purposes of any accelerated method of depreciation is to reduce income and consequently tax in the early years of the life of the asset and to postpone that tax to the later years. One does not pay less tax by using accelerated depreciation but only defers payment of the tax.

Because money does have a time value the method by which one depreciates an asset can be of great consequence as we will illustrate by the following examples.

Example 6. Given a depreciable asset costing \$15 million with no salvage value and a useful life of 5 years, assume that income for each of the 5 years is \$18 million before depreciation. Calculate taxable income and tax for the 5 years using a tax rate of 50%. Calculate depreciation using the straight-line method.

Solution:

Year	Income Before Depreciation	Depreciation	Taxable Income	Tax
1	$18,000,000	$ 3,000,000	$15,000,000	$ 7,500,000
2	18,000,000	3,000,000	15,000,000	7,500,000
3	18,000,000	3,000,000	15,000,000	7,500,000
4	18,000,000	3,000,000	15,000,000	7,500,000
5	18,000,000	3,000,000	15,000,000	7,500,000
		15,000,000		37,500,000

Note that in Example 6, accumulated depreciation is $15,000.000 and total tax paid is $37,500,000 for the five year period.

Example 7. Using the information given in Example 6 determine net income after taxes and the present value of the cash flow for the five year period using a discount rate of 12%. Remember that depreciation is a non-cash expenditure and must be added to net income after taxes to determine cash available.

Solution:

(1) Year	(2) Net Income After Taxes	(3) Cash Flow	(4) Present Value of $1.00	(5) Discounted Cash Flow
1	$7,500,000	$10,500,000	.8929	$ 9,375,450
2	7,500,000	10,500,000	.7972	8,370,600
3	7,500,000	10,500,000	.7118	7,473,900
4	7,500,000	10,500,000	.6355	6,672,750
5	7,500,000	10,500,000	.5674	5,957,700
				$37,850,400

Column (3) is column (2) plus depreciation calculated in Example 6.
Column (4) is from Table A-5 using a rate of 12%.
Column (5) is the product of column (3) and column (4).

Example 8. Assume the same information given in Example 6, only use the sum-of-the-years-digits method to calculate depreciation.

Solution:

Year	Income Before Depreciation	Depreciation	Taxable Income	Tax
1	$18,000,000	$ 5,000,000	$13,000,000	$ 6,500,000
2	18,000,000	4,000,000	14,000,000	7,000,000
3	18,000,000	3,000,000	15,000,000	7,500,000
4	18,000,000	2,000,000	16,000,000	8,000,000
5	18,000,000	1,000,000	17,000,000	8,500,000
		15,000,000		37,500,000

Note as in Example 6 accumulated depreciation is $15,000,000 and total tax paid is $37,500,000 for the five year period.

Example 9. Using the information given in Example 8 determine net income after taxes and the present value of the cash flow for the five year period using a discount rate of 12%.

Solution:

Year	Net Income After Taxes	Cash Flow	Present Value of $1.00	Discounted Cash Flow
1	$6,500,000	$11,500,000	.8929	$10,268,350
2	7,000,000	11,000,000	.7972	8,769,200
3	7,500,000	10,500,000	.7118	7,473,900
4	8,000,000	10,000,000	.6355	6,355,000
5	8,500,000	9,500,000	.5674	5,390,300
				$38,256,750

Comparing the result of Example 7 and Example 9, we see that one can realize a savings of $406,350 by making use of the sum-of-the-years-digits method of depreciation. Again we emphasize that the same total tax is paid under both methods and that the savings are a result of the time at which the taxes are paid.

SUMMARY OF EXAMPLES 7 AND 9

Discounted net income using sum-of-the-years-digits method of depreciation (Example 9)	$38,256,750
Discounted net income using straight-line method of depreciation (Example 7)	37,850,400
Savings	$ 406,350

PROBLEMS

1. The Western Corporation acquired an asset on January 1, 19*xx* at a cost of $55,000. The asset has an estimated useful life of 10 years and no salvage value. Compute the depreciation for each of the first three years assuming the following depreciation methods: (a) straight-line and (b) sum-of-the-years-digits.

2. Calculate the sum-of-the-years-digits for Problem 1 using the formula for the sum of the first n terms of an arithmetic progression.

3. The Computer Corporation bought a machine on January 1, 19*xx* at a cost of $100,000. The asset has an estimated useful life of 5 years and an estimated salvage value of $25,000. Compute the depreciation for each year of the asset's life assuming the following depreciation methods: (a) straight-line and (b) sum-of-the-years-digits.

4. Using the information given in Problem 3, plot the book value of the asset (cost − accumulated depreciation) on the vertical axis and time on the horizontal axis for each method of depreciation.

5. If an asset is acquired for $120,000 on January 1, 19*xx* and has a useful life of 10 years and an estimated salvage value of $10,000, what would be the accumulated depreciation at the end of the fourth year if:
(a) the straight-line method of depreciation is used,
(b) the sum-of-the-years-digits method is used.

6. Given a depreciable asset costing $30 million with no salvage value and a useful life of 5 years. Assume that income for the 5 years before depreciation is estimated at $26 million for each of the first three years and $28 million for each of the two remaining years. Calculate the present value of the cash flow for the five year period using a discount rate of 10 percent, and assume a tax rate of 50 percent. Calculate depreciation using the straight-line method.

7. Redo Problem 6 using the sum-of-the-years-digits method of depreciation.

8. Compare the results of Problems 6 and 7. What are the implications?

9. The sum-of-the-years-digits method involves the sum $\sum_{i=1}^{n} i$. Use (4) of this section to obtain a general formula for $\sum_{i=1}^{n} i$.

10. Use the result of Problem 9 above to show that $\sum_{i=1}^{n} i = \frac{1}{2}n(n+1)$.

11. Apply the result of Problem 10 to Example 5 to obtain the same value for s_{40}.

12. Derive $\sum_{i=1}^{n} i = \frac{1}{2}n(n+1)$ for the special case of n, an even integer. The following hints may be helpful:
(a) Group the terms as follows:

$$(1+n)+(2+n-1)+, \text{ etc.}$$

(b) Consider the value of each group and the number of groups.

13. Derive $\sum_{i=1}^{n} i = \frac{1}{2}n(n+1)$ for the case of n, an odd integer. *Hint:* Use the result of Problem 12.

6-7 SUMMARY AND APPLIED PROBLEMS

The topics covered in this chapter will perhaps have more impact on you as an individual or businessperson than any other covered in this text. At this time there should be a realization that money changes in value with the passage of time.

We began this chapter with a discussion of arithmetic and geometric progressions and they can be summarized as follows:

Arithmetic Progressions	Geometric Progressions
d = common difference	r = common ration
a = first term	a = first term
a_n = nth term	a_n = nth term
$a_n = a + (n-1)d$	$a_n = ar^{n-1}$
$a_n - a_{n-1} = d$	$a_n \div a_{n-1} = r$
s_n = sum of first n terms of progression	
$s_n = na + \dfrac{n(n-1)d}{2}$ or $s_n = \dfrac{(a+a_n)n}{2}$	$s_n = \dfrac{a(r^n - 1)}{r-1}$
An application of arithmetic progressions: sum-of-the-years-digits method of depreciation	Applications of geometric progressions: compound interest, present value, and annuities

Applications of the time value of money encountered were compound interest, annuities, computation of depreciation and its after tax effect on cash flow, and capital budgeting. The applications and formulas can be summarized as follows:

Application	Formula
Compound Interest	$V_n = P(1+i)^n$
Present Value	$P_n = V(1+i)^{-n}$
Amount of an Annuity	$S_n = Rs_{\overline{n}\rceil i},\ s_{\overline{n}\rceil i} = \dfrac{(1+i)^n - 1}{i}$
Present Value of an Annuity	$A_n = Ra_{\overline{n}\rceil i},\ a_{\overline{n}\rceil i} = \dfrac{1-(1+i)^{-n}}{i}$

where i = interest rate,
n = number of periods,
P = dollar amount on deposit now,
V = dollar amount to be received in the future,
R = dollar amount of an annuity.

Values for $(1+i)^n$, $(1+i)^{-n}$, $s_{\overline{n}|i}$, and $a_{\overline{n}|i}$ can be found in Tables A-4 through A-7, respectively, in Appendix A.

PROBLEMS

1. In the state of Massachusetts there is a state-run lottery. One of the games a person can play is called the Million Dollar game. The winner is paid \$50,000 a year for twenty years. Taking into consideration the time value of money, what would be a more appropriate title for this game? Assume an interest rate of 8 percent and that the payments are made at the end of each year.

2. Road Crew Associates is considering buying a new bulldozer which is expected to have a ten-year useful life, have zero disposal value, and contribute \$10,000 per year to profit. The cost of the bulldozer is \$48,332. What is the time-adjusted rate of return on this project?

3. Your father-in-law, who is 50 years old, is planning on depositing into a retirement fund \$2,000 at the end of every year for the next 10 years. At that time he plans on being semi-retired for 5 years. During this period he will not contribute or withdraw from his retirement fund. Assuming an interest rate of 8 percent, he has asked you what the worth of his retirement fund will be at the end of this 15 year period. What would be his annual income from this retirement fund if he did not withdraw from the fund?

4. Many companies are now using leases to finance properties. When one uses the lease-purchase method it is necessary to assign a value to the lease. The value to be assigned to the lease is the discounted value at an appropriate interest rate of all future rents. If Mr. Manley, president of a large company, enters a noncancelable lease contract of \$10,000 annual rent payable at the beginning of the year for ten years, what would be the proper entry to record the value of the lease correctly? Assume an interest rate of 8 percent. Note that the payments are made at the beginning of the year, not the end of the year.

5. Fill in the blanks given in the following progressions.

(a) $-2, 0, 2, \ldots, a_n, \ldots$

Type of progression ____________

$a =$ ____________

$d =$ ____________

$a_5 =$ ____________

$s_{10} =$ ____________

$s_3 =$ ____________

(b) $1, 1.06, (1.06)^2, \ldots, a_n, \ldots$

Type of progression ____________

$a =$ ____________

$r =$ ____________

$a_4 =$ ____________

$s_4 =$ ____________

6.* In 1969 the Archibald Freight Company negotiated and closed a long-term lease contract for newly constructed truck terminals and freight storage facilities. The buildings were erected to the Company's specifications on land owned by the Company. On January 1, 1970 Archibald Freight Company took possession of the leased properties. On January 1, 1970 and 1971 the Company made cash payments of $1,200,000 which were recorded as rental expenses.

Although the terminals have a composite useful life of 40 years, the noncancelable lease runs for 20 years from January 1, 1970 with a favorable purchase option available upon expiration of the lease. You have determined that the leased properties and related obligation should be accounted for as an installment purchase. Internal Revenue Service agents have indicated that purchase accounting will be required for tax purposes and that an assessment will be made for deficiencies in 1970 income taxes.

The 20-year lease is effective for the period January 1, 1970 through December 31, 1989. Advance rental payments of $1,000,000 are payable to the lessor on January 1 of each of the first 10 years of the lease term. Advance rental payments of $300,000 are due on January 1 for each of the last 10 years of the lease. The company has an option to purchase all of these leased facilities for $1 on December 31, 1989. It also must make annual payments to the lessor of $75,000 for property taxes and $125,000 for insurance. The lease was negotiated to assure the lessor a 6% rate of return.

Required: (round all computations to the nearest dollar)

Prepare a schedule to compute for Archibald Freight Company the discounted present value of the terminal facilities and related obligation at January 1, 1970.

Selected present value factors are as follows:

Periods	For an Ordinary Annuity of $1 at 6%	For $1 at 6%
1	.943396	.943396
2	1.833393	.889996
8	6.209794	.627412
9	6.801692	.591898
10	7.360087	.558395
19	11.158117	.330513
20	11.469921	.311805

7.* Given below are the present-value factors for $1.00 discounted at 8% for one to five periods. Each of the following items is based on 8% interest compounded annually from day of deposit to day of withdrawal.

Periods	Present Value of $1 Discounted at 8% Per Period
1	0.926
2	0.857
3	0.794
4	0.735
5	0.681

A. What amount should be deposited in a bank today to grow to $1,000 three years from today?

(a) $\frac{\$1{,}000}{0.794}$

(b) $1,000 × 0.926 × 3.

(c) ($1,000 × 0.926) + ($1,000 × 0.857) + ($1,000 × 0.794).

(d) $1,000 × 0.794.

B. What amount should an individual have in his bank account today before withdrawal if he needs $2,000 each year for four years with the first withdrawal to be made today and each subsequent withdrawal at one-year intervals? (He is to have exactly a zero balance in his bank account after the fourth withdrawal.)

(a) $2,000 + ($2,000 × 0.926) + ($2,000 × 0.857) + ($2,000 × 0.794).

(b) $\frac{\$2{,}000}{0.735} \times 4.$

(c) ($2,000 × 0.926) + ($2,000 × 0.857) + ($2,000 × 0.794) + ($2,000 × 0.735).

(d) $\frac{\$2{,}000}{0.926} \times 4.$

C. If an individual put $3,000 in a savings account today, what amount of cash would be available two years from today?

(a) $3,000 × 0.857.

(b) $3,000 × 0.857 × 2.

(c) $\frac{\$3{,}000}{0.857}.$

(d) $\frac{\$3{,}000}{0.926} \times 2.$

D. What is the present value today of $4,000 to be received six years from today?

(a) $4,000 × 0.926 × 6.

(b) $4,000 × 0.794 × 2.

(c) $4,000 × 0.681 × 0.926.

(d) Cannot be determined from the information given.

* Problem 7 from the Uniform CPA Examinations, Copyright © (1974) by the American Institute of Certified Public Accountants, Inc., is reprinted with permission.

8.* The Baxter Company manufactures toys and other short-lived fad-type items.

The research and development department came up with an item that would make a good promotional gift for office equipment dealers. Aggressive and effective effort by Baxter's sales personnel has resulted in almost firm commitments for this product for the next three years. It is expected that the product's value will be exhausted by that time.

In order to produce the quantity demanded, Baxter will need to buy additional machinery and rent some additional space. It appears that about 25,000 square feet will be needed. 12,500 square feet of presently unused, but leased, space is available now. (Baxter's present lease with 10 years to run costs $3.00 a square foot.) There is another 12,500 square feet adjoining the Baxter facility which Baxter will rent for 3 years at $4.00 per square foot per year if it decides to make this product.

The equipment will be purchased for about $900,000. It will require $30,000 in modifications, $60,000 for installation and $90,000 for testing; all of these activities will be done by a firm of engineers hired by Baxter. All of the expenditures will be paid for on January 1, 1973.

The equipment should have a salvage value of about $180,000 at the end of the third year. No additional general overhead costs are expected to be incurred.

The following estimates of revenues and expenses for this product for the three years have been developed.

	1973	1974	1975
Sales	$1,000,000	$1,600,000	$800,000
Material, Labor and Incurred Overhead	400,000	750,000	350,000
Assigned General Overhead	40,000	75,000	35,000
Rent	87,500	87,500	87,500
Depreciation	450,000	300,000	150,000
	$ 977,500	$1,212,500	$622,500
Income Before Tax	$ 22,500	$ 387,500	$177,500
Income Tax (40%)	9,000	155,000	71,000
	$ 13,500	$ 232,500	$106,500

Required:

A. Prepare a schedule which shows the incremental, after tax, cash flows for this project.

B. If the company requires a two-year payback period for its investment, would it undertake this project? Show your supporting calculations clearly.

C. Calculate the after tax accounting rate of return for the project.

D. A newly hired Business School graduate recommends that the company consider the use of the net present value analysis to study this project. If the company sets a required rate of return of 20%

* Problem 8 adapted from the CMA examination.

after taxes will this project be accepted? Show your supporting calculations clearly. (Assume all operating revenues and expenses occur at the end of the year.)

Discount Factors for 20% (Rounded off)

Period	Present Value of $1.00	Present Value of $1.00 per Period Received at End of Period	Accumulated Value of $1.00	Accumulated Value of $1.00 per Period Received at End of Period
1	.83	.83	1.20	1.00
2	.69	1.52	1.44	2.20
3	.58	2.10	1.73	3.64
4	.48	2.58	2.07	5.37

9. Edwards Corporation is a manufacturing concern that produces and sells a wide range of products. The company not only mass produces a number of products and equipment components, but also is capable of producing special-purpose manufacturing equipment to customer specifications.

The firm is considering adding a new stapler to one of its product lines. More equipment will be required to producc the new stapler. There are three alternative ways to acquire the needed equipment: (1) purchase general-purpose equipment, (2) lease general-purpose equipment, (3) build special-purpose equipment. A fourth alternative, purchase of the special-purpose equipment, has been ruled out because it would be prohibitively expensive.

The general-purpose equipment can be purchased for $125,000. The equipment has an estimated salvage of $15,000 at the end of its useful life of ten years. At the end of five years the equipment can be used elsewhere in the plant or be sold for $40,000.

Alternatively, the general-purpose equipment can be acquired by a five-year lease for $40,000 annual rent. The lessor will assume all responsibility for taxes, insurance, and maintenance.

Special-purpose equipment can be constructed by the Contract Equipment Department of the Edwards Corporation. While the department is operating at a level which is normal for the time of year, it is below full capacity. The department could produce the equipment without interfering with its regular revenue-producing activities.

The estimated departmental costs for the construction of the special-purpose equipment are

Materials and parts	$ 75,000
Direct labor	60,000
Variable overhead (50% of DL$)	30,000
Fixed overhead (25% of DL$)	15,000
Total	$180,000

Corporation general and administrative costs average 20 percent of labor dollar content of factory production.

Engineering and management studies provide the following revenue and cost estimates (excluding lease payments and depreciation) for producing the new stapler depending upon the equipment used:

	General-Purpose Equipment		Self-Constructed Equipment
	Leased	Purchased	
Unit selling price	$5.00	$5.00	$5.00
Unit production costs:			
Materials	$1.80	$1.80	$1.70
Conversion costs	1.65	1.65	1.40
Total unit production costs	$3.45	$3.45	$3.10
Unit contribution margin	$1.55	$1.55	$1.90
Estimated unit volume	40.000	40.000	40,000
Estimated total contribution margin	$62,000	$62,000	$76,000
Other costs:			
Supervision	$16,000	$16,000	$18,000
Taxes and insurance	—	3,000	5,000
Maintenance	—	3,000	2,000
Total	$16,000	$22,000	$25,000

The company will depreciate the general-purpose machine over ten years on the sum-of-the-years-digits (S-Y-D) method. At the end of five years the accumulated depreciation will total $80,000. (The present value of this amount for the first five years is $62,100.) The special-purpose machine will be depreciated over five years on the S-Y-D method. Its salvage value at the end of that time is estimated to be $30,000.

The company uses an after-tax cost of capital of 10 percent. Its marginal tax rate is 40 percent.

Required:

A. Calculate the net present value for each of the three alternatives that Edwards Corporation has at its disposal.

B. Should Edwards Corporation select any of the three options, and if so, which one? Explain your answer.

Problem 9 adapted from the CMA examination.

Discount Factors for 10% (Rounded)

Period	Present Value of $1	Present Value of $1.00 per Period Received at End of Period
1	.91	.91
2	.83	1.74
3	.75	2.49
4	.68	3.17
5	.62	3.79
6	.56	4.35
7	.51	4.86
8	.47	5.33
9	.42	5.75
10	.39	6.14

10. You have worked out the following arrangement with a "rich" uncle to help pay for a four-year college program. He agrees to loan you $4,000; $1,000 at the beginning of each college year, say September 1. The interest rate is 4% with annual compounding. No payments (principal or interest) are to be made on this loan until fifteen months after graduation, i.e., September 1. At that time you are to begin paying back principal and accumulated interest in 5 equal annual payments. The interest rate is 6% charged on the unpaid balance during the amortization period. Determine the size of the 5 equal annual payments needed to just amortize (pay back) this loan.

11. An *amortization schedule* shows exactly how much of each payment goes for interest and how much goes toward the repayment of the principal. It shows the allocation of each payment into interest and principal as well as the outstanding balance after each payment. This suggests using four columns headed: *payment number*, *payment on interest*, *payment on principal*, and *balance of loan after payment has been made.* Amortization schedules for any amount, interest rate, and term of payment may be obtained from Financial Publishing Company at Boston, Mass., for example.

Prepare an amortization schedule for a loan of $1,000 to be paid back in 12 equal monthly payments. The interest rate is 8% and interest is charged only on the unpaid balance.

A monthly payment of $86.99 should do it. Use $87.00 as monthly payment and adjust the last payment accordingly.

12. Determine the size of the equal monthly payments needed to amortize a loan of $15,000 in 15 years at an annual interest rate of 8%, the interest being charged on the unpaid balance.

REFERENCES

Hart, William L. *Mathematics of Investment.* Lexington, Mass.: D. C. Heath & Co., 1958.

Johnson, Robert W. *Financial Management*, fourth ed. Boston: Allyn and Bacon, Inc., 1971.

Kent, Frederick C., and Kent, Maude E. *Compound Interest and Annuity Tables.* New York: McGraw-Hill Book Co., Inc., 1954.

Lee, T. H. *Mathematics of Finance.* Homewood, Illinois: Richard D. Irwin, Inc., 1957.

Spiegel, Murray R. *Mathematical Handbook of Formulas and Tables.* New York: McGraw-Hill Book Co., Inc., 1968.

Weston, Fred J., and Brigham, Eugene F. *Managerial Finance.* New York: Holt, Rinehart and Winston, Inc., 1972.

SEVEN

Introduction to probability

Sets are used extensively in the study of modern mathematics. A *set* can be defined intuitively as a collection of objects. The objects in a set are called its elements and are said to *belong* to the set. Braces, { }, are normally used to enclose the elements of a set. The notation, $x \in S$, is used to indicate that x is an element of the set S.

The *set* of all possible outcomes of an experiment is called the *sample space* for the experiment. Any *subset* of a sample space is called an *event*. Two events are said to be mutually *exclusive* if their *intersection* is empty.

Sets, subsets, and set operations are treated in the first two sections of this chapter. The remainder of the chapter deals with probability. An understanding of sets and set operations is essential to the study of probability as presented in the next two chapters of this text.

7-1 SETS AND SUBSETS

The brace notation for sets is illustrated by the next four examples.

Example 1. $\{1, 3, 5, 7\}$

Example 2. $\{x_1, x_2, x_3, x_4, x_5\}$

Example 3. {product 1, product 2, product 3}

Example 4. {(head, head), (head, tail), (tail, head), (tail, tail)}

In Example 2, x_1, x_2, x_3, x_4, and x_5 are said to be *elements* of the given set.

The symbol $\in$ is used to indicate membership in a set. The expression $x \in A$, read "x belongs to A," means that x is an element of the set A. The symbol $x \notin A$ means that "x does not belong to A."

Examples using $A = \{1, 2, 3\}$ and $B = \{x, y, z, w\}$:

$2 \in A$ and $x \in B$

$4 \notin A$ but $w \in B$

The set of all elements in a given discussion or situation is called the *universal* set.

DEFINITION 7.1. *Two sets A and B are said to be equal if every element of A is an element of B and every element of B is an element of A.*

Definition 7.1 is illustrated by the next four examples.

Example 5. $\{1, 2, 4\} = \{1, 2, 4\}$

Example 6. $\{1, 3, 7\} \neq \{1, 4, 7\}$

Example 7. $\{2, 5, 8\} = \{8, 5, 2\}$ *order is not important*

Example 8. $\{1, 4, 10\} \neq \{1, 4, 10, 13\}$ *13 is in the set on the right but not in the set on the left*

DEFINITION 7.2. *A set A is said to be a subset of a set B if every element of A is also an element of B. A is a subset of B is denoted by* $A \subset B$.

Definition 7.2 is illustrated by the next four examples.

Example 9. $\{2, 3, 5\} \subset \{1, 2, 3, 5, 8\}$

Example 10. $A = \{1, 2, 5\}$ is not a subset of $\{1, 4, 5\} = B$ *2 is in A but not in B*

Example 11. $\{a, b, c\} \subset \{c, d, a, b\}$

Example 12. $\{x, y, z\} \subset \{x, y, z\}$, i.e., if $A = B$, then $A \subset B$ and $B \subset A$.

It is useful to consider the set with no elements. This set is called the *null* set, or the *empty* set, and is frequently denoted by the symbol $\varnothing$.

DEFINITION 7.3. *Let Ω be the universal set and A be a subset of Ω. Then the set of all x such that $x \in \Omega$ and $x \notin A$ is called the complement of A.*

The complement of A is denoted by A'.

Definition 7.3 is illustrated by the next two examples.

Example 13. If $\Omega = \{1, 2, 3, 4, 5, 6, 7, 8, 9, 10\}$ and $A = \{2, 4, 6, 8, 10\}$, then $A' = \{1, 3, 5, 7, 9\}$.

Example 14. If the universal set consists of the letters of the alphabet and C is the set of consonants, then the complement of C is the set of vowels, i.e., $C' = \{a, e, i, o, u\}$.

Much insight can be gained into set relationships by representing sets by shaded regions in a plane. To illustrate this, diagrams called *Venn diagrams* are used. Some examples of Venn diagrams are given in Figure 7-1.

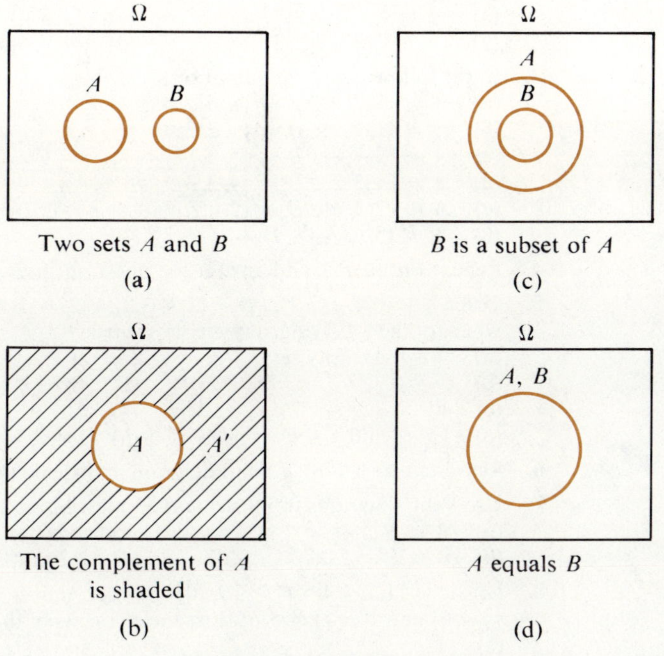

FIGURE 7-1

SUMMARY

1. Sets are commonly denoted by capital letters.
2. Braces, { }, are used in both the roster and the descriptive methods.
3. Membership in a set is indicated by $\in$.
4. Failure to be a member of a set is indicated by $\notin$.
5. Two sets, A and B, are considered equal, written $A = B$, if and only if they both contain exactly the same elements, the order in which the elements are written being immaterial.
6. The fact that two sets A and B are not equal is indicated by $A \neq B$.
7. The set of all elements that could possibly enter a given discussion is called the *universal* set. The universal set is commonly denoted by U or Ω.
8. It is convenient to make use of the set with no elements called the *null* set or the *empty* set. The empty set is denoted by $\varnothing$.
9. Consider a set A all of whose elements are members of a given universal set U. The set of elements in U that fail to be members of A is called the complement of A. The complement of A is denoted by A'.
10. If every element of a set A is also an element of another set B, then A is said to be a *subset* of B. A is a subset of B is denoted by $A \subset B$.

PROBLEMS

1. Give some examples of sets.

2. Which of the following sets are equal?
(a) $\{2, 6, 9\}$, $\{6, 2, 9\}$
(b) $\{5, 7, 3\}$, $\{5, 7, 2, 3\}$
(c) $\{2, 5, 4\}$, $\{2, 5, 4, 3\}$
(d) $\{4, 9, 6\}$, $\{9, 6, 8\}$

3. In the following, is A a subset of B?
(a) $A = \{5, 2, 9\}$, $B = \{8, 6, 2, 4, 5, 9\}$
(b) $A = \{1, 2, 3\}$, $B = \{1, 2, 3, 4\}$
(c) $A = \{5, 7, 9\}$, $B = \{5, 9, 11\}$
(d) $A = \{4, 3, 2\}$, $B = \{4, 2, 3\}$
(e) $A = \{2, 4, 6\}$, $B = \{2, 4, 6\}$
(f) $A = \{5, 6, 7, 8\}$, $B = \{8, 6, 7\}$

4. Repeat Problem 3, and answer the question: Is B a subset of A?

5. Given $A = \{1, 3, 5, 9\}$, $B = \{2, 4, 6, 8\}$, $C = \{1, 4, 5, 8\}$, and $D = \varnothing$, which of the following statements are true?
(a) D is the empty set
(b) $3 \in A$
(c) $4 \in A$
(d) $1 \in A$ and $1 \in C$
(e) $A \subset C$
(f) $C \subset B$
(g) $6 \in B$
(h) $2 \notin A$

6. Represent a set B by a Venn diagram and shade its complement.

7. Use Venn diagrams to illustrate the following:
(a) $A \subset B$
(b) $B \subset A$
(c) A is not a subset of B
(d) B is not a subset of A

8. Let $U = \{1, 2, 3, 4, 5, 6, 7, 8, 9, 10\}$ be the universal set and let $A = \{2, 4, 6\}$ and $B = \{1, 4, 5, 6\}$. List the elements of the following sets.
(a) A'
(b) B'

(c) $\{x: x \in A \text{ and } x \in B\}$
(d) $\{x: x \in A \text{ or } x \in B\}$

9. What is the complement of the universal set U? The empty set, $\varnothing$?

10. What can be said about the sets A and B if A is a subset of B and B is a subset of A?

11. Let $A = \{a, b, c\}$. Form all possible subsets of A. How many subsets did you get if both the empty set, $\varnothing$, and A itself are included in the count?

12. Complete the following table for the number of subsets of $A = \{a, b, c, d, e\}$.

Number of elements in subset	Number of subsets in this class
zero	
one	
two	
three	10
four	
five	
Total	____

13. Repeat Problem 5 with $A = \{a, b, c, d\}$.

14. Based on your experience with Problems 5, 6, and 7, can you suggest a formula for the number of subsets of a set A with k elements.

7-2 INTERSECTIONS AND UNIONS

DEFINITION 7.4. *Let A and B be sets. The set $A \cap B$, which is read A intersect B, consists of all the elements that belong to both A and B.*

Example 1. $\{2, 3, 4, 6\} \cap \{1, 4, 5, 6\} = \{4, 6\}$

Example 2. $\{a, b, c\} \cap \{d, e\} = \varnothing$, the empty set

Example 3. {plant 1, plant 2, plant 3} $\cap$ {plant 2, plant 3, plant 4} = {plant 2, plant 3}

Definition 7.4 is illustrated further by the Venn diagrams in Figure 7-2.

In this section, we introduce two important set operations called *union* and *intersection*. The union of two sets is defined in terms of the English word *or* and the intersection of two sets is defined in terms of the English word *and*.

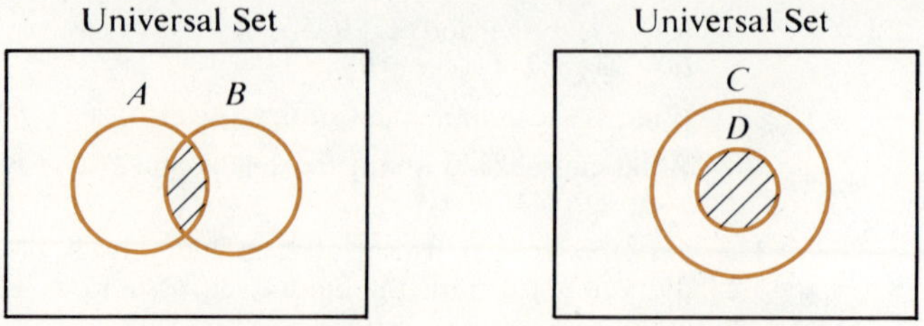

$A \cap B$ is shaded

(a)

$D \subset C$, $C \cap D = D$
$C \cap D$ is shaded

(b)

FIGURE 7-2

DEFINITION 7.5. *Let A and B be sets. The set $A \cup B$, which is read A union B, consists of the elements that belong to A or B or both.*

Example 4. $\{1, 4, 7\} \cup \{3, 5, 7, 8\} = \{1, 3, 4, 5, 7, 8\}$. Note that 7 is a member of both sets, but is only listed once in the union.

Example 5. $\{a, b, c\} \cup \varnothing = \{a, b, c\}$

Example 6. {product 1, product 2} $\cup$ {plant 1, plant 2} = {product 1, product 2, plant 1, plant 2}

Definition 7.5 is illustrated further by the Venn diagrams in Figure 7-3.

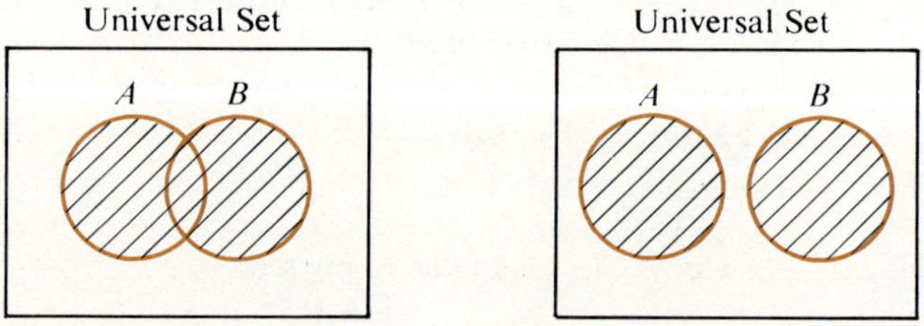

$A \cap B \neq \varnothing$
$A \cup B$ is shaded

(a)

$A \cap B = \varnothing$
$A \cup B$ is shaded

(b)

FIGURE 7-3

The following set notation is standard. $A = \{x\colon x \text{ has property } p\}$. This would read, "$A$ is the set of x such that x has property p." Note that the colon is read "such that."

SUMMARY

If A and B are arbitrary subsets of a universal set Ω, then:

1. $A \cup B = \{x: x \in A \text{ or } x \in B\}$ (Union)
2. $A \cap B = x: x \in A \text{ and } x \in B$ (Intersection)
3. $A \subset A \cup B$ and $B \subset A \cup B$
4. $A \cap B \subset A$ and $A \cap B \subset B$

PROBLEMS

1. Let A, B, and U, the universal set, be given as follows. $A = \{2, 4, 6\}$, $B = \{1, 4, 5, 6\}$ and $U = \{1, 2, 3, 4, 5, 6, 7, 8, 9, 10\}$. List the elements of each of the following sets.

(a) A' **(c)** $A \cap B$
(b) B' **(d)** $A \cup B$

2. Using the sets defined in Problem 1, list the elements of each of the following sets.

(a) $A' \cap B'$ **(c)** $(A \cup B)'$
(b) $A' \cup B'$ **(d)** $(A \cap B)'$

3. **(a)** Which of the sets in Problem 2 are equal?
(b) Can you suggest two possible set identities from the facts of part (a)? If so, what are they?

4. Show that $(A \cup B)' = A' \cap B'$ (De Morgan's law). *Hint:* Let $x \in (A \cup B)'$ and show that $x \in A' \cap B'$. Then let $x \in A' \cap B'$ and show that $x \in (A \cup B)'$. This proves that the two sets are equal.

5. Use the method described in Problem 4 to prove $(A \cap B)' = A' \cup B'$. (De Morgan's law.)

6. Illustrate the identity $(A \cap B)' = A' \cup B'$ with Venn diagrams. *Hint:* Draw two identical Venn diagrams side by side both of which include A and B. Then shade $(A \cap B)'$ in the left-hand diagram and shade $A' \cup B'$ in the right-hand diagram. The same area will be shaded in both diagrams thus "proving" the identity.

7. Given $A = \{\text{product 1, product 2, product 3}\}$ and $B = \{\text{plant 1, plant 2, plant 3, plant 4}\}$, find

(a) $A \cap B$ **(b)** $A \cup B$

8. Let $A_1 = \{x: x \text{ has property } p_1\}$ and $A_2 = \{x: x \text{ has property } p_2\}$. Using this format define A_1, $A_1 \cap A_2$ and $A_1 \cup A_2$.

9. Let

$$A_1 = \{x: (x-2)(x-3)(x-6) = 0\}$$

and

$$A_2 = \{x: (x-1)(x-3)(x-6)(x-8) = 0\}.$$

(a) Identify the properties p_1 and p_2 of Problem 8 for these two sets.
(b) List the elements of A_1 and A_2.
(c) List the elements of $A_1 \cap A_2$. $A_1 \cup A_2$.
(d) Express $A_1 \cap A_2$ in the format used to describe A_1 and A_2.
(e) Repeat part (d) for $A_1 \cup A_2$.

10. Let $A = \{a, b, c\}$, $B = \{e, f, g, h\}$ and $C = \{g, h, i, j\}$.

(a) Determine $n(A)$, $n(B)$, and $n(C)$ where $n(S) =$ the number of elements in the set S.

(b) Determine $A \cup B$, $A \cup C$, and $B \cup C$.

(c) Determine $A \cap B$, $A \cap C$, and $B \cap C$.

(d) Determine $n(A \cap B)$, $n(A \cap C)$, and $n(B \cap C)$.

(e) Determine $n(A \cup B)$, $n(A \cup C)$, and $n(B \cup C)$.

(f) Can you suggest a counting formula for the number of elements in the union of two finite sets? If so, what is it?

11. Let A and B be two finite sets and suppose $n(A) = a$ and $n(B) = b$, where $n(S)$ denotes the number of elements in the set S.

(a) What is the largest possible value of $n(A \cup B)$?

(b) What is the smallest possible value of $n(A \cup B)$?

(c) Under what conditions would $n(A \cup B) = a+b$?

(d) Under what conditions would $n(A \cup B) = 0$?

(e) Under what conditions would $n(A \cap B) = 0$?

(f) Under what conditions would $n(A \cup B) = a$?

12. There are exactly 1000 students enrolled at XYC, a school of business and economics. There are three types of fraternities available to the students of XYC; namely, honorary, professional, and social. Let the following information be given.

Type of Fraternity	Set of Members	Number of Members
Honary	H	100
Professional	P	400
Social	S	600

The following formula is valid:

$$n(A \cup B \cup C) = n(A) + n(B) + n(C) - n(A \cap B) - n(A \cap C) - n(B \cap C) + n(A \cap B \cap C)$$

where $n(A)$ equals the number of elements of A, etc. Two hundred of the students at XYC do not belong to any fraternity. Additional data are:

$$n(H \cap P) = 41, \quad n(H \cap S) = 22 \quad \text{and} \quad n(P \cap S) = 255$$

(a) Determine $n(H \cup P \cup S)$.

(b) Determine $n(H \cap P \cap S)$.

13. For a certain town with 8000 residents, the following hypothetical data are available relating to possession of Master Charge cards and Visa.

Category	Number Residents
Have Master Charge	3000
Have Bank Americard	2500
Have both Master Charge and Bank Americard	
Have neither type of card	1500

How many residents possess both types of charge cards? *Hint:* $n(A \cup B) = n(A) + n(B) - n(A \cap B)$ is a valid counting formula.

14. Let $\Omega = \{1, 2, 3, \ldots, 10\}$

$A = \{1, 2, 5\}$

$B = \{3, 5, 9\}$

Find

(a) A' | **(c)** $A' \cap B$

(b) B' | **(d)** $A \cup B'$

The set denoted by $(A - B) \cup (B - A)$ or $(A \cup B) - (A \cap B)$ is called the *symmetric difference* of the sets A and B, where $A - B = \{x\colon x \in A, x \notin B\}$.

15. Shade $(A - B) \cup (B - A)$ in the following Venn diagram.

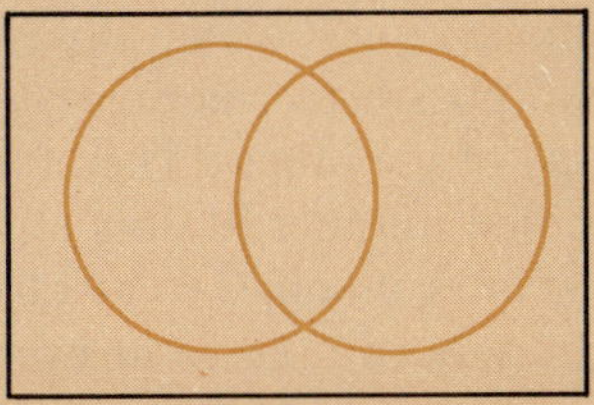

16. Shade $(A \cup B) - (A \cap B)$ in the following Venn diagram.

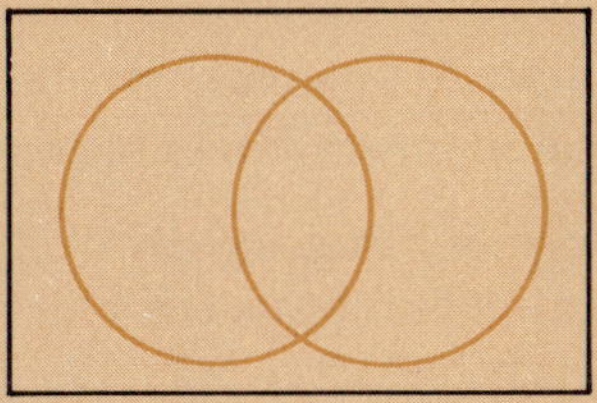

17. Compare the shading in Problems 9 and 10.

18. List the elements of the symmetric difference of $A = \{1, 3, 4, 5, 7, 8, 9\}$ and $B = \{2, 4, 6, 8, 10, 12\}$.

7-3 INTRODUCTION TO PROBABILITY

We begin this section with some questions to start you thinking about probability.

You and your family are planning to go on a picnic. Your favorite weather man comes on and announces that the *chance* of rain during the next few hours is 7 in 10, i.e., the *probability* of rain is 70 percent or 0.7. The picnic area where you are going has absolutely no shelter from rain.

Question 1. Would you cancel the picnic based on this weather forecast?

Answer: This is a subjective question and only you can answer it.

You are shown a cube (die) with dots on the faces. There are six faces numbered 1 through 6. You are told that the die is *fair* but you are not allowed to toss the die even once.

Question 2. What probability would you choose for the *event* that a six would be on top after the die is tossed? What are the chances of a two being on top?

Answer: It seems reasonable (doesn't it?) to say that the probability of a six being on top is 1/6 and the probability of a two is also 1/6.

In answering this question we have reasoned that there are 6 ways that the die could possibly come to rest. Each of these ways seems to be as *likely* as any of the others. Therefore, we argue, there is 1 chance in 6 of any specified number of dots (1, 2, 3, 4, 5 or 6) appearing on the top of the die after it comes to rest from an arbitrary toss.

The thought pattern used to answer this question is summarized by the following *classical definition of probability*. Let A be an event such as $A =$ "a two or a three appears on the die" and let $p(A)$ be the probability of A. Then

$$p(A) = \frac{\text{number of outcomes favorable to } A}{\text{total number of possible outcomes}}$$

all possible outcomes being equally likely.

The classical definition is the oldest definition of probability. It was used extensively by early workers such as Blaise Pascal (1623–1662) in connection with games of chance.

You are enrolled in a Finite Math course and are asked to consider the two possible outcomes, pass or fail. Assuming that you go on to finish the course, you will either pass it with a grade of D or higher, or (perish the thought) you will receive a grade of F.

Question 3. For you, are the two outcomes, $P =$ pass, $F =$ fail, equally likely? For a typical student in the class, are P and F equally likely?

Answer: This is also a subjective question, but in general the assumption of equal liklihood in this situation would be very questionable.

Question 4. Can the classical definition of probability be used when the possible outcomes are not equally likely?

Answer: No, categorically.

Now that you are curious about the probability of passing the Finite Math course, you ask the professor for some data. He grants your request and points out that during the past five years he has taught ten sections of the course to 284 students. Of the 284 students taking the course, 263 passed and 21 of them failed.

Question 5. Can you suggest a probability, p, that a student chosen at random from your class will pass the Finite Math course?

Answer: The data suggest $p = 263/284$ as a reasonable choice for p if no other information about the student is available.

The mode of thought used to answer the last question is based on the *frequency interpretation of probability.* From this point of view probability is defined as *relative frequency in the long run.*

The classical definition and the relative frequency definition of probability are both used in the study and applications of probability. A third definition of probability, introduced in the twentieth century, consists of a set of axioms. The axiomatic approach paved the way for an intense mathematical study of the logical structure of probability theory.

All three of these approaches to probability are discussed in this chapter. The axiomatic definition is introduced next, in Section 7-4.

PROBLEMS

1. List the possible outcomes of the act (experiment) of flipping a single coin.
2. Define the term *a priori.* Use a standard dictionary as your source.
3. List the possible outcomes that could result from flipping two coins simultaneously.
4. Define the term *a posteriori.* Use a standard dictionary.
5. For the experiment of Problem 3 with two fair coins, are the *events* "zero heads", "one head", "two heads" all equally likely? Explain.
6. Give the names and dates of two men other than Blaise Pascal (1623–1662) who were associated with the early study and use of probability. Use an encyclopedia as your source.
7. The Pioneer Collection Agency has been successful in 85 percent of the cases attempted over the past two years. If a pending account is chosen at random from the files of PCA, what is the probability that they will be successful in this case?
8. Consider the experiment of flipping a fair coin. Then $p(H) = p(T) = 1/2$ from the classical definition of probability. What do you think the relative frequency of heads might be if the coin is flipped 20 times? Try it and see.

7-4 SETS AND PROBABILITY

The primary purpose of this section is to discuss *probability* in terms of a set of axioms for a *probability space.* The reader can expect to become familiar with the following technical terms that are basic to modern probability theory. They are *sample space, sample point, event, elementary event*, and *compound event.*

A familiarity with sets, subsets, complements of sets, intersections of sets, and unions of sets is assumed at this point. Material about these topics can be found in Sections 7-1 and 7-2.

The term *sample space* refers to the *set* of all possible outcomes of an experiment. Some examples of an experiment are tossing a coin and observing the outcome (heads or tails), rolling a die, drawing a card from a deck of cards, and observing the lifetime of a battery. Any particular outcome of an experiment, i.e., an *element* of the sample space, is called a *sample point.* Any *subset* of the sample space in question is called an *event.* Next, we pause for an example.

Example 1. Consider the experiment of tossing a coin and rolling a die. Describe the sample space for this experiment and give four examples of an event.

Solution: The sample space, S, consists of twelve ordered pairs:

$$S = \{(H, 1), (H, 2), \ldots, (T, 5), (T, 6)\}.$$

Four examples of events are (there are many more, of course),

$$A = \{(H, 1), (H, 6), (T, 4)\}, \qquad B = \{(T, 3)\}$$

$$C = \varnothing, \qquad D = S$$

This example illustrates the fact that an event may be the empty set, the whole sample space, a single sample point, or a proper subset of the sample space, where the event has more than one sample point. The term *elementary event* is used for an event that contains exactly one element. All nonempty events that are not a single element are called *compound events.*

As was mentioned in the previous section, three ways of defining the notion of probability are the classical definition of probability, the "frequency interpretation of probability," and a set of axioms. All three of these approaches help one to understand the important and useful concept of probability. The axiomatic approach introduced by A. N. Kolmogorov in the twentieth century exhibits the logical structure of probability theory and it opens the way for intense mathematical study of the subject. The older definitions are still used to solve certain types of applied problems. Examples of their use can be found in later sections of this text.

Next, we list the Kolmogorov axioms and state three very important and fundamental consequences of these axioms.

DEFINITION 7.6. *Let S be a sample space. S is called a* probability space *if for every event A in S, there is a number $p(A)$ with the following three properties.*

a. $0 \leqslant p(A)$
b. $p(S) = 1$
c. $p(A \cup B) = p(A) + p(B)$, *whenever A and B are two events with $A \cap B = \varnothing$.*

In this definition, p is called a *probability function* and $p(A)$ is called the probability of the event A. If S is finite (see Problems 12 and 13), no additional axioms are needed to define probability. In case S is infinite, some additional conditions need to be imposed. Since it is not our purpose to proceed that far into the mathematical theory of probability, the additional axioms are not included in this text.

The reader should be aware of the fact that any consequences derivable from axioms *a*, *b*, and *c* are automatically true in the extended system. Hence, the three fundamental results that follow are true for both finite and infinite sample spaces.

Next, we include three important theorems that can be proven from the Kolmogorov axioms.

THEOREM 7.1. *If A is any event in S, then $p(A') = 1 - p(A)$, where A' denotes the complement of the set A relative to the universal set S.*

Proof: *A and A' are disjoint by definition, i.e. $A \cap A' = \varnothing$. Therefore, by axiom c, $p(A \cup A') = p(A) + p(A')$. But $A \cup A' = S$, also by definition of complement. Hence $p(A \cup A') = p(S) = 1$ by axiom b. These results together give $p(A) + p(A') = 1$ or $p(A') = 1 - p(A)$ as was to be shown.*

THEOREM 7.2. $p(\varnothing) = 0$, *where $\varnothing$ denotes the empty set.*

Proof: $S' = \varnothing$. $p(\varnothing) = p(S')$

$p(S') = 1 - p(S)$ *by Theorem 9.1.*

$p(S) = 1$ *by axiom b.*

These results taken together give

$$p(\varnothing) = p(S') = 1 - p(S) = 1 - 1 = 0$$

Therefore, $p(\varnothing) = 0$ as was to be shown.

THEOREM 7.3. *Let A and B be any two events in S, where A and B are not necessarily disjoint as required by axiom c. Then*

$$p(A \cup B) = p(A) + p(B) - p(A \cap B).$$

A proof of this theorem can be found in several of the references listed at the end of this chapter. No proof is given here.

Another result not proven here, and sometimes included as part of axiom a is $p(A) \leqslant 1$ for any event A in S.

The relationships indicated above both by the axioms and by the theorems are presented next in a useful summary.

PROBABILITY RELATIONSHIPS

$0 \leqslant p(A) \leqslant 1$ P-1

$p(\varnothing) = 0$ P-2

$p(S) = 1$ P-3

$p(A') = 1 - p(A)$ P-4

$p(A \cup B) = p(A) + p(B) \quad \text{if } A \cap B = \varnothing$ P-5

$p(A \cup B) = p(A) + p(B) - p(A \cap B)$ P-6

Example 2. Consider the events A and B of Example 1. Use P-5 or P-6 to find an expression for $p(A \cup B)$.

Solution: $A = \{(H, 1), (H, 6), (T, 4)\}$ and $B = \{(T, 3)\}$.
$A \cap B = \varnothing$. Therefore, P-5 applies here.
$p(A \cup B) = p(A) + p(B)$ in this case.

PROBLEMS

1. List the elements of the sample space for the experiment of flipping two coins.
2. List the elements of the sample space for the experiment of flipping three coins.
3. List the elements of the set $K = \{x : x \text{ is a letter in the name Kolmogorov}\}$.
4. Let $S = \{x_1, x_2, x_3, \ldots, x_{10}\}$,

   $$A = \{x_1, x_3, \ldots, x_9\} \quad \text{and} \quad B = \{x_2, x_4, \ldots, x_{10}\}$$

   (a) Describe $A \cap B$ and $A \cup B$.
   (b) List the elements of A'.
   (c) Determine $p(A \cap B)$ and $p(A \cup B)$.
5. Define the term *compound event* as it is used in probability theory.
6. Complete the following table for the number of subsets of $A = \{a, b, c, d, e\}$.

Number of elements in subset	Number of subsets in this class
zero	
one	
two	
three	10
four	
five	
Total	—

7. Repeat Problem 6 with $A = \{a, b, c, d\}$.

8. Based on your experience with Problems 5, 6, and 7, can you suggest a formula for the number of subsets of a set A with k elements.

9. Three light bulb sockets are mounted on a piece of board. Five colored bulbs (red, green, blue, yellow, and organge) are available for use in these three sockets. List and count the possible color "combinations." The order in which the three bulbs are placed in the sockets is not considered important in this problem—order is not regarded.

10. Repeat Problem 9 if there are only two sockets and five bulbs of the same colors.

11. A *one-to-one correspondence* between two sets consists of a "pairing" rule of the elements of one set A with the elements of another set B. Furthermore, each element of A has a unique correspondent from B, and each element of B has a unique correspondent from A. For example, $a \leftrightarrow 1$, $b \leftrightarrow 2$, $c \leftrightarrow 3$ is a one-to-one correspondence between $\{a, b, c\}$ and $\{1, 2, 3\}$. List the other possible one-to-one correspondences between these two sets.

12. A nonempty set A is said to be finite if there exists a one-to-one correspondence (see Problem 11) between A and the set $\{1, 2, \ldots, k\}$ for some k. In this case we say that A contains k elements and write $n(A) = k$.

    Remark: The empty set, $\varnothing$, is also considered finite and $n(\varnothing) = 0$. Determine k for each of the following sets:
    (a) $A = \{a, b, c, \ldots, x, y, z\}$
    (b) $B = \{x\colon x \text{ is a letter in your last name}\}$
    (c) $C = \{x\colon 2x = 10\}$
    (d) $D = \{x\colon (x-8)(x-3) = 0, x \text{ a real number}\}$

13. A definition of finite set is given in Problem 12. A set is said to be *infinite* if it is not finite. Determine which of the following sets are finite and which are infinite. Use your intuition to do this problem.
    (a) $A = \{x\colon 0 \leqslant x \leqslant 1, x \text{ a real number}\}$
    (b) $B = \{x\colon 0 \leqslant x \leqslant 1, x \text{ a rational number}\}$
    (c) $C = \{x\colon x \text{ is root of } x^5 + x^4 + x^3 + x^2 + x = 0\}$

14. Determine if the following sample spaces are finite or infinite. See Problems 12 and 13 for required definitions.
    (a) The set of possible outcomes of flipping 20 coins simultaneously.
    (b) The set of possible distances traveled by a shot put at a future athletic event.
    (c) The sample space for the experiment of a world-wide team of observers looking for grains of sands on the surface of the earth.

15. Let A and B be two nonempty finite sets. The notation $n(A)$ is defined in Problem 12. What can be said about $n(A)$ and $n(B)$ under the following conditions?
    (a) $A = B$
    (b) A is a subset of B
    (c) B is a proper subset of A, i.e., $B \neq A$ and $B \subset A$.
16. Does $n(A) = n(B)$ imply that $A = B$? See Problem 12 for a definition of $n(A)$.
17. (Based on results of Problems 9 and 10) Set up a one-to-one correspondence between the set of color combinations in Problem 9 and the set of color combinations in Problem 10.

7-5 SAMPLE SPACES AND ASSIGNMENT OF PROBABILITIES

The term *sample space* refers to the set of all possible outcomes of an experiment. Each possible outcome is called a *sample point*. The determination and description of the sample space for an experiment is an important first step in the solution of any related probability problem.

The next four examples illustrate the notion of a sample space.

Example 1. Consider the experiment of tossing a coin. What is the sample space for this experiment?

Solution: Considering the outcome of landing on its edge as impossible, we have the set $\{H, T\}$, where H stands for head and T stands for tails, as the sample space.

Example 2. A sample space for the experiment, which consists of setting up a given machine is $\{C, I\}$, where C denotes a correct set-up and I denotes an incorrect set-up.

Example 3. Consider the experiment of choosing an account from an accounts receivable file where the maximum balance is \$1000. List the sample points for this experiment.

Solution:

$$x_1 = 0,\ x_2 = \$0.01, \ldots, x_{101} = \$1.00, \ldots, x_{100{,}000} = \$1000$$

Example 4. The thickness of a tablet is known to be between 0.4 and 0.6 centimeter. Discuss the sample space consisting of the possible thicknesses of this tablet.

Solution: Let x denote a possible thickness. Then the resulting sample space is $S = \{x\colon 0.4 \text{ cm} \leqslant x \leqslant 0.6 \text{ cm}\}$.

As illustrated by the previous examples, a sample space may be *finite* or *infinite*. All finite sample spaces and some infinite sample spaces are said to be *discrete*. An infinite sample space is called *discrete* if its points can be arranged in a sequence; otherwise, it is called *continuous*.

Once the sample space for an experiment has been described, one becomes interested in the probabilities of the various sample points (outcomes).

To illustrate the process of assigning probabilities, we consider the simple experiment of tossing a coin. The sample space for this experiment has two points, $x_1 = H$ and $x_2 = T$. Intuitively, we would assign a probability of $\frac{1}{2}$ to x_1 and a probability of $\frac{1}{2}$ to x_2. Similarly, if the experiment consists of covering a pea with one of three cups, we would assign a probability of $\frac{1}{3}$ to each sample point.

The assignment of the probabilities in the previous experiments is based on the assumption that the coin is fair and the person covering the pea is honest. The assignment is also based on the judgment that all of the outcomes are equally likely. Note that the probabilities were found by dividing 1 by the number of sample points.

Example 5. Consider an experiment with sample space $x_1, x_2, \ldots, x_{12}$. Assign a probability to each sample point assuming equal likelihood.

Solution:

$$p(x_1) = \tfrac{1}{12},\ p(x_2) = \tfrac{1}{12},\ \ldots,\ p(x_{12}) = \tfrac{1}{12}$$

In general, for experiments with n sample points, the assumption of equal likelihood leads to $p(x_i) = 1/n$ for each i between 1 and n.

In the discussion so far we were forced to make some assumptions or judgments about the outcomes.

Furthermore, the probabilities were all assigned before the experiments were run. This method of assigning probabilities *prior to experimentation* is called the *a priori* method (known by reason alone versus known through experience). The *a priori* method of assigning probabilities to events is often used in conjunction with the classical definition of probability.

Next, we discuss an objective or empirical method of assigning probabilities. To illustrate this method, we ask the reader to imagine flipping a coin 10 times. Suppose this is done with the following outcomes: $H, T, H, H, T, H, T, H, T, H$. Recording these results in tabular form, we have the information shown in Table 7-1.

TABLE 7-1

Outcome	Number of Times
H	6
T	4

Sample Point	Frequency
x_1	6
x_2	4

Let $f(x_i)$ denote the number of times x_i occurs. We have

Sample Point	$f(x_i)$
x_1	6
x_2	4
	10

The following ratios can be considered estimates of $p(H)$ and $p(T)$ respectively. They are used to complete Table 7-2.

$$p(x_1) = \frac{f(x_1)}{f(x_1)+f(x_2)} \quad \text{and} \quad p(x_2) = \frac{f(x_2)}{f(x_1)+f(x_2)}$$

$$p(x_1) = \frac{6}{6+4} \quad \text{and} \quad p(x_2) = \frac{4}{6+4}$$

$$p(x_1) = 0.6 \quad \text{and} \quad p(x_2) = 0.4$$

TABLE 7-2

x_i	$f(x_i)$	$p(x_i)$
x_1	6	0.6
x_2	4	0.4

In general,

$$p(x_i) = \frac{f(x_i)}{f(x_1)+f(x_2)+\cdots+f(x_n)}$$

When applying this formula to experiments with a large number of sample points, it is convenient to use the summation notation, $\sum_{i=1}^{n} f(x_i)$, where $\sum_{i=1}^{n} f(x_i) = f(x_1) + f(x_2) + \cdots + f(x_n)$.

The symbol $\sum_{i=1}^{n} f(x_i)$ would be read "summation of $f(x_i)$ from $i = 1$ to $i = n$."

The results $p(x_1) = 0.6$ and $p(x_2) = 0.4$ determined from measurements of relative frequencies are examples of *a posteriori* probabilities. *A posteriori* refers to reasoning from observations or facts, i.e., from experience.

The interpretation of $p(x_1) = 0.6$ and $p(x_2) = 0.4$ involves the frequency interpretation of probability in which case, probability is defined as *relative frequency in the long run.* The frequency definition of probability is expressed by the mathematical expression

$$p(x_i) = \lim_{n \to \infty} \frac{f(x_i)}{n}$$

where $f(x_i)$ denotes the number of times (frequency) that the sample point x_i occurs in n trials or runs of the experiment. This definition suggests the approximation of probabilities by empirical methods, i.e., it motivates the use of *a posteriori* probabilities.

The hypothetical outcome $H, T, H, H, T, H, T, H, T, H$ used in the above discussion, which leads to $p(H) = 0.6$ and $p(T) = 0.4$, may seem inconsistent with the *a priori* probabilities, $p(H) = 0.5$ and $p(T) = 0.5$. This apparent discrepancy does not mean that the coin is not fair, but results from the small number of times that the coin was tossed. If the reader cared to flip a coin 1000 times and record the results, he would expect the resulting *a posteriori* probabilities to be much closer to $p(H) = 0.5$ and $p(T) = 0.5$.

Example 6. Express the sum $f(x_1) + f(x_2) + \cdots + f(x_7)$. Use summation notation.

Solution:

$$f(x_1) + f(x_2) + \cdots + f(x_7) = \sum_{i=1}^{7} f(x_i)$$

PROBLEMS

1. Consider an experiment with sample space $\{x_1, x_2, \ldots, x_6\}$. Assign a probability to each sample point. Assume equal likelihood.
2. Flip a coin 20 times and record your results.
3. Use the data obtained in Problem 2 to calculate the probability of a head or a tail.

4. *Group Problem.* Gather the probability of a head and the probability of a tail as calculated by each student in Problem 2. Add them and divide by the number of students. If you have a class of 20 students, the result obtained is equivalent to flipping a coin 400 times.
5. Write the following in expanded form:
   (a) $\sum_{i=1}^{5} f(x_i)$ **(c)** $(\sum_{i=1}^{4} x_i)^2$
   (b) $\sum_{i=1}^{4} x_i^2$ **(d)** $\sum_{i=1}^{5} x_i f(x_i)$
6. Use summation notation to express the following:
   (a) $x_1^2 f(x_1) + x_2^2 f(x_2) + x_3^2 f(x_3)$
   (b) $[x_1 f(x_1) + x_2 f(x_2) + x_3 f(x_3)]^2$
7. Consider the experiment of tossing a fair coin and a fair die both at the same time. Use the classical definition of probability to determine the probability of each sample point.
8. Are the probabilities of Problem 7 *a priori* or *a posteriori* probabilities?
9. Obtain an *a posteriori* probability that a thumb tack will land on its head (point up) after an arbitrary toss.
10. Explain briefly the meaning of $\lim_{n \to \infty} f(n)$. Use a calculus text (topic: the limit of a sequence) as your source.
11. Determine the following *limits*. Use your intuition to obtain these limits.
    (a) $\lim_{n \to \infty} (4 + (2/n))$
    (b) $\lim_{n \to \infty} (3 + (\frac{1}{2})^n)$
    (c) $\lim_{n \to \infty} (2n/(n+1))$
12. Discuss the meaning of $\lim_{n \to \infty} (f(x_i)/n)$ as it is used in the frequency definition (interpretation) of probability.
13. List the elements of the sample space for the experiment of tossing two coins.
14. List the elements of the sample space for the experiment of tossing three coins. [*Hint:* List the triples (H, T, H), etc.]
15. A class of students received letter grades A, B, C, D, and F on a certain exam. List the elements of the sample space for the experiment of picking a student's name at random from the professor's roll book and recording the student's grade.
16. A certain battery manufacturer produces batteries advertised to last at least 10 hours under continuous use. Consider the experiment of testing and recording the lifetime of every 100th battery from the production line.
    (a) List three possible outcomes, in your judgment, of this experiment.
    (b) Is the sample space of this experiment continuous or discrete?
17. Describe the sample space S for the experiment of tossing a pair of dice.
18. A number x is chosen at random from the interval $[-1, 2]$, where $[-1, 2] = \{x: -1 \leqslant x \leqslant 2;\ x \text{ real}\}$, and the value of $|x|$ is "observed." Describe the sample space for this "experiment."
19. Is the sample space of Problem 18 finite or infinite?
20. Redo Problem 18 if we "observe" x^2 instead of $|x|$.
21. A coin is flipped until a head comes up (if ever), and the number of flips is recorded. Describe the sample space for this experiment.
22. Is the sample space of Problem 21 finite or infinite?
23. Can you see any difference between the sample space of Problem 18 and the sample space of Problem 21? Please explain.

7-6 EVENTS

This section is designed to acquaint the reader with the notion of *event* as it is used in probability theory and practice. Any subset of a sample space is called an event. The sample space itself is an event as is the empty set. For a finite sample space with n sample points, there are exactly 2^n distinct events.

Example 1. Let $S = \{x_1, x_2, x_3, x_4, x_5\}$. Define three events.

Solution:

$$A = \{x_2\}$$

$$B = \{x_1, x_3\}$$

$$C = \{x_2, x_4, x_5\}$$

Next we will define the terms *exhaustive* and *mutually exclusive*.

DEFINITION 7.7. *Two events A and B are said to be exhaustive if $A \cup B = S$, where S is the sample space.*

DEFINITION 7.8. *Two events A and B are said to be mutually exclusive if $A \cap B = \varnothing$, where $\varnothing$ denotes the empty set.*

Example 2. Describe the events A, B, and C of Example 1 using the terms exhaustive and mutually exclusive.

Solution:

$$A = \{x_2\}, \qquad B = \{x_1, x_3\}, \qquad \text{and} \qquad C = \{x_2, x_4, x_5\}$$

Events B and C are exhaustive since

$$B \cup C = \{x_1, x_3\} \cup \{x_2, x_4, x_5\}$$
$$= \{x_1, x_2, x_3, x_4, x_5\}$$
$$\therefore\ B \cup C = S$$

Events A and B are mutually exclusive since

$$A \cap B = \{x_2\} \cap \{x_1, x_3\}$$
$$\therefore\ A \cap B = \varnothing$$

The probability of a sample point is a number between zero and one, inclusive.

DEFINITION 7.9. *Let A be an event. Then the probability of A denoted by $p(A)$ is defined to be the sum of the probabilities of all of the sample points in A.*

Example 3. Consider the experience of drawing a single card from a standard deck of playing cards. Let K be the event of drawing a king. Find $p(K)$.

Solution: $K = \{K_h, K_s, K_c, K_d\}$. The probability of each sample point in K is $\frac{1}{52}$.
Therefore,

$$p(K) = p(K_h) + p(K_s) + p(K_c) + p(K_d)$$
$$= \tfrac{1}{52} + \tfrac{1}{52} + \tfrac{1}{52} + \tfrac{1}{52}$$
$$p(K) = \tfrac{4}{52} = \tfrac{1}{13}$$

In general, if $A = \{a_1, a_2, \ldots, a_n\}$is an event, then $p(A) = p(a_1) + p(a_2) + \cdots + p(a_n)$ by Definition 7.9. Next, we ask the question: Can probabilities be added if the events are not just sample points? This question is answered below.

Let A and B be two *mutually exclusive* events. Then $p(A \cup B) = p(A) + p(B)$. See Probability Relationship P-5, Section 7-4.

Example 4. Let Q be the event of drawing a queen and K be the event of drawing a king from a standard deck of playing cards. Find $p(K \cup Q)$.

Solution:

$$K \cap Q = \varnothing$$

Therefore, K and Q are mutually exclusive.

$$p(K \cup Q) = p(K) + p(Q)$$
$$p(K \cup Q) = \tfrac{4}{52} + \tfrac{4}{52}$$
$$p(K \cup Q) = \tfrac{2}{13}$$

The probability of drawing a king or a queen is $\frac{2}{13}$.

The probability relationships P-5 and P-6 of Section 7-4 are illustrated in the next two examples. These relationships are repeated below for your convenience.

$$p(A \cup B) = p(A) + p(B)$$

if A and B are mutually exclusive. P-5

$$p(A \cup B) = p(A) + p(B) - p(A \cap B) \qquad \text{P-6}$$

in general.

Example 5. Illustrate P-6 with a Venn diagram using a sample space with 10 sample points.

Solution:

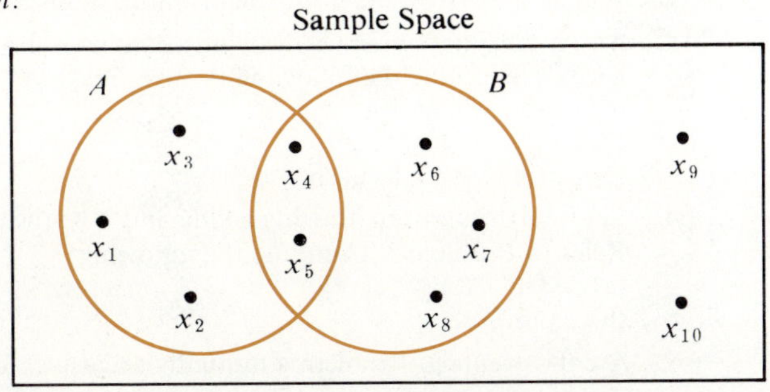

FIGURE 7-4

$$p(A) = p(x_1) + p(x_2) + p(x_3) + p(x_4) + p(x_5)$$

$$p(B) = p(x_4) + p(x_5) + p(x_6) + p(x_7) + p(x_8)$$

$$p(A \cap B) = p(x_4) + p(x_5)$$

$$p(A \cup B) = p(A) + p(B) - p(A \cap B)$$

$$p(A \cup B) = [p(x_1)+p(x_2)+p(x_3)+p(x_4)+p(x_5)] + [p(x_6)+p(x_7)+p(x_8)]$$

Remark. The point of Example 5 is that the subtraction of $p(A \cap B)$ from $p(A)+p(B)$ in P-6 compensates for "double counting."

Example 6. Consider the experiment of drawing numbers between 1 and 6 from a hat. Let $A = \{3, 4, 6\}$ and $B = \{1, 4, 5, 6\}$. Compute $p(A \cup B)$.

Solution: The sample space has 6 elements. If we assume equal likelihood, we have $p(1) = \frac{1}{6}$, $p(3) = \frac{1}{6}$, etc.

$$p(A) = \tfrac{3}{6}, \qquad p(B) = \tfrac{4}{6}, \qquad p(A \cap B) = p\{4, 6\} = \tfrac{2}{6}$$

Note that $p(A)+p(B) = \frac{7}{6}$, which clearly could not be $p(A \cup B)$, since the probability of an event never exceeds one. If we apply the correct formula, we have

$$p(A \cup B) = p(A) + p(B) - p(A \cap B)$$

$$p(A \cup B) = \tfrac{3}{6} + \tfrac{4}{6} - \tfrac{2}{6}$$

$$p(A \cup B) = \tfrac{5}{6}$$

PROBLEMS

1. Consider an experiment with sample space $S = \{x_1, x_2, \ldots, x_5\}$. Which of the following pairs of events are mutually exclusive?

(a) $A = \{x_1, x_3, x_5\}$
(b) $B = \{x_2, x_4\}$
(c) $D = \{x_2, x_4, x_5\}$
(d) $E = \{x_2, x_5\}$

2. Which pairs of events in Problem 1 are exhaustive?
3. An experiment consists of drawing a card from a standard deck of playing cards. Determine the following probabilities:
   (a) p(drawing a 10)
   (b) p(drawing a king)
   (c) p(drawing a face card)
4. Let K be the event of drawing a king and A be the event of drawing an ace. Refer to Problem 3. Compute the following:
   (a) $p(A \cap K)$
   (b) $p(A \cup K)$
5. Are the events in Problem 4 mutually exclusive? Exhaustive?
6. A number between 1 and 10, inclusive, is drawn from a hat. Events A, B, C, and D are defined as follows: $A = \{1, 3, 5, 7, 9\}$, $B = \{2, 4, 6, 8, 10\}$, $C = \varnothing$, and $D = \{3, 4, 5\}$. Assume equal likelihood. Compute the probability of the following events.
   (a) $p(A)$
   (b) $p(A \cap B)$
   (c) $p(C)$
   (d) $p(A \cup B)$
   (e) $p(A \cap D)$
   (f) $p(A \cup D)$
   (g) $p(C \cup A)$
   (h) $p(C \cup D)$
7. Compute the following probabilities based on events from Problem 6.
   (a) $p(A')$
   (b) $p[(A \cap B)']$
   (c) $p[(A \cup B)']$
   (d) $p(D')$
   (e) $p[(A \cap D)']$
   (f) $p[(A \cup D)']$

   Hint: Recall P-4 of Section 7-2.

7-7 AN APPLICATION

A final exam was given to a class of 33 students and the grades were as follows. From an alphabetical listing of students, the grades are

86	70	67	73	69	80	78	79	73	67	94
73	73	70	78	70	79	77	78	82	77	95
93	65	63	69	77	65	73	70	69	77	67

Arranging the grades in ascending order, we form the following table.

TABLE 7-3

x_i	$f(x_i)$	$p(x_i)$	x_i	$f(x_i)$	$p(x_i)$
63	1	$\frac{1}{33}$	79	2	$\frac{2}{33}$
65	2	$\frac{2}{33}$	80	1	$\frac{1}{33}$
67	3	$\frac{3}{33}$	82	1	$\frac{1}{33}$
69	3	$\frac{3}{33}$	86	1	$\frac{1}{33}$
70	4	$\frac{4}{33}$	93	1	$\frac{1}{33}$
73	5	$\frac{5}{33}$	94	1	$\frac{1}{33}$
77	4	$\frac{4}{33}$	95	1	$\frac{1}{33}$
78	3	$\frac{3}{33}$		33	1

To assign grades *A*, *B*, *C*, and *D*, we group the data into 4 mutually exclusive events. This results in the following table.

TABLE 7-4

Grade	Event	Mark	$f(E_i)$	$p(E_i)$
A	E_1	90–100	3	$\frac{3}{33}$
B	E_2	80–89	3	$\frac{3}{33}$
C	E_3	70–79	18	$\frac{18}{33}$
D	E_4	60–69	9	$\frac{9}{33}$
			33	1

A sample calculation for $p(E_4)$ follows:

$$E_4 = \{63, 65, 67, 69\}$$

$$p(E_4) = p(63) + p(65) + p(67) + p(69)$$

$$p(E_4) = \frac{1}{33} + \frac{2}{33} + \frac{3}{33} + \frac{3}{33}$$

$$p(E_4) = \frac{9}{33} \qquad \textit{as shown in the table above}$$

It is not by coincidence that this number can also be obtained by dividing $f(E_4)$ by 33. In fact, the following formula is valid for *a posteriori* probabilities:

$$p(E_i) = \frac{f(E_i)}{f(E_1)+f(E_2)+\cdots+f(E_n)}$$

Or, using summation notation, we get

$$p(E_i) = \frac{f(E_i)}{\sum_{i=1}^{n} f(E_i)}$$

Sample calculation for $p(E_3)$ using the above formula is

$$p(E_3) = \frac{f(E_3)}{f(E_1)+f(E_2)+f(E_3)+f(E_4)}$$

$$p(E_3) = \frac{18}{3+3+18+9}$$

$$p(E_3) = \frac{18}{33}$$

PROBLEMS

1. A sample of 20 accounts was taken from an accounts receivable file. The balances of these 20 accounts are as follows:

\$191	\$ 61	\$ 22	\$ 35	\$186
\$ 90	\$ 16	\$115	\$ 74	\$ 83
\$146	\$159	\$ 67	\$125	\$135
\$ 48	\$104	\$163	\$176	\$ 79

Arrange the balances in ascending order.

2. Classify the sample points in Problem 1 into four events and record the frequency of each event (i.e., 0–49, 50–99, 100–149, 150–199).
3. Calculate the probability of the events in Problem 2.
4. The height, rounded to the nearest inch of 15 students in a class, is as follows:

75	68	71	67	62
69	64	61	73	67
65	73	67	74	71

   Arrange the heights in ascending order and present them in a table that includes the $f(x_i)$.

5. Calculate the probability of each of the sample points in Problem 4.
6. Classify the sample points of Problem 4 into 3 events and record the frequency of each event (i.e., 61–65, 66–70, 71–75).
7. Calculate the probability of each event in Problem 6 two different ways.

7-8 COUNTING FORMULAS

Suppose three students, Jim, Jack, and Mary, are instructed to be seated in three chairs. The chairs are in a row. We are asked to determine the total number of possible seating arrangements. Since there are only three people involved, we can simply list the possible ways as follows:

Jim, Jack, Mary
Jim, Mary, Jack
Jack, Jim, Mary
Jack, Mary, Jim
Mary, Jim, Jack
Mary, Jack, Jim

We see that there are six possible seating arrangements in this case. It will be shown later in this section that if there are five people instead of three, then the total number of possible ways of seating them is 120.

Example 1. Refer to the seating arrangement problem of Jim, Jack, and Mary and randomly select one of the possible seating arrangements. Determine the probability of Jim getting to sit next to Mary.

Solution: Of the 6 possible seating arrangements, Jim will be seated next to Mary in 4 of them. Therefore, the desired probability is $\frac{2}{3}$.

The four counting formulas derived in this section are based on the following counting principle. *Suppose an act or procedure can be performed in N_1 different ways and a second act or procedure can be performed in N_2 different ways. Then the total number of ways that the two procedures can be performed consecutively is given by the product $N_1 N_2$.*

Imagine a set of n distinguishable objects (real or mathematical) arranged in some definite order. The given arrangement or any possible arrangement of the n objects is called a *permutation* of the objects involved. It follows from the counting principle, stated above, that the total number of possible permutations or arrangements of n objects is given by $n!$, where $n! = n(n-1)\cdots(2)1$. The symbol $n!$ is read n factorial. To see this we note that the first position can be filled in n ways, the second position can be filled in $n-1$ ways, and so on. The next to the last position can be filled in 2 ways and the last position can be filled with the one remaining object. Therefore, it follows from a repeated application of the counting principle that the total number of ways that these n procedures can be performed sequentially is given by $n!$

Example 2. How many ways can five people be seated in five chairs? The chairs are in a row.

Solution: Five objects of any kind can be arranged in five positions in $5! = 120$ ways.

When dealing with permutations, we have n objects to be arranged in the same number of positions. Suppose now that there may be fewer available positions than there are objects, say n objects and k positions with $k \leqslant n$. It follows from the counting principle that the number of possible arrangements of n objects in k positions, where order is regarded, is given by $n(n-1)\cdots(n-k+1)$. This formula can be written in factorial form as $n!/(n-k)!$. If we introduce the notation $p(n,k)$ to indicate the number of possible arrangements of n objects taken k at a time, we have

$$p(n, k) = \frac{n!}{(n-k)!} \tag{1}$$

We know from above that

$$p(n, n) = n! \tag{2}$$

Furthermore, if we adopt the usual definition $0! = 1$, we see that Equation (2) is a special case of Equation (1).

Frequently, it is of little interest to determine all possible arrangements of n objects taken k at a time. The matter of interest is the number of ways (disregarding order) that k objects can be chosen from a set of n distinguishable objects ($k \leq n$). Each possible choice is called a *combination*. We use the notation $c(n, k)$ to denote the number of combinations, or k element subsets, that can be formed from n objects. The formula for $c(n, k)$, the number of combinations of n objects taken k at a time, is as follows:

$$c(n, k) = \frac{n!}{k!(n-k)!} \tag{3}$$

Proof of (3): It follows from the counting principle and the definitions of $p(n, k)$, $c(n, k)$, and $p(k, k)$ that

$$p(n, k) = c(n, k)\,p(k, k)$$

Therefore, $n!/(n-k)! = c(n, k)k!$ and the result follows.

The notation $c(n, k) = \binom{n}{k}$ is also common.

The preceding formulas can be used for counting the possible combinations or permutations of distinguishable objects. These important *combinatoric* formulas are illustrated further in the next examples.

Example 3. Determine the number of permutations of the letters of the word SAMPLE.

Solution: The word SAMPLE has 6 distinct letters. Therefore, there are $6! = 720$ permutations of the 6 letters involved.

Example 4. Determine the number of 3-element subsets (combinations taken 3 at a time) of letters that can be chosen from the letters of the word SAMPLE.

Solution: We seek the number of combinations of 6 objects taken 3 at a time. In this case $n = 6$ and $k = 3$. Hence, by (3), there are $c(6, 3) = 60$ such combinations.

Example 5. Consider an urn containing 12 balls numbered 1 to 12 so that they are distinguishable. You are asked to *sample without replacement* from this urn. Furthermore, you are to extract an *ordered sample* of size 4 from the urn. The question is, how many ordered samples of size 4 are possible in this case?

Solution: Sampling without replacement means that the ball is not returned to the urn (population) before the next ball is chosen. In an ordered sample, the order in which the balls occur is considered important (i.e., regarded).

This problem can be done using (1) or it can be done directly from the counting principle as follows.

The first ball can be chosen in 12 ways, the second in 11 ways, the third in 10 ways, and the fourth in 9 ways. Therefore, by repeated application of the counting principle there are $12 \cdot 11 \cdot 10 \cdot 9 = 11{,}880$ possible ordered samples.

To use (1), let $n = 12$ and $k = 4$. This gives $p(12, 4) = 12!/8! = 11{,}880$ as above.

Next, we consider the case of arrangements of objects some of which are indistinguishable from each other, i.e., permutations of objects containing repetitions.

Example 6. Determine the number of distinct (distinguishable) arrangements of the letters of the word NEEDLE.

Solution: In contrast to the word SAMPLE which has 6 distinct letters, the word NEEDLE has only 4 distinct letters. This means that many of the $6! = 720$ arrangements indicated by Equation (2) would be indistinguishable from each other. Hence, the correct answer will be something less than 720. In this case, using (4) below, we must divide 720 by $6 = 3!$. This gives 120 distinct arrangements.

Consider the case of n objects of which n_1 are alike (indistinguishable), n_2 are alike, ..., and n_r are alike, where $n_1 + n_2 + \cdots + n_r = n$. Then the number of distinct (distinguishable) permutations of these n objects is given by

$$p(n, n; n_1, n_2, \ldots, n_r) = \frac{n!}{n_1!, n_2!, \ldots, n_r!} \tag{4}$$

Proof of (4): Let x denote the number of distinct arrangements. Then $n! = x(n_1!, n_2!, \ldots, n_r!)$. This follows from the counting principle and (2). Solving for x gives $x = n!/(n_1!, n_2!, \ldots, n_r!)$. Hence, the desired result is established.

Remark: In the special case of all n objects being alike, we have $r = 1$ and $n_1 = n$. This gives $p(n, n; n_1) = n!/n! = 1$, which agrees with common sense: All $n!$ permutations would be indistinguishable giving only one distinct arrangement.

Example 7. You are given a Christmas tree light set with 18 sockets along with 4 white bulbs, 6 green bulbs, 5 red bulbs, and 3 blue bulbs. Your job is to count the number of possible distinct linear patterns that can be made from the available bulbs. Hopefully, a counting formula can be used.

Solution: Use (4) with $n = 18$, $n_1 = 4$, $n_2 = 6$, $n_3 = 5$, and $n_4 = 3$. This gives

$$p(18, 18; 4, 6, 5, 3) = \frac{18!}{4!6!5!3!}$$

$$= 514{,}594{,}080$$

as the count! Does anyone care to try this experimentally?

BASIC COMBINATORIC FORMULAS

Permutations of n objects taken k at a time

$$p(n, k) = \frac{n!}{(n-k)!} \qquad \textit{(order is regarded)} \qquad \text{C-1}$$

Permutations of n objects taken all at a time

$$p(n, n) = n! \qquad \textit{(order is regarded)} \qquad \text{C-2}$$

Combinations of n objects taken k at a time

$$c(n, k) = \frac{n!}{k!(n-k)!} \qquad \textit{(order is not regarded)} \qquad \text{C-3}$$

Alternative notation: $\binom{n}{k}$ also called a binomial coefficient.

Permutations of n objects with repetitions present and taken all at a time.

$$p(n, n; n_1, n_2, \ldots, n_r) = \frac{n!}{n_1!, n_2!, \ldots, n_r!} \qquad \textit{(order is regarded)}$$

C-4

Alternative notation: $\binom{n}{n_1, n_2, \ldots, n_{r-1}}$ also called a multinomail coefficient.

PROBLEMS

1. Compute the following:

(a) $6!$ **(c)** $7!$

(b) $\frac{}{3!\,2!}$ **(d)** $\frac{}{2!\,5!}$

2. Use logs to compute $5!$

3. Compute the following:

(a) $\binom{5}{2} = c(5, 2)$ **(c)** $\binom{4}{1}$

(b) $\binom{6}{0} = c(6, 0)$ **(d)** $\binom{4}{0} + \binom{4}{1} + \binom{4}{2} + \binom{4}{3} + \binom{4}{4}$

4. How many 3-element subsets can be formed from a set containing 8 elements?

5. If a part has to be worked on in five different departments and can be worked on in any order, how many different ways are there for the part to be produced?

6. How many committees of four can be formed from a class of 25 students?

7. A committee of four is to be selected in a completely random fashion from the class of 20 students. The professor has the students listed in his roll book in alphabetical order. What is the probability that the first four students listed in the roll book will be chosen for the committee?

8. How many slates of officers (p, v, s, t) can be formed from a class of 15 students? Assume that no student may hold more than one office.

9. A 12-member club elects its officers (p, v, s, t) by a completely random process and no club member can hold more than one office at a time. Determine the probability that a given slate of officers will be elected.

10. Solve Example 5 if the sampling is to be with replacement.

11. Solve Example 5 if just samples instead of ordered samples are required.

12. Compute the following:

(a) $p(9, 9; 2, 3, 4) = \binom{9}{2, 3}$ **(b)** $p(14, 14; 4, 3, 2, 5) = \binom{14}{4, 3, 2}$

13. Determine the number of distinct patterns (signals) that can be formed with 2 red, 2 white, and 3 blue flags. A signal is formed by placing 7 flags in a row one above the other.

14. In how many ways can a 5-card hand be drawn from a deck of 52 cards?

15. In how many ways can a 13-card bridge hand be drawn from a deck of 52 cards?

16. A deal of cards in bridge involves dividing 52 cards into four subsets of 13 cards each. It can be shown that the total number of deals is given by $p(52, 52; 13, 13, 13, 13)$. Use this fact to compute the number of possible deals in bridge. Compare the answer with the answer for Problem 15.

17. Determine the number of distinct arrangements of the letters of the name KOLMOGOROV.

7-9 APPLICATIONS OF COUNTING FORMULAS

This section employs the classical definition of probability. The reader will recall that according to this definition, the probability, p, of an event, A, is taken to be

$$p(A) = \frac{\text{number of outcomes favorable to } A}{\text{number of possible outcomes}}$$

all outcomes being equally likely. Sometimes it is convenient to use the lower case letter of the given event to denote the number of elements favorable to that event and n to denote the number of possible outcomes: $p(A) = a/n$, $p(B) = b/n$, etc.

An appropriate counting technique is used to determine a and n in each example.

Example 1. A department consists of 14 members in addition to the department head. Five members are women and 9 are men. It has been decided that a 3-person committee will be elected from the 14 members to represent them on all matters of concern to the department. It is further required that the committee consist of one woman and two men. Determine the probability that a committee chosen at random will satisfy this condition.

Solution: Let W denote the event, "A three-person committee consisting of 1 woman and 2 men is chosen."

The number of possible outcomes is given by

$$\binom{14}{3} = 364$$

The number of outcomes favorable to W is given by

$$\binom{5}{1}\binom{9}{2} = 5(36) = 180$$

Therefore, the desired probability is

$$p = \frac{180}{364} = \frac{45}{91} = 0.4945$$

The next two examples illustrate the use of combination formulas in connection with urn problems. Assume sampling without replacement and that the balls are well mixed after each draw. The sample size is 4 in each case and we are not concerned about the order in which the balls appear in the sample.

Example 2. An urn contains 8 red balls, 7 green balls and 5 blue balls. Compute the probability, $p(R)$, of drawing 4 red balls, i.e. all red balls.

Solution: The sample space contains $\binom{20}{4}$ sample points, i.e. there are $\binom{20}{4}$ possible outcomes. The number of outcomes favorable to R is given by $\binom{8}{4}$. Therefore,

$$p(R) = \frac{\binom{8}{4}}{\binom{20}{4}} = \frac{70}{4845} = 0.0144$$

Example 3. An urn contains 8 red balls, 7 green balls, and 5 blue balls as in Example 2. Compute the probability of obtaining a sample of size 4 with 2 red balls and 2 non-red balls. The sampling procedure is as described prior to Example 2.

Solution: As in Example 2, there are $\binom{20}{4}$ possible outcomes.

Define the event $T =$ "the sample contains exactly two red balls." Some examples of outcomes that are favorable to T are (R, R, G, B), (G, G, R, R), and (G, R, B, R). The number of outcomes favorable to T is given by $\binom{8}{2}\binom{12}{2}$. This follows from the fact that we must choose 2 red balls in the sample from 8 red balls in the population and 2 non-red balls (green or blue) in the sample from 12 non-red balls in the population.

Finally

$$p(T) = \frac{\binom{8}{2}\binom{12}{2}}{\binom{20}{4}}$$

Hence,

$$p(T) = \frac{(28)(66)}{4845} = \frac{1848}{4845} = 0.3814$$

Next, we consider the case of n independent experiments or *trials* each of which results in two possible outcomes. Some typical names that are used for the possible outcomes are: pass, fail; success, failure; defective, nondefective; zero, one; and heads, tails. If the probability of success, say, is assumed constant from trial to trial, then the binomial distribution of Section 7-2 is applicable. The above mentioned process is illustrated by tossing a coin n times and observing the number of heads and tails.

It is our purpose here to illustrate the use of counting techniques as they relate to an analysis of n independent trials with two possible outcomes.

Example 4. Consider the case of tossing a coin 5 times (or equivalently tossing 5 coins simultaneously) and observing the number of heads and tails. Let the six events A, B, C, D, E, and F be defined as follows. A-0 heads, B-1 head, C-2 heads, D-3 heads, E-4 heads, and F-5 heads. Our job is to enumerate (count) the sample points in each of these events.

Solution: A and F consist of one sample point each, (T, T, T, T, T) and (H, H, H, H, H), respectively. The five outcomes favorable to B are (H, T, T, T, T), (T, H, T, T, T), (T, T, H, T, T), (T, T, T, H, T), and (T, T, T, T, H). Similarly, there are five outcomes favorable to E.

To count the outcomes favorable to C and D, we will use the counting formula for the number of permutations of 5 objects containing repetitions. Consider the outcome (H, H, T, T, T) which is favorable to C. The number of all outcomes favorable to C is given by $p(5, 5; 2, 3) = 5!/(2!3!)$.

This expression is also equal to $\binom{5}{2}$.

For D-3 heads, we need $p(5, 5; 3, 2) = 5!/(3!2!)$, which also equals $\binom{5}{3}$.

Denoting the number of outcomes favorable to A by a, the number of outcomes favorable to B by b, etc., we have the following results.

$$a = 1 = \binom{5}{0} \qquad \text{A-0 heads}$$

$$b = 5 = p(5, 5; 1, 4) = \binom{5}{1} \qquad \text{B-1 head}$$

$$c = 10 = p(5, 5; 2, 3) = \binom{5}{2} \qquad \text{C-2 heads}$$

$$d = 10 = p(5, 5; 3, 2) = \binom{5}{3} \qquad \text{D-3 heads}$$

$$e = 5 = p(5, 5; 4, 1) = \binom{5}{4} \qquad \text{E-4 heads}$$

$$f = 1 = \binom{5}{5} \qquad \text{F-5 heads}$$

The total number of sample points is given by $2^5 = 32$ or equivalently, $1+5+10+10+5+1 = 32$.

PROBLEMS

1. Do Example 1 again if there are 6 women and 10 men in the department.
2. State the classical definition of probability.
3. An urn contains 10 red, 12 white, and 8 blue balls. A sample of size 6 is drawn without replacement. Determine the probability of obtaining a sample with 2 balls of each color.
4. Compute the probability of obtaining a sample with 4 red balls and 2 white balls from the urn of Example 3.

5. How many 5-card hands can be formed from a standard deck of 52 cards?
6. Compute the probability of drawing a 5-card hand with 3 hearts from a well-shuffled deck of 52 cards.
7. Consider the experiment of tossing 4 coins. Let the events A, B, C, D, and E be defined as follows: A-0 heads, B-1 head, C-2 heads, D-3 heads, and E-4 heads. Compute the probability of each of these events. *Hint:* It can be shown that the probability of each sample point in this sample space is 1/16, assuming fair coins.
8. Show that $p(n, n; k, n-k) = \binom{n}{k}$. See Section 7-7 for definition of symbols.
9. Consider an urn containing 3 red balls and 7 green balls. A sample of size 6 is drawn with replacement so that the probability of drawing a red ball on any trial is 0.3. Determine the probability of obtaining a sample with 4 red balls. *Hint:* The probability of any sample point with 4 red balls can be shown to be 0.003969.
10. How many different samples of size 5 can be chosen from a lot of 104?
11. Given that we have a file of accounts receivable containing 5 accounts, we want to choose a sample of 3 accounts to estimate the average balance of the file. (Note that, in practice, when working with a small population one would not take a sample.) The balances in the five accounts are as follows:

Account	Balance
001	$100.00
002	160.00
003	120.00
004	150.00
005	130.00

(a) How many samples of size 3 can be drawn from the file? [*Note:* Order is not important.]
(b) List the possible samples that can be chosen.
(c) Compute the arithmetic average for each sample in part b.

Problems 12 and 13 are based on the following definition from statistics.

Mean of sampling distribution is denoted by $\mu_{\bar{x}}$. Therefore,

$$\mu_{\bar{x}} = \frac{\sum_{i=1}^{c(n,k)} (\text{sample mean})_i}{c(n, k)}$$

where $n =$ the population size
$k =$ the sample size

12. Calculate $\mu_{\bar{x}}$ for Problem 11.
13. Find the upper index of the summation symbol in $\mu_{\bar{x}}$ for a population size of 12 and a sample size of 4.
14. A question of interest to many managers is—How many subordinates can a supervisor handle effectively in a given situation? One of the variables to be considered is the number of interpersonal relationships inherent in a given situation. Graicunas's law states that there are three kinds of relationships:

A. Direct single relationships, i.e., supervisor with subordinate $= n$

B. Cross relationships, i.e., subordinate with subordinate $= n(n-1)$

C. Direct group relationships, i.e., supervisor with subordinate, modified by presence of another subordinate $= n\left(\frac{2^n}{2} - 1\right)$

where n = number of persons being supervised. The total number of relationships is

$$f = A + B + C \qquad \text{or} \qquad f = n\left(\frac{2^n}{2} + n - 1\right)$$

Calculate f for $n = 1, 2, 3, 4, 5, 6, 7, 8, 9, 10$.

15. Show that

$$n + n(n-1) + n\left(\frac{2^n}{2} - 1\right) = n\left(\frac{2^n}{2} + n - 1\right)$$

Refer to Problem 14.

16. The Braille alphabet using the six-dot cell was developed by Louis Braille in about 1829. The six dots of the Braille cell are arranged and numbered thus:

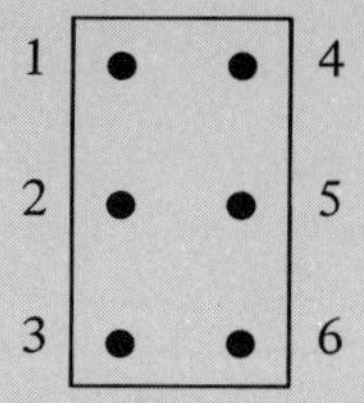

The arrangement can be a single dot in any position, two dots in any position, etc. To represent k, the dots numbered 1 and 3 appear.

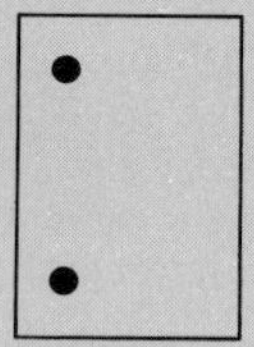

Note that the absence of all dots is not a valid character. In how many possible combinations can we arrange these dots? In other words, how many characters are there in this touch-alphabet?

17. Telephones are advertised as being available in 60 combinations of colors and styles. If all styles are available in one of five colors, how many styles are there?

REFERENCES

Hamburg, Morris. *Statistical Analysis for Decision Making*. New York: Harcourt Brace and World, Inc., 1970.

Harris, Bernard. *Theory of Probability*. Reading, Massachusetts: Addison-Wesley, 1966.

Hays, William L., and Winkler, Robert L. *Statistics: Probability, Inference, and Decision*. New York: Holt, Rinehart and Winston, Inc., 1971.

Lipschutz, Seymour. *Theory and Problems of Finite Mathematics*. New York: Schaum's Outline Series, McGraw-Hill Book Co., Inc., 1966.

Lipschutz, Seymour. *Theory and Problems of Probability*. New York: Schaum's Outline Series, McGraw Hill Book Co., Inc., 1968.

Mood, Alexander M., and Graybill, Franklin A. *Introduction to the Theory of Statistics*. New York: McGraw Hill Book Co., Inc., 1963.

Simone, Albert J. *Probability: An Introduction with Applications*. Boston: Allyn and Bacon, Inc., 1967.

Springer, Clifford H., Herlihy, R. E., Mall, Robert T., and Beggs, R. I. *Statistical Inference*. Homewood, Illinois: Richard D. Irwin, Inc., 1965.

Wani, J. K. *Probability and Statistical Inference*. New York: Appleton-Century-Crofts, 1971.

EIGHT

Applications of probability

8-1 UNIVARIATE vs. BIVARIATE DATA

In each example of the previous chapter there was only one characteristic of interest. A case in point is the grade problem, where the grade is the characteristic or variable of interest.

If we repeat Table 7-4 from Section 7-7, we have

TABLE 8-1

Grade	Event	$f(E_i)$	$p(E_i)$
A	E_1	3	$\frac{3}{33}$
B	E_2	3	$\frac{3}{33}$
C	E_3	18	$\frac{18}{33}$
D	E_4	9	$\frac{9}{33}$

We can read directly from this table $p(A) = \frac{3}{33}$, $p(B) = \frac{3}{33}$, etc. These probabilities are called marginal probabilities.

These students could also be classified according to I.Q. Such a classification is given in Table 8-2.

TABLE 8-2

Event	I.Q.	$f(F_i)$	$p(F_i)$
F_1	low $= L$	11	$\frac{1}{3}$
F_2	medium $= M$	11	$\frac{1}{3}$
F_3	high $= H$	11	$\frac{1}{3}$

This means $p(L) = \frac{1}{3}$, $p(M) = \frac{1}{3}$, and $p(H) = \frac{1}{3}$.

Although information classified according to one variable is useful, it is more significant to classify the data according to two or more variables. A set of data is said to be *univariate* when classified according to one variable and *bivariate* when classified according to two variables. In the next section the students are classified according to two variables: grades and I.Q. Once this has been done, we can determine, for example, the probability of a student having a medium I.Q. and a grade of *B*.

A typical format for displaying bivariate data is shown in Table 8-3.

TABLE 8-3

Grade \ I.Q.	F_1	F_2	F_3	Total
E_1				3
E_2				3
E_3				18
E_4				9
Total	11	11	11	33

To fill in the remaining entries in Table 8-3 we must go through the list of students and take note of both characteristics of interest at the same time. This can be done by forming a separate tally sheet and then transferring the totals from the tally sheet to the table. The tally sheet must have the same number of lines as there are blank entries in the table.

TABLE 8-4 TALLY SHEET

I.Q.	Grade	Tally Marks	Total
L	*A*	/	1
L	*B*	/	1
L	*C*	~~////~~ /	6
L	*D*	///	3
M	*A*	/	1
M	*B*	/	1
M	*C*	~~////~~ /	6
M	*D*	///	3
H	*A*	/	1
H	*B*	/	1
H	*C*	~~////~~ /	6
H	*D*	///	3

Using the information from the tally sheet, we have the following bivariate frequency table.

TABLE 8-5

Grade \ I.Q.	*L*	*M*	*H*	Total
A	1	1	1	3
B	1	1	1	3
C	6	6	6	18
D	3	3	3	9
Total	11	11	11	33

PROBLEMS

1. Given the following table of employees at the XYZ Company:

TABLE 8-6

Employees	Earnings	$f(E_i)$
E_1	1.50–2.49	15
E_2	2.50–3.49	20
E_3	3.50–4.49	30
E_4	4.50–5.49	20
E_5	5.50–6.49	15
		100

(a) Calculate all marginal probabilities.
(b) What is the probability that an employee chosen at random earns between \$2.50 and \$3.49?
(c) Calculate $p(E_1 \cup E_2)$.
(d) Explain in words the probability calculated in (c).

2. If the same group of 100 employees that work for the company in Problem 1 are classified according to skill level, we have the following table:

TABLE 8-7

Employees	Skill Level	
F_1	unskilled	17
F_2	semi-skilled	50
F_3	skilled	33
		100

(a) Calculate the marginal probabilities.
(b) What is the probability that a given employee chosen at random is a skilled worker?
(c) Calculate $p(F_2 \cup F_3)$.
(d) Explain in words the probability calculated in (c).
(e) Would it be more meaningful if these employees were classified by earnings and skill level simultaneously?

3. Given the following tally sheet for the 100 employees of the XYZ Company, set up the bivariate frequency table. (U = unskilled, SS = semi-skilled, and S = skilled).

TABLE 8-8

Earnings	Type of Worker	Total
1.50–2.49	U	10
1.50–2.49	SS	5
1.50–2.49	S	0
2.50–3.49	U	6
2.50–3.49	SS	11
2.50–3.49	S	3
3.50–4.49	U	1
3.50–4.49	SS	22
3.50–4.49	S	7
4.50–5.49	U	0
4.50–5.49	SS	10
4.50–5.49	S	10
5.50–6.49	U	0
5.50–6.49	SS	2
5.50–6.49	S	13
		100

4. The 10,500 employees of a certain company are classified according to sex and type of work, i.e., sales or production. Six thousand employees are male production workers and 3500 workers are in sales. Of the 3500 workers engaged in sales, 1500 are females.

(a) Set up the bivariate frequency table.
(b) How many males work at this company?
(c) How many females are production workers?
(d) How many females or production workers are employed?

5. Given the following table:

TABLE 8-9

		A_1 Freshmen	A_2 Sophomores	A_3 Juniors	A_4 Seniors	Total
B_1	Have financial assistance	—	40	60	70	200
B_2	No financial assistance	270	—	—	—	800
	Total	300	250	250	200	1000

(a) Complete the table.
(b) How many elements are in A_1?
(c) How many elements are in B_2?
(d) How many elements are in the population?
(e) What is $p(A_1)$?
(f) What is $p(B_2)$?

8-2 JOINT AND MARGINAL PROBABILITIES

Converting Table 8-5 of the previous section to a table of probabilities by dividing each entry by 33, we have

TABLE 8-10

Grade \ I.Q.	L	M	H	Total
A	$\frac{1}{33}$	$\frac{1}{33}$	$\frac{1}{33}$	$\frac{3}{33}$
B	$\frac{1}{33}$	$\frac{1}{33}$	$\frac{1}{33}$	$\frac{3}{33}$
C	$\frac{6}{33}$	$\frac{6}{33}$	$\frac{6}{33}$	$\frac{18}{33}$
D	$\frac{3}{33}$	$\frac{3}{33}$	$\frac{3}{33}$	$\frac{9}{33}$
Total	$\frac{11}{33}$	$\frac{11}{33}$	$\frac{11}{33}$	1

This is a typical probability table for bivariate data. The probabilities in the total row and total column are called *marginal* probabilities. The other entries are called *joint* probabilities.

The marginal probabilities correspond to a univariate classification of the given data and the joint probabilities correspond to a bivariate classification of the data. For example, $p(A) = \frac{3}{33}$, $p(M) = \frac{11}{33}$, etc. Also $p(A \cap M) = \frac{1}{33}$, $p(C \cap H) = \frac{6}{33}$, etc.

Example 1. Compute $p(A \cup H)$ using Table 8-10.

Solution:

$$p(A \cup H) = p(A) + p(H) - p(A \cap H)$$

$$= \tfrac{3}{33} + \tfrac{11}{33} - \tfrac{1}{33}$$

$$p(A \cup H) = \tfrac{13}{33}$$

In other words, 13 of the 33 students in this class have either a grade of A or a high I.Q. or both a grade of A and a high I.Q.

Next we include a bivariate classification of a standard deck of playing cards.

TABLE 8-11

		D_1	D_2	D_3	D_4	D_5	
		2–10	J	Q	K	A	Total
E_1	Diamonds	9	1	1	1	1	13
E_2	Spades	9	1	1	1	1	13
E_3	Hearts	9	1	1	1	1	13
E_4	Clubs	9	1	1	1	1	13
	Total	36	4	4	4	4	52

Example 2. Use the data in Table 8-11 to find $p(E_2)$, $p(D_3)$, $p(E_1 \cap D_4)$, and $p(E_4 \cup D_1)$.

Solution:

$$p(E_2) = \tfrac{13}{52} = \tfrac{1}{4}$$

$$p(D_3) = \tfrac{4}{52} = \tfrac{1}{13}$$

$$p(E_1 \cap D_4) = \tfrac{1}{52}$$

$$p(E_4 \cup D_1) = p(E_4) + p(D_1) - p(E_4 \cap D_1) = \tfrac{13}{52} + \tfrac{36}{52} - \tfrac{9}{52}$$

$$p(E_4 \cup D_1) = \tfrac{40}{52}$$

We point out that the sum of the joint probabilities in a given row or column equals the marginal probability of that row or column.

Also, the marginal probabilities in the bottom row and right-hand column both add up to one. These remarks are true because we classify our data into exhaustive and mutually exclusive events.

Example 3. A sample space S is classified into exhaustive and mutually exclusive events A_1 and A_2. It is also classified into exhaustive and exclusive events B_1 and B_2. It is known that $p(A_1) = 0.6$, $p(B_2) = 0.3$, and $p(A_1 \cap B_1) = 0.5$. Set up a complete probability table from this information.

Solution:

Step 1: Record the given information.

TABLE 8-12(a)

	A_1	A_2	Total
B_1	0.5		
B_2			0.3
Total	0.6		1

Step 2: After performing the required subtractions, we have the following complete table:

TABLE 8-12(b)

	A_1	A_2	Total
B_1	0.5	0.2	0.7
B_2	0.1	0.2	0.3
Total	0.6	0.4	1

PROBLEMS

1. Convert Table 8-11 to a table of joint and marginal probabilities.
2. What rules should be followed in classifying data?
3. An experiment consists of tossing a coin and a six-sided die.
   (a) Set up a bivariate probability table for this experiment.
   (b) Determine the joint probability of a head on the coin and a 4 on the die.
   (c) Determine the joint probability $p(T \cap 5)$.
   (d) Compute $p(H \cup 3)$.
4. Given the data in Table 8-13:

TABLE 8-13. ANNUAL INCOME IN THOUSANDS

	B_1 \$5,000–10,000	B_2 \$10,000–15,000	B_3 \$15,000–20,000	Total
A_1 owns a camping trailer	80	60	10	150
A_2 does not own a camping trailer	420	340	90	850
Total	500	400	100	1000

Explain the following symbols in words:

(a) $p(A_1)$
(b) $p(B_2)$
(c) $p(A_1 \cap B_2)$
(d) $p(A_2 \cup B_1)$

5. Compute the probabilities in Problem 4.

6. Which of the probabilities in Problem 5 are marginal probabilities? Which are joint probabilities?

7. Given the data in Table 8-14:

TABLE 8-14

Car Ownership \ Grade Average	B_1 A	B_2 B	B_3 C	B_4 D	Total
A_1 owns a car	30	100	350	80	500
A_2 does not own a car	70	100	250	20	500
Total	100	200	600	100	1000

Express the following symbolically:

(a) The probability that a student has an A average.
(b) The probability that a student owns a car.
(c) The probability that a student has a C average or owns a car.
(d) The probability that a student has a B average and owns a car.

8. Calculate the probabilities in Problem 7 if the student was chosen at random.

9. Which of the probabilities in Problem 7 are marginal probabilities? Which are joint probabilities?

8-3 CONDITIONAL PROBABILITY

In this section the concept of conditional probability is introduced in connection with the bivariate data given in Table 8-10 of the previous section. This table is repeated below for convenience.

TABLE 8-15

Grade I.Q.	L	M	H	Total
A	$\frac{1}{33}$	$\frac{1}{33}$	$\frac{1}{33}$	$\frac{1}{11}$
B	$\frac{1}{33}$	$\frac{1}{33}$	$\frac{1}{33}$	$\frac{1}{11}$
C	$\frac{2}{11}$	$\frac{2}{11}$	$\frac{2}{11}$	$\frac{6}{11}$
D	$\frac{1}{11}$	$\frac{1}{11}$	$\frac{1}{11}$	$\frac{3}{11}$
Total	$\frac{1}{3}$	$\frac{1}{3}$	$\frac{1}{3}$	1

The reader will recall that this table was obtained by dividing the entries in the bivariate frequency Table 8-5 by the total population of 33. Table 8-15 is a table of joint and marginal probabilities.

The probability of a student with a low I.Q. getting an A is an example of a *conditional probability*. This would be written symbolically as $p(A/L)$ and is read "the probability of A given L."

The following procedure can be used to obtain a table of conditional probabilities. The first column of the table, which gives $p(A/L)$, $p(B/L)$, $p(C/L)$, and $p(D/L)$, is shown below.

L COLUMN OF TABLE 8-16

	L
A	$\frac{1}{11}$
B	$\frac{1}{11}$
C	$\frac{6}{11}$
D	$\frac{3}{11}$
Total	$\frac{11}{11} = 1$

(Table 8-16 Partial)

Note that the frequencies were divided by the reduced population of 11 rather than the total population of 33. That is, we divided by the number of students with low I.Q. to get the entries in the L column. The remaining columns of a complete conditional probability table are obtained by dividing by the number of students with medium I.Q. and the number of students with high I.Q. Both of these numbers happen to also be 11 in this problem. The complete table is shown as Table 8-16.

TABLE 8-16

	L	M	H
A	$\frac{1}{11}$	$\frac{1}{11}$	$\frac{1}{11}$
B	$\frac{1}{11}$	$\frac{1}{11}$	$\frac{1}{11}$
C	$\frac{6}{11}$	$\frac{6}{11}$	$\frac{6}{11}$
D	$\frac{3}{11}$	$\frac{3}{11}$	$\frac{3}{11}$
Total	1	1	1

Example 1. Determine $p(B/M)$ and $p(D/H)$ from Table 8-16.

Solution:

$$p(B/M) = \frac{1}{11} \quad \text{and} \quad p(D/H) = \frac{3}{11}$$

From Table 8-16 we can read the probability of getting a certain grade subject to the condition that the student has a given I.Q. The next table shows the probabilities that result if the condition is placed on grades rather than I.Q.

TABLE 8-17

	L	M	H		
A	$\frac{1}{3}$	$\frac{1}{3}$	$\frac{1}{3}$	1	(*dividing by* 3)
B	$\frac{1}{3}$	$\frac{1}{3}$	$\frac{1}{3}$	1	(*dividing by* 3)
C	$\frac{6}{18}$	$\frac{6}{18}$	$\frac{6}{18}$	1	(*dividing by* 18)
D	$\frac{3}{9}$	$\frac{3}{9}$	$\frac{3}{9}$	1	(*dividing by* 9)

Example 2. Determine $p(M/B)$ and $p(H/D)$ from Table 8-17.

Solution:

$$p(M/B) = \tfrac{1}{3} \quad \text{and} \quad p(H/D) = \tfrac{3}{9} = \tfrac{1}{3}$$

To preview the next section, we point out that $p(L) = \frac{1}{3}$ (from Table 8-15) and $p(L/D) = \frac{1}{3}$ (from Table 8-17). In other words, $p(L/D) = p(L)$ in this case. The significance of this will be explained in Section 8-4.

Before giving the final example of this section, we develop the following important formula. Let A and B be two events in a sample space S. Then $p(A/B) = [p(A \cap B)]/p(B)$.

To establish this formula for the case of a finite sample space where all sample points are equally likely, we recall that $p(A/B)$ is computed by dividing the number of elements in the event $A \cap B$ by the number of elements in B. If we denote the number of elements in $A \cap B$ by h and the number of elements in B by b, this gives $p(A/B) = h/b$. Dividing numerator and denominator of this fraction by the number of elements n in the sample space gives

$$p(A/B) = \frac{h}{b} = \frac{h/n}{b/n}$$

But,

$$\frac{h}{n} = p(A \cap B) \qquad \text{and} \qquad \frac{b}{n} = p(B)$$

Therefore,

$$p(A/B) = \frac{p(A \cap B)}{p(B)}$$

In general, $p(A/B)$, the conditional probability of A given B, is defined by the formula

$$p(A/B) = \frac{p(A \cap B)}{p(B)}$$

Example 3. The sample space of a certain experiment is classified two different ways into pairs of exhaustive events A_1, A_2 and B_1, B_2. Given that $p(A_1) = 0.3$, $p(B_1) = 0.4$, and $p(A_1/B_1) = 0.6$, complete the joint and conditional probability tables.

Solution: The following tables reflect the given information.

TABLE 8-18

	B_1	B_2			B_1	B_2
A_1			0.3	A_1	0.6	
A_2				A_2		
	0.4		1		1	1

We know that

$$p(A_1/B_1) = \frac{p(A_1 \cap B_1)}{p(B_1)} = 0.6$$

Then

$$p(A_1 \cap B_1) = p(A_1/B_1)p(B_1)$$
$$0.24 = (0.6)(0.4)$$

With this information we can complete the joint probability table as follows:

TABLE 8-19

	B_1	B_2	
A_1	0.24	0.06	0.3
A_2	0.16	0.54	0.7
	0.4	0.6	1.0

We can now calculate

$$p(A_1/B_2) = \frac{p(A_1 \cap B_2)}{p(B_2)}$$
$$0.10 = \frac{0.06}{0.6}$$

With this information we can complete the conditional probability table as follows:

TABLE 8-20

	B_1	B_2
A_1	0.6	0.1
A_2	0.4	0.9
	1.0	1.0

PROBLEMS

1. Given the experiment of randomly drawing a single card from a standard deck:
   (a) What is the probability of drawing a jack of hearts?
   (b) What is the probability of drawing a king or a queen?
   (c) What is the probability of drawing a spade?
   (d) What is the probability of drawing a heart given that the card is a king?
   (e) What is the probability of drawing an ace given that the card is a spade?
2. Express the probabilities in Problem 1 symbolically.
3. Determine if the probabilities in Problem 1 are marginal, joint, or conditional.

4. Given the following bivariate table, set up the joint and conditional probability tables:

TABLE 8-21

	A_1	A_2	A_3	
B_1	10	5	15	30
B_2	10	5	5	20
B_3	5	5	40	50
	25	15	60	100

5. Given the following information: $p(A_1) = 0.7$, $p(B_1) = 0.5$, and $p(B_1/A_1) = 0.4$, complete the joint and marginal probability table. Assume the pairs A_1, A_2 and B_1, B_2 are exhaustive.

6. Given the following partial tables:

TABLE 8-22(a)

	A_1	A_2	A_3	
B_1				0.4
B_2				
		0.3	0.2	1

TABLE 8-22(b)

	A_1	A_2	A_3	
B_1				1
B_2		0.3		1

TABLE 8-22(c)

	A_1	A_2	A_3
B_1			0.2
B_2			
	1	1	1

Complete the tables. Assume A_1, A_2, A_3 and B_1, B_2 are both collectively exhaustive.

7. Given Table 8-23:

TABLE 8-23

	A_1 Freshmen	A_2 Sophomores	A_3 Juniors	A_4 Seniors	Total
B_1 Have financial assistance	30	40	60	70	200
B_2 No financial assistance	270	210	190	130	800
Total	300	250	250	200	1000

Calculate the following:

(a) $p(B_2)$ (d) $p(A_2/B_1)$
(b) $p(A_3)$ (e) $p(B_2/A_1)$
(c) $p(A_4 \cap B_2)$ (f) $p(A_4 \cup B_2)$

8. Explain the probabilities calculated in Problem 7 in words.

9. Which of the probabilities in Problem 7 are joint probabilities? Which are conditional probabilities?

10. Given that $p(A) = 0.3$, $p(B') = 0.2$, and $p(A \cap B') = 0.1$. Calculate

(a) $p(A')$ (d) $p(A \cup B')$
(b) $p(B)$ (e) $p(B'/A)$
(c) $p(A/B')$

11. Given that $p(A) = 0.4$, $p(B'/A) = 0.3$, and $p(A \cup B') = 0.6$. Calculate

(a) $p(A \cap B')$ (d) $p(A')$
(b) $p(B')$ (e) $p(A/B')$
(c) $p(B)$

8-4 INDEPENDENCE

In the previous section it was pointed out that $p(L/D) = p(L)$ for a certain class of students, where L stands for low I.Q. and D is a grade. Whenever two events are related in this way, they are said to be independent.

DEFINITION 8.1. *Two events, A and B, are said to be independent if* $p(A/B) = p(A)$.

In words, two events are independent if the conditional probability, $p(A/B)$, is equal to the corresponding marginal probability, $p(A)$.

Next we illustrate this definition using the I.Q. example. From Table 8-15, we have $p(A) = \frac{1}{11}$, $p(B) - \frac{1}{11}$, $p(C) = \frac{6}{11}$, and $p(D) = \frac{3}{11}$. This information and probability Table 8-16 follow:

TABLE 8-24

	F_1 L	F_2 M	F_3 H	$p(E_i)$
$E_1 = A$	$\frac{1}{11}$	$\frac{1}{11}$	$\frac{1}{11}$	$\frac{1}{11}$
$E_2 = B$	$\frac{1}{11}$	$\frac{1}{11}$	$\frac{1}{11}$	$\frac{1}{11}$
$E_3 = C$	$\frac{6}{11}$	$\frac{6}{11}$	$\frac{6}{11}$	$\frac{6}{11}$
$E_4 = D$	$\frac{3}{11}$	$\frac{3}{11}$	$\frac{3}{11}$	$\frac{3}{11}$
	1	1	1	

It is clear from Table 8-24 that the conditional probabilities $p(A/L)$, $P(A/M)$, and $p(A/H)$ are all equal to the marginal probability

$p(A)$. Therefore, by Definition 8.1, the three pairs of events A, L; A, M; and A, H are all independent. Similar statements are true for the remaining three rows of Table 8-24.

Since $p(E_i/F_j) = P(E_i)$ for $i = 1, 2, 3$, and $j = 1, 2, 3$, it follows that grades are independent of I.Q. for this particular example.

Example 1. Given the following data, determine whether or not the variables x and y are independent.

TABLE 8-25

	x_1	x_2	x_3	Total
y_1	0.3	0.1	0.2	0.6
y_2	0.3	0.1	0	0.4
Total	0.6	0.2	0.2	1.0

Solution:

$$p(y_1/x_1) = \frac{0.3}{0.6} = 0.5$$

$$p(y_1) = 0.6 \neq 0.5$$

Therefore, x and y are not independent.

Example 2. Form the conditional probability table with x as the given variable from the joint probability table given in Example 1.

Solution:

TABLE 8-26

	x_1	x_2	x_3
y_1	0.5	0.5	1
y_2	0.5	0.5	0
	1	1	1

dividing the x_1 column by 0.6, and so forth

We point out that the nonequal entries 0.5, 0.5, and 1 in the y_1 row are an indication that the variables x and y are not independent. Since $p(y_1/x_1) \neq p(y_1/x_3)$, it follows that at least one of them and possibly both are not equal to $p(y_1)$. Conversely, if a table of conditional probabilities is formed and the entries within each row are equal, then the two variables are independent.

By Definition 8.1, if A and B are independent, then $p(A/B) = p(A)$. Next we show that this relationship is equivalent to $p(A \cap B) = p(A)p(B)$. To see this we recall that $p(A/B) = p(A \cap B)/p(B)$. Multiplying both sides by $p(B)$ gives $p(A \cap B) = p(A/B)p(B)$. But if A and B are independent, then $p(A/B) = p(A)$. By substitution, we have $p(A \cap B) = p(A)p(B)$, whenever A and B are independent. Conversely, if $p(A \cap B) = p(A)p(B)$, then $p(A/B) = p(A)$ and the events A and B are independent in the sense of Definition 8.1.

From now on, we can test pairs of events for independence either by Definition 8.1 or by comparing the joint probability $p(A \cap B)$ with the product $p(A)p(B)$. The two events are independent if and only if $p(A \cap B) = p(A)p(B)$.

Example 3. Test the two events x_2 and y_2 given in Example 1 for independence by the method described.

Solution: $p(x_2 \cap y_2) = 0.1$, $p(x_2) = 0.2$, $p(y_2) = 0.4$, and $p(x_2)p(y_2) = 0.08 \neq p(x_2 \cap y_2)$. Therefore, x_2 and y_2 are not independent.

PROBLEMS

1. Let the following table of joint probabilities be given:

TABLE 8-27

$y \setminus x$	x_1	x_2	x_3	Total
y_1	0.2	0.15	0.15	0.5
y_2	0.2	0.05	0.25	0.5
Total	0.4	0.2	0.4	1.0

Which of the following pairs of events are independent? Use the method of comparing the product of the marginals with the joint probability.

(a) y_1, x_1 **(d)** y_2, x_1
(b) y_1, x_2 **(e)** y_2, x_2
(c) y_1, x_3 **(f)** y_2, x_3

2. Construct a conditional probability table from Table 8-27.

3. Check the pairs of events listed in Problem 1 for independence. Use Definition 8.1 and the conditional probability table calculated in Problem 2.

4. Are the variables x and y of Problem 1 independent? Explain briefly.

5. Under what conditions are two variables independent in the probability sense?

6. Given that the two variables A and B are independent, complete Table 8-28.

TABLE 8-28

$B \setminus A$	A_1	A_2	
B_1			0.4
B_2			
	0.3		1.0

7. Given that the two variables A and B are independent, complete Table 8-29.

TABLE 8-29(a)

	A_1	A_2	
B_1	0.12		
B_2			
			1.0

TABLE 8-29(b)

	A_1	A_2
B_1	0.3	
B_2		
	1.0	1.0

8. Given Table 8-30:

TABLE 8-30

$B \setminus A$	A_1 Freshmen	A_2 Sophomores	A_3 Juniors	A_4 Seniors	Total
B_1 Have financial assistance	30	40	60	70	200
B_2 No financial assistance	270	210	190	130	800
Total	300	250	250	200	1000

Are the variables A and B independent?

9. A student must take an exam with five questions. He believes that his success on any one problem does not depend on his success or failure on the other problems. He assigns a probability of 0.7 to answering a particular question correctly. Determine the probability of getting

(a) all five correct
(b) none correct
(c) exactly two correct
(d) exactly three correct
(e) at least two correct
(f) at least three correct

10. A fair coin is tossed 6 times. Determine the probability of
(a) six heads
(b) exactly three heads
(c) six tails
(d) at least three heads

11. An experiment consists of n independent trials. Each trial can result in a success with probability p or a failure with probability $g = 1 - p$. Let X denote the number of successes in n trials. Show that

$$p(X = x) = \binom{n}{x} p^x (1-p)^{n-x}, \quad \text{where } \binom{n}{x} = \frac{n!}{x!(n-x)!}$$

and that the formula applies for $x = 0, 1, 2, \ldots, n$.

12. Verify the formula in Problem 11 by applying it to Problems 9 and 10.

13. The manager of a department store in a suburban mall in Connecticut is interested in learning more about the shopping habits of Massachusetts residents. He is particularly interested in how much they spent, how long they stayed in the store, and their plans for future visits to the store. In order to increase the percentage of questionnaires returned, he offers to send each respondent a gift certificate. This same questionnaire was used the previous year with a sample of 50. A tabulation of these questionnaires is given in the following table:

	B_1 (gift certificate offered)	B_2 (no gift certificate offered)	Total
A_1 (returned)	35	25	60
A_2 (not returned)	15	25	40
Total	50	50	100

(a) Complete the following table by inserting the theoretical or expected frequencies assuming no relationship (i.e., variables are independent).

	B_1	B_2	Total
A_1			60
A_2			40
Total	50	50	100

(b) Are the variables independent?

14. Last year when the new credit policies were introduced, the vice president of sales challenged them on the contention that collections are dependent upon the size of the buying firm. In order to support the policies in the face of this challenge, the decision has been made to include in the presentation a discussion of the statistical independence of a bad debt occurrence and size of the buying organization. These two events are said to be statistically independent if

(a) the probability of a bad debt is equal to the sum of the probability that the buyer is a small company plus the probability that the buyer is a large company.
(b) the probability of a bad debt plus the probability that the buyer is a small company equals the probability of a bad debt plus the probability that the buyer is a large company.
(c) the probability of a bad debt given that the buyer is a small company equals the probability of a bad debt divided by one minus the probability that the buyer is a large company.
(d) the probability of a bad debt given that the buyer is a small company equals the probability of a bad debt given that the buyer is a large company equals the probability of a bad debt.
(e) none of the above.

8-5 TREE DIAGRAMS

Consider the experiment of tossing a coin and drawing a card from a deck that contains one jack, one queen, and one king. The sample space for this experiment is

$$S = \{(J, H), (J, T), (Q, H), (Q, T), (K, H), (K, T)\}$$

The subjective probabilities, assuming equal likelihood, are displayed in Table 8-31.

TABLE 8-31

	H	T	Total
J	$\frac{1}{6}$	$\frac{1}{6}$	$\frac{1}{3}$
Q	$\frac{1}{6}$	$\frac{1}{6}$	$\frac{1}{3}$
K	$\frac{1}{6}$	$\frac{1}{6}$	$\frac{1}{3}$
Total	$\frac{1}{2}$	$\frac{1}{2}$	1

It is sometimes meaningful to display the sample points in a so-called *tree diagram.* This method is illustrated as Figure 8-1.

The "total path" $O \to J \to H$ corresponds to the pair (J, H), etc. The number of sample points can still be counted by counting the number of "total paths." In this case, there are six such paths or, equivalently, six sample points. The probability of each "total path" is $\frac{1}{6}$, since in this case all "total paths" are equally likely.

Problem 14 adapted from the CMA examination.

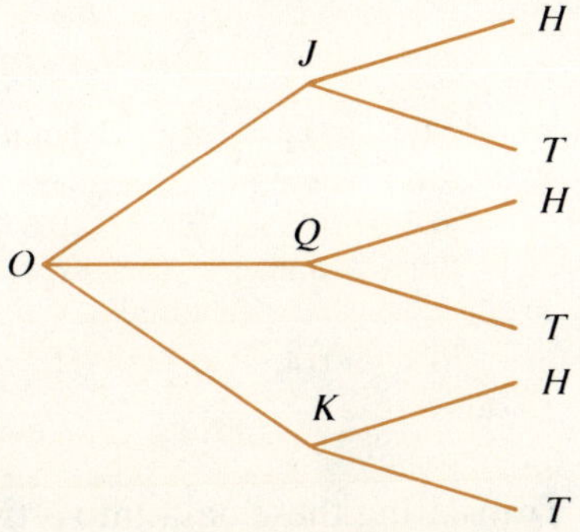

FIGURE 8-1

Example 1. Construct a tree diagram for the experiment that consists of tossing 3 coins and determine the number of sample points.

Solution:

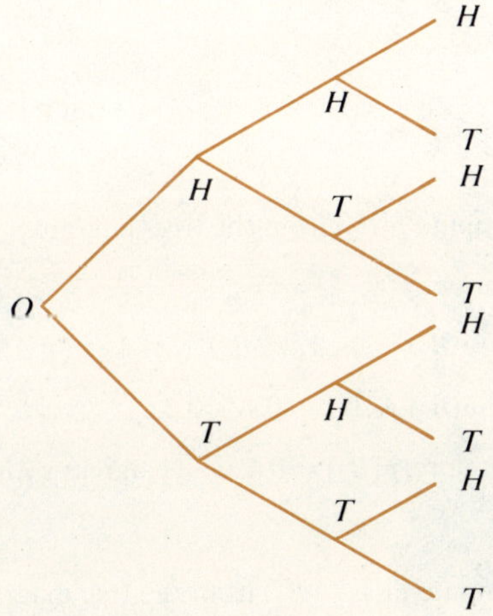

FIGURE 8-2

There are 8 "total paths" and 8 sample points.

Next we relate the concepts of joint, marginal, and conditional probabilities to tree diagrams. Let the following probability table be given.

TABLE 8-32

	A_1	A_2	A_3	Total
B_1	0.2	0.3	0.1	0.6
B_2	0.1	0.1	0.2	0.4
Total	0.3	0.4	0.3	1

Translating these data into a tree diagram format, we have

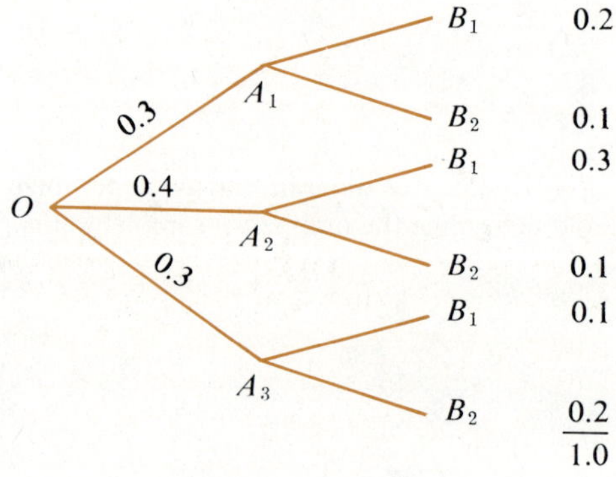

FIGURE 8-3

Example 2. Compute $p(B_2)$ from the tree diagram.

Solution:

$$p(B_2) = p(A_1 \cap B_2) + p(A_2 \cap B_2) + p(A_3 \cap B_2)$$

$$p(B_2) = 0.1 + 0.1 + 0.2$$

Therefore, $p(B_2) = 0.4$, which agrees with the information given in Table 8-32.

Example 3. Compute $p(A_1 \cup B_2)$ from the tree diagram and the results of Example 2.

Solution:

$$p(A_1 \cup B_2) = p(A_1) + p(B_2) - p(A_1 \cap B_2)$$

$$= 0.3 + 0.4 - 0.1$$

$$p(A_1 \cup B_2) = 0.6$$

Figure 8-4 illustrates the probability interpretation of the various branches and *total paths* of a typical tree diagram.

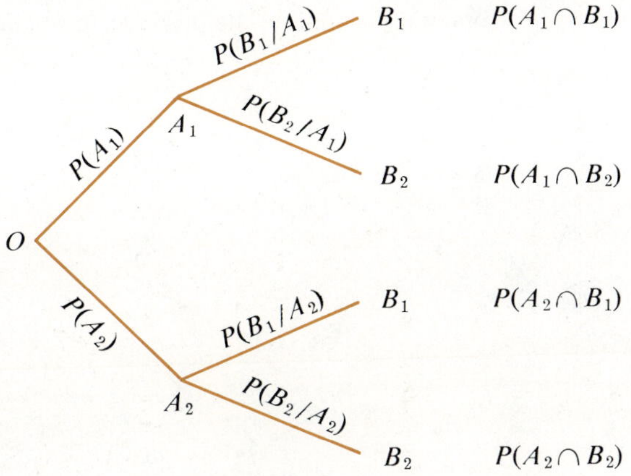

FIGURE 8-4

Finally, we solve an applied problem using a tree diagram approach.

Example 4. The probability that a machine is correctly set up is assumed to be 0.8. Furthermore, experience has shown that if the machine is correctly set up, then the probability of obtaining a good part is 0.90. However, if the machine is not correctly set up, the probability of a good part is only 0.2. Given the fact that the first part produced was tested and found to be satisfactory, determine the probability that the machine is set up correctly.

Solution: We must determine the conditional probability, p(correct set-up/first part is satisfactory) $= p(C/S)$. Figure 8-5 illustrates a partial tree diagram that includes the obvious information.

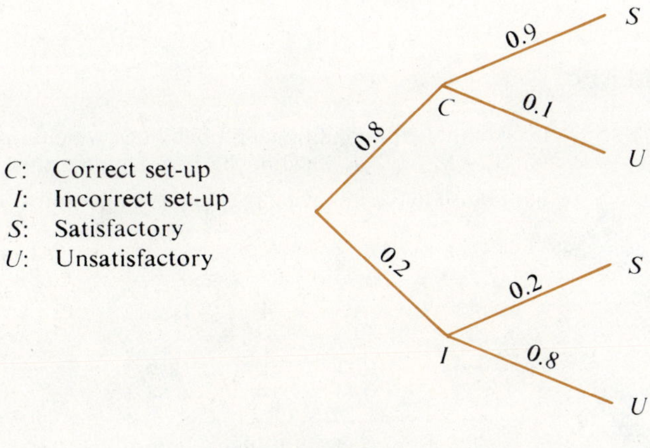

FIGURE 8-5

Compute the appropriate products to obtain the 4 joint probabilities.

$$p(S \cap C) = p(S/C)p(C)$$
$$= (0.9)(0.8)$$
$$p(S \cap C) = 0.72, \text{ etc.}$$

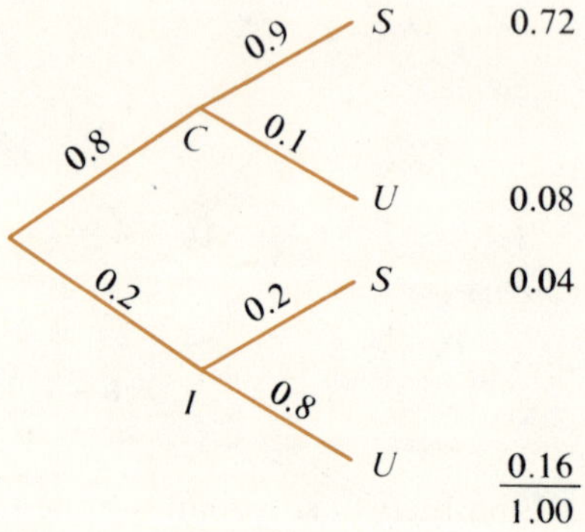

FIGURE 8-6

$$p(C/S) = \frac{p(C \cap S)}{p(S)}$$
$$p(S) = p(C \cap S) + p(I \cap S)$$
$$= 0.72 + 0.04$$
$$p(S) = 0.76$$

Therefore,

$$p(C/S) = \frac{0.72}{0.76} = \frac{18}{19}$$
$$p(C/S) = 0.95$$

PROBLEMS

1. Compute the conditional probabilities, which correspond to the branches $A_1 - B_1$, $A_1 - B_2$, etc. in the tree diagram, that result from Table 8-32.
2. Translate the following probability tables into a tree diagram.

TABLE 8-33(a)

$B \setminus A$	A_1	A_2	Total
B_1	0.11	0.05	0.16
B_2	0.44	0.4	0.84
Total	0.55	0.45	1.0

TABLE 8-33(b)

	A_1	A_2
B_1	$\frac{1}{5}$	$\frac{1}{9}$
B_2	$\frac{4}{5}$	$\frac{8}{9}$
	1	1

3. Translate the following tree diagram into a joint and a conditional probability table.

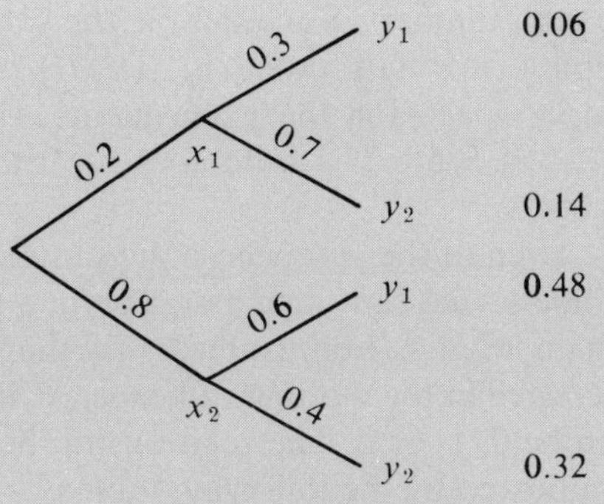

FIGURE 8-7

4. Determine the remaining probabilities in the following tree diagram.

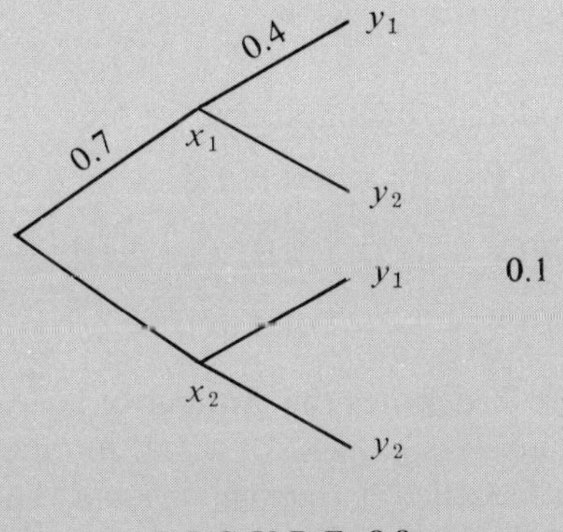

FIGURE 8-8

5. Set up a tree diagram showing all probabilities for the experiment of tossing a coin and rolling a 6-faced die.
6. Use the tree diagram found in Problem 5 to determine the following probabilities:
   (a) $p(H)$
   (b) $p(3)$
   (c) $p(T \cap 4)$
   (d) $p(H \cup 5)$
7. Given the data in Example 4 of this section, calculate $p(C/U)$.
8. Do Problem 9 of Section 8-4 using tree diagrams.
9. Do Problem 10 of Section 8-4 using tree diagrams.

8-6 RANDOM VARIABLES AND EXPECTED VALUE

The points in a sample space are the possible outcomes of the associated experiment. Sometimes, but not always, the points of a sample space are real numbers. For example the sample space for the experiment of flipping two fair coins is $\{(H, H), (H, T), (T, H), (T, T)\}$, while the sample space for the experiment of tossing a fair six-sided die is $\{1, 2, 3, 4, 5, 6\}$. The outcomes for the coin experiment are not real numbers.

Even in the case where the sample points themselves are not real numbers, one can often associate in a natural way a real number with certain events. For example, in the coin experiment we might be interested in the number of heads, X, in the outcome. Clearly, X could then be 0, 1, or 2. The relationship between X and the sample space is indicated by the following table:

TABLE 8-34

X	Event
0	$\{(T, T)\}$
1	$\{(H, T), (T, H)\}$
2	$\{(H, H)\}$

Introducing probabilities, we have

$$P(X = 0) = P\{(T, T)\} = \tfrac{1}{4}$$

$$P(X = 1) = P\{(H, T), (T, H)\} = \tfrac{1}{2}$$

$$P(X = 2) = P\{(H, H)\} = \tfrac{1}{4}$$

where X denotes the number of heads in the outcome.

The assignment of a real number to each outcome of an experiment is called a *random variable*. The formal definition of a random variable involves the concept of a real valued function. The reader may wish to review this notion which is discussed in Section 1-1.

DEFINITION 8-2. *A random variable, X, on a sample space, S, is a function from S into the set of real numbers.*

The following remarks about the definition of a random variable are in order.

1. X is a function whose domain is the sample space, S.
2. The range of X is a subset of the set of real numbers.

In many experiments the outcomes themselves are real numbers, in which case the outcome itself can be taken as a random variable. Some typical examples of random variables are: the sum of points on a pair of dice, inches of rainfall during a given month, height of a person, time taken for an electronic part to fail, number of successes in n trials of a two-outcome success or failure process.

Next, we introduce some standard functional notation that is used in connection with random variables, X. $P(X = x)$ is denoted by $f(x)$. $P(X \leqslant x)$ is denoted by $F(x)$, where X denotes a random variable and x is a particular value taken on by X.

Example 1. Complete the following table for the random variable X.

x	$P(X = x)$	$f(x)$	$F(x)$
1	0.2		
2	0.3		
3	0.4		
4	0.1		

Solution:

x	$P(X = x)$	$f(x)$	$F(x)$
1	0.2	0.2	0.2
2	0.3	0.3	0.5
3	0.4	0.4	0.9
4	0.1	0.1	1.0

In the above usage, $f(x)$ is called the probability function of x or the distribution of X. Upper case F, $F(x)$ is called the cumulative distribution function (c.d.f.) for X. Finally, we introduce the notion of the *expected value* of a random variable X. Expected value is used later in the text. It is denoted by $E(X)$.

DEFINITION 8.3. *The* expected value *of a random variable* X, $E(X)$, *is defined by* $E(X) = \sum xP(X = x)$, *where the summation is over all values of* x.

The concept of expected value is illustrated by the next example.

Example 2. Find the expected value of the random variable of Example 1.

Solution: $E(X) = 1P(X = 1) + 2P(X = 2) + 3P(X = 3) + 4P(x = 4)$

$$= 1(0.2) + 2(0.3) + 3(0.4) + 4(0.1)$$
$$= 0.2 + 0.6 + 1.2 + 0.4$$
$$= 2.4$$

The expected value of a random variable, X, is also called the *mean* of X.

To calculate the expected value of a random variable, X, it is necessary to multiply each value of X by the probability that X has that value and then add these products.

In previous sections we used relative frequencies to arrive at relevant probabilities based on past experience. In the next two sections we use known (from theoretical considerations) or assumed probability distributions to obtain the needed probabilities.

PROBLEMS

1. Let X be the number that shows on top when a fair die with six faces is rolled.
   (a) Can X be considered a random variable?
   (b) Determine the possible values of X.
   (c) Find $P(X = 2)$, $P(X = 6)$, $P(2 \leqslant X \leqslant 4)$, and $P(X \leqslant 4)$.
   (d) Set up a table for $f(x) = P(X = x)$, $x = 1, 2, \ldots, 6$.
   (e) Set up a table for $F(x) = P(X \leqslant x)$, $x = 1, 2, \ldots, 6$.
2. Use a result from Problem 1 to compute the expected value of X, $E(X)$, for the random variable defined there.
3. As usual let $f(x) = P(X = x)$. Given the following data, compute the expected value of X, i.e., $E(X)$.

   (a)

x	0	1	2	3	4
$f(x)$	0.1296	0.3456	0.3456	0.1536	0.0256

   (b) Your answer for part (a) equals $4P$ for what value of P?
4. As usual, let $f(x) = P(X = x)$. Given the following data:

x	0	1	2	3	4	5
$f(x)$	0.1681	0.3602	0.3087	0.1323	0.0284	0.0024

   (a) Compute the expected value of X, $E(X)$.
   (b) For what value n does $n(0.3)$ equal your answer for part (a)?
5. A fair coin is tossed three times. List the possible outcomes, (H, H, H) etc. of this experiment. In other words, describe the sample space.
6. Consider the coin experiment described in Problem 5. Define a random variable, X, so that X equals the number of heads.
   (a) What are the possible values of X?
   (b) Describe the event that corresponds to each possible value of X.

(c) Choose a binomial coefficient that gives the number of sample points in each of the events of part (b).

Hint: $c(n, x) = \binom{n}{x}$ which gives the number of combinations of n objects taken x at a time is also called a binomial coefficient.

7. Expand $(a+b)^n$ for the following values of n, where

$$(a+b)^n = \sum_{x=0}^{n} \binom{n}{x} a^{n-x} b^x \quad \text{and} \quad \binom{n}{x} = \frac{n!}{(n-x)!\,x!}$$

(a) $n = 2$, i.e., $(a+b)^2$
(b) $n = 3$, i.e., $(a+b)^3$
(c) $n = 5$, i.e., $(a+b)^5$

8. At a certain switchboard 54 calls are received on the average between 10 A.M. and 11 A.M. weekdays. Let X denote the number of calls that are received between 10 A.M. and 10:05 A.M. on a given day.
(a) Is X a random variable?
(b) Could X be zero?
(c) Could X be 10?
(d) Is there a largest possible value that X could take on? If so, what is it?

8-7 THE BINOMIAL DISTRIBUTION

In the previous section it was pointed out that probabilities of a discrete random variable can be assigned by using known probability distributions. In this section we will look at one of the most important discrete probability distributions, namely the binomial probability distribution.

To get a better insight into this distribution we will look at a situation which most of you will find quite familiar. Assume that you are about to take an exam which will contain three separate unrelated questions. We can look at this exam as a series of experimental trials which are characterized as follows:

(a) There are two mutually exclusive outcomes on each trial. That is, one can either pass or fail the question.
(b) The probability of passing or failing a question is constant.
(c) The trials or questions are independent.

In summer then, the exam represents a sequence of unrelated trials where n denotes the number of trials, i.e., $n = 3$.

1. The outcome state for each trial is pass, fail.
2. The probability of passing denoted by p is constant from trial to trial, consequently.
3. The probability of failure $1-p$ or q is also constant from trial to trial.
4. Success on any one question is independent of success on any of the other questions.

Example 1. Use a tree diagram to summarize the situation outlined above.

Solution:

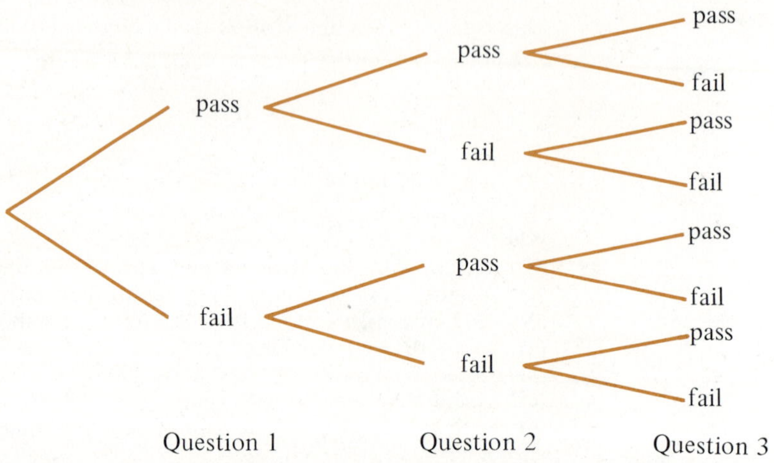

FIGURE 8-9

We can see from the previous example that there are eight points in the sample space for this experiment. Now, upon taking the exam the student can pass all three questions, pass two questions, pass one question, pass none of the questions. Next, we define an appropriate random variable, X, for this problem. Let $X =$ the number of questions passed. Then the possible values of X are 0, 1, 2, 3.

x	Events		
0	$\{F, F, F\}$		
1	$\{P, F, F\}$,	$\{F, P, F\}$,	$\{F, F, P\}$
2	$\{P, P, F\}$,	$\{P, F, P\}$,	$\{F, P, P\}$
3	$\{P, P, P\}$		

Please observe that there is only one way of passing none of the questions and only one way of passing all of the questions. However, there are three ways of passing one or two questions.

Example 2. Given that the probability of passing any one question is 80 percent, i.e., $p = 0.8$, what is the probability distribution for X, where X represents the number of questions passed?

Solution:

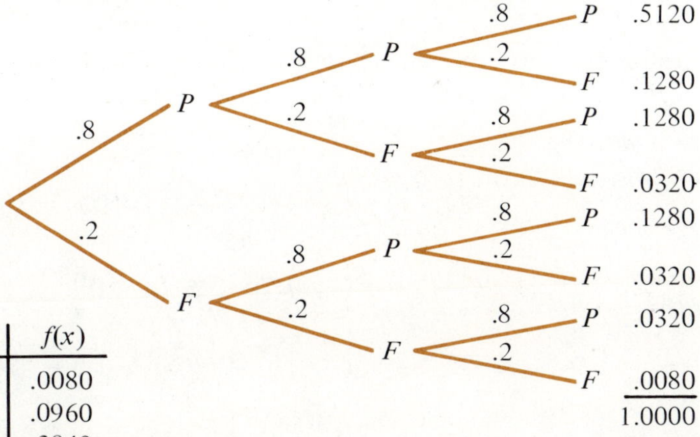

x	$f(x)$
0	.0080
1	.0960
2	.3840
3	.5120
	1.0000

FIGURE 8-10

Sample calculation:

$$\begin{aligned} f(1) = P(X = 1) &= P\{(P, F, F), (F, P, F), (F, F, P)\} \\ &= 0.0320 + 0.0320 + 0.0320 \\ &= 3(0.0320) \\ &= 0.0960 \end{aligned}$$

Note that in arriving at the 8 probabilities in the tree diagram we simply multiplied the probabilities. This is allowable because of independence. That is, the probability of passing question 2 is independent of whether or not question 1 was passed. The reader should recall that to obtain $f(1)$ we had to add $.0320 + .0320 + .0320$ because $X = 1$ can occur in three different ways. To obtain the probability distribution if there were 10 questions or $n = 10$ would be extremely tedious.

The following formula should seem reasonable in light of the previous discussion and example:

THEOREM 8.1. *Let X = the number of successes in n independent trials with constant probability, p, of success. Then*

$$f(x) = P(X = x) = \binom{n}{x} p^x q^{n-x}, \qquad \textit{where}$$

$$q = 1 - p \qquad \textit{and} \qquad \binom{n}{x} = \frac{n!}{x!(n-x)!} \qquad \textit{and}$$

$$x = 0, 1, 2, \ldots, n$$

DEFINITION 8.4. *X in Theorem 8.1 is said to have a binomial distribution.*

Example 3. Use the formula given in Theorem 8.1 to verify the result of Example 2, where $n = 3$, $p = .8$, $q = .2$, and $x = 0, 1, 2, 3$.

Solution:

$$P(X = 0) = \binom{3}{0}.8^0\ .2^3 = (1)(1)(.008) = .0080$$

$$P(X = 1) = \binom{3}{1}.8^1\ .2^2 = (3)(.8)(.04) = .0960$$

$$P(X = 2) = \binom{3}{2}.8^2\ .2^1 = (3)(.64)(.2) = .3840$$

$$P(X = 3) = \binom{3}{3}.8^3\ .2^0 = (1)(.512)(1) = .5120$$

Although the formula proves to be very handy if one is interested in a particular value of the random variable, i.e., $x = 2$, a lot of calculation would be necessary to arrive at the probability distribution when $n = 10$. For this reason Table A-8 is included in the Appendix. The table has values for $n = 1$ to 12, and $p = .1$ to .5. In addition to giving values of $f(x)$ it also gives the values of the cumulative distribution function $F(x)$, where $F(x) = P(X \leqslant x)$.

In the above problem, $p = .8$. Since $p > .5$ we will not be able to use the table unless we redefine the random variable. In our case we could let the random variable be number of questions failed rather than number of questions passed. So that rather than finding the probability of 2 questions passed where the probability of passing a question is .8, we will simply reword the problem as the probability of 1 question failed where the probability of failure is .2. The reader should use Table A-8 to verify the solution to Example 3.

Example 4. Given $n = 4$, $p = .3$, and X is a random variable that is binomially distributed, use Table A-8 to find $f(x)$ and $F(x)$ for $x = 0, 1, 2, 3, 4$.

Solution:

x	$f(x)$	$F(x)$
0	.2401	.2401
1	.4116	.6517
2	.2646	.9163
3	.0756	.9919
4	.0081	1.0000

PROBLEMS

1. Is the binomial distribution discrete or continuous?

2. What conditions need be imposed on the trials of an experiment before one can consider applying the binomial distribution?

3. An unprepared student is faced with a 10 question quiz. Given that the quiz is true or false:
   (a) What is the probability of answering 7 questions correctly?
   (b) What is the probability of answering 6 or more questions correctly?

4. Given that an unprepared student is faced with a 10 question quiz that is multiple choice with 5 possible answers for each question:
   (a) What is the probability of answering 7 questions correctly?
   (b) What is the probability of answering 6 or more questions correctly?

5. An urn contains 2 red and 8 white balls. Five balls are selected at random and with replacement from the urn. Let X equal the number of red balls drawn in these 5 trials.
   (a) What are the possible values of X?
   (b) Compute $f(x)$ for $x = 0, 1, 2, 3, 4, 5$.
   (c) What are the possible values of X if one had sampled without replacement?

6. Given $n = 3$ and $p = .2$, and that X is a random variable binomially distributed:
   (a) Use Table A-8 to find $f(x)$ for $x = 0, 1, 2, 3$.
   (b) Calculate the expected value of X.
   (c) Does the product, np, yield the same result as part (b)?

7. Given that your company writes 10,000 invoices per month and has on the average 5 invoices that are not properly prepared per month. Assuming that the binomial distribution applies, what is the probability of 5 invoices being not properly prepared? Set up only.

8. Of all the persons passing an entrance exam, 80 percent have completed the curriculum satisfactorily. Of the 8 persons who have just passed the exam what is the probability that 5 will successfully complete the curriculum, assuming the binomial distribution is applicable?

9. Ninety percent of all parts produced in a given department are not defective—they work properly. Given that a sample of 10 parts is taken from that department:
   (a) What is the probability that there will be 2 defective parts in the sample?
   (b) What is the probability that there will be 2 or less defective parts in the sample?

10. Mr. Manley owns and operates a small machine shop that has employed the same 10 employees for 5 years. It is very important that the machines are not left idle. Therefore he has compiled information on the number of employees absent on Monday for the past one hundred weeks. The results are as follows:

Number of Employees Absent	*Number of Weeks*
0	35
1	40
2	18
3	5
4	1
5	1
	100

Would the binomial distribution fit this set of data?

11. The Modern Day Magazine Company relies on the use of telephone calls to obtain subscriptions. An average of 100 calls are made by you every day. The probability that any call will result in a sale is .05. What is the probability that the 100 calls you make tomorrow, will result in
(a) exactly 4 subscriptions?
(b) more than 4 subscriptions?

8-8. THE HYPERGEOMETRIC DISTRIBUTION

In Section 8-7 we looked at the binomial probability distribution, where the probability of a success, p, was constant from trial to trial and the trials were independent.

Assume that the professor mentioned in Section 8-7, rather than give a written exam with three questions, will choose five students at random in the class and ask them a question to see if they are knowledgable in the material being presented. This is quite hypothetical, but businessmen often make decisions on the basis of sample information. This is especially true of the auditing profession. In any case returning to our hypothetical situation, we have $N =$ size of population, i.e., size of class $= 50$; $M =$ number of successes in the population, i.e., people in class who have mastered subject matter $= 30$; $n =$ size of sample, i.e., students selected for oral questions $= 5$. In addition the professor said that if all 5 who were selected had mastered the material, then there would be no final. Assuming that you are interested in starting work early this summer what is the probability that there will be no final? That is, $X =$ number of successes in the sample, i.e., people who were selected for oral exam and have mastered the material $= 5$.

Example 1. Use a tree diagram to arrive at your answer.

Solution:

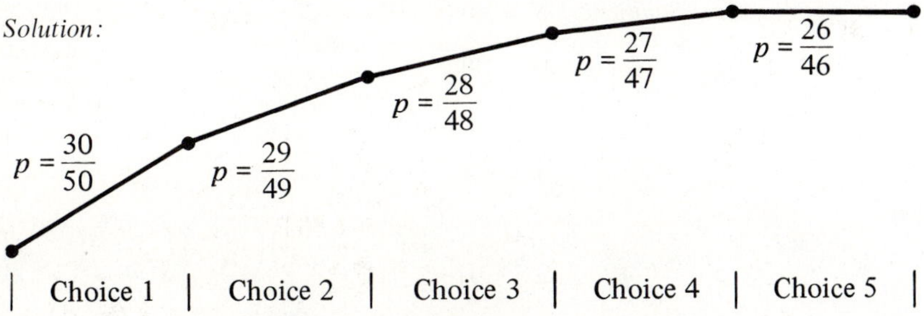

FIGURE 8-11

$$\text{Probability that there will be no final} = \left(\frac{30}{50}\right)\left(\frac{29}{49}\right)\left(\frac{28}{48}\right)\left(\frac{27}{47}\right)\left(\frac{26}{46}\right)$$

$$= (.6)(.5918)(.5833)(.5745)(.5652)$$

$$= .0673$$

Note: In the above example the probability that the first student chosen will perform successfully is 30/50. Given that the first person chosen was one that has mastered the material, then there will only be 29 people left that are knowledgable out of 49 possible when choice 2 is to be made. Note that this is an example of sampling without replacement. That is, once a student has been chosen for an oral question he or she will not be chosen again. Similarly, when sampling for an audit one would not select the same account twice during the same audit. The reader should see the importance then of the following distribution. If instead we asked the probability of 3 or more performing well on the oral exam, the method used in Example 1 would be extremely lengthy.

DEFINITION 8.5. *A random variable X is said to have a hypergeometric distribution if*

$$f(x) = \frac{\binom{N-M}{n-x}\binom{M}{x}}{\binom{N}{n}}$$

where N = size of the population

M = number of successes in the population

x = number of successes in the sample

n = size of the sample

Example 2. Verify that the above formula will give the same result obtained in Example 1.

Solution: $N = 50$

$M = 30$

$x = 5$

$n = 5$

$$P(x = 5) = f(5) = \frac{\binom{50-30}{5-5}\binom{30}{5}}{\binom{50}{5}} \frac{\binom{20}{0}\binom{30}{5}}{\binom{50}{5}}$$

$$= \frac{\dfrac{20!}{0!20!}\dfrac{30!}{3!25!}}{\dfrac{50!}{5!45!}} = \frac{\dfrac{30\cdot 29\cdot 28\cdot 27\cdot 26}{5\cdot 4\cdot 3\cdot 2\cdot 1}}{\dfrac{50\cdot 49\cdot 48\cdot 47\cdot 46}{5\cdot 4\cdot 3\cdot 2\cdot 1}}$$

$$= \frac{30\cdot 29\cdot 28\cdot 27\cdot 26}{50\cdot 49\cdot 48\cdot 47\cdot 46} = 0.0673$$

We see that the formula given does in fact yield the solution obtained in Example 1.

Example 3. What is the probability that there will be no final if 3 or more people do well in the oral exam?

Solution:

$$\begin{array}{lll} \quad x\ \ N\ \ n\ \ M & \quad x\ \ N\ \ n\ \ M & \quad x\ \ N\ \ n\ \ M \\ H(3{:}\ 50, 5, 30) & +\,H(4{:}\ 50, 5, 30) & +\,H(5{:}\ 50, 5, 30) \\ f(3) & +\,f(4) & +\,f(5) \end{array}$$

$$= \frac{\binom{20}{2}\binom{30}{3}}{\binom{50}{5}} + \frac{\binom{20}{1}\binom{30}{4}}{\binom{50}{5}} + \frac{\binom{20}{0}\binom{30}{5}}{\binom{50}{5}}$$

$$= \frac{(190)(4060) \quad + (20)(27{,}405) \quad + (1)(142{,}506)}{2{,}118{,}760}$$

$$= .3641 \quad + .2587 \quad + .0673$$

$$= 0.6901$$

If $N \geqslant 10n$, then one can use the binomial distribution to arrive at an approximation of the hypergeometric distribution. This is often necessary because of the tedious computation that occurs, as verified in Example 2. In our case, $N = 50$ and $n = 5$ or, $50 \geqslant 10(5)$, so this condition is met.

Example 4. Use the binomial distribution to obtain an approximation to the result obtained in Example 2.

Solution: Given $n = 5$, $p = 30/50 = .60$, we must compute $P(X = x)$ for $x = 3, 4,$ and 5 and add the results. If we are going to use Table A-8, recall that only values up to $p = .5$ are given so we must redefine the random variable, which will result in the following:

$$f(x) = \binom{5}{x} .4^x .6^{5-x} \qquad \text{for } x = 0, 1, 2$$

$$= f(0) + f(1) + f(2)$$

$$= .0778 + .2592 + 3456$$

$$= .6826$$

Note that .6826 is not a bad approximation to .6901. In fact the discrepancy results from the fact that we are using $(.6)^5$ rather than $(.6)(.5918)(.5833)(.5745)(.5652)$ when $x = 0$, for example.

PROBLEMS

1. Given that you have an urn with 6 red balls and 4 white balls and that you are to pick 3 balls from the urn without replacement:
   (a) Illustrate the selection process using a tree diagram.
   (b) What is the probability of choosing 3 red balls? Arrive at the answer intuitively.
   (c) Use the formula presented in Definition 8.5 to arrive at the answer obtained in part (b).

2. **(a)** If you let X equal the number of red balls drawn in Problem 1, then what are the possible values of X?
   (b) Compute $f(x)$ for $x = 0, 1, 2, 3$.

3. What determines whether one should use the hypergeometric probability distribution versus the binomial probability distribution?

4. Making reference to Example 2 in this section, what do the following terms in the numerator and denominator represent?

$$\frac{\binom{20}{0}\binom{30}{5}}{\binom{50}{5}}$$

5. Given a population of 1000 accounts from which we are going to draw a sample of 50 for the purpose of balance confirmation. Past experience leads one to hypothesize that of the 1000 accounts, 50 are in error. Also assume that sampling is done without replacement.
   (a) What is the probability that no accounts are in error? Set up only.
   (b) Use the binomial probability distribution to approximate part (a). Set up only.

6. **(a)** Given the information in Problem 5, the probability distribution is approximated as follows:

x	0	1	2	3	4	5	6	7
$f(x)$	0.0821	0.2052	0.2565	0.2138	0.1336	0.0668	0.0278	0.0099

(b) If when one took the sample of 50, it was found that 10 accounts were in error, should the original hypothesis that there are 50 accounts that are incorrect in the population of 1000 be questioned? Do you think it's more or less?

7. Given a lot of 10,000 parts. Past experience indicates that 100 of these parts are defective. If for quality control purposes one takes a sample of 60 parts without replacement:

(a) What is the probability that 3 parts will be defective? Set up only.
(b) What is the binomial approximation of (a)? Set up only.
(c) Is it likely that more than 3 of the 60 parts chosen will be defective?

8. Jurors are selected from a list of registered voters that is made up of 100,000 people, 10,000 of which can be classified as belonging to minority groups.

(a) Of the 12 jurors selected, what is the probability that there will be no minority groups represented? Set up only.
(b) Use the binomial distribution to approximate part (a).

8-9 PROBABILITY VECTORS, MARKOV ANALYSIS

In this section we will illustrate the combined use of probabilities expressed as vectors and introduce the concept of a *Markov process*.

A vector $\mathbf{p} = (p_1, p_2, \ldots, p_n)$, $\mathbf{p}\begin{pmatrix} p_1 \\ p_2 \\ \vdots \\ p_n \end{pmatrix}$ is called a *probability vector* if the components are nonnegative and their sum is 1.

Example 1. Which of the following vectors are probability vectors?

$$\mathbf{a} = (\tfrac{1}{3}, \tfrac{1}{3}, \tfrac{1}{3}) \qquad \mathbf{b} = (0.4, -0.2, 0.8)$$

$$\mathbf{c} = (0.4, 0.2, 0.3, 0.2) \qquad \mathbf{d} = \begin{pmatrix} 0.5 \\ 0.3 \\ 0.2 \end{pmatrix}$$

Solution: **a** is a probability vector, **b** is not a probability vector because the second component is negative. **c** is not a probability vector because the sum of the components is greater than 1. **d** is a probability vector.

A square matrix with nonnegative elements such that the sum of the elements in each row is 1 is called a *transition matrix*.

Example 2. Which of the following are transition matrices?

$$A = \begin{pmatrix} 0.8 & 0.2 \\ 0.6 & 0.4 \end{pmatrix}, \quad B = \begin{pmatrix} 0.3 & 0.6 & 0.1 \\ 0.2 & 0.7 & 0.1 \end{pmatrix}, \quad C = \begin{matrix} 0.3 & 0.2 & 0.5 \\ 0.2 & 0.4 & 0.3 \\ 0.5 & 0.4 & 0.1 \end{matrix}$$

Solution: A is a transition matrix. B is not a transition matrix because it is not a square matrix. C is not a transition matrix because the sum of the elements of row two is not equal to 1.

One of the primary concerns of the business manager is his market share. The *Markov process* is a method of analyzing present market share and the future of that share. A very important factor in the determination of future share is customer loyalty. In the remainder of the section we will use a transition matrix to depict purchase probabilities. Once this is done we will predict the market share of future periods and finally determine the market equilibrium.

Example 3. Three possible texts can be used for a given course. The three texts are *EP*, *XY*, and *WZ*. If the *EP* text is presently being used, the probability of it being used again is 0.8 and the probability that users of *EP* switch to *XY* or *WZ* is 0.1 and 0.1. If the *XY* text is in use, the probability of it being continued is 0.5 and the probability that users will switch to *EP* or *WZ* is 0.4 and 0.1, respectively. If the *WZ* text is in use the probability of it being continued is 0.3 and the probability that users will switch to *EP* or *XY* is 0.4 and 0.3, respectively. Use a transition matrix to display these probabilities.

Solution:

		Next Purchase		
		EP	*XY*	*WZ*
	EP	0.8	0.1	0.1
Presently in Use	*XY*	0.4	0.5	0.1
(Previous Purchase)	*WZ*	0.4	0.3	0.3

The reader should note that a transition matrix is equivalent to a table of conditional probabilities where the condition imposed is the previous purchase. At present, 200 schools are using the *EP* text, 400 are using the *XY* text, and 400 are using the *WZ* text, i.e., the current market share is 20 percent, 40 percent, and 40 percent, respectively. Assume no new customers and no old customers leaving the market. What will be the market share in the next period if the transition matrix of Example 3 applies? Of the 200 *EP* users, 0.8(200) or 160 will purchase *EP*, and so forth.

TABLE 8-35

		Purchase of		
		EP	*XY*	*WZ*
Purchase by	*EP*	160	20	20
	XY	160	200	40
	WZ	160	120	120
		480	340	180

The market share expressed as probabilities will be 0.48, 0.34, and 0.18, respectively.

We could have used the tree diagram in Figure 8-12 to obtain the same result.

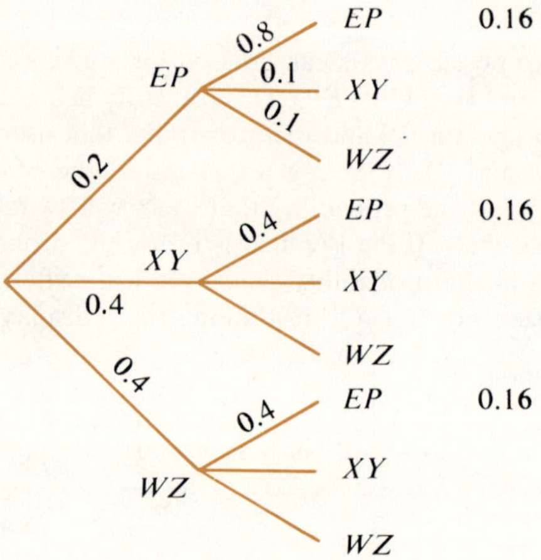

FIGURE 8-12

The market share of $EP = 0.16 + 0.16 + 0.16$, or 0.48. The reader should complete the tree diagram and verify the results.

Example 4. Use matrix multiplication to find the market share of period 2, as calculated above.

Solution: Define $\mathbf{s} = (s_1, s_2, s_3)$, where s_1 is the market share expressed as a probability of *EP*, and so forth. Then

$\mathbf{s} = (0.2, 0.4, 0.4) =$ market share in period 1 expressed as a probability vector

$$T = \begin{pmatrix} 0.8 & 0.1 & 0.1 \\ 0.4 & 0.5 & 0.1 \\ 0.4 & 0.3 & 0.3 \end{pmatrix} = \text{transition matrix}$$

$\mathbf{s}' = \mathbf{s}T =$ market share in the next period

$\mathbf{s}' = (0.48, 0.34, 0.18)$

A sample calculation follows.

$$s_1' = t_{11}s_1 + t_{21}s_2 + t_{31}s_3$$

$$s_1' = (0.8)(0.2) + (0.4)(0.4) + (0.4)(0.4)$$

$$s_1' = 0.48$$

The increase in the market share for the *EP* text was considerable. However, at some point in time the gains will be exactly offset by losses and the market will achieve an equilibrium. When this happens the following will be the case:

Share next period		=	Share this period
EP	s_1'	=	s_1
XY	s_2'	=	s_2
WZ	s_3'	=	s_3

Expanding,

$$t_{11}s_1 + t_{21}s_2 + t_{31}s_3 = s_1$$

$$t_{12}s_1 + t_{22}s_2 + t_{32}s_3 = s_2$$

$$t_{13}s_1 + t_{23}s_2 + t_{33}s_3 = s_3$$

Also, because $\mathbf{s}$ is a probability vector, we know that $s_1 + s_2 + s_3 = 1$. Using the values in the transition matrix we have

$$0.8s_1 + 0.4s_2 + 0.4s_3 = s_1 \qquad (1)$$

$$0.1s_1 + 0.5s_2 + 0.3s_3 = s_2 \qquad (2)$$

$$0.1s_1 + 0.1s_2 + 0.3s_3 = s_3 \qquad (3)$$

and

$$s_1 + s_2 + s_3 = 1 \qquad (4)$$

Using equations (1), (2), and (4) to solve for s_1, s_2, and s_3, we have

$$-0.2s_1 + 0.4s_2 + 0.4s_3 = 0$$

$$0.1s_1 - 0.5s_2 + 0.3s_3 = 0$$

$$s_1 + \quad s_2 + \quad s_3 = 1$$

If we write the above as a matrix equation, we have

$$\begin{pmatrix} -0.2 & 0.4 & 0.4 \\ 0.1 & -0.5 & 0.3 \\ 1.0 & 1.0 & 1.0 \end{pmatrix} \begin{pmatrix} s_1 \\ s_2 \\ s_3 \end{pmatrix} = \begin{pmatrix} 0 \\ 0 \\ 1 \end{pmatrix}$$

Multiplying each side by T^{-1} will give a solution.

$$\begin{pmatrix} s_1 \\ s_2 \\ s_3 \end{pmatrix} = \begin{pmatrix} \frac{-0.80}{0.48} & 0 & \frac{0.32}{0.48} \\ \frac{0.20}{0.48} & \frac{-0.20}{0.48} & \frac{0.10}{0.48} \\ \frac{0.50}{0.48} & \frac{-0.20}{0.48} & \frac{0.06}{0.48} \end{pmatrix} \begin{pmatrix} 0 \\ 0 \\ 1 \end{pmatrix} = \begin{pmatrix} 0.67 \\ 0.21 \\ 0.12 \end{pmatrix}$$

If we verify this result, we see that the market share in the next period is the same as the market share in this period.*

$$\mathbf{s}' = (0.67, 0.21, 0.12) \begin{pmatrix} 0.8 & 0.1 & 0.1 \\ 0.4 & 0.5 & 0.1 \\ 0.4 & 0.3 & 0.3 \end{pmatrix} = (0.668, 0.208, 0.124)$$

Therefore, we have an equilibrium.

PROBLEMS

1. Which of the following are probability vectors?

(a) $\mathbf{p} = \begin{pmatrix} 0.3 \\ 0.4 \\ 0.3 \end{pmatrix}$ **(c)** $\mathbf{p} = (0.2, 0.3, 0.1, 0.3)$

(b) $\mathbf{p} = (-0.1, 0.8, 0.3)$ **(d)** $\mathbf{p} = (\frac{2}{3}, \frac{1}{6}, \frac{1}{6})$

2. In words, write the meaning of the following transition matrix:

$$\begin{array}{c} \\ A \\ B \\ C \end{array} \begin{array}{c} \begin{array}{ccc} A & B & C \end{array} \\ \begin{pmatrix} 0.5 & 0.3 & 0.2 \\ 0.1 & 0.6 & 0.3 \\ 0.2 & 0.2 & 0.6 \end{pmatrix} \end{array}$$

3. Which of the following are transition matrices?

(a) $$\begin{array}{c} \\ A \\ B \\ C \end{array} \begin{array}{c} \begin{array}{ccc} A & B & C \end{array} \\ \begin{pmatrix} 0.5 & 0.2 & 0.3 \\ \frac{1}{3} & \frac{1}{3} & \frac{1}{3} \\ 0.2 & 0.2 & 0.6 \end{pmatrix} \end{array}$$

(b) $$\begin{array}{c} \\ A \\ B \\ C \end{array} \begin{array}{c} \begin{array}{ccc} A & B & C \end{array} \\ \begin{pmatrix} 0.4 & 0.4 & 0.2 \\ 0.3 & 0.5 & 0.2 \\ 0.3 & 0.1 & 0.5 \end{pmatrix} \end{array}$$

4. Do the calculations necessary to obtain Table 8-35.

* There is a slight discrepancy due to rounding.

5. Use the transition matrix and the result of Example 4 to find the market share of *EP*, *XY*, and *WZ* in period 3.

6. Show that the long run market share is $\frac{1}{3}$ for all brands given the following transition matrix:

$$\begin{array}{c} \\ X \\ Y \\ Z \end{array} \begin{array}{c} \text{Switched to} \\ \begin{array}{ccc} X & Y & Z \\ 0.8 & 0.1 & 0.1 \\ 0.1 & 0.8 & 0.1 \\ (0.1 & 0.1 & 0.8) \end{array} \end{array}$$

7. Using the transition matrix given in Problem 6 and given that the present market share is 0.4, 0.2, and 0.4 for *X*, *Y*, and *Z*, respectively, what will the market share be in the next period for *X*, *Y*, and *Z*?

8. Given the following transition matrix.

$$\begin{array}{c} \\ A \\ B \\ C \end{array} \begin{array}{c} \text{Switched to} \\ \begin{array}{ccc} A & B & C \end{array} \\ \begin{pmatrix} 0.6 & 0.2 & 0.2 \\ 0.2 & 0.6 & 0.2 \\ 0.2 & 0.2 & 0.6 \end{pmatrix} \end{array}$$

(a) What is the long run market share for brands *A*, *B*, and *C*?
(b) Does the answer to part (a) depend on the starting market share?
(c) Does the number of periods required to reach the equilibrium depend on the starting market share?

9. Assume the following breakdown of the annual $350 million dentifrice market:

Crest	38 percent
Colgate MFP	24 percent
Closeup	10 percent
Gleem	7 percent
Colgate Ultra Brite	7 percent

Given the above, "other" dentifrices must account for what percent of the market?

Given that total annual sales next year will be $360 million, what will annual sales for each brand be, assuming no change in market share?

10. Assume the following as regards low-priced pantyhose being sold in supermarkets:

Hosiery Maker	Brand	Annual Sales (in millions)
Hanes	L'eggs	$200
Burlington	Activ	$100
Kayser Roth	No Nonsense	$100

Given the following transition matrix, find the market share for next period and the market share at equilibrium.

		L'eggs	Activ	No Nonsense
Previous Purchase	L'eggs	.5	.2	.3
	Activ	.2	.4	.4
	No Nonsense	.3	.1	.6

REFERENCES

Bierman, Harold, Bonini, C. P., and Hausman, W. H. *Quantitative Analysis for Business Decisions.* Homewood, Ill.: Richard D. Irwin, Inc., 1969.

Kemeny, John G., Schleifer, A., Snell, J. L., and Thompson, G. L. *Finite Mathematics with Business Applications.* Englewood Cliffs, N.J.: Prentice-Hall, Inc., 1964.

Levin, Richard I., and Kirkpatrick, C. A. *Quantitative Approaches to Management.* New York: McGraw-Hill Book Co., Inc., 1968.

Lipschutz, Seymour. *Theory and Problems of Probability.* New York: Schaum's Outline Series, McGraw-Hill Book Co., Inc., 1968.

Neter, John, Wasserman, William, and Whitmore, G. A. *Fundamental Statistics for Business and Economics,* fourth ed. Boston: Allyn and Bacon, Inc., 1973.

Scheerer, Anne E. *Probability on Discrete Sample Spaces with Applications.* Scranton, Pennsylvania: International Textbook Company, 1969.

Schellenberger, Robert E. *Managerial Analysis.* Homewood, Ill.: Richard D. Irwin, Inc., 1969.

Schlaifer, Robert. *Probability and Statistics for Business Decisions.* New York: McGraw-Hill Book Co., Inc., 1959.

Simone, Albert J. *Probability: An Introduction with Applications.* Boston: Allyn and Bacon, Inc., 1967.

Theodore, Chris A. *Applied Mathematics: An Introduction; Mathematical Analysis for Management.* Homewood, Illinois: Richard D. Irwin, Inc., 1965.

NINE

Descriptive statistics

When talking about statistics one usually thinks of batting averages or other data related to sports, one more casualty added to an already too long list of automobile fatalities over a long holiday weekend, or, on the brighter side, the number of points your stocks have increased. On a broader scale one can define statistics as the science that deals with the collection, classification, summarization, and presentation of numerical data. This definition is applicable when talking about descriptive statistics. On the other hand when one talks about statistical inference or inductive statistics he is interested in making an inference about the population or universe (i.e., all students at your college, all of the employees employed by an industry) on the basis of sample information (a sample being a subset of the population such as commuters at your college, employees of one firm in a given industry). In this chapter the subject matter of descriptive statistics only will be discussed.

9-1 COLLECTION AND CLASSIFICATION OF THE DATA

The actual collection of the data for statistical analysis can in fact be the most difficult and time consuming task depending on whether one has to go outside his firm or organization for the data or if this information can be obtained internally. Also important at this stage is whether the data are already available from some other study (secondary data) or if they must be generated for the purpose at hand (primary data). If for example one is interested in the commissions for the month of September for the forty salesmen of the New England sales territory, one would simply have to search through the individual payroll records which are arranged alphabetically.

Example 1. List the monthly commissions of the forty salesmen from New England.

Solution:

J.A.	$ 670	P.E.	$1240	B.L.	$ 810	O.Q.	$ 560
R.A.	1080	R.E.	1060	R.L.	1200	D.R.	1170
P.B.	890	S.E.	1550	T.L.	600	E.S.	980
R.B.	1440	C.F.	1420	M.M.	810	M.S.	760
B.C.	1290	S.F.	610	S.M.	800	S.T.	970
G.C.	450	P.G.	830	C.N.	1030	J.V.	1110
M.C.	1630	H.I.	1220	N.O.	910	J.W.	540
A.D.	840	J.I.	1050	M.P.	780	R.W.	960
C.D.	630	A.K.	1350	S.P.	1010	P.Y.	760
E.D.	350	M.K.	1030	V.P.	930	T.Z.	740

Note: Monthly commissions were rounded.

It would be most difficult to work with the data as they appear in Example 1.

In order to facilitate studying the data in Example 1, it would be most helpful to arrange the data in ascending order. If this list were much longer, this would be a fairly difficult task if the files were not computerized. For our purpose, one would simply search the list for the lowest dollar value for monthly commission, enter it as the first entry on the ordered list, strike it from the unordered list, and look for the lowest dollar value for monthly commission of the remaining entries, and so on.

Example 2. Arrange the monthly commissions of the forty salesmen given in Example 1 in ascending order.

Solution:

$350	$760	$ 970	$1170
450	780	980	1200
540	800	1010	1220
560	810	1030	1240
600	830	1030	1290
610	840	1050	1350
630	890	1060	1420
670	910	1080	1440
740	930	1110	1550
760	960	1150	1630

Looking at the solution to Example 2, one can easily see that the monthly commissions range from a low of $350.00 to a high of $1630.00. We will now look at how one can still improve upon the presentation of these data by grouping or arranging the data into classes.

Rather than having to handle numerous individual pieces of data, it might be helpful if one arranged or grouped the data into various classes. Some general rules that must be observed here are—

1. that the classes be mutually exclusive (i.e., only one class for each observation or data point).
2. that the classes be exhaustive (i.e., at least one class for each observation or data point).
3. that there be neither too few nor too many classes.
4. that the class intervals be equal.
5. that the end points of each class be chosen such that the class well represents the data in that class.

The reader may note that these same rules were observed when defining events in Chapter 7. These guidelines are illustrated in some of the later examples.

Example 3. Classify the ordered data of Example 2 into 7 classes. Let the width of each class interval be $200 and let $300 be the lower limit of the first class.

Solution:

Intervals	*Number of Observations*
$ 300 and under $ 500	2
500 and under 700	6
700 and under 900	9
900 and under 1100	11
1100 and under 1300	7
1300 and under 1500	3
1500 and under 1700	2
	40

The reader should note that all the rules outlined were observed. That is—

1. Each observation is in only one interval.
2. There is an interval for each observation.
3. There are 7 classes, which is quite managable.
4. The class widths are all equal to $200.
5. The data do not tend to gather at the end points of the interval, which would cause problems when doing calculations in future sections.

One might wonder how to treat an observation of say $2400. (Assume for the present that the 40th observation was $2400 rather than $1630.) That is would you have 3 classes with 0 observations and a last interval with one observation? This would appear as follows.

Intervals	*Number of Observations*
\$ 300 and under \$ 500	2
500 and under 700	6
700 and under 900	9
900 and under 1100	11
1100 and under 1300	7
1300 and under 1500	3
1500 and under 1700	1
1700 and under 1900	0
1900 and under 2100	0
2100 and under 2300	0
2300 and under 2500	1
	40

The proper way to treat a situation such as this is to make use of what is called an open-ended interval. This can be done for the first interval as well as the last. In this case, we would have:

Intervals	*Number of Observations*
\$ 300 and under \$ 500	2
500 and under 700	6
700 and under 900	9
900 and under 1100	11
1100 and under 1300	7
1300 and under 1500	3
1500 or more	2
	40

The use of open-ended intervals should be avoided unless absolutely necessary, as in this case because of computational problems in future sections.

PROBLEMS

1. What rules must be observed when grouping the data into classes?

2. When should open-ended intervals be used?

3. What is the disadvantage of using open-ended intervals?

4. Assume that you asked the twenty-eight students in a class to count the change they had on them and that the results were as follows:

\$ .84	.60	.56	.85	.00
.35	.75	.42	.45	.80
2.26	2.49	1.03	.14	.00
.00	.73	.00	.09	.25
1.90	.58	.55	1.44	
.27	1.23	1.20	.83	

Group the observations into intervals and note the number of observations in each interval.

5. As owner of the Nice and Cozy Oil Company you are interested in the number of gallons of Number 2 fuel oil delivered to your customers. Looking at the deliveries for November 2nd and ordering the data you have the following:

40	91	202	420
80	92	210	450
81	94	220	480
81	95	231	485
82	96	235	490
83	98	238	495
88	150	240	498
90	189	241	

Group the data into classes. Assume the first class is 00 and under 100 gallons.

6. Comment on the following sets of class intervals:

(a)	(b)	(c)
50–100	60–70	16–24
101–200	70–80	30–38
201–300	80–90	40–48
301–500	90–100	50–58

9-2 SUMMARIZATION OF THE DATA—MEASURES OF CENTRAL TENDENCY

It is sometimes desirable to summarize a set of data points or observations by a single number. The three most common measures of *central tendency* are the *mean* or arithmetic average, the *median*, and the *mode*. The computations required for these measures depend on whether the data are grouped or ungrouped, that is, whether they are like Example 2 or Example 3 respectively in the previous section. If the data are ungrouped, that is in the form of Example 2 in the previous section, one simply adds all of the observations and divides by the number of observations to obtain the mean. The formula for the mean of a set of ungrouped data is

$$\overline{X} = \frac{\sum_{i=1}^{n} X_i}{n}$$

where X_i = ith observation
n = the number of observations and
$\overline{X}$ = mean or arithmetic average.

Example 1. Calculate $\overline{X}$ for the data given in Example 2 of the previous section.

Solution: $X_1 = \underline{\$350}$, $X_2 = \underline{\$450}$, ..., $X_{40} = \underline{\$1630}$

$n = 40$

$$\overline{X} = \frac{\sum_{i=1}^{n} X_i}{n} = \frac{\$38{,}400}{40} = \underline{\$960.00}$$

We see that the average of the ungrouped data is $\underline{\$960.00}$. This is the true mean of the data. If the data are grouped as in Example 3 of the previous section, the mean that is calculated is valid only if one classified the data properly. Recalling the grouped data of Example 3 in Section 11-1, we have:

Intervals	*Number of Observations*
\$ 300 and under \$ 500	2
500 and under 700	6
700 and under 900	9
900 and under 1100	11
1100 and under 1300	7
1300 and under 1500	3
1500 and under 1700	2
	40

Looking at the first interval we see that there are two observations between \$300 and \$500. We assume that the 2 observations are \$400 or the midpoint of the interval. Implicit in this assumption is that the mean of the observations in this class is approximately equal to the midpoint of this class. If this is not the case, then when one calculates the mean using grouped data, the result will significantly differ from the true mean. To calculate the mean or arithmetic average when the data are grouped, one simply:

1. multiplies the midpoints of each interval by the frequency of that interval;
2. adds the products of step 1;
3. and divides the result of step 2 by the total number of observations, n.

The formula for the mean of a set of grouped data is

$$\overline{X} = \frac{\sum_{j=1}^{k} m_j f_j}{n}$$

where m_j is the midpoint of the jth class
f_j is the number of observations in the jth class
k is the number of classes
n is the total number of observations.

Example 2. Calculate $\overline{X}$ for the following set of grouped data.

Intervals	*Midpoint* m_j	*Frequency* f_j	$m_j f_j$
\$ 300 and under \$ 500	\$ 400	2	\$ 800
500 and under 700	600	6	3600
700 and under 900	800	9	7200
900 and under 1100	1000	11	11000
1100 and under 1300	1200	7	8400
1300 and under 1500	1400	3	4200
1500 and under 1700	1600	2	3200
		40	\$38400

Solution:

$$\overline{X} = \frac{\sum_{j=1}^{k} m_j f_j}{n} = \frac{\$38,400}{40} = \$960.00$$

Note that $\sum_{j=1}^{k} f_j = 40 = n$.

Comment: This result is exactly equal to the true mean which indicates that the classification was done properly. In general, one should not expect these two results to be exactly equal but rather only approximately equal.

A characteristic of the mean is that it is that value with the property that the sum of the deviations about it will equal zero. This will be verified in the next section. The reader should observe the complications that would arise if the list in Example 2 had an open-ended interval. That is what value would be used as the midpoint.

Next, we define another common measure of central tendency called the median. The median is preferred over the mean when the set of data has an exceptionally high or low value. In effect what we're saying is that when we have an open-ended interval which occurs when there is an exceptionally high or low value, we need not be concerned with the computational difficulty mentioned earlier because then the median is a more relevant measure of central tendency. This is so because it is not influenced by extreme high or low values.

The median is a number with the characteristic that 50 percent of the observations lie above it and 50 percent of the observations lie below it regardless of the size of the observations. When working with ungrouped data, one first determines the item that is the median as follows.

$$\text{Item that is the median} = \frac{n+1}{2}$$

where n is the number of observations.

Example 3. Using the data given in Example 2 of the previous section, determine which item is the median.

Solution:

$$\text{Item that is the median} = \frac{40+1}{2} = 20.5$$

Note: 20.5 is not the median, but the 20.5th item is the median. To locate the median we must go through the list *which is in ascending order* and find the 20.5th item, i.e., find the average of the 20th and 21st observations.

Example 4. Locate the 20.5th item in Example 2 of the previous section.

Solution:

$$\frac{960+970}{2} = 965$$

Comment: This value of $965 has the characteristic that 50 percent of the observations lie below this value and 50 percent of the observations lie above this value.

When determining the median for ungrouped data there is a distinction between the cases (1) an odd number of observations and (2) an even number of observations. Assume that the observations have been arranged in ascending order: $x_1, x_2, \ldots, x_n$, where $x_1 \leqslant x_2 \leqslant x_3 \leqslant \cdots \leqslant x_n$.

In the odd number case, the median is the observation

$$x_{(n+1)/2}$$

In the even number case, the median is given by

$$\frac{x_{n/2} + x_{(n+2)/2}}{2}$$

When working with grouped data, the procedure is somewhat more complicated. As before, one must locate the item that is the median which is given by the following formula.

$$\text{Item that is median} = \frac{n}{2}$$

Example 5. For the data in Example 3 of the previous section which observation is the median?

Solution:

$$\text{Item that is the median} = \frac{40}{2} = 20\text{th}$$

One must now locate the 20th item in the following table.

Interval	*Number of Observations* (f_j)
\$ 300 and under \$ 500	2
500 and under 700	6
700 and under 900	9
900 and under 1100	11
1100 and under 1300	7
1300 and under 1500	3
1500 and under 1700	2
	40

The twentieth item is clearly in the interval \$900 and under \$1100. To locate it one first makes the assumption that the observations in this class are evenly distributed within the class or interval.

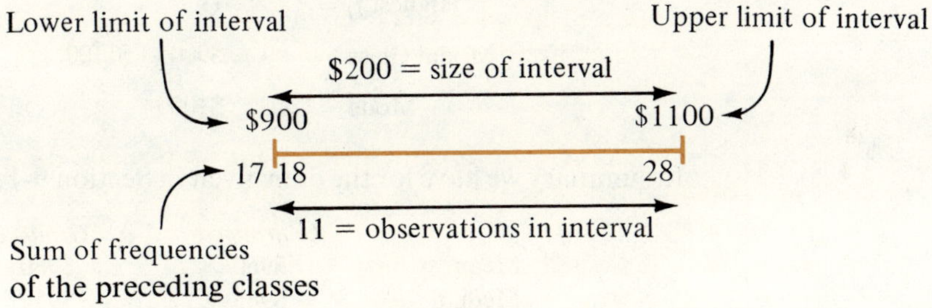

FIGURE 9-1

Using linear interpolation a general formula for calculating the median is

$$\text{Median} = LL_j + \left(\frac{\dfrac{n}{2} - \sum\limits_{i=1}^{j-1} f_i}{f_j} \right) c$$

where LL_j is the lower limit of the interval that contains the median and n is the number of observations.

$\sum_{i=1}^{j-1} f_i$ is the sum of the frequencies of the classes preceding the median class.

f_j is the number of observations in the median class.

c is the size of the intervals.

Example 6. Find the median of the grouped data.

Solution:

$$\text{Median} = \$900 + \left(\frac{20 - 17}{11} \right) 200$$

$$= \$954.54$$

Note that we are simply saying that the median is $900 + 3/11 of $200.

The last measure of central tendency to be covered is the mode. The mode is simply that observation that occurs most frequently. It is most applicable to grouped data and can simply be found by searching for the highest f_j and then noting the corresponding interval. This interval or class is called the modal class. If one wants a single value for the mode, the midpoint of the modal class is often used.

Example 7. Calculate the mode for the grouped data given in Example 3 of Section 9-1.

Solution:

Highest f_j = 11

Modal Class = $900 – $1100

Mode = $1000

In summary we have for the data given in Section 9-1 the following.

	Ungrouped	*Grouped*
Mean	$960	$960
Median	$965	$954
Mode	Not Applicable	$1000

When the mean for the grouped data is approximately equal to the mean for the actual data, then one probably did a good job of grouping the data.

Also when the mean and median are approximately equal, we can say that the data are "normal" or not "skewed." We will discuss this further in Section 9-4.

PROBLEMS

1. Verify that the mean of the observations in each class (see Example 2 of Section 9-1) is approximately equal to the midpoints of the intervals used in Example 2 of Section 9-2.
2. The ages of the students in a class are as follows: 24, 19, 19, 18, 20, 23, 19, 19, 18, 24, 19, 20, 21, 40, 21, 20, 22, 19, 19.
   (a) Calculate the mean.
   (b) Calculate the median.
   (c) Calculate the mode.
   (d) Do you arrive at the same answer for **(a)**, **(b)**, and **(c)**?
   (e) Does this have any implications?
3. Group the data given in Problem 2 and calculate the mean.

4. A student's final course grades in English, Mathematics, Economics, Speech, Sociology, and Introductory Physics are as follows: 70, 95, 80, 60, 70, and 90. If the respective credits received for these courses are 3, 3, 3, 1, 3, and 4, determine an appropriate grade for the student.

5. For the following set of numbers: 6, 3, 9, 2, 3, 4, 3 find the **(a)** mean, **(b)** median, **(c)** mode.

6. Given the following data which represent deliveries of fuel oil in gallons delivered on November 2nd:

40	91	202	420
80	92	210	450
81	94	220	480
81	95	231	485
82	96	235	490
83	98	238	495
88	150	240	498
90	189	241	

(a) Are these data grouped or ungrouped?
(b) Calculate the mean.
(c) What item is the median?
(d) What is the median?

7. Given that the data of Problem 6 are grouped as follows:

Interval	*Frequency*
0 and under 100	14
100 and under 200	2
200 and under 300	8
300 and under 400	0
400 and under 500	7

(a) Calculate the mean.
(b) What item is the median?
(c) What is the median?

8. The frequency distribution of the weekly wages of 65 part-time student assistants at G.U.T. College is as follows:

Weekly Wage	*Number of Employees*
\$ 5.00 and under \$15.00	10
15.00 and under 25.00	10
25.00 and under 35.00	20
35.00 and under 45.00	10
45.00 and under 55.00	10
55.00 and under 65.00	5
	65

(a) What is the midpoint of the second class?
(b) What are the class limits of the third class?
(c) What is the frequency of the sixth class?
(d) Which class interval has the largest frequency?
(e) Calculate the mean average weekly wage for the preceding data.
(f) Calculate the median weekly wage for the preceding data.

9. Calculate the mean, median, and mode for the data given in Problem 4 of Section 9-1.

10. What assumption is made when one calculates the mean of grouped data?

11. What assumption is made when one calculates the median of grouped data?

9-3 SUMMARIZATION OF THE DATA—MEASURES OF DISPERSION

There is a common comment about the person who drowned in a stream which on the average is only 3 feet deep. We have a similar situation here. Even though we know, using summary measures, that the salesmen of the previous examples are making on the average around $1000 a month, it is important to summarize how the salesmen fall or group around this average. That is, are twenty salesmen making $300 a month and the other twenty $1700 a month? The measures of dispersion that are covered in this section are used to answer questions such as this. The most common measures of dispersion or variability are the *range*, the *variance*, the *standard deviation*, and the *coefficient of variation.*

For ungrouped data, the range, the simplest measure of dispersion, is merely the highest value minus the lowest value. Recalling the ordered data in Example 2 of Section 9-1 we have $1630 − $350 = $1280. For grouped data it is impossible to calculate the exact range simply because of the fact that in the grouping process one has lost the distinct high and low values. Two common methods for determining the range when data are grouped are to subtract the lower limit of the lowest class from the upper limit of the highest class and to calculate the difference between the midpoints of the highest and the lowest class.

Example 1. Calculate the range for the following grouped data using the difference between the lower limit of the lower class and the upper limit of the highest class.

Interval	*Number of Observations* (f)
$ 300 and under $ 500	2
500 and under 700	6
700 and under 900	9
900 and under 1100	11
1100 and under 1300	7
1300 and under 1500	3
1500 and under 1700	2

Solution: The reader should be aware that the following solution may in fact overestimate the range of the actual data.

Range = $1700 − $300 = $1400

The range has a serious limitation in that its value depends solely on the lowest and highest observations. It could be that these two values are not representative and therefore could be misleading. In spite of these limitations it is a widely used measure of dispersion as can be evidenced by any one who is engaged in buying or selling a stock. That is, for each stock one is not only informed of the high and low value of the stock for that day but for the year as well.

Because of the limitations of the range, other measures of variability that concern themselves with all of the observations are needed. The average deviation, which is calculated by averaging the absolute value of the deviations from the mean as follows,

$$\text{Average deviation} = \frac{\sum_{i=1}^{n} |x_i - \bar{x}|}{n}$$

overcomes this limitation but is infrequently used and will not be discussed further.

One of the most widely used methods of measuring dispersion is the variance. It is calculated by first determining the difference (i.e., discrepancy between the actual x observation and the expected value or mean) of all of the x observations. Note that the sum of these differences would be zero by definition of the mean. One way to get around this problem is by squaring each difference before they are summed. This sum is then averaged by dividing by $n-1$. The fact that $n-1$ rather than n is used can be explained in a course on statistical inference. As a practical matter, whether one divides by 40 or 39, for example, will not result in a very different answer anyway. We can then summarize by giving the formula for the *variance* for *ungrouped* data. Note that the symbol s^2 is used to denote the variance.

$$s^2 = \frac{\sum_{i=1}^{n} (x_i - \bar{x})^2}{n-1}$$

Example 2. Calculate the variance for the set of data presented in Example 2 of Section 9-1.

Solution:

x_i	$(x_i - \bar{x})$	$(x_i - \bar{x})^2$
\$ 350	\$ −610	\$ 372,100
450	−510	260,100
540	−420	176,400
560	−400	160,000
600	−360	129,600
610	−350	122,500
630	−330	108,900
670	−290	84,100

x_i	$(x_i - \bar{x})$	$(x_i - \bar{x})^2$
$ 740	$−220	$ 48,400
760	−200	40,000
760	−200	40,000
780	−180	32,400
800	−160	25,600
810	−150	22,500
830	−130	16,900
840	−120	14,400
890	−70	4,900
910	−50	2,500
930	−30	900
960	00	000
970	10	100
980	20	400
1010	50	2,500
1030	70	4,900
1030	70	4,900
1050	90	8,100
1060	100	10,000
1080	120	14,400
1110	150	22,500
1150	190	36,100
1170	210	44,100
1200	240	57,600
1220	260	67,600
1240	280	78,400
1290	330	108,900
1350	390	152,100
1420	460	211,600
1440	480	230,400
1550	590	348,100
1630	670	448,900
		$3,513,800

$$s^2 = \frac{\$3{,}513{,}800}{39} = \$90{,}097.44$$

In most situations $\bar{x}$ will not be an integer; that is to say, the calculations involved in finding the difference, squaring and summing could become lengthy. Therefore, an alternative formula for the variance is given as follows:

$$s^2 = \frac{\sum_{i=1}^{n} x_i^2 - ((\sum_{i=1}^{n} x_i)^2/n)}{n - 1}$$

Note that this formula applies to ungrouped data and can be derived simply by expanding $\sum_{i=1}^{n}(x_i - \bar{x})^2$. The proof of this will be left as a problem for this section. Note that this calculation requires that one only use two work columns rather than three as in Example 2. This is an advantage, computationally.

If the data given in Example 2 were grouped as follows

Intervals	*Midpoint*, m_j	f_j
\$ 300 and under \$ 500	400	2
500 and under 700	600	6
700 and under 900	800	9
900 and under 1100	1000	11
1100 and under 1300	1200	7
1300 and under 1500	1400	3
1500 and under 1700	1600	2
		40

one would simply multiply the frequency by the difference that results from subtracting $\bar{x}$ from the midpoint. For example, we simply assume we have two values (or observations) of 400, six values of 600, etc. The resulting formula for the *variance* when the data are *grouped* is

$$s_2 = \frac{\sum_{j=1}^{k}(m_j - \bar{x})^2 f_j}{n-1}$$

where k = the number of classes.

Example 3. Calculate the variance for the set of grouped data just described.

Solution:

Intervals	*Midpoint*, m_j	f_j	$m_j - \bar{x}$	$(m_j - \bar{x})^2$	$(m_j - \bar{x})^2 f_j$
\$ 300 and under \$ 500	\$ 400	2	\$−560	\$313,600	\$ 627,200
500 and under 700	600	6	−360	129,600	777,600
700 and under 900	800	9	−160	25,600	230,400
900 and under 1100	1000	11	40	1,600	17,600
1100 and under 1300	1200	7	240	57,600	403,200
1300 and under 1500	1400	3	440	193,600	580,800
1500 and under 1700	1600	2	640	409,600	819,200
					\$3,456,000

$$s^2 = \frac{\$3{,}456{,}000}{39} = \$88{,}615$$

Again, there is an alternative formula if one expands the numerator. It is as follows:

$$s^2 = \frac{\sum_{j=1}^{k} m_j^2 f_j - ((\sum_{j=1}^{k} m_j f_j)^2/n)}{n-1}$$

You may have realized that we have yet to interpret the variance.

It is in units of squared deviations or squares of the units of the original data, in this case (dollars)^2. To obtain a measure of dispersion in terms of dollars the square root of the variance is taken, and this result is called the *standard deviation.* We then have the following formula for standard deviation when the data are grouped.

$$s = \sqrt{\frac{\sum_{j=1}^{k} (m_j - \bar{x}) f_j}{n-1}}$$

Example 4. Find the standard deviation for the data given in Example 3.

Solution:

$$s = \sqrt{s^2}$$

$$s = \sqrt{\$88{,}615}$$

$$s = \$297.68$$

This standard deviation of \$297.68 is clearly a measure of dispersion. It has many uses, which will not be presented here. One extremely important use is that when working with so called "normal" distributions one will be able to indicate the percentage of observations that will fall between ± 1, ± 2, and ± 3 standard deviations. In general most all observations will fall within 3 standard deviations if the data are normally distributed. In our situation

$$\$960 \pm 3(\$300) \quad \text{using} \quad s = \$300.00$$

$$\$960 \pm \$900$$

$$\$60 \leftrightarrow \$1860.$$

Looking back at the original data in Example 2 of Section 9-1, we see that all of the observations fall within this range.

SUMMARY OF FORMULAS FOR CALCULATING THE STANDARD DEVIATION

Data Ungrouped

$$s = \sqrt{\frac{\sum_{i=1}^{n} (x_i - \bar{x})^2}{n-1}} = \sqrt{\frac{\sum_{i=1}^{n} x_i^2 - ((\sum_{i=1}^{n} x_i)^2/n)}{n-1}}$$

Data Grouped

$$s = \sqrt{\frac{\sum_{j=1}^{k} (m_j - \bar{x})^2 f_j}{n-1}} = \frac{\sum_{j=1}^{k} m_j^2 f_j - ((\sum_{j=1}^{k} m_j f_j)^2/n)}{n-1}$$

Note that a similar summary could be made for formulas for variance, since the variance is simply the standard deviation squared.

So far in this section we have seen how one can use the variance or standard deviation as an absolute measure of dispersion about a mean. You will recall that the mean and standard deviation of the commissions paid to the salesmen in the New England region calculated in Sections 9.2 and 9.3 were \$960 and \$297.68 respectively. If one wanted to compare the standard deviation for the salesmen in the New England region (i.e., \$297.68) to that of the salesmen in the Southeastern section of the country (i.e., \$250), one could first obtain a relative measure by expressing the standard deviation as a percentage of the arithmetic mean. This result is referred to as the coefficient of variation.

$$\text{Coefficient of variation} = \frac{s}{\bar{x}}$$

Example 5. Calculate the coefficient of variation of monthly commissions for the New England and Southeastern regions given the following:

New England	*Southeast*
$\bar{x} = \$960$	$\bar{x} = \$800$
$s = \$298$	$s = \$250$

Solution:

New England	*Southeast*
$\text{C.V.} = \frac{s}{\bar{x}} = \frac{\$298}{\$960} = .31$	$\text{C.V.} = \frac{s}{\bar{x}} = \frac{\$280}{\$800} = .35$

From the result of Example 5 one can conclude that the monthly commissions are more variable in the Southeast or, stated differently, the commissions in New England are more uniform even though the standard deviations of the commissions indicate just the opposite. That is, when making comparisons, one should use a relative measure rather than an absolute measure.

PROBLEMS

1. Calculate the variance for the data given in Example 2 using the formula:

$$s^2 = \frac{\sum_{i=1}^{n} x_i^2 - ((\sum_{i=1}^{n} x_i)^2/n)}{n-1}$$

2. Show that $\sum_{i=1}^{n}(x_i - \bar{x})^2 = \sum_{i=1}^{n} x_i^2 - ((\sum_{i=1}^{n} x_i)^2/n)$.

3. Calculate the variance for the data given in Example 3 using the formula

$$s^2 = \frac{\sum_{j=1}^{n} m_j^2 f_j - ((\sum_{j=1}^{n} m_j f_j)^2/n)}{n-1}$$

4. Compute the mean, variance and standard deviation for the following data: 25, 17, 33, 19, 10, 15, 22, 20, 18, 21.

5. Compute the variance and standard deviation for the following:

Class Limits	*Frequency*
100 and under 200	2
200 and under 300	3
300 and under 400	4
400 and under 500	5

6. Calculate the variance and standard deviation of the weekly wages of 65 part-time assistants at G.U.T. College given the following:

Weekly Wage	*Frequency*
\$ 5.00 and under \$15.00	10
15.00 and under 25.00	10
25.00 and under 35.00	20
35.00 and under 45.00	10
45.00 and under 55.00	10
55.00 and under 65.00	5
	65

7. Given the following breakdown of beginning salaries for graduating seniors from the School of Business:

Starting Salaries	*Accounting*	*Management*	*Marketing*	*Total*
\$ 7000 and under \$ 8000	10	50	10	70
8000 and under 9000	20	150	100	270
9000 and under 10000	50	50	50	150
10000 and under 11000	10	30	30	70
11000 and under 12000	10	20	10	40
Total	100	300	200	600

Calculate the mean, standard deviation and coefficient of variation for a graduate of the School of Business Administration as well as for the individual majors.

8. Given that one gathers data on the closing prices of two stocks for a year and arrives at the following result:

Stock	$\bar{x}$	s
ABC	\$500	\$25
XYZ	\$ 5	\$ 1

(a) Calculate the coefficient of variation of the closing prices for both stocks.
(b) Which stock price is more variable?
(c) Given that you purchased 2 shares of ABC stock with the \$1,000 you made last summer and the price increased one standard deviation. What would your investment now be worth? What would be your return on investment?
(d) If instead of buying two shares of ABC, you purchased 200 shares of XYZ and the price increased one standard deviation, what would your investment now be worth? What would be your return on investment?

9-4 PRESENTATION OF THE DATA—FREQUENCY DISTRIBUTIONS

In the two previous sections it was shown that one could summarize how the data tended to group together by using a measure of central tendency and how they fell around this number by using a measure of dispersion. Once the data have been summarized one can make use of a *frequency distribution* to present the data in a more graphic and informative way.

When constructing a frequency distribution the original observations, which we have called x, are plotted on the horizontal axis. In most instances the original data will have been grouped into class intervals. The frequency of the values of x in each class interval is plotted on the vertical axis. The most common frequency diagram is a *histogram*, or bar chart.

Example 1. Construct a histogram for the following set of data.

Interval	*Number of Observations*, f_j
\$ 300 and under \$ 500	2
500 and under 700	6
700 and under 900	9
900 and under 1100	11
1100 and under 1300	7
1300 and under 1500	3
1500 and under 1700	2

Solution:

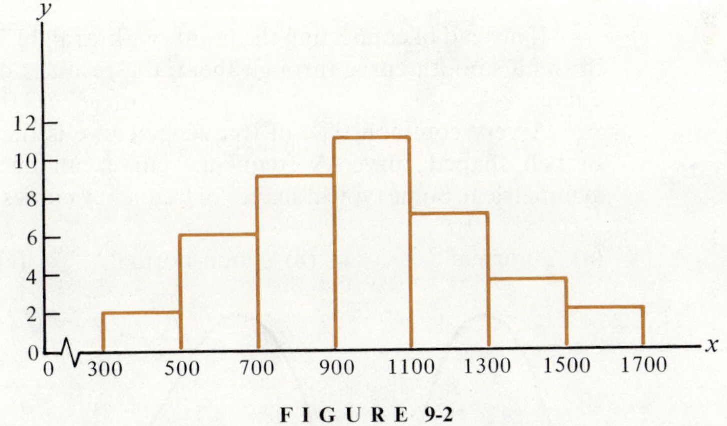

FIGURE 9-2

In Example 1 note that if the x axis does not start at zero, one breaks the scale. The y axis will always begin at zero, however. Also when one uses equal class intervals the areas of the set of vertical bars are proportional to the respective frequencies. The tallest bar indicates the modal class; in this example it is \$900–1100.

The variables that one works with can be either discrete or continuous and in some instances one treats discrete data as continuous. Another common frequency diagram is the *polygon* or line chart where one plots the frequency versus the midpoint of the class intervals and connects the points by straight line segments.

Example 2. Construct a frequency polygon for the data given in Example 1.

Solution:

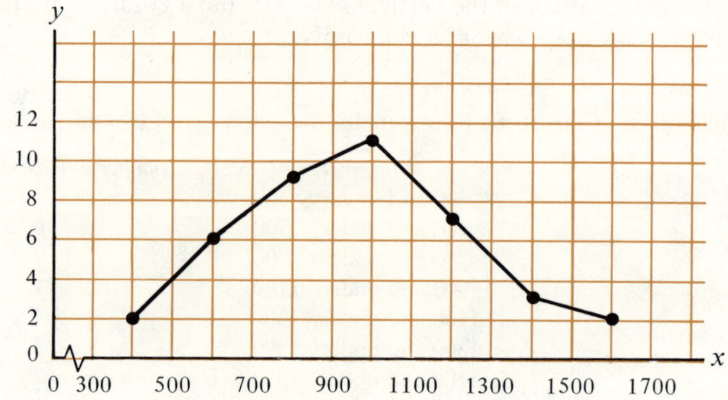

FIGURE 9-3

If instead of connecting the points with straight line segments, one draws a smooth curve through them, the result is called a frequency curve.

A very common type of frequency curve is the so-called normal or bell shaped curve. A frequency curve can be skewed, i.e., not symmetrical. Some typical shapes of frequency curves are shown below.

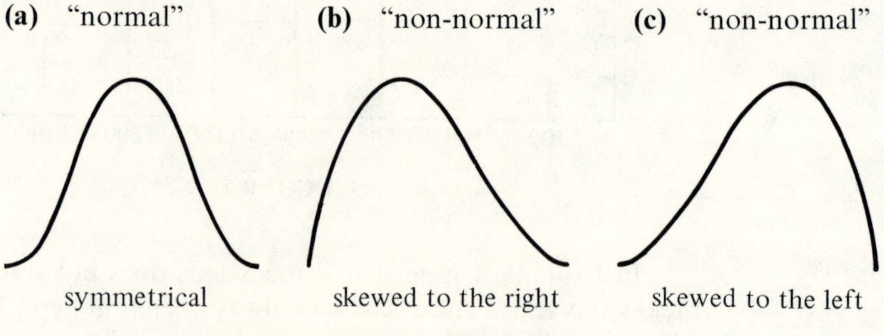

FIGURE 9-4

The values of the mean and median can be used to help determine the shape of a distribution. That is, because the mean is influenced by extreme high or low values it will be pulled away from the median. If data contain unusually high observations, the mean will be pulled to the right and the distribution is then said to be skewed to the right. A method used to determine skewness is Pearson's Coefficient of Skewness which is defined as follows:

$$\text{Skewness} = \frac{3(\text{mean} - \text{median})}{\text{standard deviation}}$$

One can observe from the formula that when the distribution is skewed to the right, i.e., mean > median, the distribution is said to be positively skewed, whereas if the distribution is skewed to the left, i.e., median > mean, then the distribution is said to be negatively skewed.

Example 3. Calculate Pearson's Coefficient of Skewness for the distribution given in Example 2. The mean is equal to \$960, the median is equal to \$954 and the standard deviation is equal to \$297.

Solution:

$$\text{Skewness} = \frac{3(960 - 954)}{297} = \frac{+18}{297} = +0.06$$

In Example 3, we see that the distribution is slightly skewed to the right.

Another graph which is of importance is called an *ogive*. It is a *cumulative* frequency distribution and can be of the less than or more than variety. When constructing a cumulative less than ogive, one simply plots the cumulative less than frequencies versus the upper limit of each class. When constructing a more than ogive one plots the cumulative more than frequencies versus the lower limit of the class.

Example 4. Construct a cumulative more than ogive for the data given in Example 1 of this section.

Solution:

Interval	f_j	*Cumulative* f_j
\$ 300 and under \$ 500	2	40
500 and under 700	6	38
700 and under 900	9	32
900 and under 1100	11	23
1100 and under 1300	7	12
1300 and under 1500	3	5
1500 and under 1700	2	2
	40	

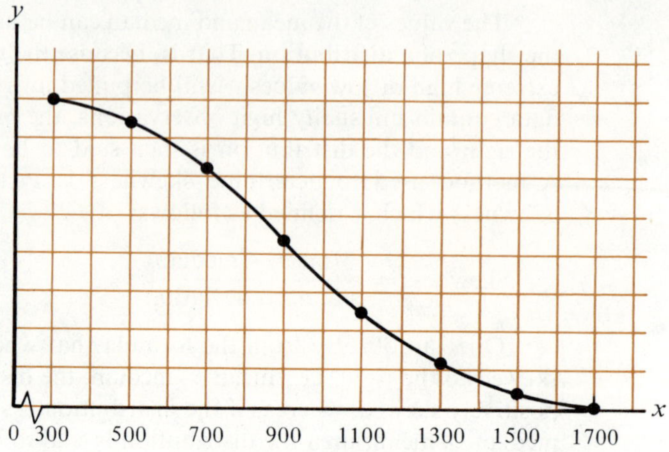

FIGURE 9-5

Interpreting the result of the previous example, we see that 40 salesmen had a commission greater than \$300, 38 had a commission greater than \$500, etc.

Summarizing the first four sections, one first collects the data, arranges the observations in some sort of sequence—usually ascending order, groups the data if there are many observations, and summarizes the data using both measures of central tendency and dispersion. Lastly, one will give a picture of how the data look, using a histogram or polygon. The end result then is a graphical presentation of the data and two numbers (one that measures central tendency, and one that measures dispersion).

Schematically, we have the following:

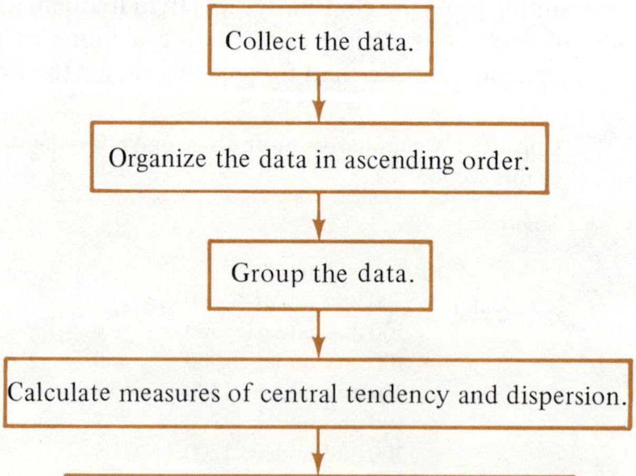

FIGURE 9-6

There are numerous library programs available that perform calculations that have been presented in the preceding sections. A sample using the hypothetical data of this chapter follows:

OPTIONAL MATERIAL

How many groups (<= 20) do you have? 7

Now, enter (A) the lower stated limit, (B) the upper stated limit, and (C) the frequency for the interval . . . one group at a time.

Group 1 :? 300, 500, 2
Group 2 :? 500, 700, 6
Group 3 :? 700, 900, 9
Group 4 :? 900, 1100, 11
Group 5 :? 1100, 1300, 7
Group 6 :? 1300, 1500, 3
Group 7 :? 1500, 1700, 2

True Limits		Mid-P	F	%	Cum F
Lower	Upper				
300	500	400	2	5	2
500	700	600	6	15	8
700	900	800	9	23	17
900	1100	1000	11	27	28
1100	1300	1200	7	18	35
1300	1500	1400	3	8	38
1500	1700	1600	2	5	40
	Total		40	101	

Mean = 960
Median = 954.545
Variance = 86400
Standard deviation = 293.939
Skewness = 0.056
Coeff. of variability = 0.306

FIGURE 9-7

Note that variance and standard deviation were calculated using n rather than $n-1$.

PROBLEMS

1. Given the following frequency distribution of weekly wages of 65 part-time student assistants:

Weekly Wage	*Number of Employees*
\$ 5.00 and under \$15.00	10
15.00 and under 25.00	10
25.00 and under 35.00	20
35.00 and under 45.00	10
45.00 and under 55.00	10
55.00 and under 65.00	5
	65

(a) Construct a histogram for the data.
(b) Construct a frequency polygon for the data.
(c) Construct a cumulative more than ogive for the data.
(d) Is the distribution positively or negatively skewed?

2. Construct a cumulative less than ogive for the following data:

Monthly Commissions	f_j	*Cumulative* f_j
\$ 300 and under \$ 500	2	
500 and under 700	6	
700 and under 900	9	
900 and under 1100	11	
1100 and under 1300	7	
1300 and under 1500	3	
1500 and under 1700	2	
	40	

3. Given the following breakdown of beginning salaries for graduating seniors from the School of Business:

Starting Salary	*Frequency*
\$ 7000 and under \$ 8000	70
8000 and under 9000	270
9000 and under 10000	150
10000 and under 11000	70
11000 and under 12000	40
	600

Construct a histogram and frequency polygon for the data.

4. Construct a cumulative more than ogive for the data given in Problem 3.

5. Calculate the coefficient of skewness for the data given in Problem 3.

6. Given the following distribution of faculty members by academic-year salaries:

Salary Intervals	*Prof.*	*Assoc.*	*Assist.*	*Instr.*	*Lecturer*	*All Ranks*
17,000 to 17,499	1					1
16,500 to 16,999	1					1
16,000 to 16,499						
15,500 to 15,999						
15,000 to 15,499						
14,500 to 14,999						
14,000 to 14,499	3	1				4
13,500 to 13,999	2					2
13,000 to 13,499	1	1				2
12,500 to 12,999	2	1				3
12,000 to 12,499	1					1
11,500 to 11,999	1	1				2
11,000 to 11,499		4	2			6
10,500 to 10,999		5	6			11
10,000 to 10,499		2	8			10
9,500 to 9,999			14	1		15
9,000 to 9,499			2	3		5
8,500 to 8,999			1			1
8,000 to 8,499			1	3		4
7,500 to 7,999				2		2

Analyze the data using the methods presented in this chapter. Include in your analysis a graphical presentation, a measure of central tendency, and a measure of dispersion.

7. The administrator of Wright Hospital has presented you with a number of service projections for the year ending June 30, 1972. Estimated room requirements for inpatients by type of service are:

		Average Number of Days in Hospital		*Percent of Regular Patients Selecting Types of Service*		
Type of Patient	*Total Patients Expected*	*Regular*	*Medicare*	*Private*	*Semi-Private*	*Ward*
Medical	2,100	7	17	10%	60%	30%
Surgical	2,400	10	15	15	75	10

Of the patients served by the hospital 10 percent are expected to be Medicare patients, all of whom are expected to select semi-private rooms. Both the number and proportion of Medicare patients have increased over the past five years. Daily rentals per patient are: $40 for a private room, $35 for a semi-private room, and $25 for a ward.

Operating room charges are based on man-minutes (number of minutes the operating room is in use multiplied by number of personnel assisting in the operation). The per man-minute charges are $.13 for inpatients and $.22 for outpatients. Studies for the current year show that operations on inpatients are divided as follows:

Type of Operation	Number of Operations	Average Number of Minutes Per Operation	Average Number of Personnel Required
A	800	30	4
B	700	45	5
C	300	90	6
D	200	120	8
	2,000		

The same proportion of inpatient operations is expected for the next fiscal year and 180 outpatients are expected to use the operating room. Outpatient operations average 20 minutes and require the assistance of three persons.

The budget for the year ending June 30, 1972, by departments, is:

General services:	
Maintenance of plant	$ 50,000
Operation of plant	27,500
Administration	97,500
All others	192,000
Revenue producing services:	
Operating room	68,440
All others	700,000
	$1,135,440

The following information is provided for cost allocation purposes:

	Square Feet	Salaries
General services:		
Maintenance of plant	12,000	$ 40,000
Operation of plant	28,000	25,000
Administration	10,000	55,000
All others	36,250	102,500
Revenue producing services:		
Operating room	17,500	15,000
All others	86,250	302,500
	190,000	$540,000

Basis of allocations:

Maintenance of plant—salaries
Operation of plant—square feet
Administration—salaries
All others—8% to operating room

Required:

Prepare schedules showing the computation of:

(a) The number of patient days (number of patients multiplied by average stay in hospital) expected by type of patients and service.

(b) The total number of man-minutes expected for operating room services for inpatients and outpatients. For inpatients show the breakdown of total operating room man-minutes by type of operation.
(c) Expected gross revenue from routine services.
(d) Expected gross revenue from operating room services.

REFERENCES

Chou, Ya-lun. *Probability and Statistics for Decision Making*. New York: Holt, Rinehart and Winston, 1972.

Hamburg, Morris. *Statistical Analysis for Decision Making*, New York: Harcourt Brace Jovanovich, Inc., 1970.

Huntsberger, David U., Billingsley, Patrick, and Croft, D. James. *Statistical Inference for Management and Economics*, Third Ed. Boston: Allyn and Bacon, Inc., 1975.

Neter, John, Wasserman, William, and Whitmore, G. A. *Fundamental Statistics for Business and Economics*, Fourth Ed. Boston: Allyn and Bacon, Inc., 1973.

Spiegel, Murray R. *Statistics.* New York: Schaum's Outline Series, McGraw-Hill Book Co., Inc., 1961.

TEN

Decision making and game theory

One area that is of interest to all, especially to the businessperson, is decision making. When a businessperson does not make the right decision, he will bear the consequence. This is usually in the form of lost sales, lost market share, lost profit, etc. What we will try to do in this chapter is to present methods for making rational decisions. The first section involves one in a decision making situation against nature or a neutral person. Strategies are given for the situation of uncertainty, risk, and certainty. The following two sections are from the body of knowledge referred to as game theory. Game theory is one of the branches of mathematics where the emphasis is on applications in the field of decision making. In the last section we look at the process of decision making in a capital budgeting situation where one has to also take into account the time value of money.

10-1 DECISION MAKING UNDER UNCERTAINTY AND RISK

In making a decision one will be faced with a situation of uncertainty, risk, or certainty. These sorts of situations are faced by businessmen, doctors, or heads of households.

To help us in the decision making process it is helpful to have some sort of structure with which we can analyze the situation. In any decision problem one will have more than one possible choice or act that can be taken; otherwise there is no problem. That is, if there is but one choice no decision need be made. Just as there is more than one possible act that can be taken, there are various outcome states that can occur. Note that unlike the acts, outcomes are not controlled by the decision maker.

If we use a table to structure the problem we can let the row headings be the acts available to the decision maker and the column headings be the various outcome states which are frequently referred to as *states of nature*. To visualize this structure, one may be trying to choose between buying a home and renting. The reason why the decision maker is faced with a problem is because he does not know if his company is going to transfer him in the next two years. Using the tabular format we have:

TABLE 10-1

	Outcomes	
	Transferred within 2 years	Will remain at present location for more than 2 years
Acts		
Buy house		
Rent		

Note that in the preceding situation the outcomes are not states of nature but are determined by the decision maker's employer. The entries that are made in the table are called payoffs. We will assume that the decision maker has calculated the following payoffs which are presented in the completed table:

TABLE 10-2

	Outcomes	
	0_1	0_2
	Transferred within 2 years	Remain at present location for more than 2 years
Acts		
A_1, Buy house	– \$2000	\$4,000
A_2, Rent	\$0	– \$1,000

The first entry can be interpreted as follows: the cost of buying and selling the house exceeds the appreciation in value by \$2,000. If the employee knew for certain that he was to be transferred within two years he would rent rather than buy a house, assuming Table 10-2 is valid. This is referred to as decision making under certainty. In general, however, he will not have this information available. At present he is faced with making a decision under *uncertainty*. Some common decision rules used when faced with uncertainty are:

THE MAXIMAX DECISION RULE: *One finds the maximum entry for each row or act in the table and enters this figure in a new column. He then chooses that act that corresponds to the highest entry in this new column.*

TABLE 10-3

	0_1	0_2	Maximum
Buy house A_1	−$2000	$4000	$4000
Rent A_2	0	−$1000	0

In this situation act A_1 or buy a house is the result of the maximax decision rule.

THE MAXIMIN DECISION RULE: *One finds the minimum entry for each row or act in the table and enters this figure in a new column. He then chooses that act that corresponds to the highest entry in this new column.*

TABLE 10-4

	0_1	0_2	Minimum
Buy house A_1	−$2000	$4000	−$2000
Rent A_2	0	−$1000	−$1000

In this situation act A_2, rent, is the result of the maximin decision rule. Note that this guarantees that no matter what outcome state occurs the very worst thing that will happen is that the decision maker will suffer a loss of $1000. In using either of these decision rules one works with payoffs. That is, we simply look at the result of a particular act or decision coupled with an outcome state. In many instances one is concerned with opportunity costs rather than actual costs. As regards the employee who is faced with a decision, he might be thinking, "If I rent (Act A_2) and the company does not transfer me I would have been $5000 better off if I had bought a house" (i.e., no loss of $1000 but rather a $4000 profit). To use the third and last rule to be presented one must first construct a regret table or "matrix." This is done by finding the largest payoff in each column and subtracting all of the other entries in that column from it. Doing this we have the following regret table or "matrix":

TABLE 10-5

	0_1	0_2
Buy house A_1	\$2000	0
Rent A_2	0	\$5000

The first column can be interpreted as follows: If outcome state one, O_1, occurs and I choose A_1, I will have regret of \$2000. (That is if I had chosen A_2 I would have been \$2000 better off.) Note that it is impossible to have a negative entry in a regret table or "matrix."

THE MINIMAX REGRET RULE: *One finds the maximum entry (regret) for each row or act in the regret table and enters this figure in a new column. He then chooses that act that corresponds to the lowest entry in this new column.*

TABLE 10-6

	Regrets		
	0_1	0_2	Maximum
Buy house A_1	\$2000	0	\$2000
Rent A_2	0	\$5000	\$5000

In this situation act A_1, buy a house, is the result of the minimax regret decision rule.

It should be pointed out that these rules are distinct alternatives, i.e., one cannot apply all three at once. The rule one chooses is in part determined by personality as well as company policy or other factors.

Once one has formalized his decision problem in the structure of a table the next step is to try to establish probabilities for each of the outcome states. This can be done by using

1. relative frequency,
2. theoretical probability distributions,
3. subjective judgment.

Assuming that the employee has examined the past actions of the company and has arrived at the following probabilities for each outcome:

$$0_1 = .8$$

$$0_2 = .2$$

Note that the sum of the probabilities is 1.

The decision rule that is commonly used under *risk* is to *maximize expected payoff.* To calculate the expected payoff of each act one multiplies the payoff by the probability of each act and sums the result for each row and enters this value in a new column. He then chooses that act that corresponds to the highest entry in this new column.

TABLE 10-7

	.8 0_1	.2 0_2	Expected payoff
Buy house A_1	– $1600	$800	– $800
Rent A_2	0	– 200	– $200

In this situation act A_2, rent, is prescribed by the maximize expected payoff decision rule. Note that the expected payoff generally will not agree with the actual results on any one decision or trial. However, this does not diminish the usefulness of this decision rule to select among alternatives.

In conclusion, if one can arrive at a probability of 1 for a given outcome state, the decision making process is quite simple. One simply selects the act that has the highest payoff. Note there is now only one outcome state. Using the preceding example, if the employee could be certain that he was to remain at this location for at least five years he would choose act A_1, i.e., buy a house.

PROBLEMS

1. What are the decision rules or criteria when faced with uncertainty?
2. What is the decision rule or criterion when one is faced with risk?
3. What is the difference between a risk situation and an uncertain situation?
4. What methods are available to establish probabilities for various outcome states?

5. Given three different situations: risk, certainty, and uncertainty. List these in the order of preference of situations you would like to be faced with.

6. Given the following payoff table in millions of dollars for the following projects and outcomes:

	0_1 Very limited energy sources	0_2 Limited energy sources	0_3 Normal energy sources
A_1 Project 1	5	8	13
A_2 Project 2	9	10	12
A_3 Project 3	12	9	8

   (a) Using the maximax decision criterion one would choose what project?

   (b) Using the maximin decision criterion one would choose what project?

7. Using the data given in Problem 6:

   (a) Construct a regret matrix.

   (b) Make a decision using the minimax regret criterion.

8. Using the data given in Problem 6 and given the following probabilities for 0_1, 0_2, and 0_3:

$$0_1 = .5, \qquad 0_2 = .3, \qquad 0_3 = .2$$

   (a) Calculate the expected value of each project.

   (b) Using the criterion of maximizing expected value one would choose which act or project?

9. Given the following payoff table:

	0_1 Move Within 5 Years	0_2 Do Not Move Within 5 years
A_1 Install Aluminum Siding	−500	1000
A_2 Paint House	300	0

   (a) Using the maximax decision criterion one would choose which act?

   (b) Using the maximin decision criterion one would choose which act?

   (c) Using the minimax regret criterion one would choose which act?

   (d) Given the following probabilities: $0_1 = .8$, $0_2 = .2$, one would choose which act?

10-2 TWO-PERSON ZERO-SUM GAMES

In the previous section we assumed that the decision maker had various acts open to him and that his choice depended on the outcome states or states of nature. It was assumed at that time that these states of nature or outcome states were neutral, and that we had a one-person game against nature. In this section we will assume that these outcome states are in fact the decisions of another player and that he also is acting rationally in his best interests. This, in effect, is the difference between one- and two-person games. We will also assume that the games (decision situations) that the players (competitors) are playing are zero-sum games. That is whatever amount A wins, B loses, or conversely whatever amount B wins, A loses. This would be the case when two companies compete for their share of the market where whatever A gets, B loses.

Game theory can be applied only to those situations where the specific sets of choices by all of the players are known in advance and numerically defined. The standard convention for displaying these choices and payoffs is a two dimensional table or matrix.

Example 1. Interpret the following two dimensional format used to represent a game being played by *A* and *B*.

		Player B	
		B_1	B_2
Player A	A_1	20	−5
	A_2	10	5

Solution: Player *A* and *B* have two strategies each. If *A* chooses strategy 1, he will win 20 points (i.e., *B* loses 20 points) or lose 5 points (i.e., *B* wins 5 points), depending on which strategy *B* chooses. Similarly, if player *B* chooses strategy 1, he will lose 20 points or lose 10 points depending on *A*'s strategy.

In the preceding example we can think of the points as being shares of the market where if *A* plays strategy 1 (i.e., say a direct sales campaign) he will gain 20 percent of the market or lose 5 percent to *B* depending on *B*'s strategy. Note that in this example one matrix is used to portray the outcomes of the game and that the numbers represent the payoffs to player *A*.

Example 2. Using the table given in Example 1 construct a table of payoffs for *B*.

Solution:

	B_1	B_2
A_1	−20	5
A_2	−10	−5

Payoffs for *B*

Note that if in Example 1 a matrix were used to illustrate the payoffs, one could simply multiply the matrix by −1 to find the payoffs for *B*.

Returning to the game presented in Example 1, we have:

		Player B	
		B_1	B_2
Player A	A_1	20	-5
	A_2	10	5

Looking at the payoff table we observe that player B can win only if he plays strategy 2. Player A knowing this will then play strategy 2. Player A can then expect to win 5 and B lose 5 each time the game is played. This game is biased in favor of A and the number 5 is referred to as the value of the game.

Example 3. Given the following payoff table determine the value of the game:

		Player B	
		B_1	B_2
Player A	A_1	6	8
	A_2	-4	5

Solution: There are two ways of approaching a solution to this problem.

1. Player A sees that best is A_1 (in fact dominates A_2) but if he does then B will play B_1 and minimize his loss (i.e., -6 rather than -8).

2. Player B can win only if he chooses strategy B_1. If he plays this strategy, then A will play A_1 and make 6 rather than lose 4.

In both approaches the value of the game is 6.

A convenient method for determining which strategy each player should choose or in effect determining the value of the game is as follows:

1. Player A knows that once he has made a choice (i.e., picked a row) that player B will make his choice so as to minimize A's points (i.e., maximize own). So in effect A should use a maximum strategy. Put a square around the minimum entry in each row.
2. Similarly, player B is faced with the same situation and should also use the same criterion. However, the table represents payoffs to A and costs to B. Rather than construct a table of payoffs for B, one can simply view the payoff matrix for A as a cost matrix for B. Then as far as B is concerned a minimum is the largest number in the column. In a sense player B is using a minimax strategy when using a table with payoffs for A. Put a circle around the maximum entry in each column.
3. If a payoff entry has both a circle and square around it, then we can say that the game is strictly determined (i.e., each player plays only one strategy) or has a saddle point. This entry is the value of the game and the players will choose the strategies that will yield this result.

Example 4. Use the above procedure to find a solution for example 3.

Solution:

	B_1	B_2
A_1	⑥	⑧
A_2	-4	5

Payoffs for A

Strategy A_1, B_1 indicated with the value of the game being 6.

Note that the procedure applies only if the game is strictly determined or has a saddle point. This occurs when the maximin strategy for both players coincides, that is when an element has both a square and circle around it. When a game has no saddle point, the players will mix their strategies. That is, players A and B will play their strategies a certain percentage of the time. This topic, and how the percentages are determined, will be discussed in the next section.

So far, we have limited ourselves to situations where each player has but two strategies or choices available to him. Before proceeding to the next section, the following example examines a case where both players have more than two strategies. Note that this is still a two-person game, however.

Example 5. Determine what strategies A and B should choose and the resulting value of the game given the following:

		Player B			
		B_1	B_2	B_3	B_4
	A_1	6	2	4	5
Player A	A_2	7	-3	-2	2
	A_3	5	-2	-2	3

Solution:

		Player B			
		B_1	B_2	B_3	B_4
	A_1	6	②	④	⑤
Player A	A_2	⑦	-3	-2	2
	A_3	5	-2	-2	3

Strategy A_1, B_2 indicated with the value of the game being 2.

Examining the table given in Example 5 we see that the act or strategy A_1 dominates A_3. For a row act to dominate any other row act all of the entries in the row have to be greater than or equal to all of the corresponding entries in another row. The row A_3 is referred to as a recessive act. Act B_2 dominates acts or strategies B_1, B_3, and B_4. For a column act to dominate any other column act all of the entries in the column have to be less than or equal to all of the corresponding entries in another column. The columns B_1, B_3, and B_4 are recessive. One should note that for a *row* to be recessive the entries will be smaller than or equal to entries in another row. For a *column* to be recessive the entries will be greater than or equal to entries in another column.

Example 6. Eliminate the recessive rows and columns from the table given in Example 5.

Solution:

	B_2
A_1	2
A_2	−3

Referring to the solution of example 6 strategies A_1, B_2 are indicated with the value of the game being 2.

PROBLEMS

1. What is meant by the term zero-sum game?
2. Interpret the following zero-sum game:

		Player Y	
		Y_1	Y_2
Player X	X_1	5	2
	X_2	−2	7

3. Using the table given in Problem 2, construct a table of payoffs for Y.
4. Is game theory applicable if the outcomes for all specific sets of choices are not numerically defined?
5. Do games that are strictly determined always have a saddle point?
6. Are the players in a two-person zero-sum game each limited to two strategies?
7. Determine the saddle point and values of the following strictly determined games.

(a)

	B_1	B_2
A_1	20	15
A_2	10	5

(b)

	B_1	B_2
A_1	−6	10
A_2	−5	−4

(c)

	B_1	B_2
A_1	5	4
A_2	2	−1
A_3	−3	3

(d)

	B_1	B_2	B_3
A_1	4	−2	−3
A_2	−1	4	−2

10-3 MIXED STRATEGIES

In the previous section all of the games were strictly determined or had a saddle point. By this we mean that the maximum strategy for both players coincided. However, in many instances, this will not be the case. Consider the following game that is not strictly determined.

		Player B	
		B_1	B_2
Player A	A_1	6	8
	A_2	10	5

The following sequence might occur.

1. Player A chooses strategy A_2.
2. Player B would then choose strategy B_2.
3. Player A would then choose strategy A_1.
4. Player B would then choose strategy B_1.
5. Player A would then choose strategy A_2.

The game is clearly biased in A's favor and the value of the game is some value between 5 and 10. The exact value of the game depends on how the players mix their strategies. The intent of this section is to determine the percentage of the time the players should choose each strategy and to determine the value of the game.

Let p_1 and p_2 be the probability that player B chooses strategy 1 and 2 respectively. Then if player A chooses strategy 1 the expected value of the game will be

$$EV = p_1(6) + p_2(8)$$

If player A chooses strategy 2 the expected value of the game will be

$$EV = p_1(10) + p_2(5)$$

Since $p_1 + p_2 = 1$, and the expected value of the game to player B is the same no matter what the opponent does, we get the following:

$$6p_1 + 8(1 - p_1) = 10p_1 + 5(1 - p_1)$$

Solving for p_1 yields the following result:

$$6p_1 + 8 - 8p_1 = 10p_1 + 5 - 5p_1$$

$$p_1(6 - 8 - 10 + 5) = 5 - 8$$

$$p_1 = \frac{5 - 8}{6 - 8 - 10 + 5}$$

$$p_1 = 3/7$$

Substituting p_1 in the equation $p_1 + p_2 = 1$ we have:

$$p_1 + p_2 = 1$$

$$3/7 + p_2 = 1$$

$$p_2 = 1 - 3/7$$

$$p_2 = 4/7$$

The expected value of the game is:

$$EV = 10(3/7) + 5(4/7)$$

$$= 30/7 + 20/7$$

$$= 50/7$$

Letting q_1 and q_2 be the probability that player A chooses strategy 1 and 2 respectively, we have:

$$EV = 6q_1 + 10q_2$$

Since $q_1 + q_2 = 1$ and $EV = 50/7$ we have:

$$50/7 = 6q_1 + 10(1 - q_1)$$

$$50/7 = 6q_1 + 10 - 10q_1$$

$$q_1(10 - 6) = 10 - 50/7$$

$$q_1 = \frac{70/7 - 50/7}{10 - 6}$$

$$q_1 = 20/28$$

Substituting q_1 in the equation $q_1 + q_2 = 1$, we have:

$$q_1 + q_2 = 1$$

$$20/28 + q_2 = 1$$

$$q_2 = 1 - 20/28$$

$$q_2 = 8/28$$

It is assumed that the opponent cannot detect a pattern in the selection of a strategy. If he can, the value of the game will be less for that player.

Letting the payoff table be as follows:

		(p_1) B_1	(p_2) B_2
(q_1)	A_1	a_{11}	a_{12}
(q_2)	A_2	a_{21}	a_{22}

we can make the following generalizations.

$$p_1 = \frac{a_{22}-a_{12}}{a_{11}-a_{12}-a_{21}+a_{22}}$$

$$p_2 = 1 - p_1$$

$$EV = a_{21}(p_1) + a_{22}(p_2)$$

or

$$EV = a_{11}(p_1) + a_{12}(p_2)$$

$$q_1 = \frac{a_{21}-EV}{a_{21}-a_{11}}$$

$$q_2 = 1 - q_1$$

The generalizations are valid only when the players have 2 strategies each. These are referred to as 2 by 2 games. If more than 2 strategies are available to one or both players, one should try to eliminate recessive rows and columns. If this fails to reduce the game to a 2 by 2 case, the simplex method can be used to find an optimal strategy. This technique is not presented here. Similarly we have not discussed nonzero sum games or games that involve more than two persons.

PROBLEMS

1. Which of the following games are strictly determined?

(a)

	B_1	B_2
A_1	−6	10
A_2	−2	−4

(b)

	B_1	B_2
A_1	3	5
A_2	−1	4

(c)

	B_1	B_2
A_1	4	2
A_2	5	−1
A_3	−1	−2

(d)

	B_1	B_2
A_1	5	2
A_2	2	−1
A_3	−3	3

2. Find the optimum strategies and the value of this game:

	B_1	B_2
A_1	-6	10
A_2	-2	-4

3. Find the optimum strategies and the value of this game:

	B_1	B_2
A_1	4	2
A_2	5	-1
A_3	-1	-2

4. Find the optimum strategies and the value of this game:

	B_1	B_2
A_1	4	2
A_2	5	-1
A_3	-1	-2

5. Find the optimum strategies and the value of this game:

	B_1	B_2
A_1	5	2
A_2	2	-1
A_3	-3	3

6. Using the methods presented in this section can you find the optimum strategies and value of the following game? Why?

	B_1	B_2	B_3
A_1	4	-2	-1
A_2	-1	4	2

10-4 AN APPLICATION OF PRESENT VALUE, MATRIX TECHNIQUES, AND PROBABILITY TO A CAPITAL BUDGETING DECISION

First we will present a capital budgeting problem with a determinate view (i.e., one cash flow for each investment alternative) and provide a simple illustration of the present value technique. Then a risk situation is presented (i.e., more than one cash flow for each investment alternative) and probabilities are introduced. All of the above is done in a matrix setting. Relaxing the assumption of a determinate world, a more complicated cash flow is then introduced where depreciation and taxes are taken into account. Because these calculations can become lengthy, the computer can be introduced. Finally, the sensitivity of the decision is examined in regard to the probabilities. To begin, let us assume that we can invest in either Machine A or Machine B, both of which require an outlay of \$5000. The following schedule shows a payoff matrix by year before discounting.

$$\begin{array}{l} \\ \textit{Machine A} \\ \textit{Machine B} \end{array} \begin{array}{c} \begin{array}{cc} \textit{Year 1} & \textit{Year 2} \end{array} \\ \begin{pmatrix} 0 & 10{,}000 \\ 10{,}000 & 10{,}000 \end{pmatrix} \end{array}$$

Example 1. Discount the payoffs using a cost of capital (i.e., discount rate) of 10%.

Solution: Using Tables A-5 and A-7 we have

$$\begin{array}{l} \\ \textit{Machine A} \\ \textit{Machine B} \end{array} \begin{array}{c} \textit{Year 0} \\ \begin{pmatrix} 8{,}264 \\ 17{,}355 \end{pmatrix} \end{array}$$

The present value of future cash inflows less investment outlay gives the following:

$$\begin{array}{l} \\ \\ \\ \textit{Machine A} \\ \textit{Machine B} \end{array} \begin{array}{c} \textit{Inflows} \\ \textit{Year 0} \\ \begin{pmatrix} 8{,}264 \\ 17{,}355 \end{pmatrix} \end{array} - \begin{array}{c} \textit{Investment} \\ \textit{outlay} \\ \textit{Year 0} \\ \begin{pmatrix} 5{,}000 \\ 5{,}000 \end{pmatrix} \end{array} = \begin{array}{c} \textit{Net excess} \\ \textit{Present} \\ \textit{value} \\ \begin{pmatrix} 3{,}264 \\ 12{,}355 \end{pmatrix} \end{array}$$

Thus far we have assumed a definite cash inflow for each investment alternative. If this assumption is valid, we would clearly invest in Machine B. Suppose, on the other hand, that the cash inflows were based on the assumption that demand is high and that, if demand just happened to be low, the following payoff matrix, before discounting, would result.

$$\begin{array}{l} \\ \textit{Machine A} \\ \textit{Machine B} \end{array} \begin{array}{c} \begin{array}{cc} \textit{Year 1} & \textit{Year 2} \end{array} \\ \begin{pmatrix} 0 & 5000 \\ 1000 & 1000 \end{pmatrix} \end{array}$$

Example 2. Discount the payoffs using a cost of capital rate of 10%.

Solution: Using Tables A-5 and A-7 as in Example 1 we have

$$\begin{array}{l} \\ \textit{Machine A} \\ \textit{Machine B} \end{array} \begin{array}{c} \textit{Year 0} \\ \begin{pmatrix} 4132 \\ 1736 \end{pmatrix} \end{array}$$

The present value of future cash inflows less investment outlay gives the following:

$$\begin{array}{l} \\ \\ \\ \textit{Machine A} \\ \textit{Machine B} \end{array} \begin{array}{c} \textit{Inflows} \\ \textit{Year 0} \\ \begin{pmatrix} 4132 \\ 1736 \end{pmatrix} \end{array} - \begin{array}{c} \textit{Investment} \\ \textit{outlay} \\ \textit{Year 0} \\ \begin{pmatrix} 5000 \\ 5000 \end{pmatrix} \end{array} = \begin{array}{c} \textit{Net excess} \\ \textit{Present} \\ \textit{value} \\ \begin{pmatrix} (868) \\ (3264) \end{pmatrix} \end{array}$$

Now that we have made life a little more complicated, we are not yet in a position to make a decision. That is, the basis for selecting an act (i.e., what machine to purchase) is a more complex issue than in the case of decision-making under certainty where each act has a simple known payoff.

Decision-making under uncertainty involves not knowing the probability that one or another of the outcome states will prevail. Three of the most popular criteria that are used for decision-making under uncertainty are the maximin, the maximax, and the minimax-regret criteria. However, decision makers frequently do have information about the probability that each outcome state will occur, even though they are uncertain as to the actual outcome state. The sources of information with regard to probability are:

1. relative frequency,
2. theoretical probability distribution,
3. subjective judgment.

Decision making, when there are a number of possible outcome states for which the probability distribution is known, is called *decision-making under risk.*

Returning to our problem, let us assume that the market research department estimates the probability of high demand to be 0.4 and low demand to be 0.6. Note that demand must be either high or low, hence 0.4 plus 0.6 equals 1.

When making a decision under risk the *criterion* used is to *maximize expected payoff.* To calculate expected payoff it is necessary to multiply each outcome (column) by its respective probability and then add the entries in each row. The probabilities can be expressed as a column vector which is then multiplied by the payoff matrix.

Example 3. Calculate the expected values of Machines A and B.

Solution:

$$\begin{array}{lcc} & \multicolumn{2}{c}{\textit{Demand}} \\ & \textit{high} & \textit{low} \\ \textit{Machine A} & \multicolumn{2}{c}{\begin{pmatrix} 3{,}300 & -900 \\ 12{,}400 & -3{,}300 \end{pmatrix}} \\ \textit{Machine B} & & \end{array} \begin{array}{c}\textit{Probabilities}\\ \textit{of outcomes}\\ \begin{pmatrix} 0.4 \\ 0.6 \end{pmatrix}\end{array} = \begin{array}{c}\textit{Expected}\\ \textit{value}\\ \begin{pmatrix} 780 \\ 2{,}980 \end{pmatrix}\end{array}$$

Note that the entries in the payoff matrix were rounded to the nearest hundredth and minus signs rather than ( ) were used to indicate a loss. Using the criterion of maximizing expected payoff, we would invest in Machine B.

Clearly the expected value will not agree with the actual results on any one trial; however, this does not diminish the usefulness of this criterion to select among alternatives.

In general, then,

$$\begin{array}{c} \\ A_1 \\ A_2 \\ \vdots \\ A_m \end{array} \begin{array}{c} \begin{matrix} O_1 & O_2 & \cdots & O_n \end{matrix} \\ \begin{pmatrix} P_{11} & P_{12} & \cdots & P_{1n} \\ P_{21} & P_{22} & \cdots & P_{2n} \\ \vdots & \vdots & \ddots & \vdots \\ P_{m1} & P_{m2} & \cdots & P_{mn} \end{pmatrix} \end{array} \begin{pmatrix} P(O_1) \\ P(O_2) \\ \vdots \\ P(O_n) \end{pmatrix} = \begin{pmatrix} EV_1 \\ EV_2 \\ \vdots \\ EV_m \end{pmatrix}$$

where A_i, O_j, and P_{ij} indicate acts, outcomes, and payoffs, respectively, and where $P(O_j)$ and EV_i indicate the probability of Outcome j and the expected value of Act i, respectively.

Going a step further, let us consider the following example.

Example 4. Given the following cash inflows which are to be discounted at a rate of 10 percent.

	Demand high		*Demand low*	
	Years 1–3	*Years 4–5*	*Years 1–3*	*Years 4–5*
Machine A	10,000	5,000	4,000	2,500
Machine B	7,500	7,000	5,000	3,000

Also assume that each of Machines A and B requires a cash outlay in Year 0 of $10,000, that the useful life of both machines is five years, and that they are depreciated using the straight-line method. A federal tax of 50 percent is also assumed. Formulate the payoff matrix.

Solution: Summarizing the information obtained by using a library program (computer) we have:

$$\begin{array}{l} \\ \textit{Machine A} \\ \textit{Machine B} \end{array} \begin{array}{c} \begin{matrix} \textit{Demand high} & \textit{Demand low} \end{matrix} \\ \begin{pmatrix} 9484 & 394 \\ 7680 & 1963 \end{pmatrix} \end{array}$$

The library program calculates an investment's annual depreciation and its after-tax effect on a stream of gross cash flows. The resulting net cash flow stream is then discounted at the user's specified rate.

Assuming the probability of demand being high as 0.4 and the probability of demand being low as 0.6, the expected value of Act 1, the purchase of Machine A, is $4030. The expected value of Act 2, the purchase of Machine B, is $4249. Since the expected values of both acts are almost equal, one would probably make a decision on a basis other than quantitative factors. Using the criterion of maximizing expected value, one would choose Machine B.

An interesting question is: Just how sensitive is our answer to these probabilities? That is, would the same decision be arrived at if the probabilities were 0.5 and 0.5 respectively?

Example 5. Using the above information, for what probabilities would we be indifferent; or stated differently, will $EV_1 = EV_2$?

Solution:

$$9484P(O_1) + 394P(O_2) = 7680P(O_1) + 1963P(O_2)$$

$$1804P(O_1) - 1569P(O_2) = 0 \qquad (1)$$

We also know that

$$P(O_1) + P(O_2) = 1 \qquad (2)$$

Writing (1) and (2) as a matrix equation we have

$$\begin{pmatrix} 1804 & -1569 \\ 1 & 1 \end{pmatrix}\begin{pmatrix} P(O_1) \\ P(O_2) \end{pmatrix} = \begin{pmatrix} 0 \\ 1 \end{pmatrix}$$

Multiplying each side by the inverse of the matrix of coefficients gives $P(O_1) = 0.47$ and $P(O_2) = 0.53$. That is, for any $P(O_1) \geqslant 0.47$ we would invest in Machine A.

In conclusion, we can say that the typical capital budgeting problem is certain to be more complex than the above examples. However, most of the complications have to do with organizing the information (cash flows in particular) into a meaningful analysis prior to the application of discounting, matrix, and decision-making techniques.

PROBLEMS

1. What would the payoff matrix in Example 4 be if one simply discounted the cash inflows using a discount rate of 10 percent?
2. Find the inverse of the matrix of coefficients presented in Example 5.
3. Use a tree diagram to summarize the information given in the solution of Example 3.
4. Use the general result of this section to show that

$$EV_i = \sum_{j=1}^{n} P_{ij}P(O_j)$$

5. Given the following information:

	Outcome 1	*Outcome 2*
Act 1	\$10,000	\$20,000
Act 2	15,000	5,000

where the entries in the table are payoffs for various acts when coupled with an outcome state. Also assume that Outcome 1 has occurred 60 percent of the time in the past, whereas Outcome 2 has occurred 40 percent of the time in the past. Calculate the expected value of both acts.

6. A company president is faced with two options, both of which require an immediate cash outlay of \$50,000. If he chooses the first option and

demand is low, he will have a cash flow of $10,000 for ten years. If demand is high, there will be a cash flow of $20,000 for ten years. If he chooses the second option and demand is low, there will be cash flow of $20,000 for five years. If demand is high, there will be a flow of $25,000 for five years.

(a) Use a discount rate of 10 percent to find the present value of these cash flows.

(b) If each outcome state is equally likely, what is the expected net present value of both options?

7. Vernon Enterprises designs and manufactures toys. Past experience indicates that the product life cycle of a toy is three years. Promotional advertising produces large sales in the early years, but there is a substantial sales decline in the final year of a toy's life.

Consumer demand for new toys placed on the market tends to fall into three classes. About 30 percent of the new toys sell well above expectations, 60 percent sell as anticipated, and 10 percent have poor consumer acceptance.

A new toy has been developed. The following sales projections were made by carefully evaluating consumer demand for the new toy:

Consumer Demand for New Toy	Chance of Occurring	Estimated Sales in Year 1	Year 2	Year 3
Above average	30%	$1,200,000	$2,500,000	$600,000
Average	60	700,000	1,700,000	400,000
Below average	10	200,000	900,000	150,000

Variable costs are estimated at 30 percent of the selling price. Special machinery must be purchased at a cost of $860,000 and will be installed in an unused portion of the factory which Vernon has unsuccessfully been trying to rent to someone for several years at $50,000 per year and has no prospects for future utilization. Fixed expenses (excluding depreciation) of a cash-flow nature are estimated at $50,000 per year on the new toy. The new machinery will be depreciated by the sum-of-the-years digits method with an estimated salvage value of $110,000 and will be sold at the beginning of the fourth year. Advertising and promotional expenses will be incurred uniformly and will total $100,000 the first year, $150,000 the second year, and $50,000 the third year. These expenses will be deducted as incurred for income tax reporting.

Vernon believes that state and federal income taxes will total 60 percent of income in the foreseeable future and may be assumed to be paid uniformly over the year income is earned.

Required:

(a) Prepare a schedule computing the probable sales of this new toy in each of the three years, taking into account the probable occurrence of above average, average and below average sales.

(b) Assume that the probable sales computed in (a) are $900,000 in the first year, $1,800,000 in the second year, and $410,000 in the third year. Prepare a schedule computing the probable net income for the new toy in each of the three years of its life.

(c) Prepare a schedule of net cash flows from sales of the new toy for each of the years involved and from disposition of the machinery purchased. Use the sales data given in part **(b)**.

(d) Assuming a minimum desired rate of return of 10 percent, prepare a schedule of the present value of the net cash flows calculated in **(c)**. The following data are relevant:

Year	Present Value of $1.00 Due at the End of Each Year Discounted at 10 percent	Present Value of $1.00 Earned Uniformly Throughout the Year Discounted at 10 percent
1	.91	.95
2	.83	.86
3	.75	.78

8. Vendo, Inc. has been operating the concession stands at the University football stadium. The University has had successful football teams for many years; as a result the stadium is always full. The University is located in an area which suffers no rain during the football season. From time to time, Vendo has found itself very short of hot dogs and at other times it has had many left. A review of the records of sales of the past five seasons revealed the following frequency of hot dogs sold.

	Total Games
10,000 hot dogs	5 times
20,000 hot dogs	10 times
30,000 hot dogs	20 times
40,000 hot dogs	15 times
	50 total games

Hot dogs sell for 50 cents each and cost Vendo 30 cents each. Unsold hot dogs are given to a local orphanage without charge.

Required:

(a) Assuming that only the four quantities listed were ever sold and that the occurrences were random events, prepare a payoff table (ignore income taxes) to represent the four possible strategies of ordering 10,000; 20,000; 30,000; or 40,000 hot dogs.

(b) Using the expected value decision rule determine the best strategy.

REFERENCES

Blackwell, David, and Girschick, M.A. *Theory of Games and Statistical Decisions*. New York: John Wiley & Sons, Inc. 1954.

Grawoig, Dennis E. *Decision Mathematics*. New York: McGraw-Hill Book Company, Inc., 1967.

Levin, Richard I., and Desjardins, Robert B. *Theory of Games and Strategies*. Scranton, Pennsylvania: International Textbook Company, 1970.

Problem 8 adapted from the CMA examination.

Luce, R. Duncan, and Raiffa, Howard. *Games and Decisions*. New York: John Wiley & Sons, Inc. 1958.

McKinsey, J. C. C. *Introduction to the Theory of Games.* New York: McGraw-Hill Book Company, Inc. 1952.

Rapoport, Anatol. *N-Person Game Theory.* Ann Arbor: The University of Michigan Press, 1970.

Vajda, S. *The Theory of Games and Linear Programming.* New York: John Wiley & Sons, Inc. 1961.

Ventsel, E. S. *An Introduction to the Theory of Games.* Boston: D. C. Heath and Company, 1963.

Von Neumann, John, and Morgenstern, Oskar. *Theory of Games and Economics Behavior*. New York: John Wiley & Sons, Inc. 1964.

Williams, J. D. *The Compleat Strategyst.* New York: McGraw-Hill Book Company, 1965.

ELEVEN

Precalculus mathematics

Our intent, as mentioned earlier, is to provide a solid foundation in mathematics as well as to help individuals use mathematics to understand real situations. The topic of graphing is ideal in highlighting the benefits of this blend of mathematics and problems taken from the real world. In many situations it is necessary to recognize the type of function from a graph and then formulate a mathematical expression or model. In addition, the experience gained in graphing these functions will be very useful as background knowledge for the study of calculus which follows this chapter.

In this chapter we will look at functions that graph as a hyperbola, and at quadratic, cubic, exponential, logarithmic, and power functions as well as the supporting topics of exponents and logarithms. In Section 11-7 the reader is introduced to the technique of graphing functions on semilog and log-log paper. We point out that each of the three types of functions introduced in Sections 11-4, 11-5, and 11-6 graphs as a straight line either on semilog paper or log-log paper. The use of logarithms for computational purposes can be found as an appendix to this chapter.

11-1 SOME IMPORTANT NONLINEAR FUNCTIONS

In this section we will look at the general form and applications of the quadratic and cubic functions and those functions that graph as a hyperbola. It seems reasonable that the mathematical relationship that exists between the price, p, of a particular commodity and the quantity demanded, x, would appear as follows:

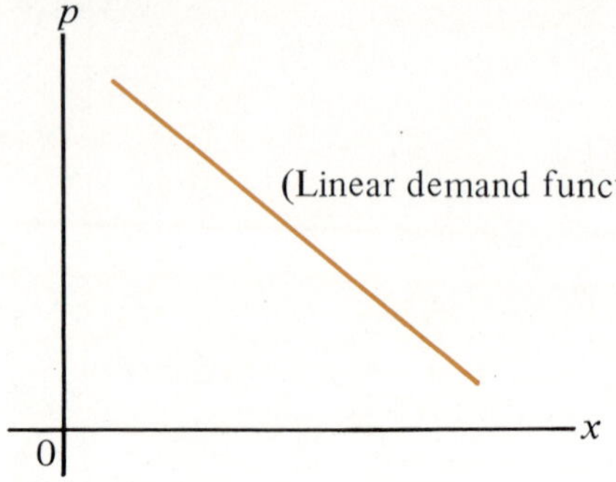

FIGURE 11-1

Verbally this graph indicates that as the price is lowered a greater quantity will be demanded. The graph appears as a straight line and the mathematical relationship could be expressed as follows:

$p = bx + c$ where b and c are constants

To obtain total revenue R one would simply multiply the price (p) by the quantity (x) to obtain $R = px$.

Total Revenue = Price × Quantity

$$R = px$$

$$R = (bx + c)x, \text{ in this case}$$

$$R = bx^2 + cx$$

The mathematical relationship

$$y = bx^2 + cx + d$$

is called a *quadratic function* and its graph is a parabola. Note that the total revenue function derived is a quadratic function and therefore graphs as a parabola. The value of d is zero in this application.

The next three examples are based on quadratic functions.

Example 1. Graph and discuss the function $y = x^2$ (quadratic).

Solution:

x	-2	-1	0	1	2
y	4	1	0	1	4

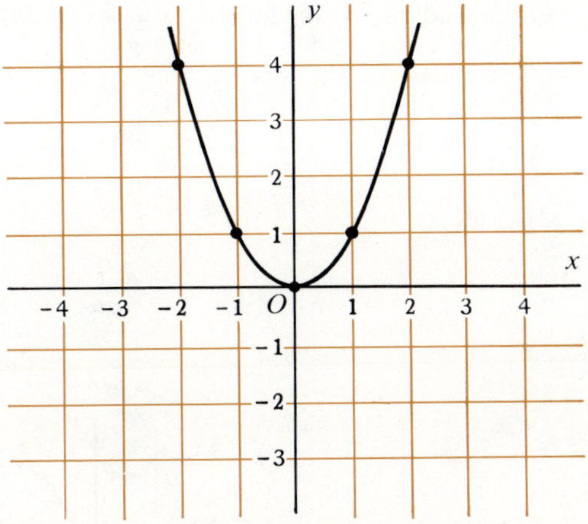

FIGURE 11-2

Discussion: $b = 1$, $c = 0$, and $d = 0$. For positive values of x, y increases at an increasing rate as x increases. The graph is a parabola opening upward with *vertex* at the origin.

Example 2. Graph and discuss the function $y = x^2 + 3x$ (quadratic).

Solution:

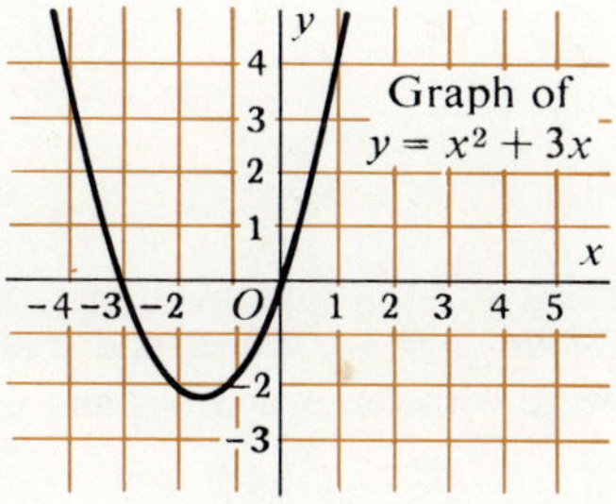

FIGURE 11-3

Discussion: $b = 1$, $c = 3$, and $d = 0$. The graph is a parabola opening upward with vertex at $x = -\frac{3}{2}$ and $y = -\frac{9}{4}$. For $x < -\frac{3}{2}$, y decreases at a decreasing rate. For $x > -\frac{3}{2}$, y increases at an increasing rate as x increases. This function is concave upward for all values of x.

Example 3. Graph and discuss the function $y = -x^2 + 3x + 2$ (quadratic).

Solution:

x	-1	0	1	2	3
y	-2	2	4	4	2

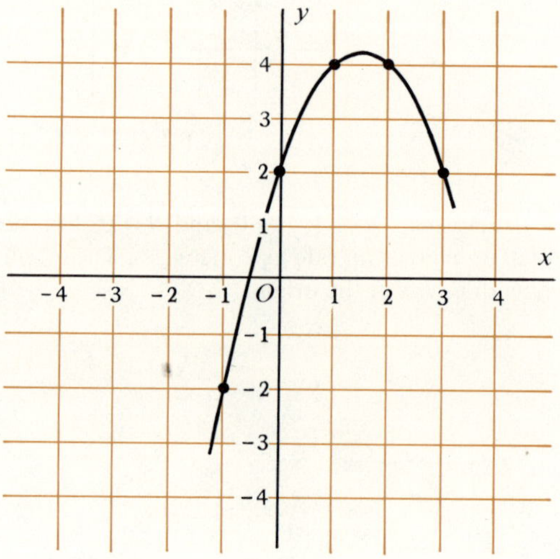

FIGURE 11-4

Discussion: $b = -1$, $c = 3$, and $d = 2$. The graph is a parabola opening downward with vertex at $x = \frac{3}{2}$ and $y = 4\frac{1}{4}$.

The following discussion shows how a cubic function can sometimes be used to model total cost versus quantity produced.

The total cost curve is often assumed to be linear. That is, it is assumed that the cost of producing an additional unit is a constant. In many cases this is not so. A more valid assumption is that at first, the total cost curve will increase at a decreasing rate (economies of scale) and then increase at an increasing rate (diseconomies of scale).

Graphically this would appear as follows:

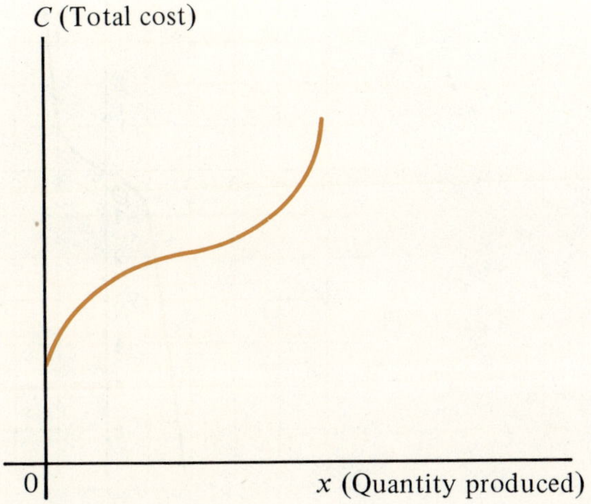

FIGURE 11-5

Some cubic functions have the same general shape as the curve shown in Figure 11-5. With this fact in mind, let us consider the class of functions of the form

$$y = ax^3 + bx^2 + cx + d$$

Functions of this form are sometimes used to model total cost.

Example 4. Graph and discuss the function $y = x^3$ (cubic).

Solution:

x	-2	-1	0	1	2	3
y	-8	-1	0	1	8	27

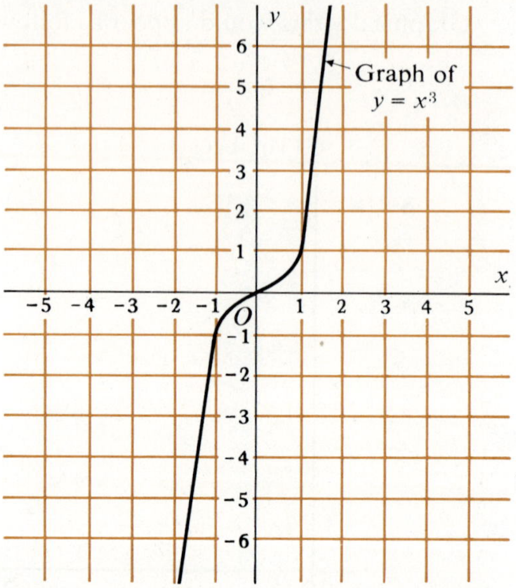

FIGURE 11-6

Discussion: $a = 1$, $b = c = d = 0$. For $x > 0$, y increases at an increasing rate. For $x < 0$, y increases at a decreasing rate as x increases. This function is concave upward for $x > 0$ and concave downward for $x < 0$.

Example 5. Graph and discuss the function $y = x^3 + 2x^2$ (cubic).

Solution:

x	-3	-2	$-\frac{4}{3}$	-1	0	1
y	-9	0	$\frac{32}{27}$	1	0	3

Discussion: $a = 1$, $b = 2$, and $c = d = 0$. This function has a relative maximum at $x = -\frac{4}{3}$ and a relative minimum at $x = 0$. The graph is concave downward for $x < -\frac{2}{3}$ and concave upward for $x > -\frac{2}{3}$. The point $x = -\frac{2}{3}$, $y = \frac{16}{27}$ is an inflection point. The concepts of *relative maximum*, *relative minimum* and *inflection point* are defined in Chapter 14.

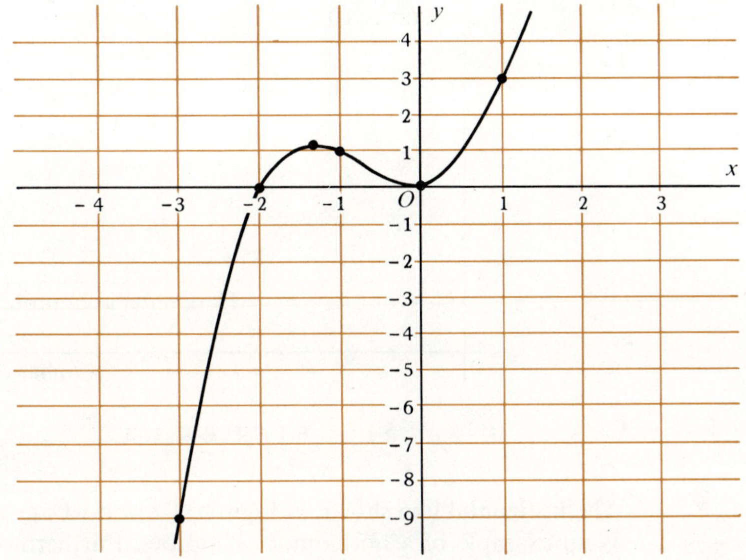

FIGURE 11-7

The next discussion illustrates the use of an equation of a hyperbola to model the relationship between *unit cost* and the *number of units produced.* A convenient form of the equation for this purpose is

$$y = \frac{a}{x-b} + c$$

where a, b, and c are constants. For this application, let y = unit cost, c = variable cost per unit, a = fixed cost, and x = the number of units produced. Under these conditions *unit cost equals a share of the fixed costs plus the variable cost per unit.* Note that the share of the fixed costs is dependent on the number of units produced.

An interesting feature of the model

$$y = \frac{a}{x-b} + c$$

is that y will approach but never reach the *limiting* value c as x takes on larger and larger values. In terms of *limits*, which are covered later in the text, it would be correct to say that y *approaches* or tends to c as x *approaches* infinity (gets arbitrarily large). One branch of a typical graph of this model is shown in Figure 11-8.

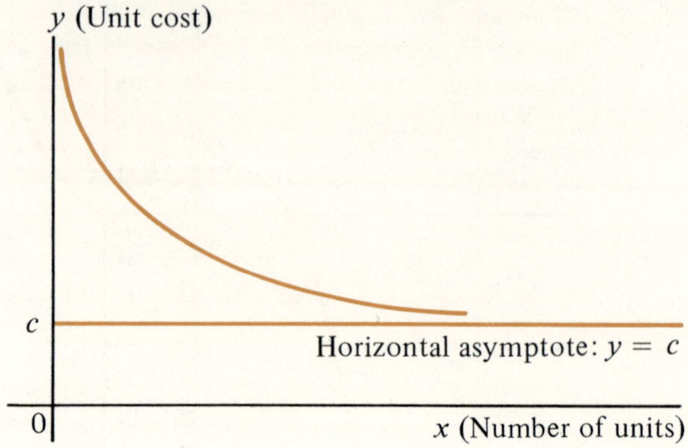

FIGURE 11-8

The horizontal line shown in Figure 11-8 and whose equation is $y = c$ is an example of a horizontal asymptote. Furthermore, in this case, the vertical axis is a vertical asymptote. In general, for the model under discussion here, an equation for the vertical asymptote is $x = b$.

A mathematical example of an equation of a hyperbola is given next.

Example 6. Graph and discuss the function $y = 2/(x - 3)$.

Solution:

x	3	3.01	4	5	7	2.99	2	1	0	-1
y		200	2	1	0.5	-200	-2	-1	$-\frac{2}{3}$	-0.5

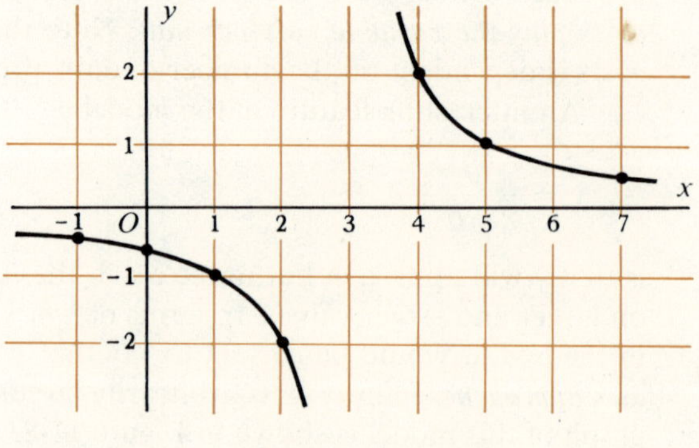

FIGURE 11-9

Discussion: $a = 2$, $b = 3$, and $c = 0$. The vertical line $x = 3$ has a special relationship to the graph of this function. It is called a *vertical asymptote.* It is possible to find points on the graph that are arbitrarily close to this vertical asymptote by choosing x close, but not equal, to 3. Note that y is undefined for $x = 3$. The horizontal line $y = 0$ has a similar relationship to the graph. It is called a *horizontal asymptote.*

Example 7. Express $(x-3)(y-4) = 2$ in the form $y = a/(x-b)+c$.

Solution:

$$y - 4 = \frac{2}{x-3}$$

$$y = \frac{2}{x-3} + 4$$

Example 8. Graph and discuss the function of Example 7.

Solution:

x	-1	1	2	3	4	5	7
y	3.5	3	2	—	6	5	4.5

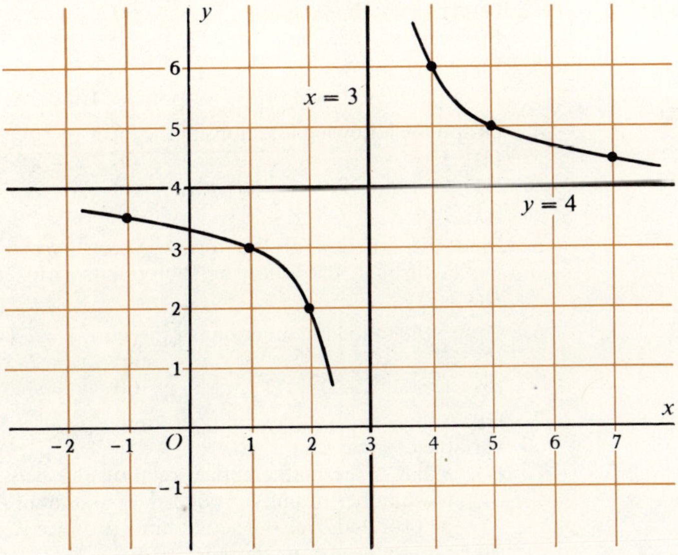

FIGURE 11-10

Discussion: $a = 2$, $b = 3$, and $c = 4$. The line $x = 3$ is a vertical asymptote. The line $y = 4$ is a horizontal asymptote. y is undefined at $x = 3$.

The following table should help summarize the information presented in this section:

Mathematical Relationship	Type of Function	Used to Model
$y = ax^3 + bx^2 + cx + d$	Cubic	Total cost
$y = bx^2 + cx + d$ (i.e., $a = 0$)	Quadratic[a]	Total revenue
$y = cx + d$ (i.e., $a = 0$, $b = 0$)	Linear[b]	Total cost
$y = \dfrac{a}{b+x} + c$	Equation of a hyperbola	Unit cost

[a] Graphs as a parabola opening upward or downward.
[b] Graphs as a straight line with slope c and y intercept d.

PROBLEMS

1. Graph the following:
(a) $y = 2x - 3$
(b) $y = -\frac{1}{2}x + 4$
(c) $2x + 3y - 6 = 0$

2. Graph the following functions of the form $y = bx^2 + cx + d$ (quadratic):
(a) $y = 3x^2$
(b) $y = x^2 - 3x$
(c) $y = x^2 + 3x$
(d) $y = x^2 + 3x + 4$

3. Find the function of the form $y = bx^2 + cx + d$ whose graph passes through the following triple of points: $A(0, 3)$, $B(1, 3)$, and $C(2, -1)$.

4. Graph the following functions of the form $y = ax^3 + bx^2 + cx + d$ (cubic):
(a) $y = -x^3$
(b) $y = x^3 - 2x^2$
(c) $y = x^3 + 2x^2 + 2$
(d) $y = x^3 + 6x$
(e) $y = x^3 - 6x$
(f) $y = -x^3 + 6x$

5. Find the function of the form $y = ax^3 + bx^2 + cx + d$ whose graph passes through the following four points: $A(0, 2)$, $B(1, 0)$, $C(2, 8)$, and $D(-1, 2)$.

6. Graph the following functions of the form $y = a/(x - b) + c$:
(a) $y = 2/x$
(b) $y = 2/(x+3)$
(c) $y = 2/(x+4)$
(d) $y = -2/(x+4)$

7. Given that the fixed cost of operating a plant is \$100,000 and that the variable cost per unit is \$10.
(a) What is the mathematical relationship between unit cost, y, and the number of units produced, x, assuming that fixed costs are to be prorated over the actual units produced?
(b) Graph the above relationship for $x = 1000, 2000, \ldots, 10{,}000, 100{,}000$.
(c) What does the product $(x)(y)$ represent?
(d) What is the function or mathematical relationship for total cost?
(e) What is the economic significance of the horizontal asymptote $y = 10$?

8. Is it logical for a total cost curve to appear as follows:

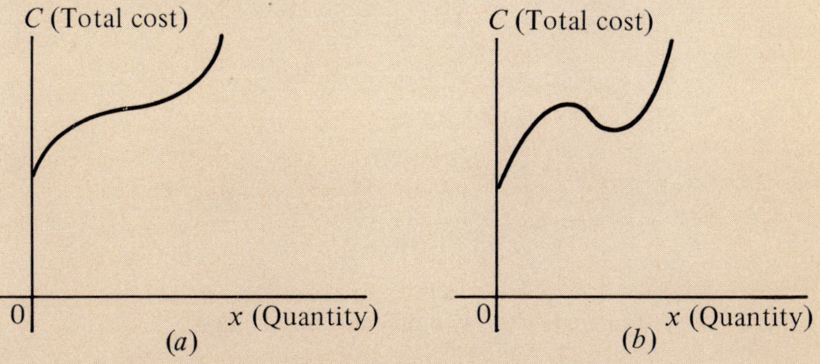

FIGURE 11-11

9. What condition must one impose on the coefficient b, to guarantee that a parabola opens downward?

10. Given that total revenue is equal to $-2x^2+120x$ where x is units sold. Graph the revenue function for $10 \leqslant x \leqslant 50$.

11. If total revenue is known to be $R = -2x^2 + 120x$ as it is in Problem 10, how does the price p depend on x? That is, determine the linear demand function that is implied by the given total revenue function. *Hint:* $R = px$.

12. Graph the linear demand function obtained in Problem 11.

13. Repeat Problems 11 and 12 if $R = 90x - 3x^2$, where R denotes total revenue and x denotes quantity.

14. What is the maximum possible revenue if $R = -2x^2 + 120x$ as it does in Problem 10? *Hint:* Look for the vertex of the parabola, which opens downward in this case.

15. What is the maximum possible revenue if $R = 90x - 3x^2$ as it does in Problem 13?

11-2 LAWS OF EXPONENTS AND LOGARITHMS

The following laws of exponents and logarithms are essential in the study of mathematics. Laws E-2, E-3, and E-4 are used in the study of logarithms and will be discussed later in this section. Laws E-5, E-7, and E-8 are used frequently in the study of calculus, which begins in the next chapter. The following laws are valid for real numbers m, n, a, and b:

LAWS OF EXPONENTS

E-1	$a^0 = 1$, if $a \neq 0$	E-5	$a^{-n} = 1/a^n$, if $a \neq 0$
E-2	$a^m a^n = a^{m+n}$	E-6	$(ab)^n = a^n b^n$
E-3	$a^m/a^n = a^{m-n}$	E-7	$a^{1/n} = \sqrt[n]{a}$, $a > 0$
E-4	$(a^m)^n = a^{mn}$	E-8	$a^{m/n} = \sqrt[n]{a^m}$, $a > 0$

Example 1. Use the laws of exponents to calculate a^5a^3 and $(a^5)^3$.

Solution:

$$a^5a^3 = a^{5+3} \qquad \text{by E-2}$$
$$= a^8$$
$$(a^5)^3 = a^{5(3)} \qquad \text{by E-4}$$
$$= a^{15}$$

Example 2. Calculate a^4/a^0 two different ways, $a \neq 0$.

Solution:

$$a^4/a^0 = a^{4-0} = a^4 \qquad \text{by E-3}$$

or

$$a^4/a^0 = a^4/1 \qquad \text{by E-1}$$
$$= a^4$$

Example 3. Express $1/x$ and $\sqrt{x}$ in exponential notation.

Solution:

$$1/x = x^{-1} \qquad \text{by E-5}$$
$$\sqrt{x} = \sqrt[2]{x} \qquad \text{(understood)}$$
$$= x^{1/2} \qquad \text{by E-7}$$

Example 4. Use laws of exponents to write a^4b^2 in a different form.

Solution:

$$a^4b^2 = (a^2)^2b^2 \qquad \text{by E-4}$$
$$= (a^2b)^2 \qquad \text{by E-6}$$

Example 5. Simplify $4^{1.5}$.

Solution:

$$4^{1.5} = 4^{3/2}$$
$$= \sqrt[2]{4^3} \qquad \text{by E-8}$$
$$= \sqrt{64}$$
$$= 8$$
$$4^{3/2} = 8$$

Logarithms are numbers that historically were used to reduce complicated multiplications and divisions to additions and subtractions, as well as to do other problems. The computational uses of logarithms can be found as an appendix to this chapter. It should be noted that calculators and computers have replaced the use of logarithms for computational efforts.

A formal definition of a logarithm follows:

DEFINITION 11.1. *Given the positive number b, where $b \neq 1$ and N is any positive real number, then the logarithm to the base b of N denoted by* $\log_b N$ *is the number that, when used as an exponent of b, gives N. In symbols,* $\log_b N = x$ *if* $b^x = N$.

Perhaps a simpler definition is to simply say that a logarithm is an exponent.

Example 6. Use Definition 11.1 to find $\log_2 8$, $\log_{16} 4$, and $\log_{1/2}(\frac{1}{16})$.

Solution:

$$\log_2 8 = 3 \quad \text{since} \quad 2^3 = 8$$

$$\log_{16} 4 = \tfrac{1}{2} \quad \text{since} \quad 16^{1/2} = \sqrt{16} = 4$$

$$\log_{1/2}(\tfrac{1}{16}) = 4 \quad \text{since} \quad (\tfrac{1}{2})^4 = \tfrac{1}{16}$$

In Example 6, we used bases 2, 16, and $\frac{1}{2}$ to illustrate Definition 11.1. Although, theoretically, any positive number not equal to one could be used as a base, the number 10 is an especially important choice. In the next section we will discuss the *base* 10 system. This system is called the *common log system.*

Example 7. Find $\log_{10} 1$, $\log_{10} 10$, $\log_{10} 100$, and $\log_{10} 1000$.

Solution:

$$\begin{aligned} \log_{10} 1 &= x \\ 10^x &= 1 \\ x &= 0 \qquad \textit{by E-1} \\ \therefore \log_{10} 1 &= 0 \end{aligned} \qquad\qquad \begin{aligned} \log_{10} 100 &= x \\ 10^x &= 100 \\ x &= 2 \\ \therefore \log_{10} 1000 &= 2 \end{aligned}$$

$$\begin{aligned} \log_{10} 10 &= x \\ 10^x &= 10 \\ x &= 1 \\ \therefore \log_{10} 10 &= 1 \end{aligned} \qquad\qquad \begin{aligned} \log_{10} 1000 &= x \\ 10^x &= 1000 \\ x &= 3 \\ \therefore \log_{10} 1000 &= 3 \end{aligned}$$

The following laws of logarithms are valid. Let M, N, and b be positive numbers with $b \neq 1$.

LAWS OF LOGARITHMS

L-1	$\log_b MN = \log_b M + \log_b N$
L-2	$\log_b M/N = \log_b M - \log_b N$
L-3	$\log_b N^P = P \log_b N$

Proof of L-1. Let $\log_b M = x$ and $\log_b N = y$. Then, by Definition 11.1, $b^x = M$ and

$$b^y = N$$

$$MN = b^x b^y$$

$$= b^{x+y} \qquad \text{by E-2}$$

Therefore,

$$\log_b MN = x + y \qquad \text{by Definition 11.1}$$

$$\log_b MN = \log_b M + \log_b N$$

(We will omit the proofs of L-2 and L-3.)

Example 8. Illustrate numerically that $\log_2 16 = \log_2 2 + \log_2 8$.

Solution:

$$\log_2 16 = 4 \qquad \text{since} \qquad 2^4 = 16$$

$$\log_2 2 = 1 \qquad \text{since} \qquad 2^1 = 2$$

$$\log_2 8 = 3 \qquad \text{since} \qquad 2^3 = 8$$

Therefore,

$$\log_2 16 = 4 = 1 + 3 = \log_2 2 + \log_2 8$$

Example 9. Show that $\log_b \sqrt[n]{M} = 1/n \log_b M$.

Solution:

$$\sqrt[n]{M} = M^{1/n} \qquad \text{by E-7}$$

$$\log_b \sqrt[n]{M} = \log_b M^{1/n}$$

$$= 1/n \log_b M \qquad \text{by L-3}$$

Therefore,

$$\log_b \sqrt[n]{M} = 1/n \log_b M$$

Example 10. Verify L-2 for the case $\log_2 \frac{16}{4} = \log_2 16 - \log_2 4$.

Solution:

$$\log_2 \tfrac{16}{4} = \log_2 4$$

$$= 2 \qquad \text{since} \qquad 2^2 = 4$$

$$\log_2 16 = 4 \qquad \text{and} \qquad \log_2 4 = 2$$

Therefore,

$$\log_2 \tfrac{16}{4} = 2 = \log_2 16 - \log_2 4$$

PROBLEMS

1. Find $\log_3 N$ for the following values of N:

   $N = 9, 81, 27, 1, \frac{1}{3}, \frac{1}{27}, \frac{1}{9}$

2. Find $\log_{1/2} N$ for the following values of N:

   $N = \frac{1}{4}, 1, 2, \frac{1}{2}, \frac{1}{8}, 16$

3. Find $\log_{10} N$ for the following values of N:

   $10^2, 10^{-3}, 10^5, 10^{1/3}, \sqrt{10}, \sqrt[3]{10}, 10^{1/3}, \sqrt[3]{100}$

4. Illustrate numerically that $\log_4 64 = \log_4 4 + \log_4 16$.
5. Illustrate numerically that $\log_2 4^3 = 3 \log_2 4$.
6. Illustrate numerically that $\log_5 \frac{625}{25} = \log_5 625 - \log_5 25$.
7. Prove L-2 and L-3.
8. Find $\log_b x$ given that $\log_b xy = 8$ and that $\log_b(x/y) = 4$.
9. Calculate or rewrite each of the following:
   (a) $\log_b(b^3)$
   (b) $\log_b 1$
   (c) $\log_b(1/b)$
   (d) $\log_b(ab^3 + b^2)$
10. Show that $\dfrac{\log_b 40 - \log_b 5}{3} = \log_b 2$.
11. Show that $\log_a N = \dfrac{\log_b N}{\log_b a}$ (change of base formula).
12. Illustrate the result in Problem 11 with $N = 16$, $a = 4$, and $b = 2$.
13. Which of the following expressions can be written as a difference of two logarithms?

    (a) $\log_{10} \dfrac{5}{3}$ **(b)** $\dfrac{\log_{10} 5}{\log_{10} 3}$

14. Express each of the following ratios as a single logarithm.

    [*Hint:* Use the result of Problem 11.]

    (a) $\dfrac{\log_{10} 5}{\log_{10} 3}$ **(b)** $\dfrac{\log_3 81}{\log_3 9}$

15. Rewrite $V_n = P(1+i)^n$ by taking logarithms of both sides.

16. Rewrite the following expressions in exponential form:

(a) $\sqrt{x}$ **(d)** $\sqrt{(xy)^4}$

(b) $\sqrt[3]{x}$ **(e)** $x\sqrt{x}$

(c) $\sqrt{x^3}$ **(f)** $\sqrt[3]{x+y}$

17. Rewrite the following expressions using radicals:

(a) $x^{1/2}$ **(d)** $x^{5/3}$

(b) $x^{3/2}$ **(e)** $(x+y)^{1/2}$

(c) $x^{3/4}$ **(f)** $x^{-1/4}$

18. Use laws of exponents to simplify the following:

(a) $3^5/3^2$ **(c)** 2^{-3}

(b) x^5/x^2 **(d)** 3^{-2}

19. Use laws of exponents to show that

(a) $\sqrt{x}\sqrt{y} = \sqrt{xy}$ **(b)** $\sqrt{x}/\sqrt{y} = \sqrt{x/y}$

20. Use the results of Problem 4 to find the following square roots:

(a) $\sqrt{a^2(b+c)^4}$ **(c)** $\sqrt{25/49}$

(b) $\sqrt{9/16}$ **(d)** $\sqrt{13/9}$

21. Solve for x.

(a) $2^x = 8$ **(d)** $\sqrt[x]{8} = 2$

(b) $3^x = 81$ **(e)** $\sqrt[3]{x} = 4$

(c) $4^x = 1/4$ **(f)** $\sqrt{3x} = 4$

22. Solve for x.

(a) $10^x = 100$ **(d)** $10^x = 0.1$

(b) $10^x = 1$ **(e)** $10^x = 0.01$

(c) $10^x = 1000$ **(f)** $10^x = 10{,}000$

23. Show that $(x^2+4)^0 = 1$ for any real number x.

24. Show that $P(1+i)^n + P(1+i)^n i = P(1+i)^{n+1}$.

11-3 THE USE OF TABLES OF LOGARITHMS

In this section we will use the base 10 system for computations. Since the base 10 is used throughout this section, we will adopt the usual convention and write $\log_{10} N$ as $\log N$.

In all of the numerical examples of the previous section, the logarithms could be obtained by intuition. Clearly, this is not the case for $\log_{10} 110 = \log 110$, for example. In other words, the answer to the question, "Ten to what power equals 110?" is not at all obvious. Fortunately, there are tables of logarithms that provide an easy answer to this type of question. One of the main purposes of this section is to acquaint the reader with the use of such tables. The reader will find a five-place table of common logarithms in Table 1 of Appendix A.

To answer the above question, we know intuitively that log 110 is between 2 and 3. The decimal part of this number is found from Table 1 to be 0.04139. Hence, $\log 110 = 2.04139$. In this example, the decimal 0.04139 is called the *mantissa* of log 110 and the integer 2 is called the *characteristic* of log 110. The terms characteristic and mantissa are illustrated further in the following examples.

Before continuing we would like to point out the fact that if one has a calculator with a log key there would be no need to make use of the table. In fact most tables are the result of the work of a computer or calculator. If one does have a calculator it would be wise to follow through the examples making use of the calculator.

Example 1. Find the characteristic and mantissa of log 101 and log 1001.

Solution: It is clear that log 101 is slightly larger than 2; i.e., $\log 100 = 2$; therefore, $\log 101 = 2 \ldots.$ The decimal needed to fill in the blank is the mantissa of log 101. From Table 1, we find that $\log 101 = 2.00432$.

The characteristic of log 101 is 2 and the mantissa of log 101 is 0.00432. Similarly, $\log 1001 = 3.00043$. $\text{Log } 1000 = 3$; therefore, the characteristic of log 1001 is 3 and the mantissa is 0.00043.

Example 2. Find the characteristic and mantissa of log 11, log 110, and log 1100.

Solution:

$$\log 11 = 1.04139$$

$$\log 110 = 2.04139$$

$$\log 110 = 3.04139$$

The characteristics of log 11, log 110, and log 1100 are one, two, and three, respectively. The mantissa, 0.04139, is the same for all of them.

Example 2 illustrates the fact that the mantissa of $\log N$ is determined by the sequence of digits that make up N with no regard to the decimal point. That is, the logarithms of two different numbers with the same sequence of digits have the same mantissa. For example, log 1.23, log 12.3, and log 123 all have mantissa 0.08991.

The characteristic of $\log N$ is determined by the location of the decimal point within the sequence of digits that make up N.

Any real number can be written with the decimal point after the first significant digit times ten to some integral power. The decimal point is then said to be in the *standard position.* This is called *scientific notation.*

Example 3. Write the numbers 0.0462 and 398 with the decimal point in standard position and find the logarithms of these numbers.

Solution:

$$0.0462 = 4.62 \times 10^{-2}$$

$$398 = 3.98 \times 10^{2}$$

$$\log 398 = \log 3.98 \times 10^{2}$$

$$= \log 3.98 + \log 10^{2} \qquad \textit{by L-1}$$

$$= \log 3.98 + 2 \qquad \textit{since } \log 10^{2} = 2 \textit{ by Definition 11.1}$$

$$= 0.59988 + 2$$

Since 3.98 is between one and ten, therefore, its logarithm has characteristic zero.

$$\log 398 = 2.59988$$

By similar reasoning,

$$\log 0.0462 = \log(4.62 \times 10^{-2})$$

$$= \log 4.62 + \log 10^{-2}$$

$$= 0.66464 - 2 \qquad \textit{since } \log 10^{-2} = -2 \textit{ by definition}$$

Example 3 illustrates the basis of a short-cut method of determining the characteristic of $\log N$. This method consists of counting the number of places the decimal must be moved to put it in standard position. If the decimal must be *moved to the left*, the characteristic is *positive*. It is *negative* if the decimal point must be *moved to the right*.

Example 4. Use the short-cut method to find log 3.4, log 9400, and log 0.69.

Solution:

$$\log 3.4 = 0.53148$$

$$\log 94 = 1.97313$$

$$\log 9400 = 3.97313$$

Finally, $\log 0.69$ has mantissa 0.83885 and characteristic -1. Negative characteristics should be written to the right of the mantissa. This gives $\log 0.69 = 0.83885 - 1$. It is also common to express such numbers as follows: $\log 0.69 = 0.83885 - 1 = 9.83885 - 10$.

The mantissa for numbers with no more than four significant digits can be found directly from Table A-1 of the Appendix. A method called *linear interpolation* can be used to approximate the mantissa for a number with five significant digits. This method treats the *nonlinear* function $y = \log x$ as if it were a *straight line* between the tabulated points. The methods of linear interpolation is illustrated numerically in the next two examples.

It is to be noted that the first two digits of the mantissa in this table are indicated for groups of mantissas in order to save space. Where an asterisk (*) appears, the first two digits of the next group of mantissas, which are located in the row below, are to be used for the mantissa.

Example 5. Use linear interpolation and Table A-1 of the Appendix to find $\log 1.8432$.

Solution: Since $\log 1.8440 = 0.26576$, and $\log 1.8430 = 0.26553$, their difference is 0.00023. Adding 2/10 of the difference, $\log 1.8440 - \log 1.8430$, to $\log 1.843$, we have

$$\begin{aligned} \log 1.8432 &= 0.26553 + 0.2(0.00023) \\ &= 0.26553 + 0.000046 \\ \log 1.8432 &= 0.265576 \end{aligned}$$

This value may be rounded off to 0.26558.

Example 6. Use Table A-1 of the Appendix and linear interpolation to find $\log 934.57$.

Solution: The mantissa for 934.6 is 0.97063 and the mantissa for 934.5 is 0.97058. Adding 7/10 of the difference, $0.97063 - 0.97058$, to 0.97058, the mantissa of the smaller tabulated number, we have

$$\begin{aligned} \text{mantissa for } 934.57 &= 0.97058 + 0.7(0.00005) \\ &= 0.97058 + 0.000035 \\ &= 0.97062 \end{aligned}$$

Therefore,

$$\log 934.57 = 2.97062$$

Note that in Example 6, the choice of 0.7 was based on the fact that the fifth digit in 934.57 is seven.

PROBLEMS

1. The logarithm of each of the following numbers is between what two integers?

(a) 84
(b) 97
(c) 154
(d) 837
(e) 1562
(f) 2437
(g) 0.23
(h) 0.0087

2. Find the common mantissa for the following numbers:
(a) 0.1234 **(e)** 0.001234
(b) 1.234 **(f)** 12340
(c) 1234 **(g)** 1,234,000
(d) 12.34 **(h)** 1.234×10^8

3. Write each of the following numbers with the decimal point in the standard position.
(a) 482 **(e)** 352×10^4
(b) 5372 **(f)** 4800
(c) 0.043 **(g)** 573,000
(d) 0.87 **(h)** 0.43×10^{-3}

4. Find the mantissa for each of the following numbers:
(a) 4832 **(e)** 0.02391
(b) 3718 **(f)** 0.8
(c) 43.92 **(g)** 712,300
(d) 257.6 **(h)** 0.001

5. Find the characteristic for each of the numbers in Problem 4.

6. Find the logarithm of each of the numbers in Problem 4. [*Hint:* Use the results of Problems 4 and 5.]

7. Use Table A-1 of the Appendix and interpolation to find the logarithm of each of the following numbers:
(a) 48217 **(d)** 8364.9
(b) 0.083175 **(e)** 0.12348
(c) 23.711 **(f)** 7.3463×10^5

8. Simplify $b = \dfrac{\log 10.10 - \log 2}{\log 3 - \log 1}$.

11-4 LOG FUNCTIONS

Functions of the form

$$y = m \log_a x + b$$

where a, m, and b are constants, will be discussed in this section.

Example 1. Draw the graph of $y = \log_3 x$.

Solution: $y = \log_3 x$ if and only if $3^y = x$.

x	1	3	9	27	$\frac{1}{3}$	$\frac{1}{9}$
y	0	1	2	3	-1	-2

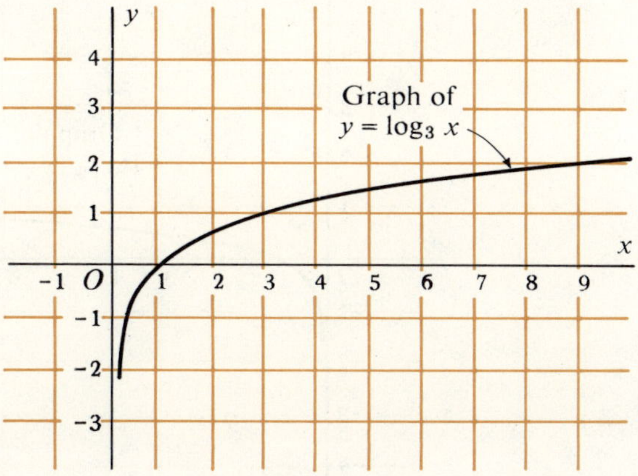

FIGURE 11-12

Analysis: (a) $m = 1$ and $b = 0$.
(b) The function does not cross the y axis and crosses the x axis at $x = 1$.
(c) The function increases at a decreasing rate.

Example 2. Draw the graph of $y = \log x$.

Solution: $y = \log x$ if and only if $10^y = x$.

x	1	2	3	4	5	10	100	$\frac{1}{10}$
y	0	?	?	?	?	1	2	-1

x	2	3	4	5
y	0.30103	0.47712	0.60206	0.69897

To complete the table the student must use Table A-1 of the Appendix.

x	1	2	3	4	5	10	100	$\frac{1}{10}$
y	0	0.30103	0.47712	0.60206	0.69897	1	2	-1

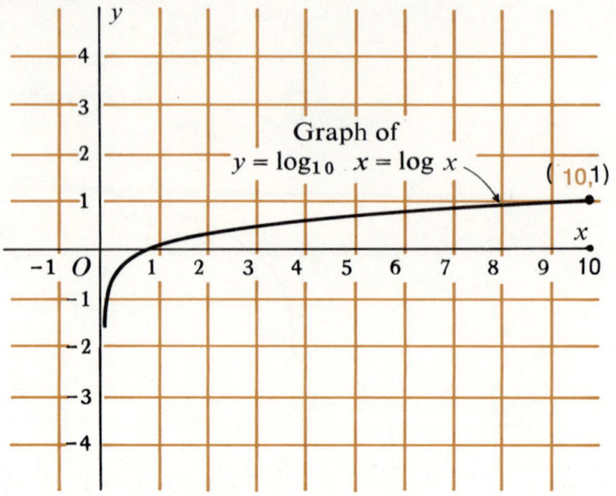

FIGURE 11-13

Analysis: (a) $m = 1$ and $b = 0$.
(b) The function does not cross the y axis and crosses the x axis at $x = 1$.
(c) The function increases at a decreasing rate.

Earlier in the text, we briefly mentioned the natural log system. This system of logs has various applications, and is especially important in advanced mathematics. The base for the natural log system is the irrational number e, where $e = 2.7183$ rounded to four decimal places. The reason for the use of this particular irrational number as a base for a log system is to be found in a study of differential calculus. For the natural log system, we write $\log_e N$ or simply $\ln N$. In other words, $\log N$ refers to the base 10 and $\ln N$ refers to the base e. In the Appendix the reader will find a table of natural logarithms.

Example 3. Discuss and graph the function $y = \ln x$.

Solution: $y = \ln x$ if and only if $e^y = x$.

x	1	2	2.7183
y	0		1

The entry for y when x is equal to 2 can be found in two ways:

Method I: Starting with $2.7183^y = 2$ and taking the log of both sides, we have

$$y \log 2.7183 = \log 2 \qquad \text{or} \qquad y = \frac{\log 2.0000}{\log 2.7183}$$

Referring to the table of logs, we get

$$y = \frac{0.30103}{0.43425} \quad \text{or} \quad y = 0.693$$

The entry when $x = 2$ is $y = 0.693$.

Method II: The second method consists of looking up the number 2 in the table of natural logs. Doing this, we get the number 0.693. From Table A-2 in the Appendix, we have

x	1	2	2.7183	3	4
y	0.000	0.693	1.000	1.099	1.386

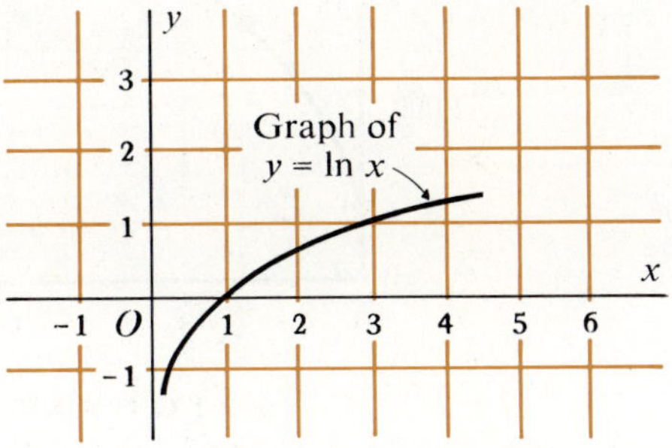

FIGURE 11-14

Analysis: (a) $m = 1$, $b = 0$, and $a = e$ (referring to the form $y = m \log_a x + b$).

(b) The function does not cross the y axis but crosses the x axis at $x = 1$.

Referring to the graphs of the log functions given in Examples 1–3 should help you to remember the following properties of logarithms.

1. There is no logarithm of a negative number.
2. The logarithm of the number 1 is zero.
3. The logarithm of a positive number less than unity is negative.
4. The logarithm of a number greater than unity is positive, and increases at a decreasing rate.

In addition we see that all logarithmic functions have the same general shape when represented graphically. The following example illustrates how one can use a log function to model cost.

Example 4. Draw the graph of $C = 1000 \ln(x+1) + 400$ where C is the cost in dollars of making buckles and x is the number of buckles made in thousands.

Solution:

x	0	2	5	9
C	400	1,499	2192	2703

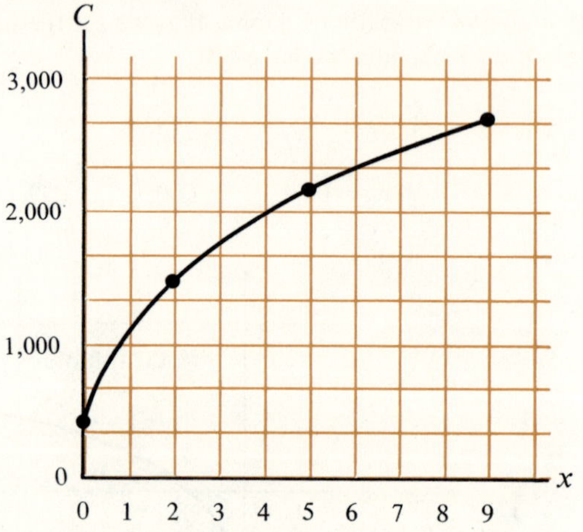

FIGURE 11-15

Analysis: The fixed cost is $400 and costs are increasing at a decreasing rate. This phenomenon is often referred to as economies of scale.

PROBLEMS

1. Draw the graph of the following functions:
   (a) $y = \log_2 x$ **(c)** $y = 2 \ln x$
   (b) $y = \log_4 x$ **(d)** $y = \log_{10} x^2$
2. Draw the graphs of the following functions:
   (a) $y = 2 \log x + 3$ **(c)** $y = \log(2x)$
   (b) $y = 3 \log x + 2$ **(d)** $y = \log(2x) + 3$
3. Draw the graph of $y = \ln x^2 + 3$.
4. Find the equation of the form $y = m \log x + b$ that passes through the points (1, 2) and (10, 4). Use the method of elimination.
5. Find the equation of the form $y = m \ln x + b$ that passes through the points (1, 3) and $(e^2, 4)$.
6. Find the equation of the form $y = m \log x + b$ that passes through the points (1, 5) and (10, 6). Use the method of elimination.
7. Use the change of base formula of Problem 11 of Section 11-2, $\log_b N / \log_b a = \log_a N$, to show that $\log N = (\log e) \ln N$.

8. Illustrate the result of Problem 7 for the following values of N:
(a) $N = 6.5$ **(b)** $N = 9.8$

Note: $\log e = 0.43429448$ accurate to eight decimal places.

9. Draw the graph of $C = 2000 \ln(x+1) + 800$ where C is the cost in dollars of manufacturing a lamp and x is the number of lamps made in thousands.

11-5 EXPONENTIAL FUNCTIONS

The expression

$$y = ab^x$$

where a and b are constants, is called an *exponential function*. Examples of such equations are $y = 3^x$, $y = e^x$, $y = 4(2^x)$, and $y = a(1+r)^x$.

Example 1. Draw the graph of $y = 2^x$.

x	-1	0	1	2	3
y	0.5	1	2	4	8

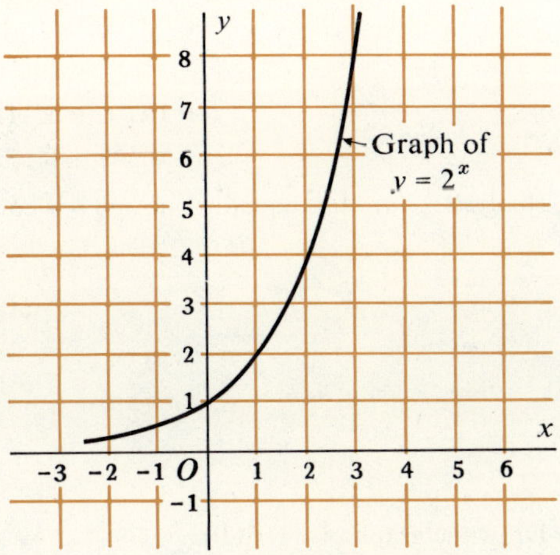

FIGURE 11-16

Analysis: (a) a is equal to one and b is equal to 2.
(b) The function crosses the y axis at $y = 1$ and does not cross the x axis.

Example 2. Draw the graph of $y = e^x$.

Solution: The entries for $x = 2$, 3, etc. can be found in Table A-3 of the Appendix.

x	0	1	2	3
y	1.0	2.72	7.39	20.1

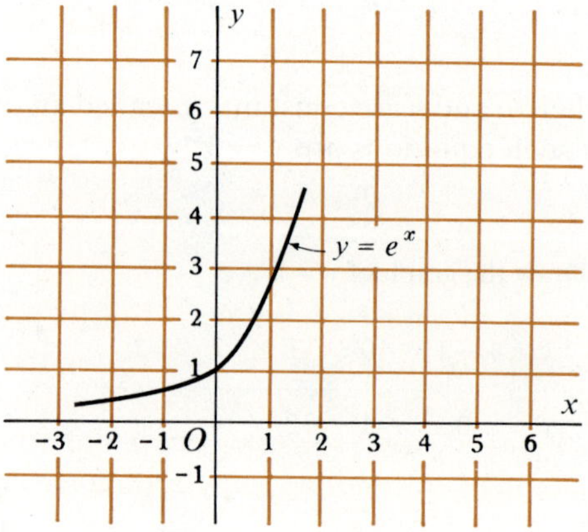

FIGURE 11-17

Analysis: (a) a is equal to one and b is equal to e.

Example 3. Draw the graph of $y = 5(2^x)$.

Solution:

x	−1	0	1	2	3
y	2.5	5	10	20	40

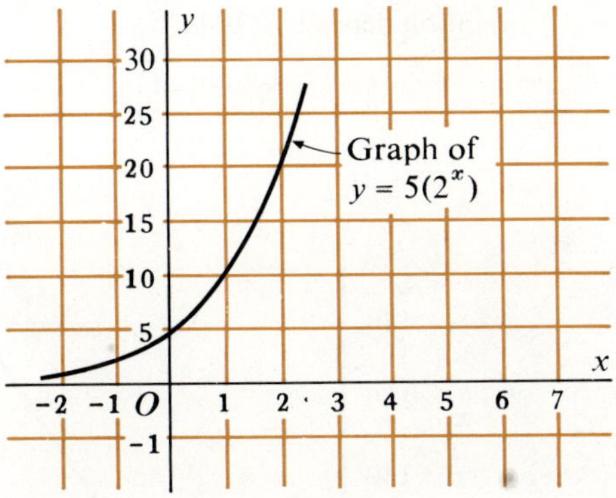

FIGURE 11-18

Analysis: (a) $a = 5$ and $b = 2$.
(b) The function crosses the y axis at $y = 5$ and does not cross the x axis.

Exponential functions are of particular interest because they can be used to model situations where there is a constant rate of change. Many physical quantities such as the biological reproduction of cells can be modeled by exponential functions. Of special interest is the formula given in Chapter 6 to determine the amount on deposit assuming continuous compounding.

Amount on deposit $= Pe^{in}$

where P = initial deposit

C = 2.7183

i = interest rate

n = number of years of deposit

Example 4. Find the amount on deposit at the end of years 1, 2, 3, 4, and 5 if \$100 is deposited in an account that pays an interest rate of 10% compounded instantaneously. Plot these points on a graph.

Solution: Amount on deposit $= 100e^{.10n}$.

n	1	2	3	4	5
Amount on deposit	\$111	122	135	149	165

The entry for $n = 1$ using Table A-3 follows.

$$\text{Amount on deposit} = 100e^{.10}$$
$$= 100(1.11)$$
$$= \$111$$

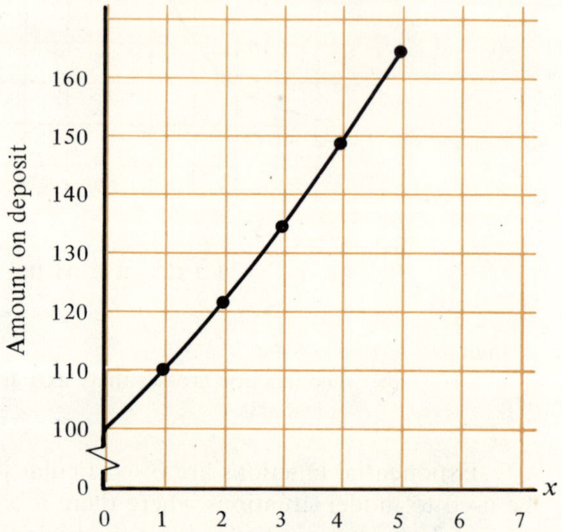

FIGURE 11-19

PROBLEMS

1. Graph the following functions:
   (a) $y = 3^x$ (c) $y = 3(4)^x$
   (b) $y = 2e^x$ (d) $y = 2(\frac{1}{2})^x$
2. Determine the function of the form $y = 3b^x$ that passes through the point (2, 12) and graph the function.
3. Determine the function of the form $y = ab^x$ that passes through the points (0, 3) and (1, 6). Use the method of elimination. Graph the resulting function.
4. Determine the function of the form $y = ab^x$ that passes through the points (0, 2) and (4, 3). Use the method of elimination. Graph the resulting function.
5. Sales for a certain industry have been growing at an annual rate of 12% for the past four years. Assume that this rate of growth remains constant over the next eight years and that current sales are $20,000,000. Estimate the expected level of sales:
   (a) five years hence (b) eight years hence

6. Sales are currently \$1,000,000 per year. Annual growth over the next five years is expected to be 10% per year.

   (a) Complete the following table:

x (year)	1	2	3	4	5
y (annual sales at year end in millions)	1.100				

   (b) Determine the function of the form $y = ab^x$ that passes through the two points corresponding to Year 1 and Year 5.

7. The manager of a branch bank knows that he can calculate the amount of money plus interest a person has on deposit after n years by using the following formula:

$$\text{Amount on deposit} = P(1+i)^n$$

He would like to know how much would be on deposit after 17 years using an interest rate, i, of 8 percent if someone just deposited \$1000(P). Normally this information is available in tables but he has misplaced them. He has asked you to help him find an answer using the formula.

8. Your client, Mr. Yvon, who works at the First National Bank, has determined (using tables) that if a customer deposits \$1000, the amount on deposit after ten years at 8 percent if interest is compounded quarterly would be \$2,208. The bank has recently introduced instantaneous or continuous compounding. Your client wishes to put an advertisement in the paper showing the difference that results in the amount on deposit when one compounds interest instantaneously versus quarterly. He has asked you to determine this for him using the above data and the following formula:

$$\text{Amount on deposit} = Pe^{in}$$

where P = initial deposit

e = mathematical constant = 2.7183 (rounded to four decimal places)

i = interest rate

n = number of years of deposit

9. Show that $y = ab^x$ can be put in the form $y = ae^{kx}$, where $k = \ln b$.

11-6 POWER FUNCTIONS

The expression

$$y = ax^b$$

where a and b are constants is called a *power function.* Examples of power functions are $y = x^3$, $y = 4x^2$, $y = x^{1.4}$, and $y = x^e$.

To look at the decision making aspect of power functions is very difficult without the knowledge of calculus. One of the objectives of calculus is to make the student adept at finding derivatives. A rule which is utilized in almost every problem in differential calculus is the rule for derivatives of power functions. The so-called power rule will be discussed in the following chapter.

Example 1. Draw the graph of $y = x^2$.

Solution:

x	0	±1	±2	±3
y	0	1	4	9

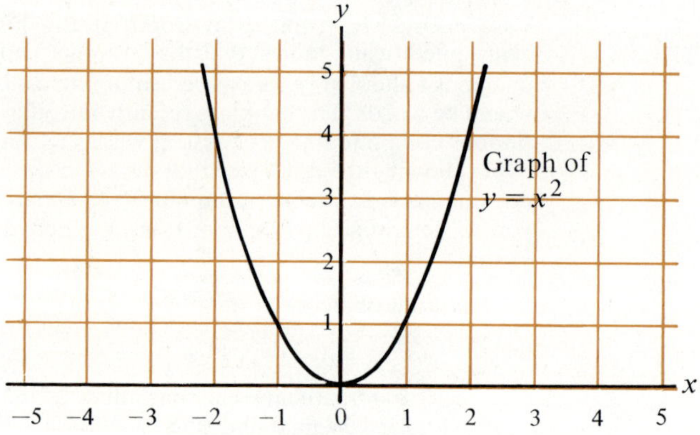

FIGURE 11-20

Analysis: a is equal to one and b is equal to 2.

Example 2. Draw the graph of $y = 2x^2$ and observe the effect of increasing the value of the coefficient a to 2.

Solution:

x	0	±1	±2	3
y	0	2	8	18

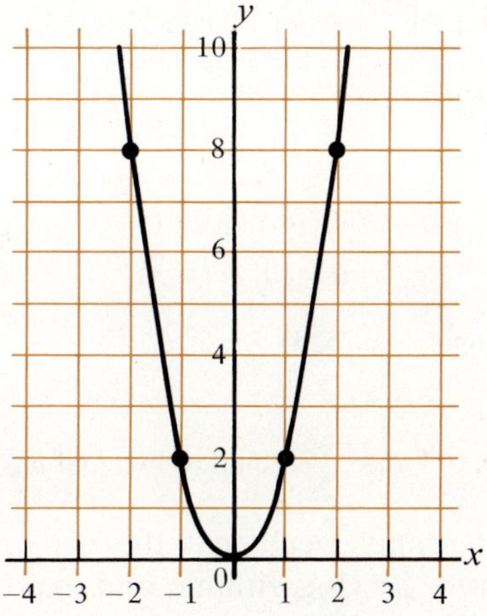

FIGURE 11-21

Analysis: The graph of $y = 2x^2$ is "tighter" or has "less spread" than the graph of $y = x^2$.

Example 3. Draw the graph of $y = 2(x^{1.5})$.

Solution:

x	0	1	2	3
y	0	2	5.65	10.39

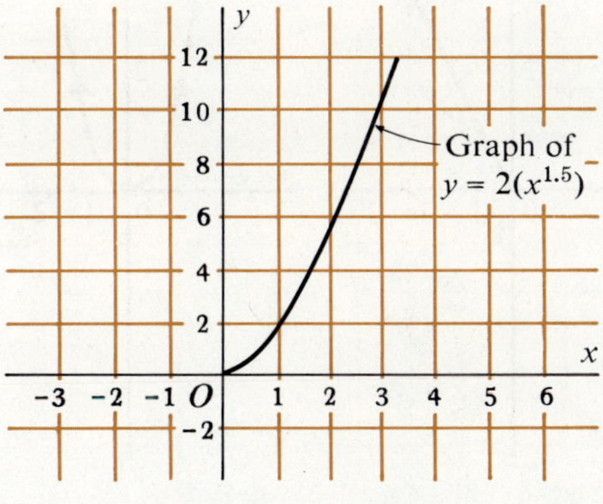

FIGURE 11-22

To find the values of y when $x = 2$, 3, etc., one must make use of logs. We have then $\log y = \log 2 + 1.5 \log x$.

When x is 2,

$$\begin{aligned} \log y &= \log 2 + 1.5 \log 2 \\ &= 0.301 + 1.5(0.301) \\ &= 0.301 + 0.4515 \\ \log y &= 0.7525 \\ y &= 5.65 \end{aligned}$$

Analysis: (a) a is equal to two and b is equal to 1.5.

One may recall that the two previous functions that we have discussed, i.e., logarithmic and exponential functions, had a general shape. This is not the case with power functions. Consider the following power functions:

$$y = 2x$$

$$y = 2x^2$$

$$y = 2x^3$$

These would graph as follows:

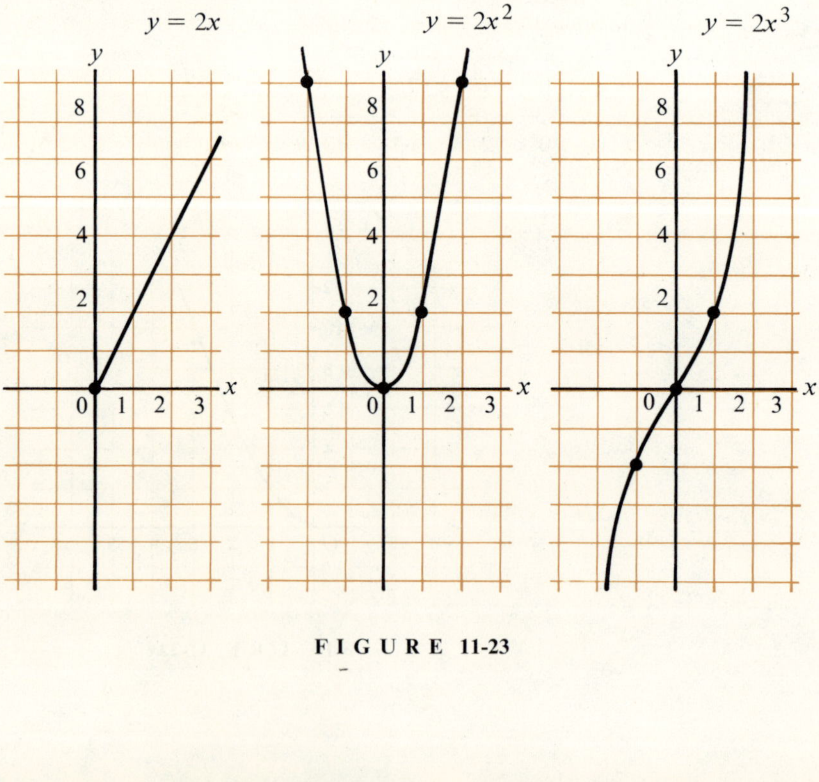

FIGURE 11-23

It should be noted that these power functions certainly do not have a general shape.

If $y = kx$, where a is a constant, y is said to be directly proportional to x. More generally, if $y = kf(x)$, where k is a constant, y is said to be *directly proportional to* $f(x)$. The statement "*y varies directly as* $f(x)$" is equivalent to the statement "y is directly proportional to $f(x)$."

Example 4. Given that w varies directly as the square of z, express w as a function of z.

Solution: w varies directly as the square of z means that w is directly proportional to z^2. Therefore, $w = kz^2$ is the desired function. Note that $f(z) = z^2$ in this example.

PROBLEMS

1. Graph the following functions:
   (a) $y = x^3$
   (b) $y = 3x^{1/2}$
   (c) $y = 0.5x^{2.5}$
   (d) $y = 2x^{1/3}$

2. Determine the function of the form $y = 2x^b$ that passes through $(2, 16)$. Graph the resulting function.

3. Determine the function of the form $y = ax^b$ that passes through the points $(1, 1)$ and $(10, 10)$. Use the method of elimination.

4. Express the following statements functionally and identify the resulting functions by name:
   (a) y is directly proportional to x^3.
   (b) y varies directly as the square root of x.
   (c) w is directly proportional to $\log z$.
   (d) v varies directly as t.
   (e) s is directly proportional to t^2.
   (f) y varies directly as 2^x.
   (g) y is directly proportional to $(1 + r)^x$.

5. Grosch's law (for computers) states that computing power $= k(\text{cost})^2$, where k is a constant.
   (a) Express this law verbally.
   (b) What type of function is this?

6. In many applications it is necessary to recognize the type of function from a graph and then formulate a mathematical expression or model. Classify the following expressions as logarithmic exponential or power functions.

Expression	*Type of Function*
$y = ab^x$	**(a)** EXPONENTIAL
$y = m \log_a x + b$	**(b)** LOG
$y = ax^b$	**(c)** POWER
$y = 6^x$	**(d)**
$y = 2x^2$	**(e)** POWER
$y = 3 \log x$	**(f)**
$y = \ln x$	**(g)**
$y = e^x$	**(h)**
$y = P(1+i)^x$	**(i)**
$y = x^e$	**(j)** POWER

11-7 SEMILOG AND LOG-LOG PAPER

Consider the equation $y = m \log x + b$, where m and b are constants. The reader will recognize this as a nonlinear function. However, if we think of y as a function of $\log x$ rather than x, we shall obtain a linear function with slope m and y intercept b. Semilog paper is constructed in such a way that one of the axes is a *logarithmic* or *ratio scale.* (See Figure 11-24.) It follows that any function of the form $y = m \log x + b$ will graph as a straight line with slope m and y intercept b on semilog paper.

Example 1. Draw the graph of $y = \log x$ on semilog paper with the x axis in logarithmic scale.

Solution: A table of values and the graph appear below.

x	1	2	3	4	5	10
$y = \log x$	0.00000	0.30103	0.47712	0.60206	0.69897	1.00000

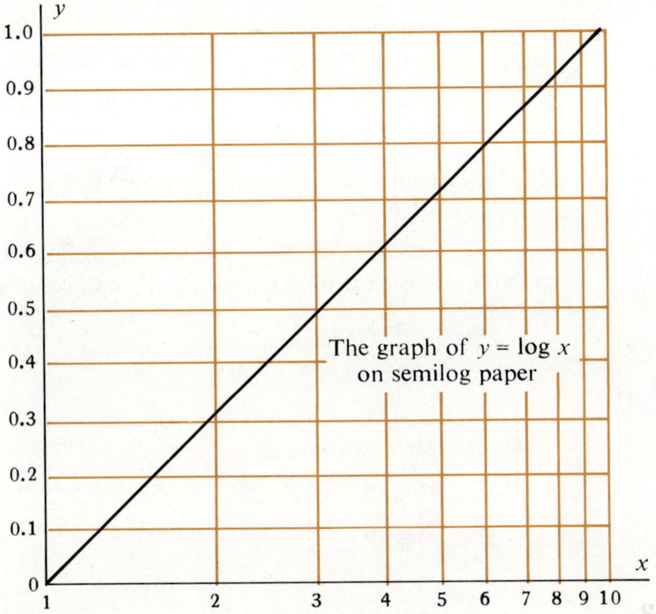

FIGURE 11-24

Example 2. Draw the graph of $y = 2 \log x + 3$ on semilog paper with the x axis in logarithmic scale.

Solution: A table of values and the graph appear below.

x	1	2	5	10
y	3	3.602	4.397	5

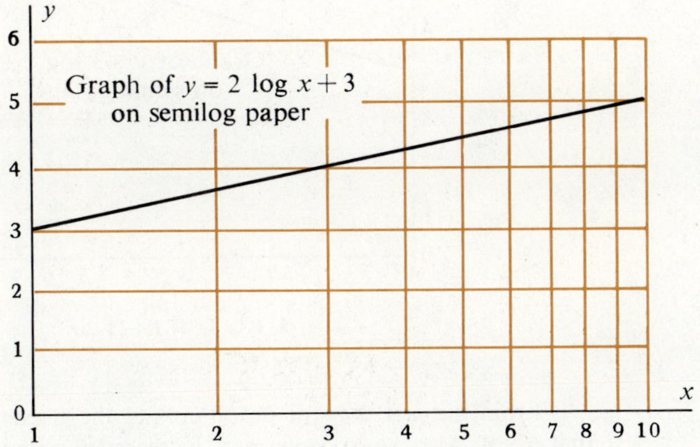

FIGURE 11-25

Analysis: The function crosses the y axis at $y = 3$, and for a one unit increase in x ($\log 10 - \log 1$), y increases by two units. Therefore, $b = 3$ and $m = 2$.

Example 3. Find the equation of the form $y = m \log x + b$ that passes through the points $A(1, 4)$ and $B(10, 7)$.

Solution: We can solve for m and b by the method of elimination or we can plot the points on semilog paper to find the values of m and b.

$$e_1\colon\ 4 = m \log 1 + b$$

$$e_2\colon\ 7 = m \log 10 + b$$

Because log 1 is equal to zero, we obtain $b = 4$. Then substituting the value of b into e_2 we have

$$7 = m \log 10 + 4$$

$$3 = m \log 10$$

$$m = \frac{3}{\log 10}$$

$$m = 3$$

The desired function is $y = 3 \log x + 4$.

Next, we obtain this function graphically using semilog paper. Plotting the two given points on semilog paper and drawing a straight line through them gives the following graph.

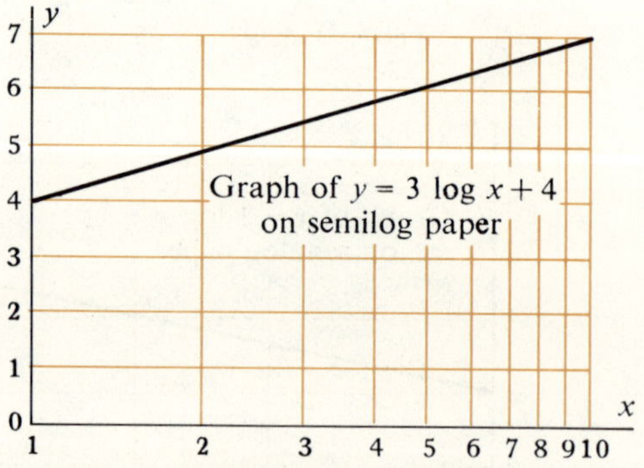

FIGURE 11-26

From the fact that the graph crosses the y axis at $y = 4$, we conclude that $b = 4$. Also, since a one unit increase in x ($\log 10 - \log 1$) results in a 3 unit increase in y, we see that the slope is 3. Therefore, $m = 3$. The desired equation is $y = 3 \log x + 4$, as we found before by the method of elimination.

Consider the equation $y = ab^x$, where a and b are positive constants. This equation is equivalent to $\log y = \log a + x \log b$. This means that $y = ab^x$ graphs as a straight line on semilog paper with the y axis in logarithmic scale. Furthermore, a equals the vertical intercept and $\log b$ equals the slope of the line.

Example 4. Draw the graph of $y = 5(2^x)$ on semilog paper with the y axis in logarithmic scale.

Solution:

x	0	1	2	3
y	5	10	20	40

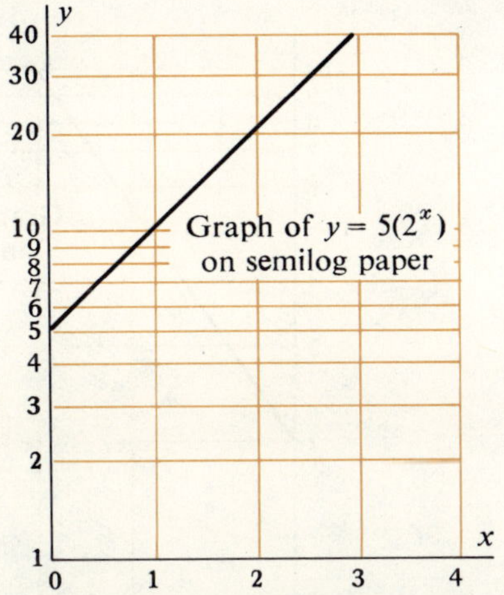

FIGURE 11-27

Analysis: (a) The vertical intercept is 5; therefore, $a = 5$.

(b) The slope is given by $\dfrac{\log 40 - \log 5}{3} = \dfrac{1.6021 - 0.6990}{3}$.

$\log b = 0.301$

Therefore,

$b = 2$

Finally, we illustrate the use of log-log paper. Consider the power function $y = ax^b$ with a, b constant and a, x positive. If we take the logarithm to the base 10 of both sides, we obtain $\log y = \log a + b \log x$. Log-log paper is constructed in such a way that both axes are in logarithmic scale. (See Figure 11-28.) It follows that any power function of the type described above graphs as a straight line on log-log paper. Furthermore, a equals the vertical intercept and b equals the slope of the line.

Example 5. Draw the graph of $y = 2(x^{1.5})$ on log-log paper.

Solution: A table of values and the graph appear below.

x	1	2	3
y	2	5.65	10.1039

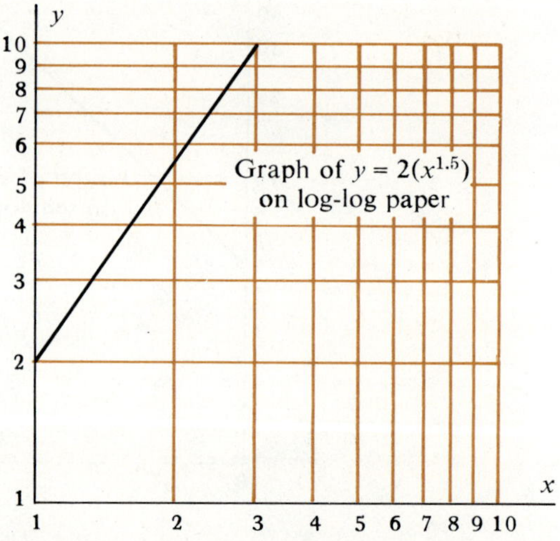

FIGURE 11-28

Analysis: (a) The vertical intercept is 2; therefore, $a = 2$.

(b) The slope is given by $b = \dfrac{\log 10.10 - \log 2}{\log 3 - \log 1}$

$$b = \frac{1.0167 - 0.301}{0.47712}$$

$$b = 1.5$$

A mathematical equation or model can be found to describe approximately certain classes of empirical data as follows:

1. If the data graph approximately as a straight line on semilog paper with the x axis in logarithmic scale, then the data can be described by a function of the form $y = m \log x + b$, a log function.
2. If the data graph approximately as a straight line on semilog paper with the y axis in logarithmic scale, then the data can be described by a function of the form $y = ab^x$, an exponential function.
3. If the data graph approximately as a straight line on log-log paper, then the data can be described by a function of the type $y = ax^b$, a power function.

The parameters a, b, and m can be estimated using the method of least squares which is normally covered in a statistics course.

PROBLEMS

1. Graph the following functions on semilog paper:
(a) $y = 2 \log x + 2$ **(c)** $y = \log(2x)$
(b) $y = 3 \log x + 2$ **(d)** $y = \log(2x) + 3$

2. Find the equation of the form $y = m \log x + b$ that passes through the points (1, 2) and (10, 4). Use semilog paper.

3. Find the equation of the form $y = m \log x + b$ that passes through the points (1, 5) and (10, 6). Use semilog paper.

4. Graph the following functions on semilog paper:
(a) $y = 3^x$ **(c)** $y = 3(4)^x$
(b) $y = 2e^x$ **(d)** $y = 2(\frac{1}{2})^x$

5. Determine the function of the form $y = ab^x$ that passes through the points (0, 3) and (1, 6). Use semilog paper.

6. Determine the function of the form $y = ab^x$ that passes through the points (0, 2) and (4, 3). Use semilog paper.

7. Graph the following functions on log-log paper:
(a) $y = x^3$ **(c)** $y = 0.5x^{2.5}$
(b) $y = 3x^{1/2}$ **(d)** $y = 2x^{1/3}$

8. Determine the function of the form $y = ax^b$ that passes through the points (1, 1) and (10, 10). Use log-log paper.

9. You are given the following figures for cost in thousands of dollars and production in hundreds of units. Find the equation of the form $y = ab^x$ that corresponds to this information.

x	0	1	2	3	4	5
y	1	2	4	8	16	32

x is in hundreds of units.
y is in thousands of dollars.

10. Given the following:

x	1	2	3	4	5	6	7	8
y	0.77	1.50	2.34	3.21	4.11	5.03	5.95	6.90

(a) Complete the table:

log x
log y

(b) Prepare a scatter diagram, i.e., plot the points, for the transformed data on ordinary graph paper.
(c) Prepare a scatter diagram on log-log paper for the given data.

11. The general linear homogeneous constant elasticity of substitution (CES) production function can be put in the following form[1]:

$$\frac{V}{L} = \left(\frac{1}{1-\delta}\right)^{1/(1+P)} \gamma^{P/(1+P)} W^{1/(1+P)} \tag{1}$$

where
V = output (value added in current dollars)
L = labor input (production man-hours worked)
W = wage rate (wage bill/man-hours)
γ = efficiency parameter
P = substitution parameter
δ = distribution parameter

(a) Express (1) in the form

$$\frac{V}{L} = AW^b \tag{2}$$

and identify A and b.
(b) Express (2) in logarithms by taking the logarithm of both sides.

12. Given Figure 11-29:

(a) What type of function would best fit these data, i.e., linear, exponential, or logarithmic?
(b) What is the general form of this function?
(c) On what kind of paper would these data graph approximately as a straight line?

Life insurance sales in the U.S. in 1972 totaled $212 billion, up from $189.2 billion in 1971, the Life Insurance Institute reports.

1. Christos C. Paraskevopoulos, "Impact of Regional Wage Differentials," *Growth and Change* IV, 2: 40–41.

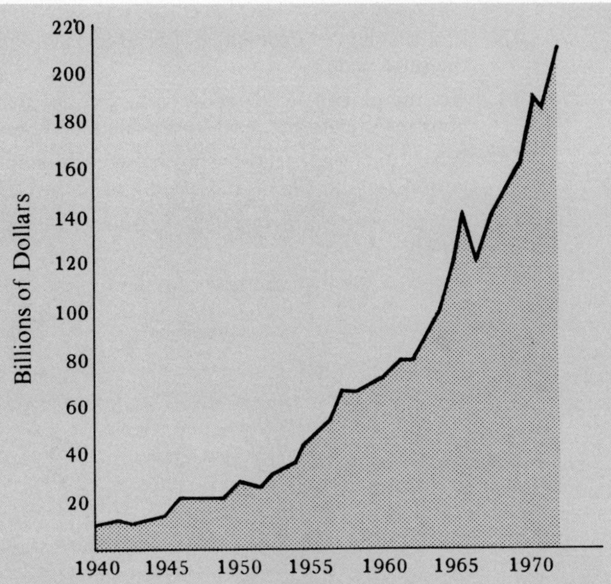

FIGURE 11-29[2]

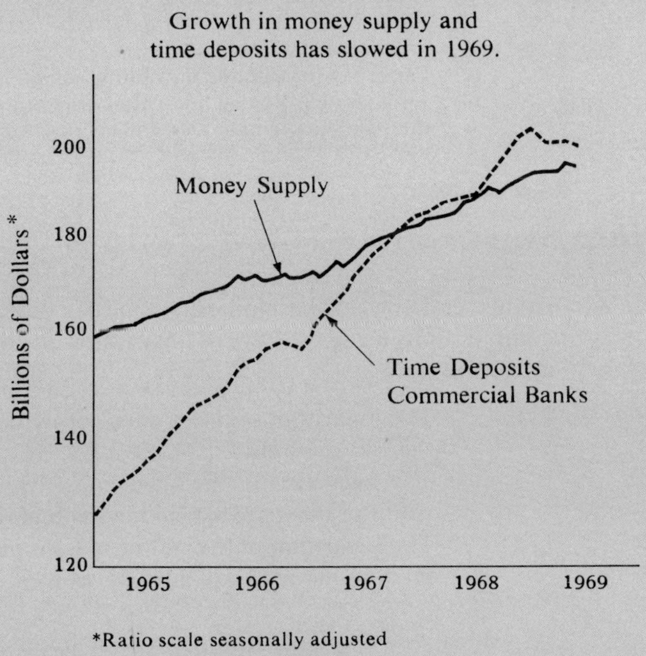

FIGURE 11-30[3]

2. *Wall Street Journal:* June 15, 1973, p. 1.

3. *Business Conditions:* June 1969, p. 3.

13. Plot the data of Problem 6 of Section 11-5 on semilog paper using y as the ratio scale.

14. Use the plot of Problem 13 to fit $y = ab^x$ to these data, and compare your result with the result of Problem 6 of Section 11-5.

15. What functional relationship between money supply and time for the period 1965 to 1969 is indicated by Figure 11-30?

16. Use the graph in Problem 15 to estimate a and b in the function money supply $= ab^{\text{time}}$.

17. Consider the hypothetical sales data in the following table.

Year	Sales
1	1,120,000
2	1,254,300
3	1,404,900
4	1,573,600
5	1,762,400
6	1,973,800

(a) Form a scatter diagram for these data on arithmetic paper.
(b) Form a scatter diagram for these data on semilog paper using log y versus x.
(c) Perform a logarithmic transformation on $S = a(1+r)^x$.
(d) Fit a straight line to the scatter diagram of (b) by eye. Use your fitted line to estimate $1+r$ and a.

APPENDIX TO CHAPTER 11—COMPUTATIONS WITH LOGARITHMS

In this section we will examine how one can perform otherwise difficult computations using the laws of logarithms given in Section 11-2. They are as follows:

L-1 The logarithm of the product of two numbers is the sum of their logarithms.

L-2 The logarithm of the quotient of two numbers is the logarithm of the dividend minus the logarithm of the divisor.

L-3 The logarithm of the power of a number is the logarithm of the number multiplied by the power.

When one uses logarithms it will be necessary to understand the concept of an antilogarithm, which is defined as follows:

DEFINITION 11.2. *The antilogarithm of a number x, denoted by antilog x, is defined by the relationship antilog* $x = N$ *if and only if* $x = \log N$.

For the base 10 system, the antilog of x is the number whose logarithm to the base 10 is x. This definition is further illustrated:

1. antilog $2 = 100$
2. antilog $3 = 1000$
3. antilog $(-1) = 0.1$

   By Definition 11.2, antilog $x = N$ if $x = \log N$.
   By Definition 11.2, $x = \log N$ if $N = 10^x$. Combining these two statements gives antilog $x = 10^x$.

The following examples illustrate the use of logarithms in certain calculations.

Example 1. Use logarithms to calculate $\sqrt{4}$.

Solution: Let $N = \sqrt{4}$.

$$N = 4^{\frac{1}{2}} \qquad \textit{by E-7}$$

$$\log N = \tfrac{1}{2}\log 4 \qquad \textit{by L-3}$$

$$= \tfrac{1}{2}(0.60206)$$

$$\log N = 0.30103$$

Looking up 0.30103 in the body of Table A-1 of the Appendix, we find that $N = 2$. The answer $N = 2$ in Example 1 is called the antilog of 0.30103.

Example 2. Calculate $(47)^{0.3}$ using logarithms.

Solution: Let $N = (47)^{0.3}$.

$$\log N = 0.3(1.67210) \qquad \textit{by L-3}$$

$$\log N = 0.50163$$

$$N = 3.174$$

Example 3. Use logarithms to calculate $\dfrac{(283)(8.1)}{2500}$.

Solution: Let $N = \dfrac{(283)(8.1)}{2500}$.

$$\log N = \log 283 + \log 8.1 - \log 2500 \qquad \textit{by L-1 and L-2}$$

$$= 2.45179 + 0.90849 - 3.39794$$

$$= 3.36028 - 3.39794$$

$$\log N = -0.03766$$

The antilog of a negative number cannot be found directly from Table A-1 of the Appendix. By adding and subtracting one to -0.03766, we obtain

$$\log N = 1 - 0.03766 - 1$$

$$\log N = 0.96234 - 1$$

$$N = 0.9169$$ *without using interpolation*

The technique of adding and subtracting a number large enough to introduce a positive decimal is the standard way of looking up antilogs of negative numbers.

The method of linear interpolation can also be used for finding antilogs. This is illustrated in the next example.

Example 4. Use Table A-1 of the Appendix and interpolation to find antilog (0.94118).

Solution:

$$\text{antilog } 0.94121 = 8.7340$$

$$\text{antilog } 0.94118 = 8.733x$$

$$\text{antilog } 0.94116 = 8.7330$$

$$\frac{x}{10} = \frac{0.94118 - 0.94116}{0.94121 - 0.94116}$$

$$\frac{x}{10} = \frac{0.00002}{0.00005} = \frac{2}{5}$$

$$x = 4$$

$$\text{antilog } (0.94118) = 8.7334$$

Example 5. Find $\sqrt[3]{0.432}$ using logs.

Solution:

$$N = (0.432)^{\frac{1}{3}}$$

$$\log N = \tfrac{1}{3} \log 0.432$$

$$\log N = \tfrac{1}{3}(0.63548 - 1)$$

$$\log N = \tfrac{1}{3}(2.63548 - 3)$$ *adding and subtracting 2 to obtain a number divisible by 3*

$$\log N = 0.87849 - 1$$

$$N = \text{antilog}(0.87849 - 1)$$

$$\sqrt[3]{0.432} = 0.75595$$ *using interpolation and Table A-1 of the Appendix in reverse*

PROBLEMS

1. Find the following antilogarithms:
   - **(a)** antilog 4
   - **(b)** antilog(−2)
   - **(c)** antilog(0)
   - **(d)** antilog(2.67025)
   - **(e)** antilog(0.71037 − 3)
   - **(f)** antilog(0.66011)
2. Use logarithms to find the following square roots:
   - **(a)** $\sqrt{9}$
   - **(b)** $\sqrt{432}$
   - **(c)** $\sqrt{8126}$
   - **(d)** $\sqrt{0.63}$
   - **(e)** $\sqrt{0.063}$
   - **(f)** $\sqrt{0.0063}$
3. Use logarithms to find the following cube roots:
   - **(a)** $\sqrt[3]{64}$
   - **(b)** $\sqrt[3]{432}$
   - **(c)** $\sqrt[3]{360}$
   - **(d)** $\sqrt[3]{0.063}$
4. Perform the following multiplications using logarithms:
   - **(a)** (483)(7543)
   - **(b)** (0.317)(3754)
   - **(c)** (48212)(672)
   - **(d)** (9.37)(856)(1432)
   - **(e)** (16427)(845.21)
   - **(f)** (23)(84)(570)(6842)
5. Use logarithms to do the following division problems:
   - **(a)** 8723/415
   - **(b)** 415/8723
   - **(c)** 0.428/642
   - **(d)** 0.03475/0.004182
6. Use logarithms to perform the indicated operations.
   - **(a)** (234)(8752)/4827
   - **(b)** $\sqrt{(572)(0.459)}$
   - **(c)** $(8432)\sqrt{954}$
   - **(d)** $\sqrt[3]{(856)/439}$
   - **(e)** $437/\sqrt{(848)(5342)}$
   - **(f)** $(0.01245)/\sqrt[5]{(6542)}$
7. Calculate the following quantities using logarithms. Expressions of this form arise in connection with compound interest calculations.
   - **(a)** $2000(1.06)^{10}$
   - **(b)** $(1+0.08)^5$
   - **(c)** $1000(1.04)^{20}$
8. Solve $\left(1+\frac{i}{4}\right)^4 = 1.08$ for i.
9. Find antilog $\left(\frac{\log 40 - \log 5}{3}\right)$.
10. Determine or simplify the following:
    - **(a)** antilog(log 2)
    - **(b)** antilog(3 log 5)
    - **(c)** antilog(log b)
    - **(d)** antilog(a log b)

TWELVE

Introduction to differential calculus

Differential calculus deals with the problem of finding the *instantaneous* or *point* rate of a change of a nonlinear function. The key concept of differential calculus is the *derivative*. Geometrically speaking, the derivative can be used to find the slope of the *tangent line* to a curve at a given point on the curve.

The notion of a *rate of change* is familiar to us from everyday language and discourse. Some familiar examples are inflation rate, unemployment rate, interest rate, attrition rate, and the rate of return on an investment. Some less familiar examples, chosen from the fields of science and engineering, are rate of decay, rate of growth, reaction rate, radiation rate, velocity, and acceleration.

Historically, the derivative has been applied in many areas of study such as physics, engineering, economics, etc. In keeping with the purpose of this text we illustrate the use of the derivative through business and economic applications.

12-1 THE RATE OF CHANGE OF A FUNCTION

The next definition formalizes the notion of average rate of change.

DEFINITION 12.1 *Let $y = f(x)$ and let $A = (x_1, y_1)$ and $B = (x_2, y_2)$ be two points on the graph of $y = f(x)$. Then the average rate of change of $f(x)$ between the points A and B is defined to be the ratio*

$$\frac{y_2 - y_1}{x_2 - x_1}$$

Interpreted geometrically, the ratio in Definition 12.1 equals the slope of the straight line passing through the points A and B. This line is called a *secant line.*

Since the average rate of change as defined depends on the function $y = f(x)$ as well as the two points A and B, we have chosen to introduce a symbol to denote the average rate of change of a function $f(x)$ between two points A and B. Our choice of a symbol for this purpose is

$$R_f(A, B)$$

In the context of Definition 12.1,

$$R_f(A, B) = \frac{y_2 - y_1}{x_2 - x_1}$$

Furthermore,

$$R_f(A, B) = \frac{\Delta y}{\Delta x}$$

where $\Delta x = x_2 - x_1$ and $\Delta y = y_2 - y_1$.

If the secant line passing through A and B is denoted by S, then

$$R_f(A, B) = \text{Slope of } S$$

The terms and relationships discussed so far are illustrated in Figure 12-1.

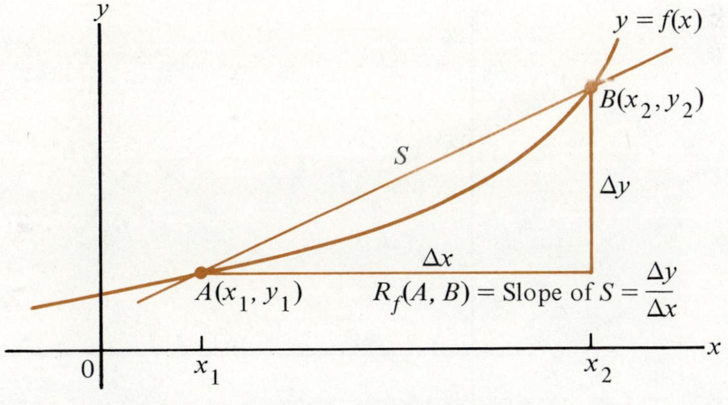

FIGURE 12-1

Example 1. Let $f(x) = x^2 + 3x - 3$, $x_1 = 1$, and $x_2 = 3$ be given. Calculate y_1, y_2, and $R_f(A, B)$.

Solution: Let $A = (x_1, y_1)$ and $B = (x_2, y_2)$ as in Definition 12.1.

$$y_1 = f(x_1) \qquad y_2 = f(x_2)$$
$$= f(1) \qquad\quad = f(3)$$
$$= 1 \qquad\quad = 15$$

Therefore, $A = (1, 1)$ and $B = (3, 15)$

$$R_f(A, B) = \frac{15-1}{3-1}$$
$$= 7$$

The terms *instantaneous* and *point* rate of change are defined next. The adjective *instantaneous* comes from the fields of physics and engineering where one is concerned with instantaneous (in contrast to average) velocity, for example. The adjective *point* comes from the field of economics where one talks about the point (in contrast to arc) elasticity, for example.

DEFINITION 12.2 *Consider the sequence of points $B_1, B_2, B_3, \ldots$ on the graph of $y = f(x)$, as shown in Figure 12-2. From the diagram it appears that the numbers $R_f(A, B_1)$, $R_f(A, B_2)$, ... are all distinct. Now it often happens in such a situation that although these average rates are all different, they get numerically close to some fixed number as the points B_i get closer and closer to the fixed point A. If such a number does exist, it is called the "instantaneous" or "point" or simply the rate of change of $f(x)$ at A.*

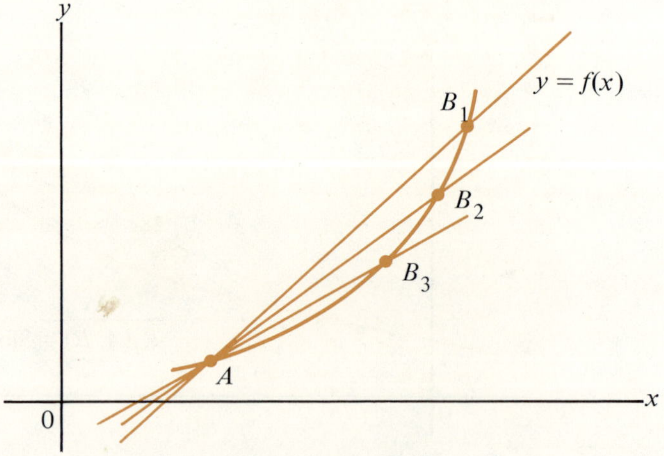

FIGURE 12-2

The problem of determining *the* rate of change of a function (in contrast to an average rate of change) involves the study of the behavior of the ratio $\Delta y/\Delta x$ as the points B_i get closer and closer to A, or equivalently as Δx gets closer and closer to zero. The somewhat imprecise terminology of "getting closer and closer to" will be explained further in the next section.

We use the notation $R_f(A)$, in contrast to $R_f(A, B)$, to denote *the* rate of change of $f(x)$ at the point A.

Example 2. $A(-1, 7)$ and $B(2, 4)$ are points of the graph of $y = 2x^2 - 3x + 2$. Calculate the slope of the secant line, S, that passes through the points A and B.

Solution: Let $f(x) = 2x^2 - 3x + 2$. Then

$$R_f(A, B) = \frac{4-7}{2+1}$$

$$= -1$$

PROBLEMS

Calculate the rate of change, $R_f(A, B)$ in Problems 1–10.

1. $f(x) = x^2 + 2x$, $x_1 = 2$, and $x_2 = 4$
2. $f(x) = x^2 + 2x$, $x_1 = 2$, and $x_2 = 2.5$
3. $f(x) = x^2 + 2x$, $x_1 = 2$, and $x_2 = 2.001$
4. $f(x) = x^3 - 3x + 1$, $x_1 = 1$, and $x_2 = 2$
5. $f(x) = x^3 - 3x + 1$, $x_1 = 1$, and $x_2 = 1.1$
6. $f(x) = x^3 - 3x + 1$, $x_1 = 1$, and $x_2 = 0.99$
7. $f(x) = 3(2^x)$, $x_1 = 0$, and $x_2 = 1$
8. $f(x) = 3(2^x)$, $x_1 = 0$, and $x_2 = .05$
9. $f(x) = \log x^2$, $x_1 = 1$, and $x_2 = \sqrt{10}$
10. $f(x) = \log x^2$, $x_1 = 1$, and $x_2 = 0.001$

Graph the secant lines, S, and find their slopes in Problems 11–14.

11. S passes through $A(1, 1)$ and $B(2, 4)$, when A and B are points on the graph of $y = x^2$.
12. S passes through $A(1, 1)$ and $B(2, 8)$, where A and B are points on the graph of $y = x^3$.
13. S passes through $A(2, y_1)$ and $B(4, y_2)$, where A and B are points on the graph of $y = 0.12^x$.
14. S passes through $A(-1, y_1)$ and $B(2, y_2)$, where A and B are points on the graph of $y = 3e^x$.

12-2 LIMITS AND CONTINUITY

You are, no doubt, familiar with the expression, "the sky's the limit." This saying expresses the thought that there are no bounds or limits to the process or situation under discussion.

Another expression that suggests the idea of a lower limit is: "the bottom line." The bottom line might represent a price below which we would not go and still be willing to sell an item.

If $y = 2^n$, where $n = 0, 1, 2, 3, \ldots$, then the sky is the limit for y. Or, as a mathematician might put it, y increases without bound as n approaches infinity.

On the other hand, suppose $y = 4 + 2^{-n}$, where $n = 0, 1, 2, \ldots$. In this case 4 is the bottom line for y. Again, our mathematician friend might say that

y approaches 4 as n approaches infinity (∞)

or

y has the limit 4 as $n \to \infty$

or

the limit of $4 + 2^{-n}$, as $n \to \infty$, equals 4

or

$$\lim_{n \to \infty} (4 + 2^{-n}) = 4$$

Limits are used extensively in the study of calculus. The notion of limit is used in this text to define:

1. continuity,
2. the derivative,
3. the definite integral.

The limit of a function is expressed by the symbol

$$\lim_{x \to a} f(x) = L$$

where L denotes the number (limit) being approached by the value of $f(x)$ as x approaches a.

The limit of a sequence is denoted by the symbol

$$\lim_{n \to \infty} S_n = L$$

where L is the number (limit) being approached by the terms of the sequence as n increases indefinitely through the natural numbers: $1, 2, 3, \ldots$

Example 1. Select the correct value for L if $\lim_{x \to 3} (x^2 + 2x) = L$.

(a) $L = 9$ **(b)** $L = 6$

(c) $L = 15$ **(d)** $L = 14.98$

(e) $L = 15.00001$.

Answer: **(c)**, $L = 15$

Example 2. Find L, where

$$\lim_{n \to \infty} \left(12 + \frac{1}{n^2}\right) = L$$

Answer: The "bottom line" is 12 for this sequence.

$$\begin{aligned} &13 \\ &12\tfrac{1}{4} \\ &12\tfrac{1}{9} \\ &12\tfrac{1}{16} \\ &\vdots \\ L = {}&12 \qquad \text{(bottom line)} \end{aligned}$$

A function $y = f(x)$ defined on an interval (a, b) is said to be *continuous* on (a, b) if its graph can be drawn completely without lifting the pencil from the paper. This is a very informal definition of continuity, but it does convey the basic intuitive idea of this important mathematical concept. Continuity of a function $y = f(x)$ at a point $x = c$ is defined next in terms of limits.

DEFINITION 12.3 *A function $y = f(x)$ is said to be continuous at $x = c$ if*

$$\lim_{x \to c} f(x) = f(c)$$

The graphical "definition" of continuity is illustrated in Figure 12-3.

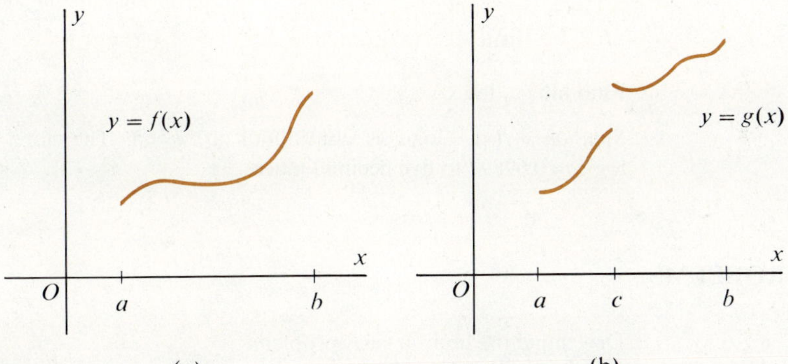

(a) This function is continuous on the interval $a < x < b$.

(b) This function fails to be continuous at $x = c$.

FIGURE 12-3

Example 3. The polynomial function $f(x) = 2x^2 - 4x + 1$ is continuous at every real value c. Use this fact to compute $\lim_{x \to 2} f(x)$.

Solution:

$$f(x) = 2x^2 - 4x + 1$$

Let $c = 2$

$$f(c) = f(2) = 2(2)^2 - 4(2) + 1$$
$$= 1$$

Since $f(x)$ is continuous at $x = 2$,

$$\lim_{x \to 2} f(x) = f(2) = 1$$

Functions of the type shown below are *continuous on* (at each value of) the indicated intervals.

SOME CONTINUOUS FUNCTIONS

1. Polynomial functions

$$y = a_n x^n + a_{n-1} x^{n-1} + \cdots + a_1 x + a_0, \text{ all real } x$$

2. Exponential functions

$$y = ab^x, \text{ all real } x, \qquad b > 0$$

3. Logarithmic functions

$$y = a \log_b x, \qquad x > 0$$

Example 4. Find $\lim_{x \to 5} \log x$.

Solution: $f(x) = \log x$ is continuous at $x = 5$. Therefore, $\lim_{x \to 5} \log x = \log 5 = 0.69897$ to five decimal places.

PROBLEMS

Determine the limit in each problem.

1. $\lim_{n \to \infty} \left(1 + \frac{1}{n}\right)$

2. $\lim_{n \to \infty} \left(1 + \frac{1}{n^2}\right)$

3. $\lim_{n\to\infty}\left(1+\frac{1}{n}\right)^2$

4. $\lim_{n\to\infty}\left(1+\frac{1}{n^2}\right)^2$

5. $\lim_{n\to\infty}\left(\frac{2n+3}{n}\right)$

6. $\lim_{n\to\infty}(5-e^{-n})$

7. $\lim_{n\to\infty}(0.4)^n$

8. $\lim_{x\to 2} 2x^3 - x + 2$

9. $\lim_{x\to -2} 2x^3 - x + 2$

10. $\lim_{x\to e} \ln x$ (*base e*)

11. $\lim_{x\to e} \ln x^2$ (*base e*)

12. $\lim_{x\to 50} \log 2x$ (*base 10*)

13. $\lim_{x\to\infty} 3 + \frac{10}{x}$

14. $\lim_{x\to\infty} \frac{4x}{x^2}$ *Hint:* divide first.

15. $\lim_{x\to 0} \frac{x^2}{4x}$ *Hint:* divide first.

12-3 TOWARD A MATHEMATICAL DEFINITION OF LIMIT

The symbolic statement

$$\lim_{x\to a} f(x) = L$$

would be read aloud as follows: the limit of $f(x)$ as x approaches a equals L. Note that a and L are constants and x is a variable. Recall that if $f(x)$ is continuous at $x = a$, then $L = f(a)$.

The meaning of $\lim_{x\to a} f(x) = L$ is conveyed somewhat informally by Definition 12.3(a).

DEFINITION 12.4(a) (*Informal Version*) $\lim_{x\to a} f(x) = L$ *if the values of* $f(x)$ *get closer and closer to* L *as* x *gets close to* a.

Example 1. Find $\lim_{x\to 2} x^2$

Solution: Form Table 12-1.

TABLE 12.1

x	$f(x)$
1.9	3.61
1.99	3.9601
1.999	3.996001

From the values of this table and Definition 8.4(a), we suspect that $\lim_{x\to 2} x^2 = 4$.

DEFINITION 12.4(b) *(Modified Version)* $\lim_{x\to a} f(x) = L$ *if* $f(x) - L$ *gets closer and closer to zero as* $x - a$ *gets close to zero.*

Example 2. Interpret the solution of Example 1 in terms of the modified version of Definition 12.4(b).

Solution:

TABLE 12-2

x	$f(x)$	$x-2$	$f(x)-4$
1.9	3.61	-0.1	$3.61-4=-0.39$
1.99	3.9601	-0.01	-0.0399
1.999	3.996001	-0.001	-0.00399

DEFINITION 12.4(c) *(Mathematical Version)* $\lim_{x\to a} f(x) = L$ *if for any positive real number* ϵ *no matter how small, there is always another positive number* δ *depending on the choice of* ϵ *such that* $|x-a| < \delta$ *(see note) insures that* $|f(x)-L| < \epsilon$.

The reader should study the three versions of Definition 12.4 from time to time until he is convinced that they all express the same idea. Mathematicians prefer version 12.3(c) because it enables them to give more convincing arguments for theorems about limits.

One such theorem which is stated next without proof is

THEOREM 12.1 *If* $\lim_{x\to a} f(x)$ *and* $\lim_{x\to a} g(x)$ *both exist, then*

(a) $\lim_{x\to a}[f(x)+g(x)] = \lim_{x\to a} f(x) + \lim_{x\to a} g(x)$

(b) $\lim_{x\to a} f(x)g(x) = \lim_{x\to a} f(x) \lim_{x\to a} g(x)$

(c) $\lim_{x\to a} cf(x) = c \lim_{x\to a} f(x)$, where c is a constant

This theorem can be summarized by the statement that the *limit of a sum* is the *sum of the limits*; the *limit of a product* is the *product of the limits*; and *any constant* can be *taken outside* the *limit symbol.* This theorem is of fundamental importance in the study of calculus, and it is used later in this text.

The next two examples provide a contrast to the method of computing $\lim_{x \to a} f(x) = L$ by noting that in the continuous case and only in the continuous case $L = f(a)$. In these examples $f(a)$ leads to the undefined expression "0/0"!

Example 3. Find

$$\lim_{x \to 0} \frac{x^2+2x}{x^2-x}$$

Solution: Substitution of $x = 0$ into the given function gets us nowhere, since

$$\frac{0^2+2(0)}{0^2-0} = \frac{0}{0}$$

is undefined. However, if we divide before taking the limit, we have

$$\frac{x^2+2x}{x^2-x} = \frac{x+2}{x-1} \qquad \text{for all } x \neq 0$$

Therefore,

$$\lim_{x \to 0} \frac{x^2+2x}{x^2-x} = \lim_{x \to 0} \frac{x+2}{x-1}$$

$$= \frac{0+2}{0-1}$$

$$= -2$$

Example 4. Find

$$\lim_{x \to 1} \frac{x^2-1}{x-1}$$

Solution: As in Example 1, substitution of $x = 1$ into the given function leads to the expression "0/0," which is undefined. However,

$$\frac{x^2-1}{x-1} = \frac{(x+1)(x-1)}{x-1} = x+1 \qquad \text{if } x \neq 1$$

Therefore,

$$\lim_{x \to 1} (x+1) = 2$$

is the desired limit.

PROBLEMS

1. Let $f(x) = x^2$. Complete the following table:

x	2.9	2.99	2.999
$f(x)$			

Using the completed table as evidence, what value would you give for $\lim_{x \to 3} x^2$?

2. Complete the following table.

x	$x-3$	$f(x) = x^2$
3.1		
3.01		
3.001		

Using the completed table as evidence, what value would you give for $\lim_{x \to 3} x^2$?

3. Using the information given in Problem 2, complete the following table.

x	$x-3$	$f(x) - L$
3.1		
3.01		
3.001		

4. Complete the following table for $f(x) = x^2 + 2x$.

x	$f(x)$
1.1	
1.01	
1.001	

What value would you give for $\lim_{x \to 1}(x^2 + 2x)$?

5. Complete the following table for $\lim_{x \to 1}(x^2 + 2x) = 3$, where $a = 1$ and $L = 3$.

x	$f(x)$	$(x-a)$	$(f(x) - L)$
1.1			
1.01			
1.001			

6. Describe the statement $\lim_{x \to 1}(x^2 + 2x) = 3$ in words. Use the modified version of Definition 12.4(b).

7. Set up a table which could be used as evidence for the statement $\lim_{x \to 4} x^2 = 16$.

8. Show that $\lim_{x \to a}(x + b) = a + b$.

9. Use Theorem 12.1 and the result of Problem 8 to show that $\lim_{x \to a} x^2 = a^2$.

10. Find $\lim_{x \to 3}(x^2+3x+6)$.

11. Find $\lim_{x \to 0}(x^2-2x+3)$.

12. Find $\lim_{\Delta x \to 0}[3a^2+3a(\Delta x)+(\Delta x)^2]$.

12-4 DEFINITION OF THE DERIVATIVE

Consider the function $y=f(x)$, the points A and B, the x values a and $a+\Delta x$, the function values $f(a)$ and $f(a+\Delta x)$, the secant line S passing through A and B, and the *increments* Δx and Δy shown in Figure 12-4.

The behavior of the ratio $\Delta y/\Delta x$ as $\Delta x \to 0$ is the object of study of this section. This ratio can be interpreted geometrically as the slope of the secant line S.

The ratio $\Delta y/\Delta x$ can also be interpreted as the average rate of change of the function $y=f(x)$ between A and B. You may recall that we denote this quantity by the symbol $R_f(A,B)$. It follows from this discussion that

$$\frac{\Delta y}{\Delta x} = \text{Slope of } S = R_f(A,B)$$

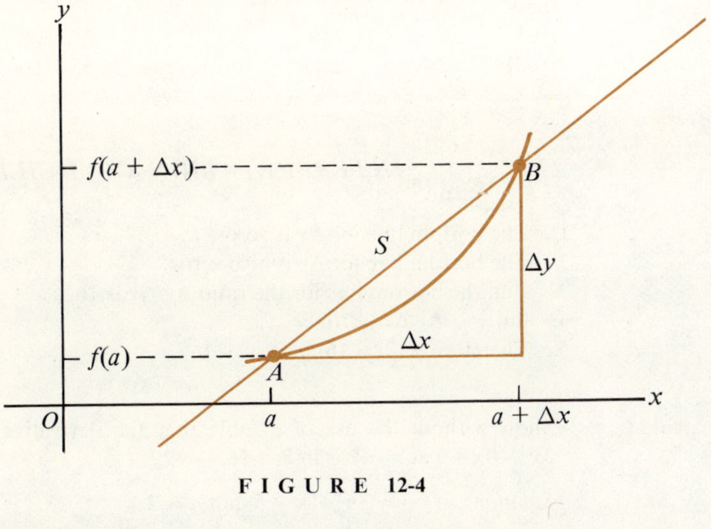

FIGURE 12-4

Now the point rate of change or simply *the rate of change* of $y=f(x)$ at the point A is defined to be

$$\lim_{B \to A} R_f(A,B) = \lim_{\Delta x \to 0} \frac{\Delta y}{\Delta x}$$

where B is constrained to be in the graph of $y = f(x)$. We use $R_f(A)$ in this text to denote the point rate of change of $f(x)$ at A.

Finally, the limit (if it exists) of the ratio $\Delta y/\Delta x$ as Δx approaches zero is called the *derivative* of $y = f(x)$ at $x = a$. This statement is also in the context of Figure 12-4.

DEFINITION 12.5 *The derivative of the function $y = f(x)$ at $x = a$ is defined to be $\lim_{\Delta x \to 0}(f(a+\Delta x)-f(a))/\Delta x$. This limit is denoted by the symbol $f'(a)$.*

A function $y = f(x)$ is said to be *differentiable* at $x = a$ if the derivative of $f(x)$ exists at $x = a$. A function $y = f(x)$ is said to be *differentiable on an interval* $a < x < b$ if $f(x)$ is differentiable at each point of this interval.

The reader should gain considerable insight into the concept of the derivative by studying the following numerical example.

For $y = 3x^2+4x+5$, we can form the following table, where $a = 2$ and Δx takes on the values 1, 0.5, 0.01, and 0.001.

TABLE 12-3 $y = 3x^2+4x+5$

Δx	$a = x_1$	$x_2 = x_1+\Delta x$	$y_1 = f(x_1)$	$y_2 = f(x_2)$	Δy	$\Delta y/\Delta x$
1	2	3	25	44	19	19
0.5	2	2.5	25	33.75	8.75	17.5
0.01	2	2.01	25	25.1603	0.1603	16.03
0.001	2	2.001	25	25.016003	0.016003	16.003

INTERPRETATION OF TABLE 12-3

1. The bottom line for Δx is zero.
2. The bottom line for Δy is also zero.
3. But, the bottom line for the ratio $\Delta y/\Delta x$ is 16.
4. $\lim_{\Delta x \to 0} \Delta y/\Delta x = 16$.
5. Therefore, $f'(2) = 16$.

Example 1. Show without the use of a table that the derivative of the function $y = 3x^2+4x+5$ at $x = 2$ is in fact 16.

Solution: $f(x) = 3x^2+4x+5$ and $a = 2$.

$$\frac{\Delta y}{\Delta x} = \frac{f(a+\Delta x)-f(a)}{\Delta x} = \frac{f(2+\Delta x)-f(2)}{\Delta x}$$

$$= \frac{3(2+\Delta x)^2+4(2+\Delta x)+5-[3(4)+4(2)+5]}{\Delta x}$$

$$= \frac{3(4+4\,\Delta x+\Delta x\,\Delta x)+8+4\,\Delta x+5-(12+8+5)}{\Delta x}$$

$$= \frac{12\,\Delta x+3\,\Delta x\,\Delta x+4\,\Delta x}{\Delta x}$$

$$\frac{\Delta y}{\Delta x} = 16 + 3\,\Delta x$$

By definition, the derivative in this case is given by $f'(2) = \lim_{\Delta x \to 0}(16+3\,\Delta x)$. Now it should be intuitively clear that this limit, and hence $f'(2)$, equals 16.

PROBLEMS

1. Complete the following table for $y = 2x^2+3x+5$.

$a = x_1$	x_2	Δx	y_1	y_2	Δy	$\Delta y/\Delta x$
2	3					
2	2.4					
2	2.02					
2	2.001					

2. From the completed table of Problem 1, what value would you give for $f'(2)$, where $f(x) = 2x^2 + 3x + 5$?

3. Use the method of Example 1 of this section to show that if $f(x) = 2x^2+3x+5$, then $f'(2) = 11$.

4. Repeat Problem 1 using $y = g(x)$, where $g(x) = x^3+2x$, $a = x_1 = 3$, and $x_2 = 4$, 3.3, 2.99, and 3.001.

5. Using the results of Problem 4, what value would you give for $g'(3)$?

6. Using the method of Example 1 of this section, show that if $g(x) = x^3+2x$, then $g'(3) = 29$. [*Hint:* $(a+b)^3 = a^3+3a^2b+3ab^2+b^3$.]

7. Given $h(x) = x^2+3x+4$, find
(a) $h'(4)$
(b) $h'(a)$

8. Complete the following table for $y = \sqrt{x}$:

$a = x_1$	x_2	Δx	y_1	y_2	Δy	$\Delta y/\Delta x$
4	5					
4	3.4					
4	4.1					

[*Hint:* Use logs if necessary.]

9. Using the results of Problem 8, what value would you give for $f'(4)$ if $f(x) = \sqrt{x}$?

10. Given the function $y = x^2$

(a) Let $A = (1, 1)$ and $B_n = [1 + 1/n, (1 + 1/n)^2]$.
Find the x and y coordinates for B_1, B_2, B_3, and B_4, i.e., $n = 1, 2, 3, 4$.

(b) Plot a large graph of $y = x^2$ between $x = 1$ and $x = 2$.

(c) Draw the straight lines between A and B_1, between A and B_2, between A and B_3, and A and B_4 on the graph of part **(b)**.

(d) Complete the following table:

n	B_n	X_n	y_n	$\Delta x = X_n - 1$	$\Delta y = y_n - 1$	$R_f(A, B) = \dfrac{\Delta y}{\Delta x}$
1	B_1	2	4			
2	B_2	1.5	2.25			
3	B_3	1.33	1.77			
4	B_4	1.25	1.56			
100	B_{100}	1.01	1.02			

(e) By examining the table in part **(d)** what can one conclude about Δx as n gets larger and larger?

(f) By examining the table in part **(d)** try to predict the point rate of change of $f(x) = x^2$ at $A(1, 1)$.

11. Compute $R_f(A, B_n)$ in general (i.e., for any positive integer, n) if $y = x^2$, $A = (1, 1)$ and $B_n = [1 + 1/n, (1 + 1/n)^2]$.

$$\textit{Hint:}\quad \frac{\Delta y}{\Delta x} = \frac{(1+1/n)^2 - 1}{1 + (1/n) - 1}$$

12. Redo Problem 10 letting $x_n = 1 + 1/n^2$ and $y_n = (1 + 1/n^2)^2$.

13. Examining the results of Problems 5 and 6, what can one conclude about $R_f(A, B_n)$ as n gets very large?

12-5 INTERPRETATIONS OF THE DERIVATIVE

The following terms from plane geometry are illustrated in Figure 12.5(a).

1. Circle
2. Radius, OA
3. Tangent, T
4. Point of tangency, A
5. Chord, AB
6. Secant, S

Remark: You may recall that the tangent to a circle at a given point A, is the line that intersects the circle in exactly one point.

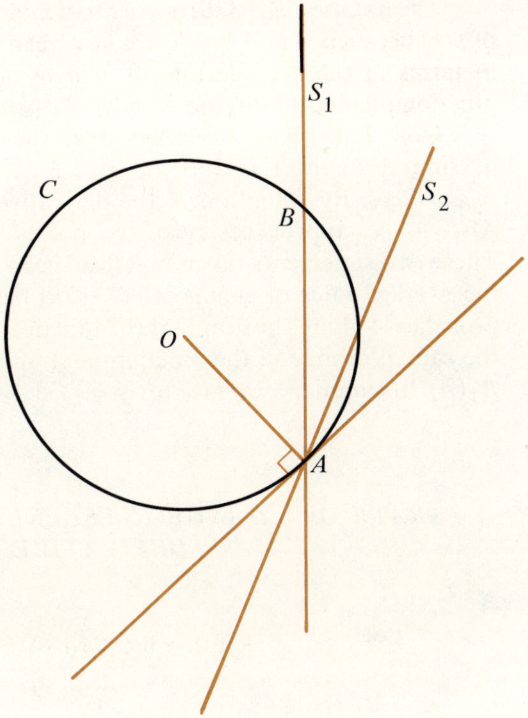

FIGURE 12-5(a)

It is also true that the tangent to a circle occupies the "limiting position of the secant lines." This notion of "tangent to" also applies to curves other than circles. See Figure 12-5(b) for an example.

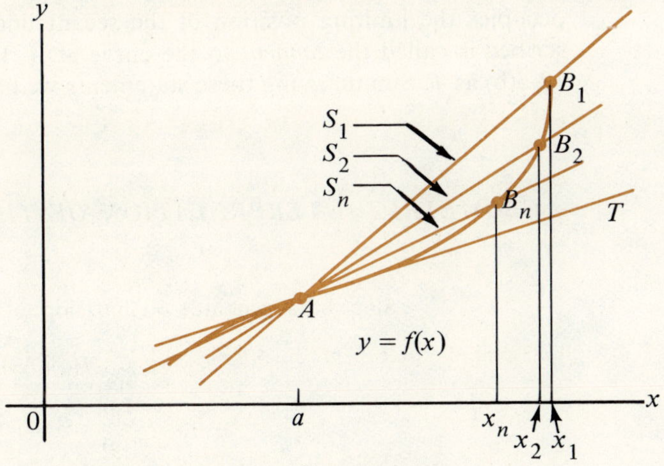

FIGURE 12-5(b)

The notation $R_f(A, B)$ is used to denote the average rate of change of $f(x)$ between A and B. We are now ready to interpret the derivative in terms of the average rate of change of a function. Consider the function pictured in Figure 12-5(b).

Now $R_f(A, B_i) = (\Delta y)/(\Delta x)$ gives the slope of the straight line passing through A and B_i $(i = 1, 2, 3, \ldots, n)$. These lines are called *secant lines*. By definition of the derivative, $\lim_{\Delta x \to 0}(\Delta y)/(\Delta x) = f'(a)$. Also, as Δx approaches zero, the points B_i approach the point A. These two statements taken together mean that $f'(a)$ gives the limit of the average rates of change of $y = f(x)$ between A and B_i as B_i approaches A along the graph of the function. This limit is called simply the rate of change of the function at A and is denoted in this text by $R_f(A)$. In summary we have for $y = f(x)$:

RATE OF CHANGE INTERPRETATION OF THE DERIVATIVE

$$\begin{aligned} R_f(A) &= \lim_{B \to A} R_f(A, B) \\ &= \lim_{\Delta x \to 0} \frac{f(a+\Delta x)-f(a)}{\Delta x} \\ &= f'(a) \end{aligned}$$

Interpreting the derivative geometrically, we see that it gives the slope of the straight line that touches the curve at $A = (a, f(a))$ and occupies the limiting position of the secant lines. The line just described is called the *tangent* to the curve at A. It is shown in Figure 12-5(b) as T. Summarizing these statements we have:

GEOMETRIC INTERPRETATION OF THE DERIVATIVE

$$\begin{aligned} \text{Slope of tangent at } A &= \lim_{\Delta x \to 0} (\text{slope of secant}) \\ &= \lim_{\Delta x \to 0} \frac{f(a+\Delta x)-f(a)}{\Delta x} \\ &= f'(a) \end{aligned}$$

Finally, the two interpretations discussed above are summarized by the following statement. The point rate of change of $y = f(x)$ at the point A equals the slope of the tangent line to the graph of $y = f(x)$ at the point A, and both are equal to $f'(a)$.

PROBLEMS

1. Consider the function $y = x^2+1$.
(a) Compute the slopes of the secant lines that pass through the point $(2, 5)$ and the points on the graph of $y = x^2+1$ with x coordinates 4, 3, 2.5, and 2.1, respectively.
(b) Sketch all 4 secant lines of (a) on the same coordinate system.

2. In Problem 1, use the derivative to find the slope of the tangent line to the given curve at the point (2, 5).

$$\left[\textit{Hint:} \quad f'(x) = \lim_{\Delta x \to 0} \frac{(x+\Delta x)^2+1-(x^2+1)}{\Delta x} \right]$$

Sketch the tangent line on the coordinate system of Problem 1(b).

3. Consider the function $y = x^2+2$.
(a) Sketch the curve.
(b) Compute the slopes of the secant lines that pass through the point $(1, 3)$ and the points on the graph of $y = x^2+2$ with x coordinates 3, 2, 1.5, and 1.1, respectively.
(c) Sketch all 4 secant lines of part (b) on the same coordinate system.

4. Use the derivative to find the slope of the tangent to the curve $y = x^2 + 2$ at the point (1, 3).

5. Given the equation $y = 6x+2$.
(a) Sketch the graph of the equation.
(b) Compute the slopes of the secant lines that pass through the point $(1, 8)$ and the points on the graph of $y = 6x+2$ with x coordinates 4, 3, and 2, respectively.
(c) What can you say about the slopes of the 3 secant lines calculated in **(b)**?
(d) Under what conditions do you think this would happen?

6. Given that $f(x) = x^3$, $A(1, 1)$, and $B = (2, 8)$, find
(a) $R_f(A, B)$
(b) $R_f(A)$ [*Hint:* $(a+b)^3 = a^3+3a^2b+3ab^2+b^3$.]

7. Consider a function $f(x)$ with the following geometric properties:
(a) The slope of the tangent at A is 4.
(b) The slope of the secant line passing through A and B is 5. What is the average rate of change of $f(x)$ between A and B? Why?

8. What is *the* rate of change of $f(x)$ at A for the function described in Problem 7? Why?

In economics we have the following definition for arc elasticity of demand:

$$\text{Elasticity} = \frac{\text{percentage change in quantity demanded}}{\text{percentage change in price}}$$

9. Given the following for a commodity:

TABLE 12-4

Price	Quantity demanded
\$1.00	10
0.90	12

(a) Calculate the percentage change in quantity demanded.
(b) Calculate the percentage change in price.
(c) Calculate elasticity.
(d) In general, if the formula for elasticity yields a value greater than one, then demand is said to be elastic. Is demand elastic in this case?
(e) Did you have any trouble doing the calculations in **(a)** or **(b)**?

Your answer to **(e)** should have been yes. In order to calculate a percentage one needs to work from a base. In calculating the percentage change in quantity, we have 2/?, where one could use 10, 12, or $(10+12)/2$ in the denominator. The same applies to the calculation of the percentage change in price. This problem is not encountered when calculating the point elasticity using the calculus.

10. Let the demand law be $x = f(P)$, where x = quantity demanded and P = price. Formulate the definition of elasticity given above using increments and/or functional notation. Use original price and original quantity as the basis of comparison for calculating the percentage change.

11. Show that the definition of elasticity given above can be expressed symbolically as

$$e_d = \frac{\Delta x}{\Delta P} \cdot \frac{P}{x}$$

12. Let $\Delta P \to 0$ in $e_d = (\Delta x/\Delta P)(P/x)$ of Problem 11 to obtain

$$e_d = \frac{P}{x} \lim_{\Delta P \to 0} \frac{\Delta x}{\Delta P}$$

13. Let $A = (P, x)$ and $B = (P + \Delta P, x + \Delta x)$, where $x = f(P)$ denotes a demand law. Show that elasticity of demand, e_d, is given by the following formula: $e_d = (P/x)\, R_f(A, B)$. In economics this quantity is called *arc* elasticity of demand.

14. Use the result of Problem 13 and let $B \to A$ to obtain $e_d = (P/x)\, R_f(A)$. In economics this quantity is called *point* elasticity of demand.

12-6 THE DERIVATIVE OF $y = x^n$

It is recalled that a real-valued function $y = f(x)$ associates with each element $x = a$ of the domain a real number that is denoted by $f(a)$. The process of finding the derivative of a function at $x = a$ associates with a another real number called $f'(a)$. Given a function $y = f(x)$, we can define a *new* function by associating with each x the unique real number $\lim_{\Delta x \to 0}(f(x+\Delta x)-f(x))/\Delta x$ if it exists. This new function is denoted by $f'(x)$ or by y'.

The notation

$$\frac{dy}{dx}, \frac{dy}{du}, \frac{dy}{dt}, \ldots$$

depending on the choice of the independent variable is also in common use.

The procedure illustrated in the next example is called "finding a derivative by definition." Many calculus teachers believe that the experience of finding derivatives by definition helps the student to *understand* the derivative as a mathematical concept.

In practice, however, most derivatives are calculated using derivative formulas rather than by definition. Several derivative formulas are discussed in this text.

Example 1. Find the derivative of $y = x^2$ by definition.

Solution: $dy/dx = \lim_{\Delta x \to 0}(\Delta y/\Delta x)$, by definition.

$$\frac{\Delta y}{\Delta x} = \frac{(x+\Delta x)^2 - x^2}{\Delta x}$$

$$= \frac{x^2 + 2x\,\Delta x + (\Delta x)^2 - x^2}{\Delta x}$$

$$= 2x + \Delta x$$

$$\therefore \frac{dy}{dx} = 2x$$

It is not too hard to show that, if $y = x^n$, where n is a positive integer, then $dy/dx = nx^{n-1}$. This means that:

1. $\dfrac{d}{dx}(x) = 1$

2. $\dfrac{d}{dx}(x^2) = 2x$

3. $\dfrac{d}{dx}(x^3) = 3x^2$ power rule

$\vdots$

n. $\dfrac{d}{dx}(x^n) = nx^{n-1}$

The power rule can be shown to be valid for any real number n.

It is stated below for future reference.

$$\frac{d}{dx}(x^n) = nx^{n-1} \qquad \text{D-2}$$

Or, in terms of the independent variable u:

$$\frac{d}{du}(u^n) = nu^{n-1}. \qquad \text{D-2}$$

Example 2. Use the power rule to find dy/dx if $y = \sqrt{x}$.

Solution:

$$y = \sqrt{x}$$

$$y = x^{1/2}, \qquad n = \tfrac{1}{2}$$

$$\frac{dy}{dx} = \left(\frac{1}{2}x\right)^{(1/2)-1} \qquad \textit{(power rule)}$$

$$dy = \frac{1}{(2x)^{1/2}}$$

$$\therefore \frac{dy}{dx} = \frac{1}{2\sqrt{x}}$$

PROBLEMS

1. Given the function $f(x) = x^2 + 3x + 4$, find
   (a) $f'(x)$
   (b) $f'(a)$
   (c) $f'(4)$
2. Using the function given in Problem 1, find
   (a) The slope of the tangent to $y = f(x)$ at the point $(1, 8)$.
   (b) The slope of the secant line passing through $(1, 8)$ and $(2, 14)$.
3. Find an equation for the tangent line in Problem 2(a).
4. Find dy/dx for the following functions:
   (a) $y = 4$
   (b) $y = x$
   (c) $y = x^2$
   (d) $y = x^3$
   (e) $y = x^4$
   [*Hint:* $(x+\Delta x)^4 = (x+\Delta x)(x+\Delta x)^3$

   and

   $$(a+b)^3 = a^3 + 3a^2b + 3ab^2 + b^3.]$$
5. Find dy/du for the following functions:
   (a) $y = u^2$
   (b) $y = u + 3$
   (c) $y = u^2 + 2u$
   (d) $y = 4 - u^2$
6. Given that $y = f(x)$, list two symbols that have the same meaning as y'.

7. Given that $w = g(u)$, list two symbols that have the same meaning as w'.
8. Given that $f'(a) = \lim_{\Delta x \to 0}((-2+\Delta x)^3 - (-2)^3)/\Delta x$, find $f(x)$ and a.
9. From Problem 14 of Section 12-5 we have $e_d = (P/x)R_f(A)$, where $x = f(P)$, $A = (P, x)$, and e_d denotes the point elasticity of demand at P. Using derivative notation, rewrite this formula for point elasticity.
10. Obtain $e_d = (P/x)(dx/dP)$ from Problem 12 of Section 12-5.
11. Find $f'(3)$ by definition if $f(x) = x^2 - 4x + 3$.

12-7 THE POWER RULE: $(d/dx)(u^n)$

In Section 12-4 we defined the derivative of $y = f(x)$ to be

$$\lim_{\Delta x \to 0} \frac{f(x+\Delta x) - f(x)}{\Delta x}$$

and this definition was used to find the derivative either exactly or approximately for some specific functions. We remind the student at this point that, in practice, the procedure for determining the derivative of a function is usually not as lengthy as is indicated by the previous section. This is because there exist many *differentiation formulas* that are quite easy to apply. Some of these will be given in this section.

For reasons that will become clear in Chapter 13, these formulas will be stated in terms of u rather than x.

The first formula that follows directly from the definition of the derivative is

$$\frac{d}{du}(c) = 0 \qquad \text{where } c \text{ is any constant} \qquad \text{D-1}$$

One of the most useful differentiation formulas is the so-called *power rule:*

$$\frac{d}{du}(u^n) = nu^{n-1} \qquad \text{D-2}$$

Formula D-2 is valid for any constant n. Before giving a proof of D-2 for n, a positive integer, we include several examples to illustrate its use.

Example 1. Differentiate $y = u^3$ with respect to u.

Solution:

$$y = u^3$$

$$\frac{dy}{du} = 3u^2$$

Example 2. Differentiate $y = u^3$ with respect to u using the definition of the derivative.

Solution: Let $y = u^3$.

$$\Delta y = (u+\Delta u)^3 - u^3$$

$$\Delta y = u^3 + 3u^2\,\Delta u + 3u(\Delta u)^2 + (\Delta u)^3 - u^3$$

$$\Delta y = \Delta u[3u^2 + 3u\,\Delta u + (\Delta u)^2]$$

$$\frac{\Delta y}{\Delta u} = 3u^2 + 3u\,\Delta u + (\Delta u)^2$$

$\text{Limit}_{\Delta u \to 0}(\Delta y/\Delta u) = 3u^2$. Therefore,

$$\frac{dy}{du} = 3u^2 \qquad \textit{as was found in Example 1 using the power rule}$$

Example 3. Differentiate $y = \sqrt{u^3}$ by the power rule.

Solution: First of all, from algebra we have $\sqrt{u^3} = u^{3/2}$.

$$y = u^{3/2}$$

$$\frac{dy}{du} = \frac{3}{2}u^{1/2} = \frac{3}{2}\sqrt{u}$$

Example 4. Differentiate $y = 1/u^4$ with respect to u.

Solution: First of all, $y = u^{-4}$.

$$\frac{dy}{du} = -4u^{-4-1}$$

$$= -4u^{-5}$$

$$= -\frac{4}{u^5} \qquad \textit{by laws of exponents}$$

The proof of the power rule for the special case where n is a positive whole number is not difficult and this proof will follow.

Proof: Let $y = u^n$, where n is a positive integer. Then

$$\Delta y = (u+\Delta u)^n - u^n$$

$$\Delta y = u^n + nu^{n+1}\,\Delta u + k_2 u^{n-2}(\Delta u)^2 + k_3 u^{n-3}(\Delta u)^3 + \cdots + (\Delta u)^n - u^n \qquad \textit{by the binomial theorem of algebra*}$$

* The binomial theorem. For any positive integer n,

$$(a+b)^n = k_0 a^n + k_1 a^{n-1}b + k_2 a^{n-2}b^2 + \cdots + k_j a^{n-j}b^j + \cdots + nab^{n-1} + b^n$$

where

$$k_j = \frac{n!}{j!(n-j)!} \qquad \text{and} \qquad j! = j(j-1)(j-2)\cdots 2\cdot 1$$

$$\Delta y = \Delta u[nu^{n-1} + k_2 u^{n-2}\Delta u + k_3 u^{n-3}(\Delta u)^2 + \cdots + (\Delta u)^{n-1}]$$

$$\frac{\Delta y}{\Delta u} = nu^{n-1} + k_2 u^{n-2}\Delta u + k_3 u^{n-3}(\Delta u)^2 + \cdots + (\Delta u)^{n-1}$$

$\text{Limit}_{\Delta u \to 0}(\Delta y/\Delta u) = nu^{n-1}$ as was to be proved. Therefore,

$$\frac{dy}{du} = nu^{n-1}$$

We emphasize the fact that even though we have proven this result for n, a positive integer, the power rule is valid for *any* real number. This fact is proven in more extensive books on calculus.

The theorem that follows from the definition of the derivative and limit, Theorem 12.1, is extremely important. A proof of Theorem 12.2(b) will be given later.

THEOREM 12.2 *Let $f(u)$ and $g(u)$ be differentiable functions and c be a constant. Then*

(a) $$\frac{d}{du}cf(u) = c\frac{d}{du}f(u)$$

(b) $$\frac{d}{du}[f(u)+g(u)] = \frac{d}{du}f(u) + \frac{d}{du}g(u)$$

(*easily extended to any finite sum*)

In words, Theorem 12.2(a) states that a constant can be taken "outside" the derivative symbol. Furthermore, by repeated application of Theorem 12.2(b), we see that the derivative of a finite sum of functions equals the sum of the derivatives of these functions. Finally, we caution the reader that the derivative of a product of functions *does not equal* the product of the derivatives.

Example 5. Differentiate the two functions $y = 3u^4$ and $y = 3$.

Solution:

$$y = 3u^4$$

$$\frac{dy}{du} = 3\frac{d}{du}(u^4) \qquad \text{by Theorem 12.2(a)}$$

$$= 3(4)u^3 \qquad \text{by D-2}$$

$$= 12u^3$$

If

$$y = 3$$

then

$$\frac{dy}{du} = 0 \qquad \textit{by D-1}$$

Note that a constant by itself is treated differently from a constant times a function.

Example 6. Find the derivative of $y = 4u$ with respect to u.

Solution: Write $y = 4u$ as $y = 4u^1$. By Theorem 12.2(a) and D-2, we have

$$\frac{dy}{du} = 4(1)u^0$$

$$= 4(1)(1) \qquad \textit{since } u^0 = 1$$

Therefore,

$$\frac{dy}{du} = 4$$

Example 7. Differentiate $y = 5u^2 + 5u - 7$ with respect to u.

Solution:

$$y = 5u^2 + 5u - 7$$

$$\frac{dy}{du} = 10u + 5 \qquad \textit{by Theorem 12.2(a), (b), D-1, and D-2}$$

Formula D-2 enables us to find $d(x^n)/dx$ or $d(u^n)/du$. Next we consider the problem of finding

$$\frac{dy}{dx} \quad \text{if} \quad y = u^n$$

and u is a function of x. In other words y is a "function of a function". In this situation, the variable u is a dependent variable relative to x, but an independent variable relative to y.

The correct formula for dy/dx in this case is the *product* of two other derivatives. The extended power rule to cover the function of a function case is:

$$\frac{d(u^n)}{dx} = nu^{n-1}\frac{du}{dx} \qquad \text{D-2}'$$

Example 8. Use D-2′ to find dy/dx if $y = (9 - 12x)^{3/4}$.

Solution: Think of the expression $9 - 12x$ as u. Then $y = u^{3/4}$.

$$\frac{dy}{dx} = \frac{3}{4}u^{-1/4}\frac{du}{dx} \qquad \text{by D-2}'$$

$$= \frac{3}{4}(9-12x)^{-1/4}(-12)$$

$$\frac{dy}{dx} = \frac{-9}{(9-12x)^{1/4}}$$

PROBLEMS

1. Find the derivative of $y = 3u^2 - 2u$ by definition.

2. Find the derivative of the two functions $f(u) = 3u^2$, $g(u) = -2u$.

3. Relate the results of Problems 1 and 2 to Theorem 12.2.

4. Let $y = f(u) = u^3$.
(a) Find dy/du by the power rule.
(b) Use the result in **(a)** to find $f'(-2)$.

5. Use the appropriate formula to find dy/du for the following functions:
(a) $y = 4$ **(d)** $y = u^4$
(b) $y = u$ **(e)** $y = u^5$
(c) $y = u^2$ **(f)** $y = u^{1/3}$

6. Find dy/du for the following functions:
(a) $y = u^{1/2}$ **(d)** $y = \sqrt[3]{u}$
(b) $y = \sqrt{u}$ **(e)** $y = u^{-1/3}$
(c) $y = u^{1/3}$ **(f)** $y = 1/\sqrt[3]{u}$

7. State the power rule for $w = z^k$.

8. Find dy/du for the following functions:
(a) $y = 3u^2$ **(c)** $y = u^2 + 10u + 13$
(b) $y = 2u + 4$ **(d)** $y = 4u^{3/2} - 2\sqrt{u}$

9. Find dy/du for the following functions:
(a) $y = 6u^4$ **(c)** $y = 4u^2 - 8u + 10$
(b) $y = 5u + 8$ **(d)** $y = 2u^2 + 2u + \sqrt[4]{u}$

10. Find dy/du for the following power functions:
(a) $y = 2u^4$
(b) $y = 1.3u^{2.7}$
(c) $y = -0.8u^{-1.4}$

Find dy/dx in Problems 11–14.

11. $y = (2x+3)^{4/3}$

12. $y = 6(10-3x)^5$

13. $y = 3\sqrt{x^2+9}$. *Hint:* $\sqrt{x^2+9} = (x^2+9)^{1/2}$

14. $y = \dfrac{4}{\sqrt{x}} + \sqrt[3]{4-x^2}$

12-8 SUMMARY OF DERIVATIVE FORMULAS

1. The derivative of a constant (by itself) equals zero.

$$\frac{d}{dx}(c) = 0 \qquad \text{D-1}$$

2. The derivative of x to the nth power equals n times x to the $(n-1)$st power.

$$\frac{d}{dx}(x^n) = nx^{n-1} \qquad \text{D-2}$$

3. The derivative of a constant times a function equals the constant times the derivative of the function.

$$\frac{d}{dx}(cu) = c\,\frac{du}{dx}$$ *(denoted by D-8 later on)*

Remark: A constant can be taken outside of the derivative symbol.

4. The derivative of a sum of functions equals the sum of the derivative of the given functions.

$$\frac{d}{dx}(u_1+u_2+\cdots+u_n) = \frac{du_1}{dx} + \frac{du_2}{dx} + \cdots + \frac{du_n}{dx}$$ (Special case $n=2$ is denoted in the next chapter.)

Remark: The statement in 4 would be false if *sum* were replaced with *product.*

5. The derivative of u to the nth power when u is a function of x is given by n times u to the $(n-1)$st power *times* the derivative of u with respect to x.

$$\frac{d}{dx}(u^n) = nu^{n-1}\frac{du}{dx} \qquad \text{D-2}'$$

PROBLEMS

Find dy/dx.

1. $y = x^2 + 2x + 7$

2. $y = x^3 - 3x^2 + x$

3. $y = 3x^{1/2} + 4x^{-1/2}$

4. $y = 3\sqrt{x} + \sqrt[3]{x}$

5. $y = 4x(x^2 - x^3)$

6. $y = \sqrt[3]{x \cdot x^4}$

7. $y = (2x+3)^2$

8. $y = (x+1)^3$

9. $y = \sqrt{x}(x-3)^2$

10. $y = (x+\sqrt{x})^2$

11. $y = \sqrt[5]{x}$

12. $y = \sqrt[6]{x^4}$

13. $y = 10$

14. $y = 4x^5 + x^{-2} + 2$

15. $y = \dfrac{10}{x^2}$

16. $y = \dfrac{4}{\sqrt{x}}$

17. $y = 0.1x^{100}$

18. $y = (1-x)^3$

19. $y = (\sqrt{x} + \sqrt[4]{x})^2 + 10x$

20. $y = \dfrac{8}{x^3} + \dfrac{x^3}{9}$

12-9 DERIVATIVES OF LOGARITHMIC AND EXPONENTIAL FUNCTIONS

Logarithmic functions are encountered frequently in economic literature. Exponential functions are used to model growth and decay processes.

Four new derivative formulas are introduced in this section. They are:

$$\frac{d}{du}(\log_a u) = \frac{1}{u}\log_a(e) \qquad \text{D-3}$$

$$\frac{d}{du}(\ln u) = \frac{1}{u} \qquad \text{D-4}$$

$$\frac{d}{du}(a^u) = a^u \ln a \qquad \text{D-5}$$

$$\frac{d}{du}(e^u) = e^u \qquad (e = 2.7183 \text{ to four decimal places}) \qquad \text{D-6}$$

These four formulas can also be extended to D-3′, D-4′, D-5′, and D-6′. For example

$$\frac{d}{dx}(e^u) = e^u \frac{du}{dx} \qquad \text{D-6}'$$

Mr. Brin has suggested that we point out the fact that formula*

* Mr. Jack Brin is a professor of mathematics at Western New England College.

D-3 can be expressed in the following, alternative form:

$$\frac{d}{du}(\log_a u) = \frac{1}{u \ln a} \qquad \text{D-3 (alternative)}$$

This form is simpler, computationally, since $\log_a x$ is generally not tabulated for values of a other than 10 or e.

Considerable insight into formula D-3 can be attained by studying the following proof.

Let $y = \log_a u$.

Then

$$\Delta y = \log_a(u+\Delta u) - \log_a u$$

$$\Delta y = \log_a\left(\frac{u+\Delta u}{u}\right)$$

$$\frac{\Delta y}{\Delta u} = \frac{1}{\Delta u}\log_a\left(1+\frac{\Delta u}{u}\right)$$

$$\frac{\Delta y}{\Delta u} = \log_a\left(1+\frac{\Delta u}{u}\right)^{1/\Delta u} \qquad \textit{using a property of logs}$$

Now to find the limit of $(\Delta y)/(\Delta u)$ as $\Delta u \to 0$ is by no means a trivial matter in this case. Some progress can be made by the following procedure.

Starting again with $\Delta y/\Delta u = 1/\Delta u \log_a[1+(\Delta u/u)]$ and writing $(1/\Delta u)\log_a[1+(\Delta u/u)]$ as $(u/u\,\Delta u)\log_a[1+(\Delta u/u)]$ leads to $\Delta y/\Delta u = (1/u) \log_a[1+(\Delta u/u)]^{u/\Delta u}$. Writing $\Delta u/u = h$ gives $\Delta y/\Delta u = (1/u)\log_a(1+h)^{1/h}$ since h approaches zero as Δu approaches zero. The converse of this is also true. It follows that the derivative is given by $dy/du = \lim_{h\to 0}(1/u)\log_a(1+h)^{1/h}$. This limit is known to exist and it is denoted by e, the base of the natural log system. In symbols, $e = \lim_{h\to 0}(1+h)^{1/h}$. The numerical value of this limit correct to four decimal places is $\lim_{h\to 0}(1+h)^{1/h} = 2.7183$.

Continuing now with the expression for dy/du, we have

$$\frac{dy}{du} = \frac{1}{u}\lim_{h\to 0}\log_a(1+h)^{1/h} \qquad \textit{by Theorem 12.1(c)}$$

$$\frac{dy}{du} = \frac{1}{u}\log_a\left[\lim_{h\to 0}(1+h)^{1/h}\right]$$

$$\frac{dy}{du} = \frac{1}{u}\log_a(e) \qquad \textit{by definition of e}$$

Example 1. Find dy/du if $y = \log_2 u$.

Solution:

$$\frac{dy}{du} = \frac{1}{u} \log_2 e$$

Example 2. Find dy/du if $y = \log_{10} u$.

Solution:

$$\frac{dy}{du} = \frac{1}{u} \log_{10} e$$

$$= \frac{1}{u}(0.43425)$$

Example 3. Find dy/du if $y = \ln u$.

Solution:

$$y = \ln u = \log_e u$$

$$\frac{dy}{du} = \frac{1}{u} \log_e e$$

$$\frac{dy}{du} = \frac{1}{u}$$

We note from Example 3 that the derivative of $y = \ln u$ is simply $dy/du = 1/u$. This explains the term "natural" as applied to the natural log system. It is "natural" in the sense that of all the bases a that could be used for a log system, the base e gives the simplest possible derivative formula for $y = \log_a u$,

$$\frac{d}{du}(\ln u) = \frac{1}{u} \qquad \text{D-4}$$

Finally, we state without proof the formula for the derivative of an exponential function.

$$\frac{d}{du}(a^u) = a^u \ln a \qquad \text{D-5}$$

Example 4. Find dy/du if $y = 2^u$.

Solution:

$$\frac{dy}{du} = 2^u \ln 2$$

Example 5. Find dy/du if $y = 10^u$.

Solution:

$$\frac{dy}{du} = 10^u \ln 10$$

$$= 10^u(2.30259)$$

Example 6. Find dy/du if $y = e^u$.

Solution:

$$\frac{dy}{du} = e^u \ln e$$

$$\frac{dy}{du} = e^u(1)$$

$$\frac{dy}{du} = e^u$$

We note from Example 6 that the derivative of $y = e^u$ is simply e^u,

$$\frac{d}{du}(e^u) = e^u \qquad \text{D-6}$$

The simplicity of D-6 explains the expression "*the* exponential function" that is used to denote the function e^u.

In particular $(d/dx)(e^x) = e^x$.

PROBLEMS

1. Find dy/du for the following functions:
   (a) $y = \log_3 u$ (c) $y = 2^u$
   (b) $y = \log_5 u$ (d) $y = 10^u$
2. Find dy/dz for the following functions:
   (a) $y = \ln z$ (c) $y = \log z$
   (b) $y = 4^z$ (d) $y = e^z$
3. Use a property of logarithms to write the following in a different form:
   (a) $\log(1+h)^h$ (c) $\log_a\left(1+\frac{1}{h}\right)^{1/h}$
   (b) $\ln\left(1+\frac{1}{h}\right)^{1/h}$ (d) $\log_a\left(1+\frac{\Delta u}{u}\right)^{u/\Delta u}$
4. Use logarithms to compute the following:
   (a) $\left(1+\frac{1}{2}\right)^2$ (c) $\left(1+\frac{1}{100}\right)^{100}$
   (b) $\left(1+\frac{1}{10}\right)^{10}$ (d) $\left(1+\frac{1}{1000}\right)^{1000}$

5. Compare your answers for Problem 4 with the value 2.7183 given for $\lim_{h\to 0}(1+h)^{1/h}$.

6. Find dy/dx for the following functions:
   (a) $y = e^x + \ln x$
   (b) $y = x^2 + 2x$
   (c) $y = x^2 + \log x$
   (d) $y = \sqrt{x} + 3x + 2$

7. Find dy/dx for the following functions:
   (a) $y = 2^x + \log_2 x$
   (b) $y = x^3 + 3^x$
   (c) $y = x^{-3} + \ln x$
   (d) $y = 2x^{-5/2} - x^{3/2} + 4^x$

The compound interest formula for compounding k times per year at i percent for n years is

$$V_n = P\left(1 + \frac{i}{k}\right)^{kn}$$

8. Show that $V_n = P[(1+i/k)^{k/i}]^{ni}$.

9. Express V_n of Problem 8 in terms of h, where $h = i/k$, and observe that h approaches zero as k approaches infinity.

10. Use the result of Problem 9 to show that for continuous compounding $V_n = Pe^{ni}$

11. Find dy/du for the following exponential functions:
   (a) $y = 100(1.06)^u$
   (b) $y = 50(1+0.08)^u$
   (c) $y = a(1+r)^u$, where a and r are constants

Find dy/dx in Problem 12–15. *Hint:* Use D-3*, D-4′, D-5′ or D-6′.

12. $y = 4 \log x^3 + \ln \sqrt{x}$

13. $y = 2^{4x} + e^{x^2}$

14. $y = 3 \ln(x^2+2x+7)$

15. $y = 7 \log_5(3x) + 6 \log_7(8x)$

Use D-3 (alternative to find dy/dx in Problems 16–20.

16. $y = \log_3 x$

17. $y = 5 \log_7 x$

18. $y = 6 \log x$ (*base 10*)

19. $y = 3 \log_{1/2} x$

20. $y = \ln x$ (*base e*)

12-10 GEOMETRIC AND ECONOMIC APPLICATIONS OF THE DERIVATIVE

We have used the notation $R_f(A, B)$ to denote the average rate of change of $y = f(x)$ between two points A and B on the graph of $y = f(x)$ and the notation $R_f(A)$ to denote the rate of change of $y = f(x)$ at the single point A. In terms of limit notation, $R_f(A) = \lim_{B\to A} R_f(A, B)$.

Interpreted geometrically, $R_f(A, B)$ is the slope of the secant line passing through the points A and B whereas $R_f(A)$ is the slope of the tangent to the graph of $y = f(x)$ at A. It should also be recalled that $R_f(A) = f'(a)$, where a is the x coordinate of A.

The notation $f'(a)$ has been used to denote the derivative of $y = f(x)$ evaluated at $x = a$. An alternative notation, which is also useful, is

$$f'(a) = \left.\frac{dy}{dx}\right|_{x=a}$$

Example 1. Find an equation of the line that is tangent to the curve $y = 3x^2$ at the point $A = (2, 12)$.

Solution: The slope $m = R_f(A)$ is given by $f'(2)$.

$$y = 3x^2$$

$$\frac{dy}{dx} = 6x$$

$$f'(2) = \left.\frac{dy}{dx}\right|_{x=2} = 6(2) = 12$$

$$m = 12$$

Writing the equation of the tangent line in the form $y = mx + b$, where m is the slope and b is the y intercept, we have

$$y = 12x + b$$

$$12 = 12(2) + b$$

$$b = -12$$

Therefore, $y = 12x - 12$ is an equation of the tangent line.

So far in this section, we have examined the geometric interpretation of the derivative. The derivative can also be interpreted as the *instantaneous* or *point* rate of change of the dependent variable with respect to the independent variable.

Instantaneous rates of change are important in applications of mathematics to economics and business administration.

Examples of this are the concepts of marginal cost and marginal revenue.

In an article entitled "Computer Graphics for Decision Making," Irvin M. Miller writes:

> To solve business problems requiring executive decisions, one must define the total problem and then assign a mathematical equation to each aspect of the problem. A composite of all equations yields a mathematical model representing the problem confronting the executive.

Mr. Miller also includes a list of eleven typical parameters used to phrase business problems.

Of these eleven parameters, the following have particular significance in this text: marginal cost (MC), marginal revenue (MR), total cost, and total revenue.*

Let $C = f(x)$ be a total cost function. Then the first derivative, $f'(x)$, is called the *marginal cost function.* We use the notation MC to denote the marginal cost function that is derived from the total cost function $C = f(x)$. In symbols, we have $MC = f'(x)$. The marginal cost at a particular point $x = a$ will be denoted by $MC|_{x=a}$.

Example 2. Given the total cost function $C = 10(2^x)$, find the marginal cost function and the marginal cost at $x = 1$.

Solution:

$$C = 10(2^x)$$

$$MC = 10(2^x)\ln 2$$

$$MC = 10(2^x)0.69315$$

The marginal cost function is $6.9315(2^x)$.
Marginal cost at $x = 1$ is given by $6.9315(2^1) = 13.863$.

PROBLEMS

1. Find an equation of the line tangent to the curve $y = 2x^2$ at point A $(3, 18)$.
2. Find an equation of the line tangent to the curve $y = 2^x$ at point A $(1, 2)$.
3. Find an equation of the line tangent to the curve $y = \ln x$ at the point $A(1, 0)$.
4. Given the total cost function $C = 3^x$, find the marginal cost function and the marginal cost at $x = 1$.
5. Given the total cost function $C = x^3$, find the marginal cost function and the marginal cost at $x = 4$.

The remaining problems in this set are based on the concept of elasticity. We quote Michael J. Brennan, *Preface to Econometrics*, for a definition and a brief description of this concept**:

One extremely important measure employed in economic theory is elasticity. Textbooks on economic theory define *price elasticity* of demand as the per-

---

* Irvin M. Miller, "Computer Graphics for Decision Making," *Harvard Business Review*, 47, p. 124.

** Michael J. Brennan, *Preface to Econometrics*. Cincinnati, Ohio: South-Western Publishing Co., 1960, pp. 134–139.

centage change in the quantity of a good demanded in response to a percentage change in its price.

He then defines *elasticity of the function* $f(x)$ as $e = \dfrac{dy}{dx} \cdot \dfrac{x^2}{y}$.

6. Consider the demand function $y = 100 - 15x$, where x denotes the price and y denotes the quantity demanded. Determine the elasticity at $x = 2$, $x = 4$, and $x = 6$.

7. Determine the elasticity of $f(x) = 24/x^4$.

8. Show that if $y = A/x^n$, where A and n are constants, then the elasticity, e, is given by $e = -n$.

Typically, one expects *demand*, x, for a product to decrease as the price, P, increases. Some demand laws frequently encountered in the literature are listed below.

$$x = \frac{a - P}{b} \qquad (1)$$

$$x = \frac{a}{P + c} - b \qquad (2)$$

$$x = \sqrt{\frac{a - P}{b}}, \qquad \text{where } a, b, \text{ and } c \text{ are positive constants} \qquad (3)$$

9. The above demand laws (1)–(3) are given in the form $x = f(P)$. For each of these demand laws express P as a function of x, i.e., in the form $P = g(x)$.

10. Illustrate (1) graphically with $a = 10$ and $b = 1$. Label the vertical axis P and the horizontal axis x.

11. By definition, the total revenue R received from x units of a product sold at price P is given by the formula $R = Px$. Using the results of Problem 9, determine the total revenue as a function of x for each of the demand laws (1)–(3).

12. Determine MR, the marginal function, for demand laws (1) and (2), where $MR = dR/dx$.

Consider the following types of cost functions:

$C = ax + b$	*linear*	(1)
$C = ax^2 + bx + c$	*quadratic*	(2)
$C = ax^3 - bx^2 + cx + d$	*cubic*	(3)
$C = ae^{bx}$	*exponential*	(4)

where a, b, c, and d are positive constants.

13. Let $a = 1$, $b = 3$, $c = 3$, and $d = 1$ in (3) of Problem 12.

(a) Graph the resulting cost function, labelling the vertical axis C.

(b) Find the marginal cost function, $MC = dC/dx$.

(c) Graph the marginal cost function of **(b)**, labelling the vertical axis MC.

14. Find $MC = dC/dx$ for the cost functions (1), (2), and (3). If price is denoted by P and output is denoted by x, then elasticity of demand, e_d, is given by the formula

$$e_d = \frac{dx}{dP} \cdot \frac{P}{x}$$

15. Obtain a formula for elasticity of demand, e_d, for Demand Law 1, $x = (a-P)/b$.

Marginal cost, as it is used in a non-calculus treatment of economics, is the change in cost, ΔC, that results from producing one more unit of output. It is the per unit change in cost. Similarly, *marginal revenue* is the per unit change in revenue.

16. Let $C = f(x)$ and $R = g(x)$ be cost and revenue functions, respectively, where x is the level of output.

(a) Express the marginal cost of the $x+1$th unit of output using functional notation.

(b) Let $A = (x, f(x))$ and $B = (x+1, f(x+1))$. Show that the marginal cost of the $x+1$th unit is equal to $R_f(A, B)$.

17. **(a)** Complete Table 12-7 adapted from hypothetical data by Campbell R. McConnell.*

TABLE 12-7

X	C	MC	MR	R	P
0	100				
1	190	90	131	131	−59
2	270	80	131	262	−8
3	340				
4	400				
5	470				
6	550				
7	640				
8	750				
9	880				
10	1030	150			

where
X = output
C = total cost
MC = marginal cost of xth unit
MR = marginal revenue of xth unit
MR = price = \$131
R = total revenue
P = profit (or loss, if negative)

* Campbell R. McConnell, *Economics.* New York: McGraw-Hill Book Co., Inc., 1966, p. 460.

(b) At what level would you produce to maximize profit? How are marginal cost and marginal revenue related at this level of output?

(c) What is the form of the cost function? (i.e., linear, cubic, etc.)

(d) What is the form of the revenue function?

18. **(a)** Use the hypothetical cost function $C = x^3 - 3x^2 + 3x + 1$ and price = \$76 to construct a table like the one in Problem 17.

(b) At what level would you produce to maximize profit? How are marginal cost and marginal revenue related at this level of output? Do you think that profit is always maximized when marginal cost equals marginal revenue?

THIRTEEN

Additional differentiation techniques and applications

INTRODUCTION

In Chapter 12 we introduced the mathematical concept of the derivative. You will recall that if $y = f(x)$, then

DEFINITION OF THE DERIVATIVE

$$\frac{dy}{dx} = \lim_{\Delta x \to 0} \frac{\Delta y}{\Delta x}$$

The derivative of $y = f(x)$ at $x = a$ gives the slope of the tangent line at $(a, f(a))$. This is the geometric meaning of the derivative. The derivative of a function can also be interpreted as the "point" or "instantaneous" rate of change of the function at a specified point.

Several convenient derivative formulas were developed in the previous chapter.

Chapter 13 includes additional derivative formulas. Furthermore, the practically indispensable *chain rule* is explained and illustrated. The chain rule combines with other derivative formulas to extend their scope.

Derivatives are used to find slopes of tangents to curves, later in this chapter. This provides additional preparation for the next chapter on max-min theory and applications. Derivatives are also used in this chapter in a *marginal analysis* setting.

The chapter ends with a brief treatment of *differentials.* Differentials can be used to approximate exact changes in a dependent variable.

13-1 CONTINUITY AND DIFFERENTIABILITY OF FUNCTIONS OF ONE VARIABLE

A function $y = f(x)$ is said to be *continuous* at a point $x = c$ if

$$\lim_{x \to c} f(x) = f(c)$$

and it is said to be *continuous on an interval* $(a, b) = \{x \mid a < x < b\}$ if it is continuous at each point of (a, b).

The notion of *continuity* can be paraphrased as follows: A function $y = f(x)$ is said to be continuous on an interval (a, b) if and only if its graph can be drawn completely without lifting the pencil from the paper.

The concept of continuity is illustrated graphically in Figures 13-1 and 13-2.

A function $y = f(x)$ is said to be *differentiable* on an interval (a, b) if the derivative $f'(c)$ exists at each point c (a, b). For example $f(x) = x^2$ is differentiable for all x, since $f'(x) = 2x$ is valid for any x.

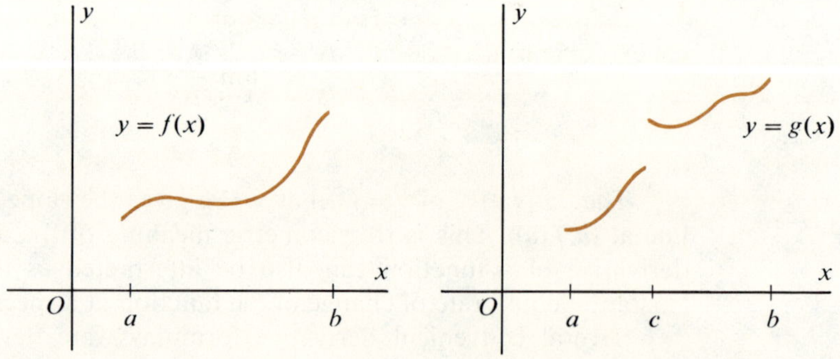

(a) This function is continuous on the interval $a < x < b$.

(b) This function fails to be continuous at $x = c$.

FIGURE 13-1

It is possible for a function to be continuous at a point, but fail to be differentiable at that point. A simple example of this point is furnished by the function $y = |x|$, where $|x|$ is called the absolute value of x. A graph of $y = |x|$ is shown below.

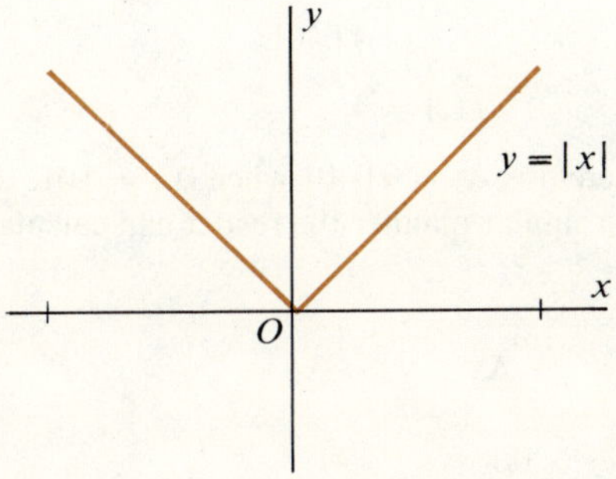

FIGURE 13-2

This function is continuous at $x = 0$, but fails to have a derivative at $x = 0$.

The next theorem can be interpreted to mean that *differentiability at $x = c$ implies continuity at $x = c$.* It can be interpreted more directly to mean that if $y = f(x)$ is differentiable at $x = c$, then Δy approaches zero as Δx approaches zero. This fact is used to prove some important differentiation formulas later on in this chapter.

THEOREM 13.1. *Let $y = f(x)$ be a function that is differentiable at $x = c$. Then $\lim_{\Delta x \to 0} [f(c + \Delta x) - f(c)] = 0$, i.e., $\lim_{\Delta x \to 0} \Delta y = 0$.*

Before giving a proof of this important theorem, we give a numerical example to illustrate it.

Example 1. Let $y = f(x) = x^2$. Determine the value of Δy at $x = 2$ for $\Delta x = 0.1$, $\Delta x = 0.01$, and $\Delta x = 0.001$.

Solution:

$$y = x^2$$

$$\frac{dy}{dx} = 2x$$

At $x = 2$,

$$\frac{dy}{dx} = 2(2)$$

$$= 4$$

Hence, by direct calculation we see that $y = x^2$ is differentiable at $x = 2$. Sample calculation for $\Delta x = 0.01$:

$$\begin{aligned} f(2+0.01) &= f(2.01) \\ &= (2.01)^2 \\ &= 4.0401 \\ f(2) &= 4 \end{aligned}$$

Therefore, $\Delta y = 0.0401$ when $\Delta x = 0.01$.
In a similar manner the reader can calculate Δy for $\Delta x = 0.1$ and 0.001.

x	Δx	Δy
2	0.1	0.41
2	0.01	0.0401
2	0.001	0.004001

Note that Δy tends toward zero as Δx takes on values 0.1, 0.01, and 0.001, respectively.

PROOF OF THEOREM 13.1. Let $y = f(x)$ be $x = c$.

Consider the identity $\Delta y = (\Delta x/\Delta x)\,\Delta y$, $\Delta x \neq 0$. Applying Theorem 12.1(b), we have $\lim_{\Delta x \to 0} \Delta y = \lim_{\Delta x \to 0} \Delta x \lim_{\Delta x \to 0}(\Delta y/\Delta x)$ at $x = c$. Since $f(x)$ is differentiable at $x = c$, $\lim_{\Delta x \to 0}(\Delta y/\Delta x)$ exists and equals the real number $f'(c)$. Therefore,

$$\begin{aligned} \lim_{\Delta x \to 0} \Delta y &= \lim_{\Delta x \to 0} \Delta x\, f'(c) \\ &= 0 f'(c) \\ &= 0 \end{aligned}$$

Continuity of $y = f(x)$ at $x = c$ can also be described as follows. A function $y = f(x)$ is continuous at $x = c$ if "small" deviations in x from $y = f(c)$. Therefore, Theorem 13.1 can be interpreted to mean that differentiability at $x = c$ implies continuity at $x = c$. This is equivalent to the statement that $y = f(x)$ is differentiable at $x = c$ *only if* $y = f(x)$ is continuous at $x = c$. In other words, continuity is a prerequisite, so to speak, for differentiability.

Example 2. Test $y = 1/(x - 2)$ for continuity.

Solution: Examination of the graph of $y = 1/(x - 2)$ shown in Figure 13-3 reveals that it fails to be continuous at, and only at, $x = 2$.

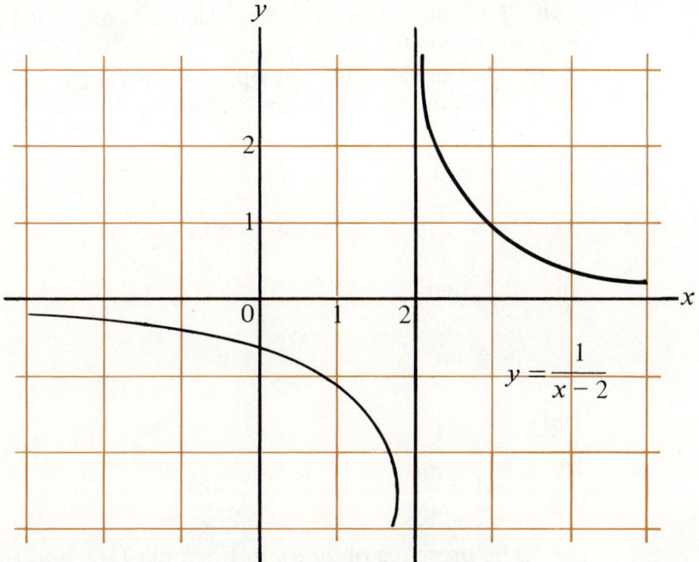

FIGURE 13-3

PROBLEMS

1. Let $y = x^3$. Determine the value of Δy at $x = 1$ for the following values of Δx:
   (a) $\Delta x = 0.1$ (c) $\Delta x = 0.001$
   (b) $\Delta x = 0.01$ (d) $\Delta x = 0.0001$
2. Repeat Problem 1 if $x = 1$ is changed to $x = 2$.
3. Are the results of Problems 1 and 2 to be expected? Why?
4. Let $f(x) = x^2$. Find $y = f(g(x))$ for the following functions:
   (a) $g(x) = x$ (d) $g(x) = 2 - x^3$
   (b) $g(x) = x^2$ (e) $g(x) = e^x$
   (c) $g(x) = x^2 + 4$ (f) $g(x) = 2^x$
5. Let $f(x) = e^x$. Find $y = f(g(x))$ for the following functions:
   (a) $g(x) = x$ (c) $g(x) = x^2 + 4$
   (b) $g(x) = x^2$ (d) $g(x) = -x^2$
6. Let $f(x) = \ln x$, $g(x) = x^2$, and $y = f(g(x))$. Compute Δy at $x = 1$ for the following values of Δx:
   (a) $\Delta x = 1$ (c) $\Delta x = 0.1$
   (b) $\Delta x = 0.2$ (d) $\Delta x = 0.01$
7. Test $y = [2/(1 - x)] + 3$ for continuity.
8. Test $y = 1 - |x|$ for continuity and for differentiability.
9. Take the derivative of $y = x^3 - 2x^2 + 3x + 7$ and thus show differentiability and hence continuity for all real x.

10. Find all values of x for which $y = 2e^x$ is differentiable and hence continuous.

11. Find all values of x for which $y = \ln x$ is continuous. *Hint:* $f(x)$ must be defined at $x = c$ to be continuous at $x = c$.

13-2 THE CHAIN RULE FOR DERIVATIVES

If $y = u^n$, then

$$\frac{dy}{du} = nu^{n-1} \qquad \text{D-2}$$

but

$$\frac{dy}{dx} = nu^{n-1}\frac{du}{dx} \qquad \text{D-2}'$$

The presence of du/dx in Formula D-2′ and the other (′) formulas is explained in this section. In many cases where a function is defined by a formula, some simpler combination of xs occur within the main formula. For example, suppose $y = 2^{(x^2)}$. In this case the main formula is $y = 2^u$, where $u = x^2$. If we agree to write 2^u as $f(u)$, then it follows that $y = f(u)$ or, better still, $y = f(u(x))$. Functions of this type are called *composite functions.*

Example 1. Given that $f(u) = \ln u$, $u(x) = 2^x$, and $y = f(u(x))$, write out y directly in terms of x.

Solution: $y = \ln u$, $u = 2^x$. Therefore, $y = \ln(2^x)$. *chain rule*

An important theorem of calculus called the *chain rule* states that if $y = f(u(x))$, where f and u are differentiable functions, then the derivative of y with respect to x equals the derivative of y with respect to u times the derivative of u with respect to x.

THEOREM 13.2. *If $y = f(u(x))$, where f and u are differentiable functions, then*

$$\frac{dy}{dx} = \frac{dy}{du}\frac{du}{dx} \qquad \text{D-7}$$

Before giving the proof of this theorem, we examine its geometric interpretation.

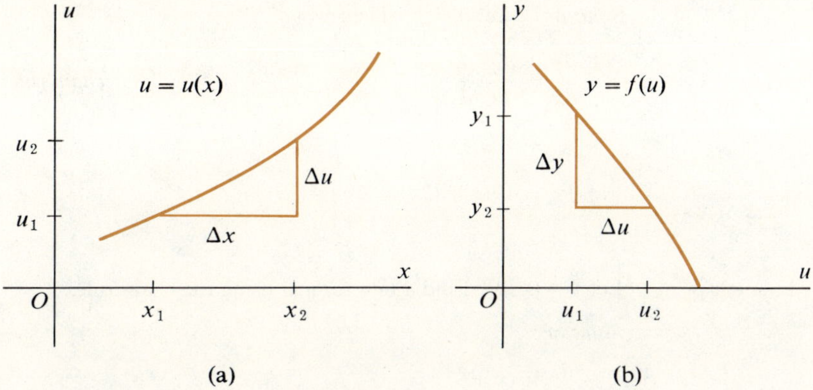

FIGURE 13-4

Since $y = f(u(x))$, a change, Δx, induces a change, Δu, in the function $u = u(x)$. This change in u, in turn, induces a change, Δy, in the function $y = f(u)$. Figure 13-4 should help the reader visualize these statements.

Proof of the Chain Rule $\dfrac{dy}{dx} = \dfrac{dy}{du}\dfrac{du}{dx}$

Consider the identity

$$\frac{\Delta y}{\Delta x} = \frac{\Delta y}{\Delta x}\frac{\Delta u}{\Delta u}$$

$$= \frac{\Delta y}{\Delta u}\frac{\Delta u}{\Delta x} \qquad \textit{algebra of fractions}$$

Taking limits, we have

$$\lim_{\Delta x \to 0} \frac{\Delta y}{\Delta x} = \lim_{\Delta x \to 0} \frac{\Delta y}{\Delta u}\frac{\Delta u}{\Delta x}$$

$$\lim_{\Delta x \to 0} \frac{\Delta y}{\Delta x} = \lim_{\Delta x \to 0} \frac{\Delta y}{\Delta u} \lim_{\Delta x \to 0} \frac{\Delta u}{\Delta x} \qquad \textit{by limit Theorem 12.1(B)}$$

$$\frac{dy}{dx} = \lim_{\Delta u \to 0} \frac{\Delta y}{\Delta u} \lim_{\Delta x \to 0} \frac{\Delta u}{\Delta x}$$ *since Δu approaches zero as Δx approaches zero by Theorem 13.1*

Therefore,

$$\frac{dy}{dx} = \frac{dy}{du}\frac{du}{dx} \qquad \textit{by definition of the derivative}$$

•

Example 2. Use the chain rule to find dy/dx if $y = (x^2)^{10}$.

Solution: Let $u = x^2$. Then $y = u^{10}$.

$$\frac{dy}{dx} = \frac{dy}{du} \cdot \frac{du}{dx} \qquad \textit{chain rule}$$

$$= 10u^9(2x)$$

$$= 10(x^2)^9(2x) = 10x^{18}(2x) = 20x^{19}$$

Example 3. Let $y = (x^2)^{10}$. Find dy/dx *without* using the chain rule.

Solution:

$$y = (x^2)^{10}$$

$$y = x^{20}$$

$$\frac{dy}{dx} = 20x^{19} \qquad \textit{by D-2}$$

The reader should compare this result with the result in Example 2.

Example 4. Use the chain rule to find dy/dx if $y = e^{bx}$; b is a constant.

Solution: Let $u = bx$. Then $y = e^u$.

$$\frac{dy}{dx} = \frac{dy}{du} \cdot \frac{du}{dx}$$

$$= e^u \frac{d(bx)}{dx}$$

$$= e^{bx}(b)$$

$$\frac{dy}{dx} = be^{bx}$$

Example 5. Let $y = e^{bx}$. Find dy/dx *without* using the chain rule.

Solution:

$$y = e^{bx}$$

$$y = (e^b)^x \qquad \textit{by E-4}$$

$$y = a^x \qquad \textit{where } a = e^b$$

$$\frac{dy}{dx} = a^x \ln a \qquad \textit{by D-5}$$

$$\frac{dy}{dx} = (e^b)^x \ln e^b$$

$$\frac{dy}{dx} = be^{bx} \qquad \textit{since } \ln e^b = b, \textit{ by definition}$$

$$\frac{dy}{dx} = be^{bx} \qquad \textit{as found in Example 4}$$

Several examples of functions that cannot be differentiated without using the chain rule are included in Section 13-4.

PROBLEMS

1. Use the chain rule to find dy/dx for the following functions:
   (a) $y = (x^3)^4$ (c) $y = (\sqrt{x})^6$
   (b) $y = (x^3)^{10}$ (d) $y = (\sqrt[3]{x})^9$
2. Find dy/dx for the functions given in Problem 1 above *without* using the chain rule.
3. Use the chain rule to find dy/dx for the following functions:
   (a) $y = e^{2x}$ (c) $y = 2^{4x}$
   (b) $y = e^{5x}$ (d) $y = 3^{5x}$
4. Find dy/dx for the functions given in Problem 3 above *without* using the chain rule.
5. Use the chain rule to find dy/dx for the following functions:
   (a) $y = e^{x^2}$ (c) 3^{x^3}
   (b) $y = e^{-x^2}$ (d) $2^{\sqrt{x}}$
6. Use the chain rule to find dy/dx for the following functions:
   (a) $y = \ln(x^2)$ (c) $y = \log(x^3)$
   (b) $y = \ln(\sqrt{x})$ (d) $y = \log(x^{3/2})$
7. Find dy/dx for the following functions:
   (a) $y = e^{x^2+3x}$ (c) $y = (x^3+2)^3$
   (b) $y = \ln(x^2+4x+7)$ (d) $y = e^{3x} + \sqrt{x^2+4}$
8. Find dy/dx for the following functions:
   (a) $y = 2^{x^2} - 3x$ (c) $y = (x^4-x)^2$
   (b) $y = \log(x^2-3x+2)$ (d) $y = \ln(5x) + \sqrt[3]{x^3+8}$
9. Let $f(x) = x^2$ and $u = e^x$.
   (a) Draw the graph of $u = e^x$ from $x = 0$ to $x = 2$.
   (b) Draw the graph of $y = f(u) = u^2$ from $u = 0$ to $u = 10$.
   (c) Show Δx, Δu, and Δy on your graphs. [*Hint:* See Figure 13-3.]

13-3 DERIVATIVES OF PRODUCTS AND QUOTIENTS

The derivative of a *sum* is the *sum* of the derivatives. Furthermore, the derivative of a *difference* is the *difference* of the derivatives. The derivative of any constant (by itself) equals zero, but the derivative of a constant *times* a function is the constant *times* the derivative of the function.

In general, the derivative of a product is *not* the product of the derivatives, and the derivative of a quotient is *not* the quotient of the derivatives. The formulas for the derivative of a product and for the derivative of a quotient are given next.

PRODUCT FORMULA FOR DERIVATIVES

$$\frac{d}{dx}(uv) = u\frac{dv}{dx} + v\frac{du}{dx} \qquad \text{D-10}$$

QUOTIENT FORMULA FOR DERIVATIVES

$$\frac{d}{dx}\left(\frac{u}{v}\right) = \frac{v\dfrac{du}{dx} - u\dfrac{dv}{dx}}{v^2} \qquad \text{D-11}$$

Example 1. Find dy/dx if $y = x \ln x$.

Solution: Let $u = x$ and $v = \ln x$. Apply D-10.

$$\frac{dy}{dx} = x\left(\frac{1}{x}\right) + \ln x(1)$$

$$\frac{dy}{dx} = 1 + \ln x$$

Example 2. Find dy/dx if $y = x^2/(1-2x)$.

Solution: Let $u = x^2$ and $v = 1-2x$. Use D-11.

$$\frac{dy}{dx} = \frac{(1-2x)(2x) - x^2(-2)}{(1-2x)^2}$$

$$\frac{dy}{dx} = \frac{2x - 2x^2}{(1-2x)^2}$$

The next example illustrates the fact that a finite sum of functions can be differentiated term by term.

Example 3. Find dy/dx if $y = x^4 + x^3 + x^2$.

Solution: Let $u = x^4$, $v = x^3$, and $w = x^2$. Then

$$y = u + v + w \qquad \text{and} \qquad \frac{dy}{dx} = \frac{du}{dx} + \frac{dv}{dx} + \frac{dw}{dx}$$

It follows that

$$\frac{dy}{dx} = \frac{d}{dx}(x^4) + \frac{d}{dx}(x^3) + \frac{d}{dx}(x^2)$$

$$\frac{dy}{dx} = 4x^3 + 3x^2 + 2x$$

Operationally, and without the explanation, we have:

$$y = x^4 + x^3 + x^2$$

$$\frac{dy}{dx} = 4x^3 + 3x^2 + 2x$$

The proof of the formula for the derivative of a product is as follows:

Proof: Let $y = u(x)v(x)$ or, more briefly, $y = uv$.

$$\Delta y = (u + \Delta u)(v + \Delta v) - uv$$

$$= uv + u\,\Delta v + v\,\Delta u + \Delta u\,\Delta v - uv$$

$$\Delta y = u\,\Delta v + v\,\Delta u + \Delta u\,\Delta v$$

$$\frac{\Delta y}{\Delta x} = u\frac{\Delta v}{\Delta x} + v\frac{\Delta u}{\Delta x} + \Delta u\frac{\Delta v}{\Delta x}$$

$$\lim_{\Delta x \to 0}\frac{\Delta y}{\Delta x} = \lim_{\Delta x \to 0}\left(u\frac{\Delta v}{\Delta x}\right) + \lim_{\Delta x \to 0}\left(v\frac{\Delta u}{\Delta x}\right) + \lim_{\Delta x \to 0}\left(\Delta u\frac{\Delta v}{\Delta x}\right)$$

recalling that the limit of a sum is the sum of the limits.

$$\frac{dy}{dx} = u\lim_{\Delta x \to 0}\frac{\Delta v}{\Delta x} + v\lim_{\Delta x \to 0}\frac{\Delta u}{\Delta x} + \lim_{\Delta x \to 0}\Delta u\lim_{\Delta x \to 0}\frac{\Delta v}{\Delta x}$$

$$\frac{dy}{dx} = u\frac{dv}{dx} + v\frac{du}{dx} + \left(\lim_{\Delta x \to 0}\Delta u\right)\frac{dv}{dx}$$

Since $u(x)$ is differentiable, it follows that $\lim_{\Delta x \to 0}\Delta u = 0$.

Hence, the term $(\lim_{\Delta x \to 0}\Delta u)\,dv/dx$ equals zero, leaving

$$\frac{dy}{dx} = u\frac{dv}{dx} + v\frac{du}{dx}$$

Example 4. Find dy/dx if $y = (x^2 + 3)(x^3)$.

Solution: Let $u = x^2 + 3$, $v = x^3$. Then $y = uv$ and $\dfrac{dy}{dx} = u\dfrac{dv}{dx} + v\dfrac{du}{dx}$.

$$\frac{dy}{dx} = (x^2+3)\frac{d}{dx}(x^3) + x^3\frac{d}{dx}(x^2+3)$$

$$\frac{dy}{dx} = (x^2+3)\,3x^2 + x^3(2x) \qquad \text{by } D\text{-}2 \text{ and } D\text{-}7$$

$$\frac{dy}{dx} = 3x^4 + 9x^2 + 2x^4$$

$$\frac{dy}{dx} = 5x^4 + 9x^2$$

or

$$y = x^5 + 3x^3$$

$$\therefore \frac{dy}{dx} = 5x^4 + 9x^2$$

Example 5. Find dy/dx if $y = xe^x$.

Solution: $y = xe^x$. Let $u = x$, $v = e^x$. Then $y = uv$.

$$\frac{dy}{dx} = u\frac{dv}{dx} + v\frac{du}{dx}$$

$$= x\frac{de^x}{dx} + e^x\frac{d(x)}{dx}$$

$$\frac{dy}{dx} = xe^x + e^x(1) \qquad \textit{by D-6 and D-2}$$

The proof of the formula for the derivative of a quotient is left as an exercise. The following two examples illustrate its use.

Example 6. Find dy/dx if $y = x^2/(x^3+x)$.

Solution:

$$y = \frac{x^2}{x^3+x}$$

$$y = \frac{x}{x^2+1}$$

Let $u = x$ and $v = x^2+1$

$$\frac{dy}{dx} = \frac{(x^2+1)(1) - x(2x)}{(x^2+1)^2}$$

$$\frac{dy}{dx} = \frac{1-x^2}{(x^2+1)^2}$$

Example 7. Find dy/dx if $y = 2^x/x^2$.

Solution: Let $u = 2^x$, $v = x^2$.

$$\frac{dy}{dx} = \frac{x^2 \frac{d}{dx}(2^x) - 2^x \frac{d}{dx}(x^2)}{(x^2)^2} \qquad \text{by D-11}$$

$$= \frac{x^2 2^x \ln 2 - 2^x 2x}{x^4} \qquad \text{by D-2 and D-5}$$

$$\frac{dy}{dx} = \frac{2^x x(x \ln 2 - 2)}{x^4}$$

$$\frac{dy}{dx} = \frac{2^x(x \ln 2 - 2)}{x^3}$$

PROBLEMS

1. Prove formula D-8. See (3) on page 458.
2. Find dy/dx for the following functions:
   (a) $y = 6x^3$ (d) $y = 6(2)^x$
   (b) $y = 3 \log x$ (e) $y = 5e^x$
   (c) $y = 5 \ln x$ (f) $y = 10(3)^x$
3. Find dy/dx for the following functions; a, c, and h are constants.
   (a) $y = 2x^h$ (c) $y = he^x$
   (b) $y = cx^h$ (d) $y = c(a)^x$
4. Differentiate the following functions with respect to x.
   (a) $f(x) = 3x^2 + 6$ (d) $f(x) = 2x + 5^x$
   (b) $f(x) = 5x^3 + 6x^2 + 6$ (e) $f(x) = x^3 + \ln x$
   (c) $f(x) = 4x + e^x$ (f) $f(x) = x^4 + 3x^2 + 5 \log x$
5. Use the product formula to find dy/dx for the following functions:
   (a) $y = (x^2)(6x)$ (d) $y = x \ln x$
   (b) $y = x^3 e^x$ (e) $y = x^2 \log x$
   (c) $y = 5(6x^2)$ (f) $y = x^5(5)^x$
6. Prove formula D-11.
7. Use the quotient formula to differentiate the following functions with respect to x.
   (a) $f(x) = \dfrac{3x^2 + 6x}{4}$ (d) $f(x) = \dfrac{\ln x}{3x}$
   (b) $f(x) = \dfrac{x^3}{2x^2}$ (e) $f(x) = \dfrac{\log x}{x^2}$
   (c) $f(x) = \dfrac{e^x + 4}{4}$ (f) $f(x) = \dfrac{5^x}{6x^3}$
8. Do Problem 7 without using the quotient formula.

9. Complete Table 13-1.

TABLE 13-1

$f(x)$ (function)	$f'(x)$ (derivative)	$f'(1)$ (derivative evaluated at $x = 1$)
$4\sqrt{x}$		
$2e^{3x}$		
$e^x + e^{-x}$		
$e^x + e^{-2x}$		
$(\ln x)/e^x$		

13-4 SUMMARY OF DIFFERENTIATION TECHNIQUES

This section includes a complete list of the differentiation formulas included in this text. Formulas D-1 through D-6 enable us to differentiate some of the most common functions that occur in the application of differential calculus to business and economics.

Formulas D-8 through D-11 enable us to differentiate sums, products, and quotients of the specific functions covered by D-1 through D-6. The reader will encounter repeated use of D-7, the chain rule.

We now give several illustrative examples, each of which requires two or more of the following formulas for its solution.

D-1 $$\frac{d}{du}(c) = 0$$ *Derivative of a constant*

D-2 $$\frac{d}{du}(u^n) = nu^{n-1}$$ *Power rule*

D-3 $$\frac{d}{du}(\log_a u) = \frac{1}{u}\log_a e$$

D-4 $$\frac{d}{du}(\ln u) = \frac{1}{u}$$

D-5 $$\frac{d}{du}(a^u) = a^u \ln a$$

D-6 $$\frac{d}{du}(e^u) = e^u$$

D-7 $$\frac{dy}{dx} = \frac{dy}{du}\frac{du}{dx}$$ *Chain rule*

D-8	$\frac{d}{dx}(cu) = c\frac{du}{dx}$	*Derivative of a constant times a function*
D-9	$\frac{d}{dx}(u+v) = \frac{dy}{dx} + \frac{dv}{dx}$	*Derivative of a sum*
D-10	$\frac{d}{dx}(uv) = u\frac{dv}{dx} + v\frac{du}{dx}$	*Derivative of a product*
D-11	$\frac{d}{dx}\left(\frac{u}{v}\right) = \frac{v\frac{du}{dx} - u\frac{dv}{dx}}{v^2}$	*Derivative of a quotient*

Example 1. Find dy/dx if $y = \ln x^2$.

Solution: Let $u = x^2$. Then $y = \ln u$.

$$\frac{dy}{dx} = \frac{1}{u}\frac{du}{dx} \qquad \textit{D-4 and chain rule}$$

$$\frac{dy}{dx} = \frac{1}{x^2}\frac{d}{dx}(x^2) \qquad \textit{substitution}$$

Or, using a property of logs, we have

$$y = \ln x^2$$

$$y = 2\ln x$$

$$\frac{dy}{dx} = \frac{2}{x}$$

$$\frac{dy}{dx} = \frac{2x}{x^2} = \frac{2}{x} \qquad \textit{D-2}$$

Example 2. Find dy/dx if $y = \sqrt{4-x^2}$.

Solution: $y = (4-x^2)^{1/2}$. Let $u = 4-x^2$. Then $y = u^{1/2}$.

$$\frac{dy}{dx} = \frac{1}{2}u^{-1/2}\frac{du}{dx} \qquad \textit{D-2 and chain rule}$$

$$\frac{dy}{dx} = \frac{1}{2\sqrt{4-x^2}}\frac{d}{dx}(4-x^2) \qquad \textit{substitution, law of exponents}$$

$$\frac{dy}{dx} = \frac{-x}{\sqrt{4-x^2}} \qquad \textit{D-1, D-2, D-9}$$

Example 3. Find dy/dx if $y = e^{2x} + x^2$.

Solution:

$$y = e^{2x} + x^2$$

$$\frac{dy}{dx} = \frac{d(e^{2x})}{dx} + \frac{d(x^2)}{dx} \qquad \textit{by D-9}$$

$$\frac{dy}{dx} = 2e^{2x} + 2x$$

$$\frac{dy}{dx} = 2(e^{2x} + x)$$

Example 4. Find dy/dx if $y = 2^{\sqrt{3x+7}}$.

Solution: Let $u = \sqrt{3x+7}$, then $y = 2^u$.

$$\frac{dy}{dx} = 2^u \ln 2 \frac{du}{dx} \qquad \textit{D-5 and chain rule}$$

Let $z = 3x + 7$. Then $u = z^{1/2}$ and

$$\frac{du}{dx} = \frac{1}{2} z^{-1/2} \frac{dz}{dx}$$

$$\frac{du}{dx} = \frac{1}{2}(3x+7)^{-1/2}(3) \qquad \textit{D-2 and chain rule}$$

$$\frac{du}{dx} = \frac{3}{2\sqrt{3x+7}}$$

$$\frac{dy}{dx} = 2^{\sqrt{3x+7}} \ln 2 \frac{3}{2\sqrt{3x+7}} \qquad \textit{substitution}$$

Example 5. Find dy/dx if $y = 2(x^2 + 4)$.

Solution:

$$y = 2(x^2 + 4)$$

$$\frac{dy}{dx} = 2 \frac{d}{dx}(x^2 + 4) \qquad \textit{D-8}$$

$$\frac{dy}{dx} = 2(2x + 0) \qquad \textit{D-1, D-2, D-9}$$

$$\frac{dy}{dx} = 4x$$

Example 6. Find dy/dx if $y = x^3e^{2x}$.

Solution: Let $u = x^3$ and $v = e^{2x}$. Then $y = uv$.

$$\frac{dy}{dx} = u\frac{dv}{dx} + v\frac{du}{dx} \qquad \textit{D-10}$$

$$\frac{dy}{dx} = x^3\frac{d}{dx}(e^{2x}) + e^{2x}\frac{d}{dx}(x^3) \qquad \textit{substitution}$$

$$\frac{dy}{dx} = x^3e^{2x}2 + e^{2x}3x^2 \qquad \textit{D-2, D-6, and chain rule}$$

$$\frac{dy}{dx} = x^2e^{2x}(2x+3)$$

Example 7. Find dy/dx if $y = 3x^2/\log_4 x^3$.

Solution: Let $u = 3x^2$ and $v = \log_4 x^3$.

$$\frac{dy}{dx} = \frac{\log_4 x^3 \dfrac{d}{dx}(3x^2) - 3x^2\dfrac{d}{dx}(\log_4 x^3)}{(\log_4 x^3)^2} \qquad \textit{D-11}$$

Let $z = x^3$. Then

$$\frac{dv}{dx} = \frac{d}{dx}\log_4 z$$

$$= \frac{1}{z}(\log_4 e)\frac{dz}{dx}$$

$$\frac{dv}{dx} = \frac{1}{x^3}\log_4 e \cdot 3x^2$$

$$\frac{d}{dx}(\log_4 x^3) = \frac{3}{x}\log_4 e$$

$$\frac{dy}{dx} = \frac{\log_4 x^3 6x - 3x^2(3/x)\log_4 e}{(\log_4 x^3)^2} \qquad \textit{substitution and D-2}$$

$$\frac{dy}{dx} = \frac{3x(2\log_4 x^3 - 3\log_4 e)}{(\log_4 x^3)^2}$$

Formulas D-2′ through D-6′ have the chain rule "built in," so to speak.

D-2′ $$\frac{d}{dx}(u^n) = nu^{n-1}\frac{du}{dx}$$

D-3′ $$\frac{d}{dx}(\log_a u) = \frac{1}{u}\log_a e\frac{du}{dx}$$

D-4′ $$\frac{d}{dx}(\ln u) = \frac{1}{u}\frac{du}{dx}$$

D-5′ $$\frac{d}{dx}(a^u) = a^u \ln a \frac{du}{dx}$$

D-6′ $$\frac{d}{dx}(e^u) = e^u \frac{du}{dx}$$

The formula D-3′ can also be written in the form

$$\frac{d}{dx}(\log_a u) = \frac{1}{u \ln a}\frac{du}{dx} \qquad \text{D-3}' \text{ (alternative)}$$

PROBLEMS

Find the derivatives of the following functions:

1. $y = x^2 + 6x$
2. $y = (x^2 + 4)^5$
3. $y = x^3 + 4$
4. $y = e^{3x}$
5. $y = e^{x^2} + 4x$
6. $y = \ln x^3$ (use chain rule)
7. $y = \ln x^3$ (do not use chain rule)
8. $y = \ln e^x$
9. $y = 6^{\sqrt{2x^2+5}}$
10. $y = \dfrac{x+3}{(x^2+4x+2)^2}$
11. $y = 6(x^2+3)^2$
12. $y = e^{3x^2}/5x^3$
13. $y = x/(1+2x)$
14. $y = (x+3)\sqrt{x^2+4}$ ← product rule
15. $y = \sqrt{(x+2)/(x-3)}$
16. $y = (ax+b)^n$
17. $y = ax^2 + bx + c$
18. $y = (e^{3x} - e^{-3x})/2$

Indicate the derivatives of the following functions in simplified form:

19. $y = [f(x)]^2$
20. $y = f(x)e^{g(x)}$
21. Complete Table 13-2.

TABLE 13-2

Function	$\left.\frac{dy}{dx}\right\|_{x=1}$	$\left.\frac{dy}{dx}\right\|_{x=2}$
$y = 3$		
$y = \frac{x^2}{2}$		
$y = x^{1/3}$		
$y = 2^x$		
$y = e^x$		
$y = xe^x$		
$y = \ln x$		
$y = \log x^2$		
$y = x^2 \ln 2x$		
$y = 2 + 3e^{-4x}$		

x = 2

$(2x^2 + 9)^7$

13-5 EQUATIONS OF TANGENT LINES AND MARGINAL ANALYSIS

The derivative, evaluated at a given point on a curve, gives the slope of the line that is tangent to the curve at the given point. Marginal cost and marginal revenue are also defined in terms of derivatives. The following examples illustrate these important points.

Example 1. Find an equation of the line that is tangent to the curve $y = x^2 + 3x - 3$ at the point $A = (1, 1)$.

Solution: The slope $m = R_f(A)$ is given by $f'(1)$.

$$y = x^2 + 3x - 3$$

$$\frac{dy}{dx} = 2x + 3$$

$$f'(1) = \left.\frac{dy}{dx}\right|_{x=1} = 2(1) + 3 = 5$$

$$m = 5$$

If we write the equation of the tangent line in the form $y = mx + b$, where m is the slope and b is the y intercept, we have

$$y = 5x + b$$

$$1 = 5(1) + b$$

$$b = -4$$

Therefore, $y = 5x - 4$ is an equation of the line in question.

Equation 2. Find an equation of the line tangent to the curve $y = \ln x^2$ at $x = 2$.

Solution: Start with $y = mx + b$ and determine the value of m. We find that m equals $(dy/dx)|_{x=2}$

$$y = \ln x^2$$

$$\frac{dy}{dx} = \frac{1}{x^2} 2x = \frac{2}{x}$$

$$\left.\frac{dy}{dx}\right|_{x=2} = \frac{2}{2} = 1$$

Therefore, $y = 1x + b$. To solve for b we must substitute values for x and y. If we remember that the tangent line at $x = 2$ touches the curve at the point of tangency, we can get the corresponding y value from the original function $y = \ln x^2$. This gives $y = \ln 2^2$ or $\ln 4$, which is 1.3863.

Solving for b gives

$$y = 1x + b$$

$$1.3863 = 1(2) + b$$

$$b = 1.3863 - 2$$

$$b = -0.6137$$

Therefore, an equation of the tangent line at $x = 2$ is $y = x - 0.6137$.

An important application of calculus to economics is the so-called *marginalist approach* to profit maximization. Using the *marginalist approach* one seeks to maximize the profit of a firm by producing at the level where the rate of change of revenue is equal to the rate of change of cost. These rates are called *marginal revenue* and *marginal cost*. Using calculus notation, we have

$$\text{Marginal revenue} = MR = \frac{dR}{dx}$$

$$\text{Marginal cost} = MC = \frac{dC}{dx}$$

where C = total cost

R = total revenue

x = quantity produced

The marginalist approach is discussed in more detail in the next chapter, and the remainder of this section deals with marginal cost and marginal revenue functions.

Example 3. Given the cost function $C = 2500 + 0.6x$, find the marginal cost of this function at $x = 50$ and at $x = 100$.

Solution:

$$C = 2500 + 0.6x$$

$$MC = \frac{dC}{dx} = 0.6$$

$$\left.\frac{dC}{dx}\right|_{x=50} = 0.6$$

$$\left.\frac{dC}{dx}\right|_{x=100} = 0.6$$

Note that the marginal cost remains constant for all values of x. This is the case because C is a linear function of x.

Example 4. Given the nonlinear cost function $C = 2500 + 0.6x + (1000/x)$, $45 < x < 1000$, find the marginal cost at $x = 50$ and at $x = 100$.

Solution:

$$C = 2500 + 0.6x + 1000x^{-1}$$

$$\frac{dC}{dx} = 0.6 - 1000x^{-2}$$

$$\frac{dC}{dx} = 0.6 - \frac{1000}{x^2}$$

$$\left.\frac{dC}{dx}\right|_{x=50} = 0.6 - \frac{1000}{2500} = 0.6 - 0.4$$

$$MC|_{x=50} = 0.2$$

$$\left.\frac{dC}{dx}\right|_{x=100} = 0.6 - \frac{1000}{10{,}000} = 0.6 - 0.1 \qquad \textit{by definition of marginal cost}$$

$$MC|_{x=100} = 0.5 \qquad \textit{by definition of marginal cost}$$

Note that the marginal cost varies with the level of production. This dependence of marginal cost on the level of production is typical of nonlinear cost functions.

Two important demand laws are $p = a - bx$ and $p = a/(x+b) - c$, where p denotes price per unit of x and a, b, and c are constants. The revenue functions, which correspond to these demand laws, are $R = ax - bx^2$ and $R = (ax)/(x+b) - cx$, respectively. This follows from the relationship $R = px$.

The following example illustrates the use of the quotient formula to compute the marginal revenue function for one of these functions.

Example 5. Find the marginal revenue function that corresponds to $R = ax/(x+b) - cx$.

Solution:

$$MR = \frac{(x+b)a - ax(1)}{(x+b)^2} - c$$

$$MR = \frac{a}{x+b} - \frac{ax}{(x+b)^2} - c$$

PROBLEMS

1. Find an equation of the line that is tangent to the curve $y = x^2 - 3x + 3$ at the point $A = (1, 1)$.
2. Find an equation of the line tangent to the curve $y = e^{x^2}$ at $x = 1$.
3. Given the cost function $C = 2500 + 0.4x + (1000/x)$, find the marginal cost at $x = 60$ and at $x = 80$.
4. Find the marginal revenue function that corresponds to the following total revenue functions:
   - **(a)** $R = 3x$
   - **(b)** $R = 5x$
   - **(c)** $R = 0.4\sqrt{x}$
   - **(d)** $R = 1.3 \ln(x+1)$
5. Given the total cost function $C = 10 + e^x + \ln x$, find the marginal cost function.
6. Given the cost function $C = x + 6$ and the revenue function $R = 6x - x^2$, find the break-even point or points.
7. Using the functions given in Problem 6, find the x value for which marginal revenue equals marginal cost.
8. Using the functions given in Problem 6, find the profit function, where profit = total revenue − total cost.
9. Using the profit function of Problem 8, compute and compare the profits at $x = 2.3$, $x = 2.5$, and $x = 2.6$. Interpret your results with respect to Problem 7 and the marginalist approach described in this section.
10. Given the total cost function $C = 2x^2 + c + 3$.
    - **(a)** Find the marginal cost function.
    - **(b)** Find the average cost function, i.e., C/x.
11. Find the marginal revenue function that corresponds to the demand law $p = a - bx$.

13-6 DEFINITION AND APPLICATION OF DIFFERENTIALS

In many problems involving functions of the type $y = f(x)$, we are interested in finding the change, Δy, in the dependent variable, y, which results from a small change, Δx, in the independent variable, x. The derivative provides an easy method of approximating this change. To see this we start with the identity

$$\Delta y = \frac{\Delta y}{\Delta x}\Delta x \qquad (1)$$

and recall that

$$f'(x) = \lim_{\Delta x \to 0} \frac{\Delta y}{\Delta x}$$

by definition of the derivative. This means that the ratio $\Delta y/\Delta x$ is approximately equal to $f'(x)$ for small values of Δx. Using this approximation in (1) gives

$$\Delta y \approx f'(x)\,\Delta x \tag{2}$$

where the symbol $\approx$ denotes the expression "is approximately equal to."

A strict equality can be obtained from (2) by introducing a slack variable q,

$$\Delta y = f'(x)\,\Delta x + q \tag{3}$$

DEFINITION 13.1. *Let $y = f(x)$ be a differentiable function. Then $dy = f'(x)\,\Delta x$.*

If we examine the above definition more closely, we see that $\Delta y = dy + q$, where q is a small quantity if Δx is small. Therefore, the differential, dy, is an approximation for the desired quantity Δy, i.e., Δy equals (approximately) $f'(x)\,\Delta x$ for small Δx. Furthermore, dy is usually easier to calculate than Δy.

Example 1. Given the linear cost function $C = 2500 + 0.72x$. Calculate both the exact change ΔC and the approximate change dC that result if the production level is changed from $x = 50$ to $x = 100$.

Solution: $C = 2500 + 0.72x$. Evaluating C at $x = 50$ and at $x = 100$, we have

$$C\,|_{x=50} = 2500 + 36 = 2536$$

$$C\,|_{x=100} = 2500 + 72 = 2572$$

Therefore, $\Delta C = 2572 - 2536$ and $\Delta C = \$36$.

By definition, $dC = (dC/dx)\,|_{x=50}$ times Δx, where $\Delta x = 100 - 50$. Therefore,

$$dC = (0.72)(50)$$

$$dC = \$36$$

We note that in this case $\Delta C = dC$. This is because linear cost functions have a constant rate of change.

Example 2. Given the nonlinear cost function $C = 0.1x^2 + 1$, where C is in hundreds of dollars and x is in hundreds of units, calculate both the exact change, ΔC, and the approximate change, dC, that result if the production level is changed from 100 units to 200 units.

Solution:

$$C|_{x=1} = 0.1(1) + 1$$
$$= 1.1$$
$$C|_{x=2} = 0.1(4) + 1$$
$$= 1.4$$
$$\Delta C = 1.4 - 1.1 = 0.3$$
$$\Delta C = \$30$$
$$dC = \left.\frac{dC}{dx}\right|_{x=1} \Delta x$$
$$= 0.2x|_{x=1}(2-1)$$
$$dC = 0.2, \qquad \text{i.e., } dC = \$20$$

The fact that dC is not a very accurate approximation of the exact change is because the cost function is nonlinear and Δx is quite large. A smaller Δx is used in the next example.

Example 3. Repeat Example 2, using a change in the production level of 10 units.

Solution: In this case, $\Delta x = 0.1$, since x is in hundreds of units.

$$\Delta C = 0.1(1.1)^2 + 1 - 1.1$$
$$= 1.121 - 1.1$$
$$\Delta C = 0.021 = \$2.10$$
$$dC = \left.\frac{dC}{dx}\right|_{x=1} \Delta x$$
$$= 0.2x|_{x=1}(0.1)$$
$$= (0.2)(0.1)$$
$$dC = 0.02, \qquad \text{i.e., } dC = \$2.00$$

To see that the approximation in Example 3 is indeed more accurate than the approximation in Example 2, we consider the percentage error involved. By *percentage error* we mean $(\Delta C - dC)/\Delta C$ times 100. The percentage error in Example 2 is 33 percent, whereas in Example 3 it is only 5 percent.

In Definition 9.1, we see that the *differential*, dy, of the dependent variable is defined to be the product of $f'(x)$ and Δx. The exact change in x, i.e., Δx, is called an increment. It is also customary to define the differential dx of the independent variable x to be equal to the increment Δx.

DEFINITION 13.2. *Let $y = f(x)$. Then $dx = \Delta x$.*

Combining Definition 9.1 and Definition 9.2, we can have

$$dy \div dx, \qquad dx = \Delta x \neq 0$$
$$= [f'(x)\,\Delta x] \div \Delta x$$
$$= f'(x)$$

In other words, the symbol dy/dx can now be interpreted either as a single symbol denoting $f'(x)$ or as the fraction (the differential of y divided by the differential of x). This dual interpretation is made possible by the above definitions.

Example 4. Transform the following derivatives into differential form.

(a) $\dfrac{dy}{dx} = 2e^{2x}$

(b) $\dfrac{dR}{dq} = 1 - 0.1q^2$

Solution:

(a) $dy = 2e^{2x}\,dx$

(b) $dR = (1 - 0.1q^2)\,dq$

PROBLEMS

1. Let $y = x^2 + 3x + 7$, $x_1 = 2$, $x_2 = 2.2$.
  (a) Compute Δy. **(b)** Compute dy.

2. Find the exact change in the function $y = x^2 - 4x$ as x changes from $x = 3$ to $x = 3.1$.

3. Use the differential, dy, to approximate the exact change found in Problem 2.

4. Given the nonlinear cost function $C = 0.2x^2 + 2$, where C is in hundreds of dollars and x is in hundreds of units, compute the following:
  (a) The exact change, ΔC, that results if the production level is changed from 100 units to 105 units.
  (b) The approximate change, dC, if the production level is changed from $x = 100$ to $x = 105$.

5. Define dw, Δw, dz, and Δz for the differentiable function $w = g(z)$.

6. Find dy for the following functions:
  (a) $y = x^2 + 3x + 4$ **(d)** $y = \ln(4x)$
  (b) $y = \sqrt{x}$ **(e)** $y = x^{-1/2}$
  (c) $y = e^{3x}$ **(f)** $y = e^{x^2}$

7. Let $y = x^2 + 2x$.
  (a) Draw the graph of $y = x^2 + 2x$ from $x = 0$ to $x = 3$.
  (b) Plot the tangent to $y = x^2 + 2x$ at $x = 1$ on your graph in (a).
  (c) Locate $x_1 = 1$, $x_2 = 3$, y_1, and y_2 on your graph.
  (d) Identify Δx, dx, Δy, and dy on your graph for (c).

FOURTEEN

Max-min theory and applications

One of the primary interpretations of the derivative is the rate of change concept. In this chapter we will see how this rate of change concept can be of aid in graphing functions as well as determining where a function has a maximum or minimum. This is the area where the calculus has many business applications. The most common application will be to find that level of activity which will maximize profits or minimize cost. In Chapter 1 we used the tools of algebra to find the maxima of quadratic functions. The calculus can be applied more generally and more efficiently.

14-1 THE FIRST DERIVATIVE TEST FOR MAXIMA AND MINIMA

Consider the function $f(x) = x^2 - 3x + 6$. A look at Table 14-1 suggests that this function has a minimum value somewhere between $x = 1$ and $x = 2$. Table 14-2 indicates that this minimum value of $f(x)$ is between $x = 1.4$ and $x = 1.6$. As we will see, the first derivative test provides a powerful tool to determine the exact location of such a point.

Now, for the given function

$$f(x) = x^2 - 3x + 6$$

$$f'(x) = 2x - 3,$$

where $f'(x)$ denotes the derivative of $f(x)$ with respect to x.

TABLE 14-1

x	$f(x)$
-1	10
0	6
1	4
2	4
3	6
4	10

TABLE 14-2

x	$f(x)$
1.0	4.0
1.4	3.76
1.5	3.75
1.6	3.76
2.0	4.0

TABLE 14-3

x	$f'(x)$
0	-3.0
1.0	-1.0
1.4	-0.2
1.5	0
1.6	0.2
2.0	1.0

Table 14-3 gives the value of $f'(x)$ at the indicated x values. We make two important observations about $f'(x)$:

1. It is zero at $x = 1.5$.
2. It changes from minus to plus at $x = 1.5$.

From the graphs of the original function $y = x^2 - 3x + 6$ and the derivative $y' = 2x - 3$ and from Table 14-3 we see that for $x < 1.5$, $f'(x)$ is negative. $f'(x)$ is positive for $x > 1.5$.

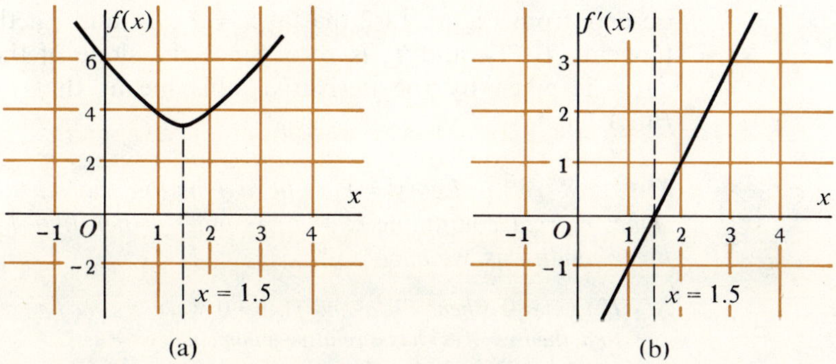

FIGURE 14-1

We are now ready to define the terms *relative minimum*, and *relative maximum*, and to state the so-called first derivative test for maxima and minima for a proper coordinate series.

DEFINITION 14.1. *Let $y = f(x)$ be a function that is defined in some interval containing $x = a$. Then $f(x)$ is said to have a relative minimum [maximum] at $x = a$ if $f(x) \geqslant f(a)$ $[f(x) \leqslant f(a)]$ for all values of x sufficiently close to a. The term local is sometimes used instead of relative.*

If $f(x) \geqslant f(a)$ $[f(x) \leqslant f(a)]$ for all x in the domain of f, then $f(x)$ is said to possess an *absolute* or *global* minimum [maximum] at $x = a$.

Example 1. Discuss the maxima and minima of the function $y = f(x)$, as shown in Figure 14-2.

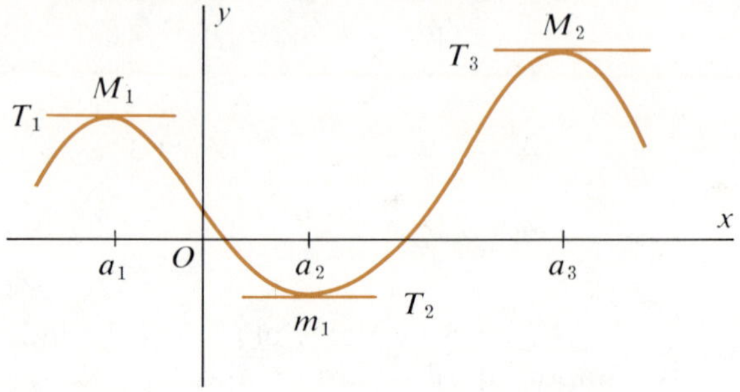

FIGURE 14-2

Solution: This function has relative maxima at $x = a_1$ and $x = a_3$. It has a relative minimum at $x = a_2$. It also has an absolute maximum at $x = a_3$ and an absolute minimum at $x = a_2$.

Before stating the first theorem about maxima and minima, we observe from Figure 14-2 that at $x = a_1$, a_2, and a_3, the slope of the tangents T_1, T_2, and T_3 is zero. Since the slope of the tangent to a curve is given by the derivative, this means that $f'(a_1) = f'(a_2) = f'(a_3) = 0$.

THEOREM 14.1. *Let $y = f(x)$ be a function that is differentiable on some interval containing $x = a$. Assume further that $f'(a) = 0$. Under these conditions, we have*

A. *If $f'(x) < 0$ when $x < a$ and $f(x) > 0$ when $x > a$, for x sufficiently close to a, then $y = f(x)$ has a relative minimum at $x = a$.*

B. *If $f'(x) > 0$ when $x < a$ and $f'(x) < 0$ when $x > a$, for x sufficiently close to a, then $y = f(x)$ has a relative maximum at $x = a$.*

Example 2. Use Theorem 14.1 to determine the exact location of the minimum value of $f(x) = x^2 - 3x + 6$.

Solution:

$$f(x) = x^2 - 3x + 6$$

$$f'(x) = 2x - 3$$

To find where $f'(x)$ is zero we set $2x - 3 = 0$. Solving for x gives

$$2x = 3$$

$$x = \tfrac{3}{2} = 1.5$$

We suspect from Table 14-2 of this section that $f(x)$ has a minimum at $x = 1.5$. To see that this is indeed the case, we apply Theorem 14.1(A). At $x = 1.4$,

$$\begin{aligned} f'(x) &= 2(1.4) - 3 \\ &= 2.8 - 3 \\ &= -0.2 < 0 \end{aligned}$$

At $x = 1.6$,

$$\begin{aligned} f'(x) &= 2(1.6) - 3 \\ &= 3.2 - 3 \\ &= 0.2 > 0 \end{aligned}$$

Therefore, $y = x^2 - 3x + 6$ has a minimum at $x = 1.5$.

Example 3. Examine the function $f(x) = 2x^3 + 3x^2 - 36x + 6$ for possible maxima and minima and illustrate these points graphically.

Solution:

$$f(x) = 2x^3 + 3x^2 - 36x + 6$$

$$f'(x) = 6x^2 + 6x - 36$$

$$6x^2 + 6x - 36 = 0$$

$$6(x^2 + x - 6) = 0$$

$$6(x+3)(x-2) = 0$$

Hence, $f'(x) = 0$ if $x + 3 = 0$ or $x - 2 = 0$. This occurs for $x = -3$ and $x = 2$.

We know that $f'(2) = 0$. To see if $f(x)$ has a maximum or minimum at $x = 2$ we evaluate $f'(x)$ at $x = 1.9$ and at $x = 2.1$.

$$\begin{aligned} f'(1.9) &= 6(1.9+3)(1.9-2) \\ &= 6(4.9)(-0.1), \end{aligned}$$

which is negative. For $x = 1.9 < 2, f'(x) < 0$

$$\begin{aligned} f'(2.1) &= 6(2.1+3)(2.1-2) \\ &= 6(5.1)(0.1), \end{aligned}$$

which is positive. For $x = 2.1 > 2$, $f'(x) > 0$. Applying Theorem 14.1, we see that $y = 2x^3 + 3x^2 - 36x + 6$ has a minimum at $x = 2$.

Similar calculations show that for $x = -3.1$, $f'(x)$ is positive and for $x = -2.9$, $f'(x)$ is negative. Therefore, this function has a maximum at $x = -3$.

Substitution into $y = 2x^3 + 3x^2 - 36x + 6$ gives

$$y = 2(2)^3 + 3(2)^2 - 36(2) + 6$$
$$= 2(8) + 3(4) - 72 + 6$$
$$= 16 + 12 - 72 + 6$$
$$y = -38 \text{ when } x = 2, \qquad y = 87 \text{ when } x = -3$$

These results are summarized in Table 14-4 and Figure 14-3.

TABLE 14-4

x	y	y'	Type
−3.1		+	
−3.0	87	0	MAX
−2.9		−	
1.9		−	
2.0	−38	0	MIN
2.1		+	

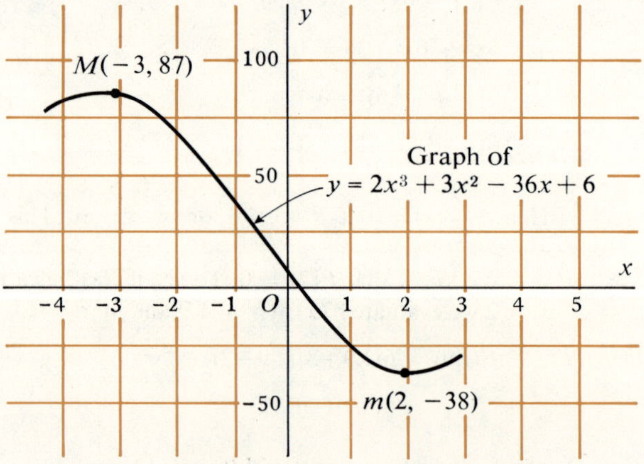

FIGURE 14-3

The next example illustrates the fact that a function may fail to have a maximum or minimum at $x = a$ even though $f'(a) = 0$.

Example 4. Test $y = x^3 + 2$ for possible maxima and minima.

Solution:

$$f(x) = x^3 + 2$$

$$f'(x) = 3x^2$$

$$\text{Set } f'(x) = 0.$$

$$3x^2 = 0$$

$$x^2 = 0$$

$$x = 0$$

This is the only solution. But for $x = 0.1$, $f'(x)$ is positive and for $x = -0.1$, $f'(x)$ is also positive. Therefore, $f(x) = x^3 + 2$ has neither a maximum nor a minimum at $x = 0$ since the first derivative does not change sign. The graph in Figure 14-4 illustrates the reason.

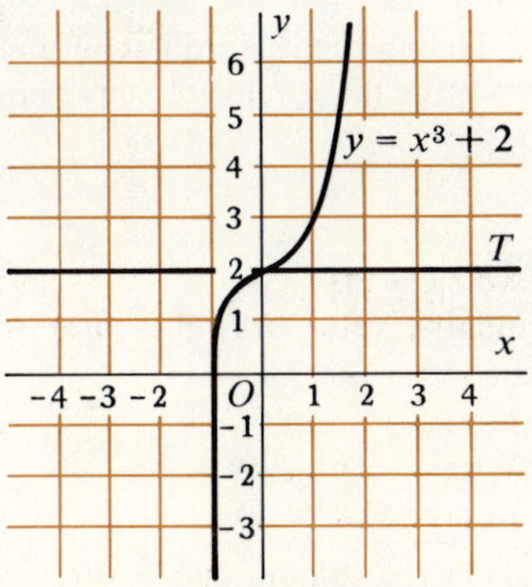

FIGURE 14-4

It is possible for a function to have a maximum or a minimum at a point where the function fails to be differentiable. In this case the derivative may be different from zero at all other points in the domain of the function. This situation is illustrated below for a function $f(x)$ that is continuous on the interval $a \leqslant x \leqslant b$, but fails to have a derivative at $x = c$.

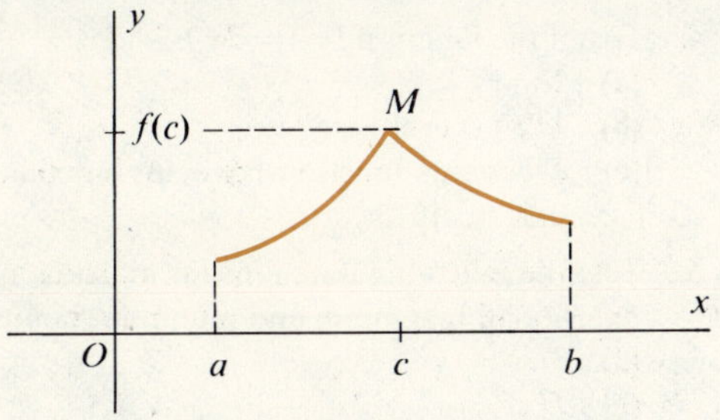

FIGURE 14-5

The above situation is *not covered* by the first and second derivative tests included in this chapter.

summary

FIRST DERIVATIVE TEST (see Theorem 14.1)

Step 1: Find $f'(x)$.

Step 2: Solve the equation $f'(x) = 0$ for x. These values of x are candidates for maxima and minima. If $f'(x) = 0$ has no solution, there are no maxima or minima.

Step 3: Determine the sign of the first derivative slightly to the left and slightly to the right of each of the x values in Step 2.

Step 4: If the sign of the derivative is negative on the left and positive on the right, the function has a minimum. If the sign of the derivative is positive on the left and negative on the right, the function has a maximum.

Step 5: To find the y coordinate of the maximum or the minimum point, substitute the x value of Step 2 into the expression $y = f(x)$.

PROBLEMS

1. Find the value or values of x that make $f'(x) = 0$ for the following functions:

(a) $f(x) = x^2 - 3x$
(b) $f(x) = x^2 + 3x + 4$
(c) $f(x) = x^3 + 2x^2$
(d) $f(x) = x^3 - 6x$
(e) $f(x) = 2x^3 + 3x^2 - 36x + 4$
(f) $f(x) = x^3 + 2x^2 - 3x + 7$

2. Sketch the graphs of $y = f(x)$ and $y = f'(x)$ if $f(x) = x^2 + 4x$ and determine the approximate location of all maxima and minima of $y = f(x)$ from your graphs. [*Hint:* Check the x intercepts of the graph of $y = f'(x)$.]

3. Using the function $f(x) = x^2 + 4x$ of Example 2,

(a) Show that $f'(x) = 0$ at $x = -2$.
(b) Calculate the sign of $f'(x)$ at $x = -3$ and $x = -1$.
(c) Use the results of **(a)** and **(b)** and the first derivative test to determine if $y = x^2 + 4x$ has a maximum or a minimum at $x = -2$. Compare this result with Problem 2.

4. Given the function $f(x) = 2x + 3$.

(a) Find $f'(x)$.
(b) Is $f'(x)$ ever zero?
(c) Does this function have any maxima or minima? Why (based on calculus)?

5. Test the following functions for maxima and minima. Find both coordinates of all maximum and minimum points.

(a) $f(x) = x^2 + 3x + 4$
(b) $f(x) = x^3 + 2x^2 - 36x + 4$
(c) $f(x) = e^{3x} - 3x$

6. Test the following functions for maxima and minima. Find both coordinates of all maximum and minimum points.
   (a) $f(x) = x^3 + 2x^2$
   (b) $f(x) = \frac{1}{4}x^4 + 3x^3 + 9x^2$
   (c) $f(x) = \ln x - 2x$
   (d) $f(x) = \ln x + 2x$

Omit for test

7. Given the function

$$f(x) = \begin{cases} x^2 & \text{if } 0 \leqslant x \leqslant 1 \\ 2 - x^2 & \text{if } 1 \leqslant x \leqslant 2 \end{cases}$$

   (a) Draw the graph of this function. Is this function continuous?
   (b) Determine $f'(x)$ at points other than $x = 1$. Draw the graph of $f'(x)$.
   (c) Is $y = f'(x)$ a continuous function? Is $y = f(x)$ differentiable at $x = 1$?
   (d) Determine the maximum and the minimum for the given function.

14-2 THE SECOND DERIVATIVE AND INFLECTION POINTS

The derivative of $y = f(x)$ constitutes a new function called $f'(x)$. This new function can be interpreted geometrically as the slope of the tangent to $y = f(x)$ at each point or, in terms of rates, as the rate of change of $y = f(x)$ at each point. We are now ready to examine the derivative of the function $f'(x)$. This is called the second derivative of $y = f(x)$.

DEFINITION 14.2. *Let $y = f(x)$. Then the second derivative of $y = f(x)$ denoted by the symbol d^2y/dx^2 is defined to be $d^2y/dx^2 = d/dx[f'(x)]$.*

Other common symbols for the second derivative are y'' and $f''(x)$. The second derivative gives the rate of change of the slope of the tangent line. If $f''(x)$ is positive on an interval, then the slope of the tangent line is increasing. This means that as x increases from left to right the tangent line rotates in a counterclockwise sense. The graph of a function that exhibits this property is said to be *concave upward*. If $f''(x)$ is negative then the slope of the tangent line is decreasing and as x increases from left to right the tangent line rotates in the clockwise sense. In this case the graph of $y = f(x)$ is said to be *concave downward*.

DEFINITION 14.3. *A point on the graph of $y = f(x)$ at which the curve changes from concave upward to concave downward or from concave downward to concave upward is called an inflection point.* The derivative must exist at that point.

When the graph of $y = f(x)$ switches from concave upward to concave downward, the sign of the second derivative switches from positive to negative. If it switches from concave downward to concave upward, $f''(x)$ switches from negative to positive. Therefore, at an inflection point, the second derivative equals zero and switches sign.

Example 1. Test the function $y = x^3$ for concavity and inflection points. Illustrate graphically.

Solution:

$$f(x) = x^3$$

$$f'(x) = 3x^2$$

$$f''(x) = 6x$$

$f''(x) = 6x$ is negative for all $x < 0$, zero at $x = 0$, and positive for all $x > 0$. Therefore, the graph of $y = x^3$ is concave downward for all $x < 0$, concave upward for all $x > 0$, and has an inflection point at $(0, 0)$.

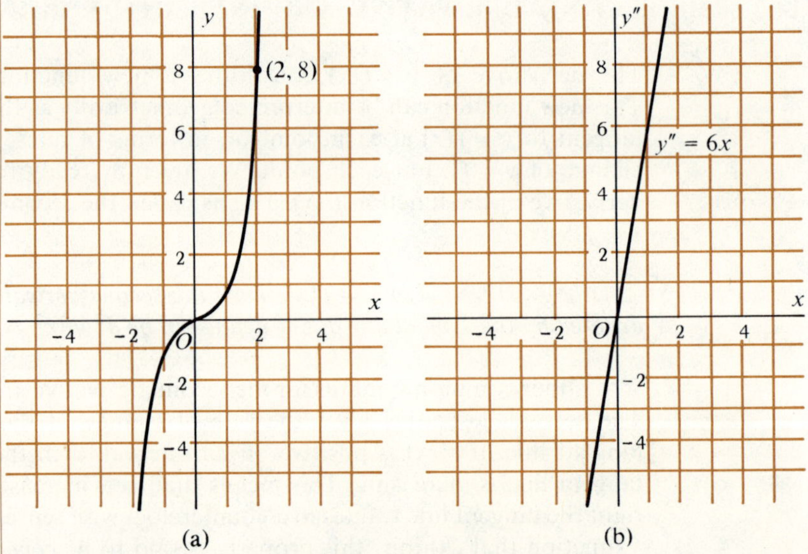

FIGURE 14-6

Example 2. Graph and discuss the function $y = x^3 - 3x^2 + 3x + 1$.

Solution:

$$y = x^3 - 3x^2 + 3x + 1$$

$$y' = 3x^2 - 6x + 3$$

$$y'' = 6x - 6$$

Let $y'' = 0$.

$$6x - 6 = 0$$

$$6x = 6$$

$$x = 1$$

Therefore, $x = 1$ is a *candidate* for an inflection point.

Furthermore, $y'' = 6x - 6$ is negative for $x < 1$ and positive for $x > 1$. This means that the graph of $y = x^3 - 3x^2 + 3x + 1$ has an inflection point at $(1, 2)$, is concave downward for all $x < 1$, and is concave upward for all $x > 1$.

Direct substitution gives the following table of values.

TABLE 14-5

x	y	$\Delta y/\Delta x$
-1	-6	
		7
0	1	
		1
1	2	
		1
2	3	
		7
3	10	

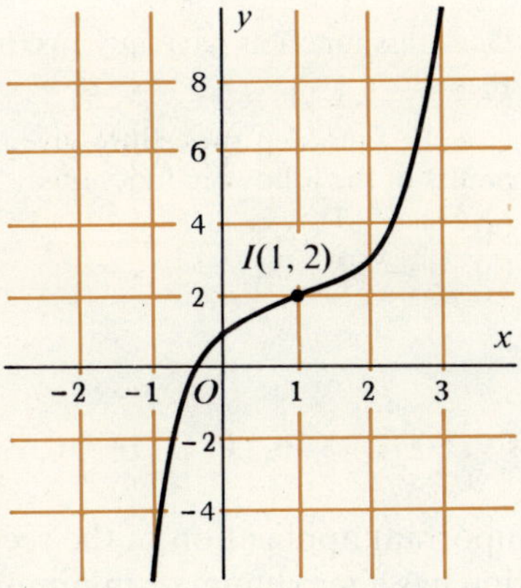

FIGURE 14-7

It can be seen from the last column of Table 14-5 that this function is increasing at a decreasing rate for $x < 1$ and increasing at an increasing rate for $x > 1$.

TEST FOR INFLECTION POINTS

Step 1: Find $f''(x)$.

Step 2: Solve the equation $f''(x) = 0$ for x. These values of x are candidates for inflection points. If $f''(x) = 0$ has no solution there are no inflection points.

Step 3: If the sign of the second derivative is negative on the left and positive on the right of one of these x values, or vice versa, then the function has an inflection point there.

Step 4: To find the y coordinate of the inflection point, substitute into $y = f(x)$.

PROBLEMS

1. Find the first and second derivative of the following functions:
   (a) $y = x^2 + 4x - 7$ (c) $z = x^3 + 3x$
   (b) $y = x^{1/2} + x^{-1/4}$ (d) $y = 3 \ln x + 4 \log x$
2. Find the first and second derivative for the following functions:
   (a) $y = 2x^3 - 6x^2 + 3x - 8$ (c) $y = x^2 - 3x + 2$
   (b) $y = x^{3/4} + 2e^x + 3 \ln x$ (d) $y = x^4 - 8x^3 + 18x^2 + 10$
3. Determine the sign of the second derivative of $y = x^3 + x^2 + x + 1$ at $x = -2, -1, 0$, and 2.
4. Use calculus to show that $y = x^4 - x$ is concave upward for all values of x.
5. Draw a graph of the function given in Problem 4.
6. Does this function have any maximum or minimum points? If so, find them.
7. Use the four-step procedure given in this section to find the inflection points of the following functions:
   (a) $y = x^5$ (c) $y = x^3 + 3x^2$
   (b) $y = x^3 + 3x$ (d) $y = x^4 + 6x^2$

14-3 THE SECOND DERIVATIVE TEST FOR MAXIMA AND MINIMA

An important application of the second derivative is to determine if a function has a maximum or minimum at a point where the first derivative is zero.

THEOREM 14.2. *Let $f''(x)$ be continuous at $x = a$.*

A. *If $f'(a) = 0$ and $f''(a)$ is positive, then the graph of $y = f(x)$ is concave upward near $x = a$ and the function has a minimum at $x = a$.*

B. *If $f'(a) = 0$ and $f''(a)$ is negative, then the graph of $y = f(x)$ is concave downward near $x = a$ and the function has a maximum at $x = a$.*

(We will omit a proof of this theorem.)

Example 1. Use Theorem 14.2 to determine the maxima and minima of $y = 2x^3 + 3x^2 - 36x + 6$.

Solution:

$$y = 2x^3 + 3x^2 - 36x + 6$$
$$y' = 6x^2 + 6x - 36$$
$$y'' = 12x + 6$$
$$6x^2 + 6x - 36 = 0$$
$$6(x^2 + x - 6) = 0$$
$$6(x+3)(x-2) = 0$$

Therefore, the first derivative is zero at $x = 2$ and at $x = -3$. Also, when $x = 2$,

$$y'' = 30$$

which is positive, indicating a minimum at $x = 2$. At $x = -3$,

$$y'' = -30$$

which is negative, indicating a maximum at $x = -3$. [*Note:* These results are the same as in Example 3 of Section 14-1.]

Example 2. Test the function $y = xe^x$ for maxima and minima. Use Theorem 14.2.

Solution:

$$y = xe^x$$
$$y' = xe^x + e^x \qquad \textit{by D-10}$$
$$y' = e^x(x+1)$$
$$y'' = e^x(1) + (x+1)e^x$$
$$y'' = e^x(x+2)$$
$$y' = 0 \qquad \textit{by setting the first derivative equal to zero}$$
$$e^x(x+1) = 0$$
$$e^x = 0 \quad \text{or} \quad x + 1 = 0$$

However, e^x is never zero. Therefore, the only value of x that makes the first derivative zero is $x = -1$. This means that $y = xe^x$ has at most one maximum or one minimum.

$$y''|_{x=-1} = e^{-1}(-1+2)$$
$$= \frac{1}{e}$$

which is positive. Therefore, $y = xe^x$ has a minimum at $x = -1$. The coordinates of this minimum point are $(-1, -1/e)$, where $e = 2.7183$ to four decimal places.

SECOND DERIVATIVE TEST (see Theorem 14.2)

Step 1: Find $f'(x)$.

Step 2: Set $f'(x)$ equal to zero and solve for x.

Step 3: Find $f''(x)$.

Step 4: Determine the sign of the second derivative at each of these points.

Step 5: If $f''(x)$ is positive at one of these x values, then the function has a minimum there. If $f''(x)$ is negative, then the function has a maximum. If $f''(x) = 0$ at one of these x values, use the first derivative test. Theorem 14.2

Step 6: As before, substitute into $y = f(x)$ to get the y coordinate.

PROBLEMS

1. Use the second derivative to show that $y = x^2+4$ is concave upward for all values of x and has a minimum at $(0, 4)$.
2. Draw the graph of the function $y = x^2+4$ given in Problem 1.
3. Use the second derivative test to find the maxima and minima of the following functions:
   (a) $y = x^3 + 2x^2 - 36x + 4$ **(b)** $y = e^{4x} - 4x$
4. Use the second derivative test to find the maximum and minimum points of $y = \frac{1}{3}x^3 + \frac{1}{2}x^2 - 6x + 8$.
5. Test $y = x \ln x$ $(0 < x)$ for maxima, minima. Use the second derivative test.
6. Draw the graph of the function in Problem 5 from $x = 0.1$ to $x = 2$.
7. Test the function $y = (x-3)^2(x-4)$ for maxima, minima. Use the second derivative test.
8. Use the second derivative test to find the maximum and minimum of the function $y = x^3+6x^2+9x+7$.
9. Let $y = ax^3+cx$, where $a \neq 0$ and $c \neq 0$. Show that
   (a) The graph of this function has x intercepts other than zero if and only if $c/a < 0$.
   (b) This function has a maximum and a minimum if and only if $c/a < 0$.

10. Let $y = ax^3 + bx^2$, where $a \neq 0$ and $b \neq 0$. Show that
 (a) The graph of this function has x intercepts at $x = 0$ and $x = -b/a$.
 (b) This function has both a maximum and a minimum.

11. Use the results of Problem 9 or Problem 10 to test the following functions for the existence of a maximum and a minimum, and illustrate graphically.
 (a) $y = x^3 + 2x^2$
 (b) $y = x^3 + 6x$
 (c) $y = x^3 - 2x^2$
 (d) $y = x^3 - 6x$
 (e) $y = x^3 + 2x^2$
 (f) $y = -x^3 + 6x$

12. **A.** The mathematical notation for the total cost of a business is $2x^3 + 4x^2 + 3x + 5$ where x equals production volume. Which of the following is the mathematical notation for the marginal cost function for this business?
 (a) $2(x^3 + 2x^2 + 1.5x + 2.5)$.
 (b) $6x^2 + 8x + 3$.
 (c) $2x^3 + 4x^2 + 3x$.
 (d) $3x + 5$.

 B. When a firm's profit model is represented by an algebraic function, one method of determining the number of units which should be produced and sold to maximize profit is by which of the following?
 (a) Integrating the function.
 (b) Differentiating the function.
 (c) Computing the inverse of the function.
 (d) Vector analysis.

 C. To find a minimum-cost point given a total-cost equation, the initial steps are to find the first derivative, set this derivative equal to zero, and solve the equation. Using the solution(s) so derived, what additional steps must be taken, and what result indicates a minimum?
 (a) Substitute the solution(s) in the first derivative equation and a positive solution indicates a minimum.
 (b) Substitute the solution(s) in the first derivative equation and a negative solution indicates a minimum.
 (c) Substitute the solution(s) in the second derivative equation and a positive solution indicates a minimum.
 (d) Substitute the solution(s) in the second derivative equation and a negative solution indicates a minimum.

14-4 SUMMARY OF MAXIMA, MINIMA, AND INFLECTION POINTS

In this section we include a sequence of examples starting with a function that has no maximum or minimum and ending with a function that has a maximum, a minimum, and an inflection point. We will also illustrate the procedure for handling the somewhat troublesome case of both y' and y'' being zero for the same value of x.

Problem 12 adapted from the Uniform CPA Examinations, with permission of the AICPA.

Example 1. Test $y = e^{2x}$ for maxima, minima, and inflection points.

Solution:

$$y = e^{2x}$$
$$y' = 2e^{2x}$$
$$y'' = 4e^{2x}$$

Since both y' and y'' are never zero, this function has no maximum, minimum, or inflection point.

Example 2. Test $y = x^3$ for maxima, minima, and inflection points.

Solution:

$$y = x^3$$
$$y' = 3x^2$$
$$y'' = 6x$$

Both y' and y'' are zero if and only if $x = 0$. This means that the second derivative test fails. By the first derivative test, we find that $y' = 3x^2$ does not change sign; therefore, $y = x^3$ has neither a maximum nor a minimum at $x = 0$. However, using the test for inflection points, we see that $y'' = 6x$ changes sign at $x = 0$; therefore, the point $(0, 0)$ is an inflection point. Some of these facts are illustrated graphically as follows.

x	-2	-1	0	1	2
$y = x^3$	-8	-1	0	1	8

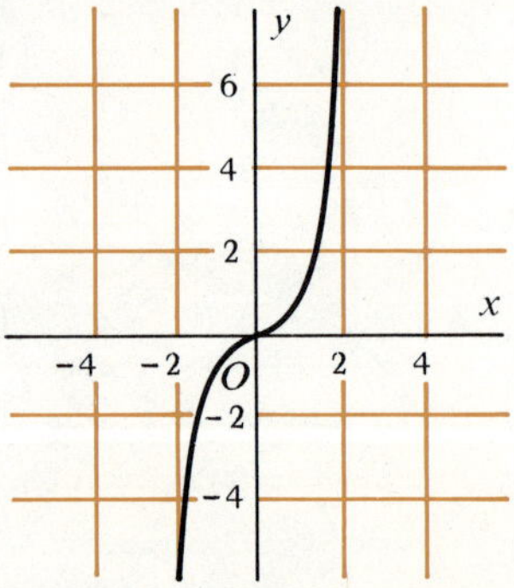

(a)

x	-2	-1	0	1	2
$y' = 3x^2$	12	3	0	3	12

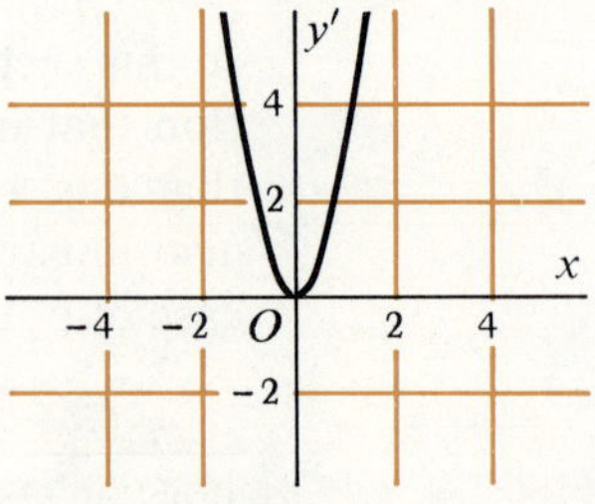

(b)

x	-2	-1	0	1	2
$y'' = 6x$	-12	-6	0	6	12

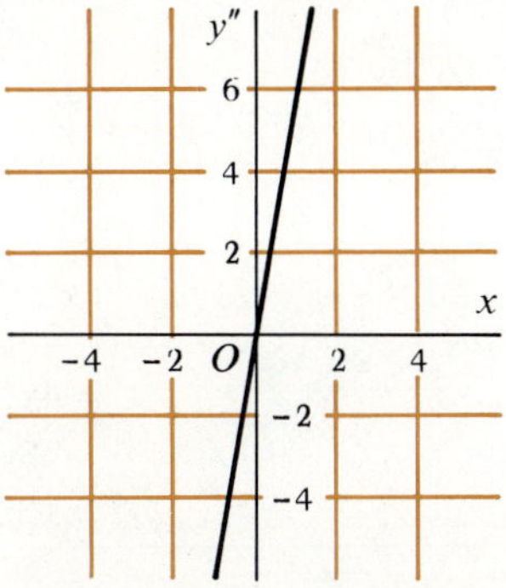

(c)

FIGURE 14-8

Example 3. Test $y = x^4$ for maxima, minima, and inflection points.

Solution:

$$y = x^4$$

$$y' = 4x^3$$

$$y'' = 12x^2$$

Again y' and y'' are both zero at $x = 0$. The first derivative switches from negative to positive at $x = 0$. Therefore, by the first derivative test, $y = x^4$ has a minimum at $(0, 0)$.

These sign relationships are also illustrated graphically as follows:

x	-2	-1	0	1	2
$y = x^4$	16	1	0	1	16

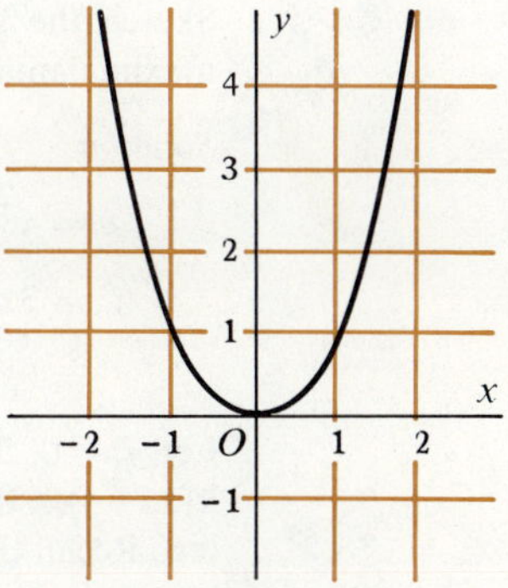

(a)

FIGURE 14-9

x	-2	-1	0	1	2
$y' = 4x^3$	-32	-4	0	4	32

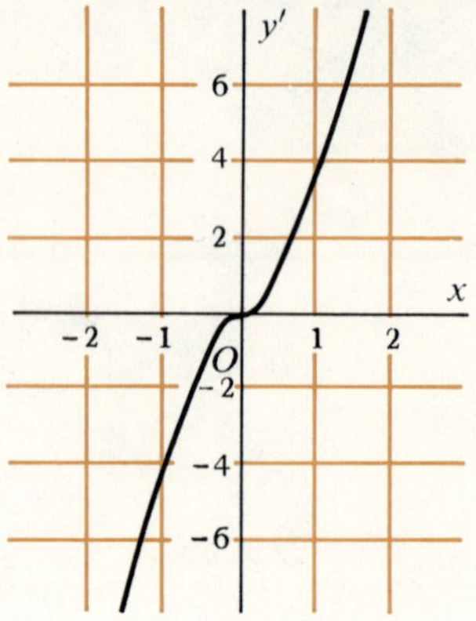

(b)

x	-2	-1	0	1	2
$y'' = 12x^2$	48	12	0	12	48

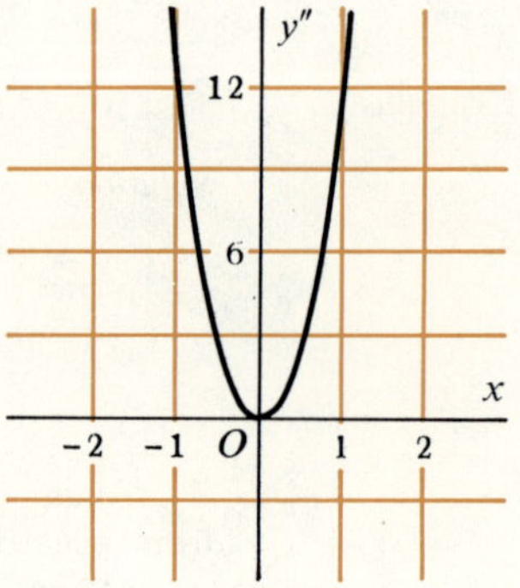

(c)

F I G U R E 14-9 *(continued)*

Example 4. Sketch the graphs of y, y', and y'' if $y = x^3 + 2x^2$. Test this function for maxima, minima, and inflection points.

Solution:

$$y = x^3 + 2x^2$$

$$y' = 3x^2 + 4x$$

$$y'' = 6x + 4$$

Let $y' = 0$. This gives the equation $3x^2 + 4x = 0$. Solving for x, we have $x(3x+4) = 0$ or $x = 0$ and $x = -\frac{4}{3}$ as solutions. Use the second derivative test. Recall that $y'' = 6x + 4$; therefore, we find that

$$y''|_{x=0} = 4 \qquad \text{and} \qquad y''|_{x=-4/3} = 6(-\tfrac{4}{3}) + 4 = -4$$

This means that $y = x^3 + 2x^2$ has a minimum at $(0, 0)$ and a maximum at $(-\frac{4}{3}, \frac{32}{27})$.

To test for inflection points, we set $y'' = 0$. This gives the equation $6x+4 = 0$. This equation has the unique solution $x = -\frac{2}{3}$. Since $y'' = 6x+4$ changes sign at $x = -\frac{2}{3}$, the point $(-\frac{2}{3}, \frac{16}{27})$ is an inflection point. Table 10-6 gives some of the values for $y = x^3+2x^2$.

TABLE 14-6

x	y	y'	y''
-2	0	4	-8
$-\frac{4}{3}$	$\frac{32}{27}$	0	-4
-1	1	-1	-2
$-\frac{2}{3}$	$\frac{16}{27}$	$-\frac{4}{3}$	0
0	0	0	4
1	3	7	10
2	16	20	16

The graphs of y, y', and y'' are shown in Figure 14-10.

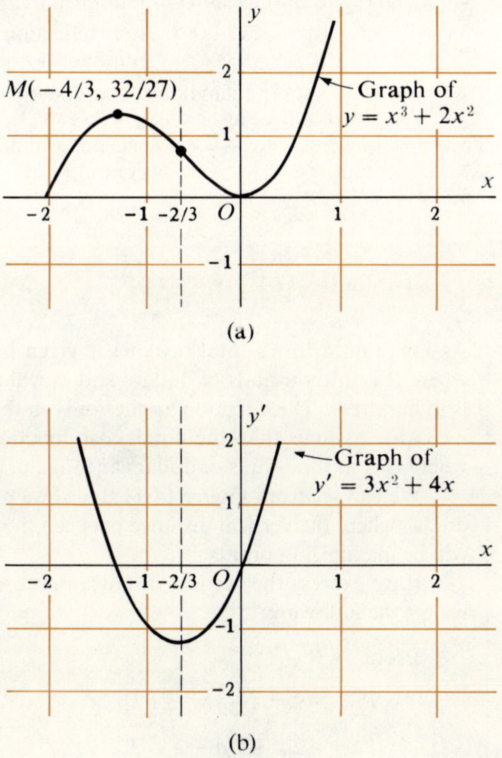

FIGURE 14-10

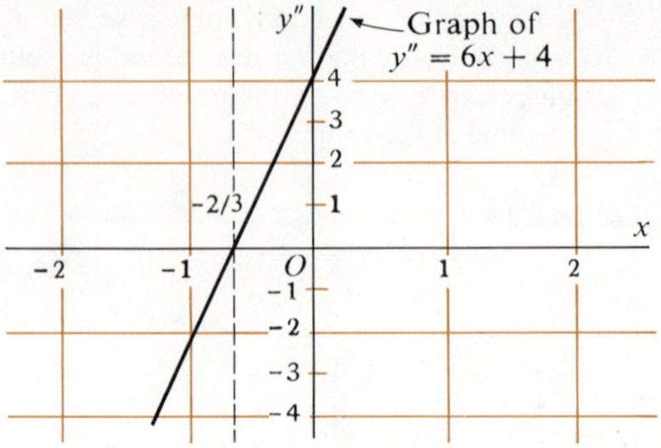

(c)

FIGURE 14-10 (*continued*)

PROBLEMS

Test the following functions for relative maxima and minima and for inflection points. Illustrate your results with tables and graphs.

1. $y = -x^4$ (one maximum)
2. $y = (1-x)^4 + 2$ (one minimum)
3. $y = x^3 - 6x^2 + 2x - 6$ (one inflection point, one maximum, and one minimum)
4. $y = xe^{-x}$ (one maximum and one inflection point)
5. $y = \frac{x^3}{3} + \frac{x^2}{2} - 6x + 7$ (one maximum, one minimum, and one inflection point)
6. $y = x^3 + 3x^2$

14-5 APPLIED MAX-MIN PROBLEMS

~~Example 1.~~ Assume that a firm's total revenue is given by the function $R = -2q^2 + 7q$, where R is in thousands of dollars and q, which is the number of units sold, is in hundreds. This function applies only in the range $0 \leqslant q \leqslant 3$.

Also assume that the total cost function for this firm is $C = q+3$, where C is in thousands of dollars and, again, q is in hundreds.

We can see from Figure 14-11 that if we produce at the q value (in hundreds), where the vertical distance between the two curves is a maximum, we will be maximizing profits.

If we express the vertical distance between the two functions for any q, we get the following:

$$\text{Profit} = R - C$$

$$P = -2q^2 + 7q - (q+3)$$

$$P = -2q^2 + 7q - q - 3$$

$$P = -2q^2 + 6q - 3$$

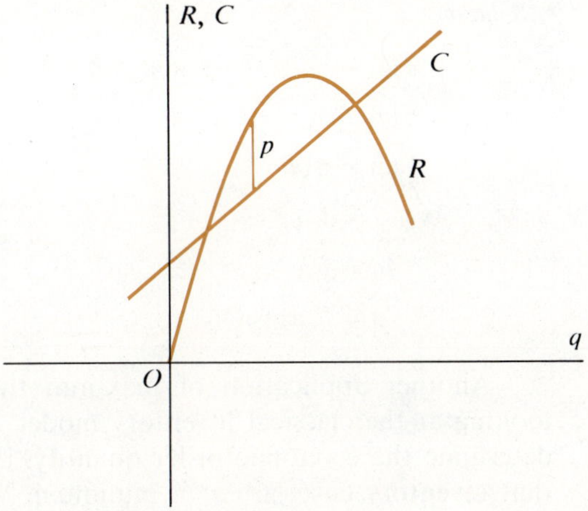

FIGURE 14-11

The first and second derivatives are

$$P' = -4q + 6$$

$$P'' = -4$$

Set P' equal to zero. This gives

$$-4q + 6 = 0$$

$$q = \tfrac{3}{2}$$

Substitution into $P = -2q^2 + 6q - 3$ gives $P_{max} = 1.5$ at $q = 1.5$. That is, the sale of 150 units will result in a profit of $1500.

The following argument justifies the marginalist approach to profit maximization, i.e., *profit is maximized (or loss is minimized) when marginal revenue equals marginal cost.*

Let $R = f(q)$ and $C = g(q)$, where f and g are differentiable functions of quantity, q.

$$\text{Profit} = \text{total revenue} - \text{total cost}$$

$$P = f(q) - g(q)$$

$$\frac{dP}{dq} = f'(q) - g'(q) = 0$$

if and only if $f'(q) = g'(q)$.

Recall that $f'(q)$ is called marginal revenue, and $g'(q)$ is called marginal cost. We see that maximum profit or minimum loss occurs when marginal revenue equals marginal cost.

Example 2. Redo Example 1 using the marginalist approach.

Solution:

$$f(q) = -2q^2 + 7q, \ g(q) = q + 3$$

$$f'(q) = -4q + 7, \ g'(q) = 1$$

$$f'(q) = g'(q)$$

$$-4x + 7 = 1$$

$$4q = 6$$

$$q = \tfrac{3}{2}, \text{ as before}$$

Another application of max-min theory can be observed by looking at the classical inventory model. We will use this model to determine the economic order quantity, i.e., the order quantity such that inventory costs are at a minimum. We denote economic order quantity by EOQ.

Inventory costs are made up of two basic factors. The first is holding cost or the cost of holding an item in inventory. If the total number of items in inventory for any period is $x/2$, or one-half the amount that is ordered, then the cost of holding would be equal to $x/2$ (i.e., the amount of items in inventory during a period) times h, the cost of holding one item. The reader should note that holding cost increases as x, the number of units ordered, is increased. A simple illustration is shown in Figure 14-12.

We see in this figure that the usage is constant and the receipt of the order occurs exactly when the inventory level falls to zero.

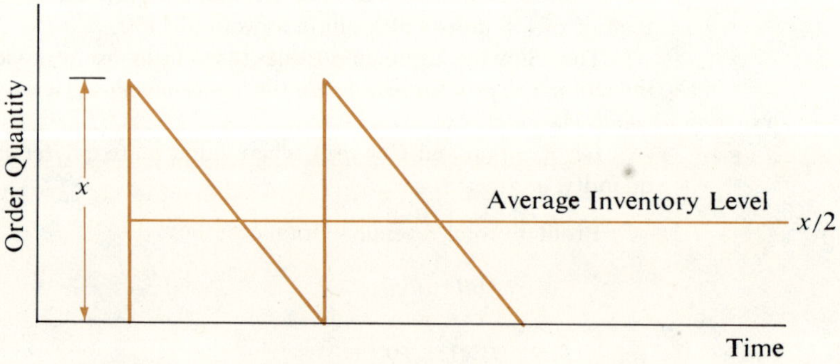

FIGURE 14-12

The other factor affecting inventory cost is the ordering cost. This cost will decrease as the number of units ordered increases. The total number of orders in a given year is equal to the annual demand, q, divided by the size of the order, x. The ordering cost would be equal to the number of orders times the cost of placing an order.

This gives the following equation:

$$C = \frac{x}{2}(h) + \frac{q}{x}(r)$$

where h = holding cost per item

x = order size

r = cost of placing an order

We see that holding cost varies directly with x, and reorder cost varies inversely with x. We are faced with a trade-off of costs such that total cost, C, will be at a minimum.

Graphically, these costs appear as shown in Figure 14-13.

We see that the economic order quantity occurs at a point where C is a minimum.

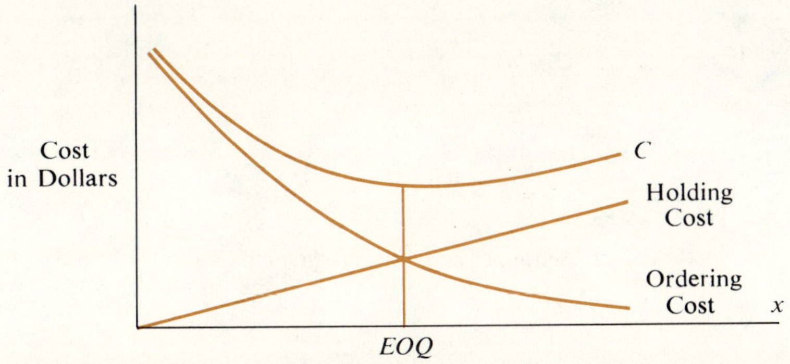

FIGURE 14-13

Example 3. Suppose that the cost of holding one item in inventory is \$0.50 and that the cost of placing an order is \$10. Also assume that the annual demand for the item is 490 units. Find the value of x at which the cost is minimum.

Solution:

$$C = \frac{x}{2}(0.50) + \frac{490}{x}(10)$$

To find where C is a minimum we find the first derivative of C with respect to x and set it equal to zero.

$$C' = 0.25 - \frac{4900}{x^2} = 0$$

$$\frac{4900}{x^2} = 0.25$$

$$x^2 = \frac{4900}{0.25}$$

$$x = \frac{70}{0.5} = 140$$

$$C'' = -(-2)\frac{4900}{x^3}$$

C'' is positive at $x = 140$, so we have a minimum.
Costs will be minimized if we order 140 units at a time.

We point out that when the two basic factors that make up C are of the form

$$y_1 = ax$$

$$y_2 = \frac{b}{x}$$

then

$$C = ax + \frac{b}{x}$$

and

$$C' = a - \frac{b}{x^2}$$

Setting C' equal to zero gives

$$\frac{b}{x^2} = a$$

$$x^2 = \frac{b}{a}$$

$$x = \sqrt{\frac{b}{a}}$$

It is interesting to note that this result can also be obtained by setting y_1 equal to y_2 and solving for x.

$$ax = \frac{b}{x}$$

$$x^2 = \frac{b}{a}$$

$$x = \sqrt{\frac{b}{a}}$$

Therefore, *in this case*, total cost is a minimum when the holding cost is equal to the ordering cost.

TABLE 14-7

Units Ordered	Ordering Costs	Cost of Carrying	Total Cost
130	37.6923	32.5000	70.1923
131	37.4045	32.7500	70.1545
132	37.1212	33.0000	70.1212
133	36.8421	33.2500	70.0921
134	36.5671	33.5000	70.0671
135	36.2963	33.7500	70.0463
136	36.0294	34.0000	70.0294
137	35.7664	34.2500	70.0164
138	35.5072	34.5000	70.0072
139	35.2518	34.7500	70.0018
140	35.0000	35.0000	70.0000
141	34.7517	35.2500	70.0017
142	34.5070	35.5000	70.0070
143	34.2657	35.7500	70.0157
144	34.0277	36.0000	70.0277
145	33.7931	36.2500	70.0431
146	33.5616	36.5000	70.0616
147	33.3333	36.7500	70.0833
148	33.1081	37.0000	70.1081
149	32.8859	37.2500	70.1359
150	32.6666	37.5000	70.1666

A numerical approach to this same problem, using a computer, is included in Table 14-7. If we let the units ordered vary from 130 to 150, we see that total cost is indeed a minimum at $x = 140$, and that holding cost is equal to ordering cost at this quantity.

PROBLEMS

1. Given the cost function $C = q^2 - 10q + 125$ and the revenue function $R = 100q - 15q^2$, where q, C, and R are in hundreds, find the value of q where marginal revenue equals marginal cost.

2. Determine if the value of q in Problem 1 gives maximum profit or minimum loss.

3. Given the cost function $C = 2500 + 60x - (1000/x)\,[1 < x \leqslant 50]$ and the revenue function $R = 160x$, find the maximum profit or minimum loss.

4. The cost of holding one item in inventory is \$1.00 and the cost of placing an order is \$5.00. Assume a yearly demand for the item of 1000 units. Find the cost function.

5. Find the value for x in Problem 4 that minimizes cost. Find the minimum cost.

6. The cost of holding one item in inventory is \$0.25 and the cost of placing an order is \$0.50. Assume a yearly demand for the item of 100 units. Find the value of x for which costs are minimized.

7. An open storage bin is to be constructed with a square base and vertical sides so as to contain 500 cu ft. Show that the cost of lining the bin with aluminum will be minimum if the depth is made half of the width.

8. Repeat Problem 7. Replace the volume, 500 cu ft, with any fixed volume, V.

9. A rectangular box is to be built with a square base and a square top. The volume is V cubic feet. The cost per square foot for the top and bottom is \$0.20 and the cost per square foot for the sides is \$0.30. Determine the dimensions that result in the minimum material cost of the box.

10. Repeat Problem 9. Use a cost of \$0.50/sq ft for the top and bottom and \$0.40/sq ft for the sides.

11. A manufacturer expects to produce 200,000 widgets during the year ending June 30, 1972 to supply a demand which is uniform throughout the year. The setup cost for each production run of widgets is \$144 and the variable cost of producing each widget is \$5. The cost of carrying one widget in inventory is \$.20 per year. After a batch of widgets is produced and placed in inventory, it is sold at a uniform rate, and inventory is exhausted when the next batch of widgets is completed.

Management wishes an equation to describe the above situation and determine the optimal quantity of widgets to produce in each run in order to minimize total production and inventory carrying costs.

Let c = Total annual cost of producing and carrying widgets in inventory.

x = Number of widgets to be produced in each production run.

Select the appropriate answers from the following.

A. The number of production runs to be made in fiscal year 1972 could be expressed as

(a) $200{,}000 + 144x$

(b) $200{,}000 + x$

(c) $200{,}000$

(d) $\dfrac{200{,}000}{x}$

(e) $\dfrac{x}{200{,}000}$

B. Total setup costs for fiscal year 1972 could be expressed as

(a) $\$144\left(\dfrac{200{,}000}{x}\right)$

(b) $\dfrac{\$200{,}000}{x}$

(c) $\$144x$

omit

(d) $\dfrac{\$144x}{200{,}000}$

(e) $\dfrac{\$144x}{200{,}000} + \$x$

C. Total cost of carrying inventory during fiscal year 1972 could be expressed as

(a) $\$.20(\$144x)$

(b) $\$.20x$

(c) $\$.20\left(\dfrac{200{,}000}{x}\right)$

(d) $\$.20\left(\dfrac{x}{2}\right)$

(e) $\$.20\left(\dfrac{\$144x}{200{,}000}\right)$

D. The derivative, dc/dx, of the equation to determine the optimal quantity of widgets which should be produced during each production run in fiscal year 1972 is

(a) $-28{,}800{,}000x^{-2} + \dfrac{.20}{2}$

(b) $-144(200{,}000)x^{-1} + \dfrac{.20}{2}$

(c) $-28{,}800{,}000x^{2} + 1{,}000{,}000 + \dfrac{.20}{2}$

(d) $-28{,}800{,}000x + \dfrac{.20}{2}$

(e) $-28{,}800{,}000x$

E. The quantity of widgets (to the nearest whole number) which should be produced in each run in fiscal year 1972 to minimize total costs is

(a) 19,000
(b) 17,000
(c) 16,000
(d) 12,480
(e) 12,000

12. MacKenzie Park sells trivets for \$.25 per unit and during 1969 reported net sales of \$500,000 and net income of \$35,000. Production capacity is limited to 15,000 trivets per day and trivets are produced 300 days each year. Variable costs are \$.10 per trivet.

The Company does not maintain an inspection system but has an agreement to reimburse the wholesaler \$.50 for each defective unit the wholesaler finds. The wholesaler uses a method of inspection which detects all defective units. The number of defective units in each lot of 300 units is equal to the daily unit production rate divided by 200.

Problem 11 adapted from the Uniform CPA Examinations, with permission of the AICPA.

Letting x = daily production in units, select the best answer for each of questions A through E.

A. The number of defective units per day could be expressed

(a) $\dfrac{x}{60{,}000}$

(b) $\left(\dfrac{200}{x}\right)\left(\dfrac{x}{300}\right)$

(c) $\dfrac{x}{500}$

(d) $\dfrac{x^2}{60{,}000}$

B. The equation to compute the maximum daily contribution to profit, including the reimbursement to the wholesaler for defective units, could be expressed

(a) $.25x - .10x - .50\left(\dfrac{x}{60{,}000}\right)$

(b) $.25x - .10x - .50\left(\dfrac{x^2}{60{,}000}\right)$

(c) $.25x - .10x - \dfrac{x^2}{60{,}000} - \dfrac{125{,}000}{300}$

(d) $.25x - .10x - \dfrac{x}{60{,}000} - 125{,}000$

C. The first derivative of the equation to determine the number of units to maximize daily profits could be expressed

(a) $.25 - .10 - \dfrac{x}{60{,}000}$

(b) $.10 - .25 - \dfrac{x^2}{60{,}000}$

(c) $\dfrac{x^2}{60{,}000}$

(d) $\dfrac{x}{(200)(300)}$

D. The second derivative of the equation to determine the number of units to be produced daily to maximize profits would be

(a) $\dfrac{1}{(200)(300)}$

(b) $\$.25 - \$.10 - \dfrac{1}{60{,}000} - \$125{,}000$

(c) $\dfrac{\$125{,}000}{60{,}000}$

(d) $-\dfrac{1}{60{,}000}$

E. To maximize profits, the results of the equation to determine the daily contribution margin of MacKenzie Park Co. should yield
(a) A negative first derivative and a positive second derivative.
(b) A zero first derivative and a negative second derivative.
(c) Negative first and second derivatives.
(d) Positive first and second derivatives.

13. A. A Company places orders for inventory with its suppliers for a certain item for which the order size is determined in advance as

$$\text{Order size} = \sqrt{\frac{2 \times \text{cost to place one order} \times \text{demand per period}}{\text{cost to hold one unit for one period}}}$$

All orders are the same size. When the policy is implemented, demand per period is only one-half what was expected when order size was computed. Consequently, actual total inventory cost will be which of the following?
(a) Larger than if the expected demand per period had occurred and larger than if the actual demand per period had been used to calculate order size.
(b) Larger than if the expected demand per period had occurred and smaller than if the actual demand per period had been used to calculate order size.
(c) Smaller than if the expected demand per period had occurred and larger than if the actual demand per period had been used to calculate order size.
(d) Smaller than if the expected demand per period had occurred and smaller than if the actual demand per period had been used to calculate order size.

B. A business determines its inventory policy using the economic-order-quantity model that allows a finite stockout cost per period and backordering. Which one of the following statements accurately describes that stockout cost?
(a) The stockout cost will have to be accumulated in a special account for comparison with expected stockout cost.
(b) The smaller the stockout cost, the more often the stockout will occur.
(c) The larger the stockout cost, the more often stockout will occur.
(d) The smaller the stockout cost, the larger will be the average amount of inventory on hand.

Expected annual usage of a particular raw material is 2,000,000 units, and the standard order size is 10,000 units. The invoice cost of each unit is $500, and the cost to place one purchase order is $80.

C. The average inventory is which of the following?
(a) 1,000,000 units.
(b) 5,000 units.
(c) 10,000 units.
(d) 7,500 units.

Problem 12 adapted from the Uniform CPA Examinations, with permission of the AICPA.

D. The estimated annual order cost is which of the following?
(a) \$16,000.
(b) \$100,000.
(c) \$32,000.
(d) \$50,000.

E. For its economic-order-quantity model a company has a cost of placing an order equal to \$10, and an annual cost of carrying one unit in stock equal to \$2. If the cost of placing an order increases by 20%, and the annual cost of carrying one unit in stock increases by 25% and all other considerations remain constant, the economic order quantity will
(a) Remain unchanged.
(b) Decrease.
(c) Increase.
(d) Either increase or decrease depending on the reorder point.

14-6 AN ADDITIONAL APPLICATION

In this section we will give cost data at four production levels for a hypothetical company. These data are used to find the third degree polynomial function that passes through the given points. Finally, this function is used in conjunction with the revenue function $R = -2q^2 + 7q$ from Example 1 of the previous section to find the level of production that gives the greatest profit.

Example 1. Given the following company data, find the third degree polynomial that passes through these four points.

Hundreds of Units	*Hundreds of Dollars*
0	1
1	2
2	3
3	10

Solution: To find the third degree polynomial function that passes through these four points, we assume

$$C = aq^3 + bq^2 + cq + d$$

When q is equal to zero, $C = 1$. Therefore, $1 = a0^3 + b0^2 + c0 + d$ and $d = 1$.

To find a, b, c, we must solve three equations in three unknowns for a, b, and c.

$$2 = a1^3 + b1^2 + c1 + 1$$

$$3 = a2^3 + b2^2 + c2 + 1$$

$$10 = a3^3 + b3^2 + c3 + 1$$

Problem 13 adapted from the Uniform CPA Examinations, with permission of the AICPA.

or

$$a + b + c = 1$$

$$8a + 4b + 2c = 2$$

$$27a + 9b + 3c = 9$$

Written in matrix form, this system of equations becomes

$$\begin{pmatrix} 1 & 1 & 1 \\ 8 & 4 & 2 \\ 27 & 9 & 3 \end{pmatrix}\begin{pmatrix} a \\ b \\ c \end{pmatrix} = \begin{pmatrix} 1 \\ 2 \\ 9 \end{pmatrix}$$

The matrix of coefficients is

$$A = \begin{pmatrix} 1 & 1 & 1 \\ 8 & 4 & 2 \\ 27 & 9 & 3 \end{pmatrix}$$

Evaluating the determinant of A by cofactors, we have

$$|A| = 1(12-18) - 1(24-54) + 1(72-108)$$

$$|A| = \quad -6 \qquad +30 \qquad -36$$

$$|A| = -12$$

The inverse can be found by dividing each element of the matrix of cofactors by $|A|$ and transposing the resulting matrix. This gives

$$A^{-1} = \begin{pmatrix} -6/-12 & 30/-12 & -36/-12 \\ 6/-12 & -24/-12 & 18/-12 \\ -2/-12 & 6/-12 & -4/-12 \end{pmatrix}^t$$

$$= \begin{pmatrix} 1/2 & -5/2 & 3 \\ -1/2 & 2 & -3/2 \\ 1/6 & -1/2 & 1/3 \end{pmatrix}^t$$

Multiplying the constant vector by the inverse to find the values of a, b, and c, we have

$$\begin{pmatrix} a \\ b \\ c \end{pmatrix} = \begin{pmatrix} \frac{1}{2} & -\frac{1}{2} & \frac{1}{6} \\ -\frac{5}{2} & 2 & -\frac{1}{2} \\ 3 & -\frac{3}{2} & \frac{1}{3} \end{pmatrix}\begin{pmatrix} 1 \\ 2 \\ 9 \end{pmatrix} = \begin{pmatrix} 1 \\ -3 \\ 3 \end{pmatrix}$$

This gives $a = 1$, $b = -3$, $c = 3$, and $C = q^3 - 3q^2 + 3q + 1$ is the desired function.

Example 2. Find the level of production that gives the maximum profit, assuming the cost function $C = q^3 - 3q^2 + 3q + 1$ from Example 1 and the revenue function $R = -2q^2 + 7q$ from Example 1 of the previous section.

Solution:

$$P = R - C$$

$$P = -2q^2 + 7q - (q^3 - 3q^2 + 3q + 1)$$

$$P = -2q^2 + 7q - q^3 + 3q^2 - 3q - 1$$

$$P = -q^3 + q^2 + 4q - 1$$

Setting the first derivative equal to zero gives

$$\frac{dP}{dq} = -3q^2 + 2q + 4$$

$$0 = -3q^2 + 2q + 4$$

Using the quadratic formula* to solve for q we have

$$q = \frac{-2 \pm \sqrt{4+48}}{-6}$$

$$= \frac{-2 \pm \sqrt{52}}{-6}$$

$$= \frac{-2 \pm 7.2}{-6}$$

$$q = \frac{-9.2}{-6} \quad \text{or} \quad q = \frac{5.2}{-6}$$

Clearly, $q < 0$ is not operationally meaningful in this example. Using the positive value for q gives $q = 1.53$ and the maximum profit occurs when 153 units are produced.

PROBLEMS

Table 14-8 is from *The Price System and Resource Allocation* by Richard H. Leftwich.†

TABLE 14-8

Quantity of X	Total Fixed Cost	Total Variable Cost	Total Cost
1	$100	$40	$140
2	100	70	170
3	100	85	185
4	100	96	196

* Equations of the type $ax^2 + bx + c = 0$ have solutions $x = (-b \pm \sqrt{b^2 - 4ac})/2a$. This expression is called the quadratic formula.

† Richard H. Leftwich, *The Price System and Resource Allocation.* New York: Holt, Rinehart and Winston, Inc., 1966, p. 131.

TABLE 14-8 (*continued*)

Quantity of X	Total Fixed Cost	Total Variable Cost	Total Cost
5	100	104	204
6	100	110	210
7	100	115	215
8	100	120	220
9	100	126	226
10	100	134	234
11	100	145	245
12	100	160	260
13	100	180	280
14	100	206	306
15	100	239	339
16	100	280	380
17	100	330	430
18	100	390	490
19	100	461	561
20	100	544	644

1. **(a)** Use price = \$25 and the data in Table 10-8 to complete Table 14-9.

TABLE 14-9

x	C	MC	MR	R	P
0	100				
1	140	70	25	25	−115
⋮					
20	644				

(b) What level of production gives the largest profit?

2. The least-squares procedure was used to fit a cubic cost function to the cost data of Problem 1(a). The resulting cost function is $C = 0.173x^3 - 3.74x^2 + 31.944x + 116$. Use this fitted cost function and the revenue function $R = 25x$ to find the production level that gives the maximum profit. How does your result compare with the result of Problem 1(b)?

3. If one can identify a function as being a quadratic, does it follow that it has a maximum or minimum? Can it have both a maximum and a minimum?

4. Once one identifies a function as being a cubic, does it necessarily follow that it has an inflection point? A maximum or minimum?

5. What is the general shape of a function that graphs as a hyperbola? Will it have a maximum or minimum? An inflection point?

6. What is the general shape of a logarithmic function? Can it have a maximum or minimum? An inflection point?

7. What is the general shape of an exponential function? Will it have a maximum or minimum? An inflection point?

8. If one is faced with the following cost and revenue functions what would the resulting profit function graph as?

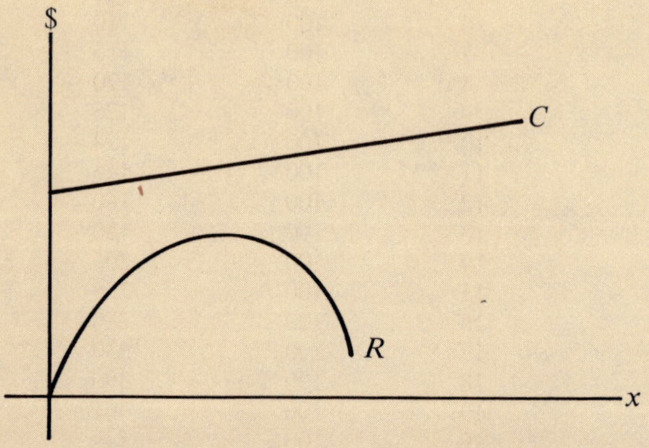

9. Would the profit function in Problem 8 have a maximum or minimum?

10. Given the following cost and revenue functions:

$$C = x + 2.5$$

$$R = -2x^2 + 7x$$

(a) Graph the functions.
(b) For what value of x is R at a maximum.
(c) Find and graph the profit function.
(d) For what value of x is P at a maximum.
(e) Do the revenue and profit functions have a maximum at a common level of production?

11. Given the following cost and revenue functions:

$$C = 1x + 40$$

$$R = -.2x^2 + 10x$$

(a) Graph the functions.
(b) For what value of x is R at a maximum.
(c) Find and graph the profit function.
(d) For what value of x is P at a maximum.
(e) Do the revenue and profit functions have a maximum at a common level of production?

FIFTEEN

Integral calculus

The subject of calculus can be divided into two main branches, differential calculus and integral calculus. The basic concept of differential calculus is the *derivative*, which has been discussed in the last few chapters. In this chapter, we will give a brief treatment of integral calculus. The basic concept of integral calculus is the *definite integral*. Both of these concepts have a natural geometric interpretation. We have seen that the derivative gives the slope of the tangent line to a curve at a given point. It will be shown in this chapter that the definite integral can be interpreted geometrically as the *area under a curve*, and evaluated by making use of the fundamental theorem of integral calculus. Lastly we will look at the indefinite integral.

15-1 DEFINITION OF THE DEFINITE INTEGRAL

The definition of the integral involves, among other things, subdivisions of certain intervals on the x axis. We will illustrate this idea before giving a definition of the integral.

Consider the interval $[a, b]$ consisting of the points x, such that $a \leqslant x \leqslant b$. Divide the interval $[a, b]$ into n subintervals of length $(b-a)/n$. This gives $n+1$ points on the x axis, each of which is the endpoint of at least one subinterval. It is customary to label these points as follows: $x_0, x_1, x_2, \ldots, x_{n-1}, x_n$, where $x_0 = a$ and $x_n = b$.

This labelling is illustrated further in Figure 15-1. This figure illustrates the special case where $n = 9$.

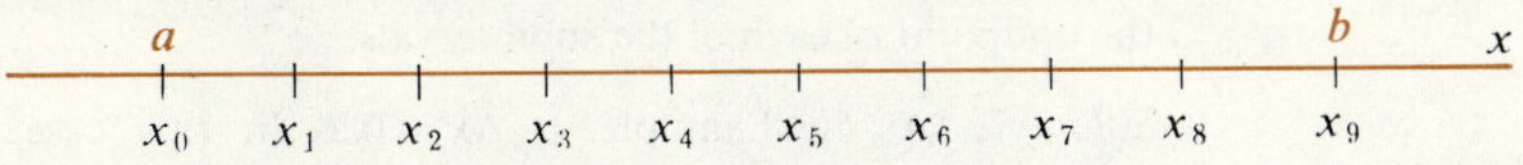

FIGURE 15-1

DEFINITION 15.1. *Let $y = f(x)$ be a function defined for all x between $x = a$ and $x = b$. Then $\int_a^b f(x)\,dx$, called the definite integral of $f(x)$ between a and b, is defined to be the following limit:*

$$\int_a^b f(x)\,dx = \lim_{n\to\infty} [f(t_1)\,\Delta x + f(t_2)\,\Delta x + \cdots + f(t_n)\,\Delta x]$$

where $\Delta x = (b-a)/n$
$=$ any point beloning to the ith subinterval, i.e., $x_{i-1} \leqslant t_i \leqslant x_i$ for $i = 1, 2, 3, \ldots, n$

By arithmetic interpretation, a definite integral is the limit of a sum of products. This concept also has an important geometric interpretation. This interpretation will be discussed later. The next few examples illustrate the various parts of Definition 15.1. A sum of products of the form $f(t_1)\,\Delta x + f(t_2)\,\Delta x + \cdots + f(t_n)\,\Delta x$ will be denoted by the symbol S_n.

$$S_n = f(t_1)\,\Delta x + f(t_2)\,\Delta x + \cdots + f(t_n)\,\Delta x$$

Example 1. Describe S_5 as fully as possible for $\int_2^3 f(x)\,dx$ if $t_1, t_2, \ldots, t_n$ are taken as the left-hand endpoints of each of the subintervals.

Solution: $a = 2$ and $b = 3$; therefore,

$$\Delta x = \frac{3-2}{5} = \frac{1}{5} = 0.2$$

The subintervals are normally labelled as follows with $a = x_0$ and $b = x_n$.

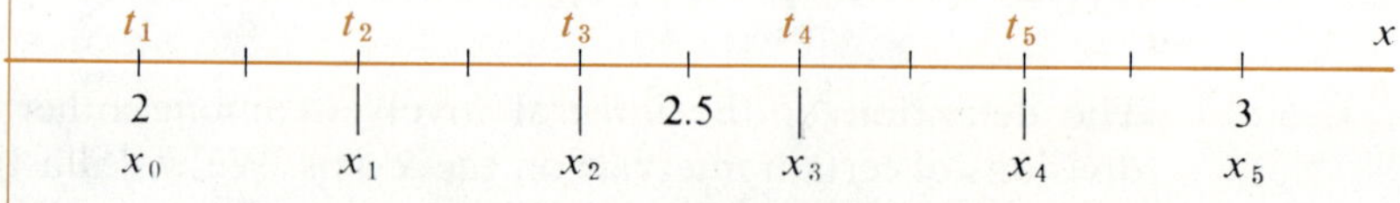

FIGURE 15-2

$$S_5 = f(t_1)\,\Delta x + f(t_2)\,\Delta x + f(t_3)\,\Delta x + f(t_4)\,\Delta x + f(t_5)\,\Delta x$$

$$S_5 = f(2)0.2 + f(2.2)0.2 + f(2.4)0.2 + f(2.6)0.2 + f(2.8)0.2$$

Example 2. Describe S_5 as fully as possible for $\int_2^3 f(x)\,dx$ if $t_1, t_2, \ldots, t_n$ are taken at the midpoint of each of the subintervals.

Solution: As in Example 1, $\Delta x = 0.2$. In this case, $t_1 = 2.1$, $t_2 = 2.3$, $t_3 = 2.5$, $t_4 = 2.7$, and $t_5 = 2.9$.

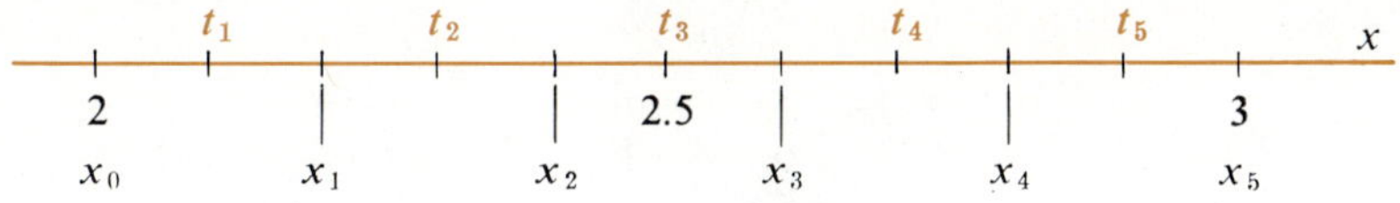

FIGURE 15-3

$$S_5 = f(t_1)\,\Delta x + f(t_2)\,\Delta x + \cdots + f(t_5)\,\Delta x$$

$$S_5 = f(2.1)0.2 + f(2.3)0.2 + f(2.5)0.2 + f(2.7)0.2 + f(2.9)0.2$$

Example 3. Calculate S_5 for Examples 1 and 2 if $f(x) = x^2$.

Solution: From Example 1,

$$S_5 = f(2)0.2 + f(2.2)0.2 + f(2.4)0.2 + f(2.6)0.2 + f(2.8)0.2$$

$$= 2^2(0.2) + (2.2)^2(0.2) + (2.4)^2(0.2) + (2.6)^2(0.2) + (2.8)^2(0.2)$$

$$= 0.2[4 + 4.84 + 5.76 + 6.76 + 7.84]$$

$$= 0.2[29.2]$$

$$S_5 = 5.84$$

From Example 2, $S_5 = f(2.1)0.2 + \cdots + f(2.9)0.2$.

$$S_5 = (2.1)^2 0.2 + (2.3)^2 0.2 + (2.5)^2 0.2 + (2.7)^2 0.2 + (2.9)^2 0.2$$

$$= 0.2[4.41 + 5.29 + 6.25 + 7.29 + 8.41]$$

$$= 0.2(26.65)$$

$$S_5 = 6.33$$

We see from Example 3 that the value of S_n for fixed n depends on the choice of $t_1, t_2, \ldots, t_n$. It is shown, however, in more advanced treatments of calculus that if $f(x)$ is continuous on $a \leqslant x \leqslant b$, then $\lim_{n\to\infty} S_n$ exists, and this limit is not dependent on the choice of $t_1, t_2, \ldots, t_n$. A proof of this fact is well beyond the scope of this text.

The value of S_n for some particular choice of $t_1, t_2, \ldots, t_n$ gives an approximate value of $\int_a^b f(x)\,dx$. The larger the value of n, the better this approximation becomes. Both of these statements follow from the fact that the definite integral is defined to be the limit as n approaches infinity of S_n.

if the curve is smooth, there will always be an area under the curve

Example 4. Find an approximation for $\int_0^2 x^3\,dx$ using $n = 4$ and taking t_i at the right-hand side of the ith subinterval.

Solution:

$$a = 0, \quad b = 2, \quad \Delta x = \frac{2-0}{4} = \frac{1}{2} = 0.5$$

$$t_1 = 0.5, \quad t_2 = 1, \quad t_3 = 1.5, \quad \text{and} \quad t_4 = 2$$

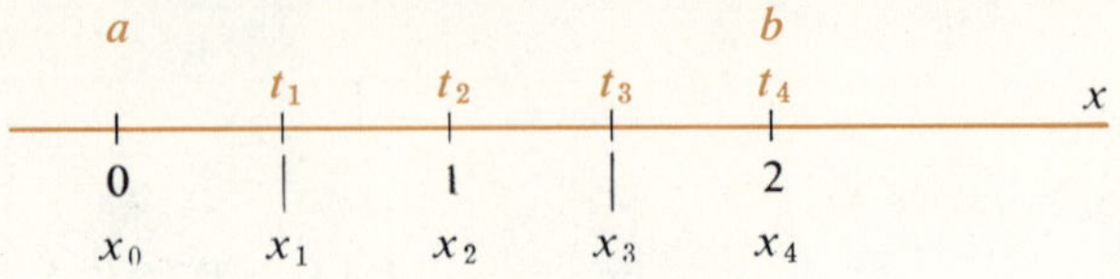

FIGURE 15-4

$$S_4 = \left(\tfrac{1}{2}\right)^3\tfrac{1}{2} + (1)^3\tfrac{1}{2} + \left(\tfrac{3}{2}\right)^3\tfrac{1}{2} + (2)^3\tfrac{1}{2}$$
$$= \tfrac{1}{2}\left(\tfrac{1}{8}+1+\tfrac{27}{8}+8\right) = \tfrac{1}{2}\left(\tfrac{25}{2}\right)$$

Therefore, $S_4 = 6\frac{1}{4}$ is a rough (since n is only 4) approximation for $\int_0^2 x^3\,dx$.

PROBLEMS

1. Consider the *definite integral* $\int_1^2 x^2\,dx$.
   (a) Calculate S_5 by taking function values at the left endpoint of each subinterval.
   (b) Calculate S_5 by taking function values at the right endpoint of each subinterval.
   (c) Calculate S_5 by taking function values at the midpoint of each subinterval.
2. Consider the *definite integral* $\int_2^4 (2x+3)\,dx$.
   (a) Calculate S_4 by taking function values at the left endpoint of each subinterval.
   (b) Calculate S_4 by taking function values at the right endpoint of each subinterval.
3. Consider the *definite integral* $\int_0^1 e^x\,dx$. Compute S_5 by taking function values at the midpoint of each subinterval.
4. Consider the *definite integral* $\int_1^2 (1/x)\,dx$. Compute S_4 by taking function values at the midpoint of each subinterval.
5. State the definition of $\int_a^b f(x)\,dx$.
6. State the definition of S_n.
7. What is the relationship of $\int_a^b f(x)\,dx$ to S_n?

15-2 A GEOMETRIC INTERPRETATION OF THE DEFINITE INTEGRAL

Let C be the graph of $y = f(x)$ between $x = a$ and $x = b$, where $f(x)$ is continuous and positive on the interval $a \leqslant x \leqslant b$. The area, A, of the part of the xy plane bounded by the vertical lines $x = a$ and $x = b$, the x axis, and the curve C is called the *area under the curve C*.

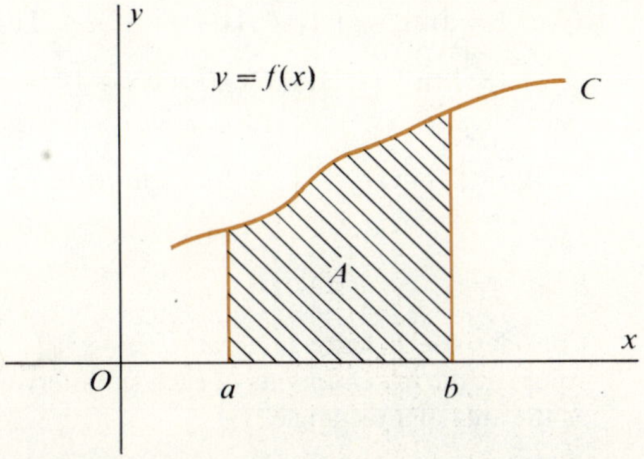

FIGURE 15-5

The area, A, is equal to $\int_a^b f(x)\,dx$.

As a first step in the explanation of the above statement, we consider the geometric interpretation of a typical S_4.

Taking the function values at the midpoints, we have

$$S_4 = f(t_1)\,\Delta x + f(t_2)\,\Delta x + f(t_3)\,\Delta x + f(t_4)\,\Delta x$$

The first term of S_4 is $f(t_1)\,\Delta x$, where t_1 is at the midpoint of the first subinterval. Now, $f(t_1)$ gives the vertical distance between the x axis and the graph of $y = f(x)$ at $x = t_1$. Therefore, $f(t_1)$ times Δx gives the area, A_1, of the first rectangle shown in Figure 15-6.

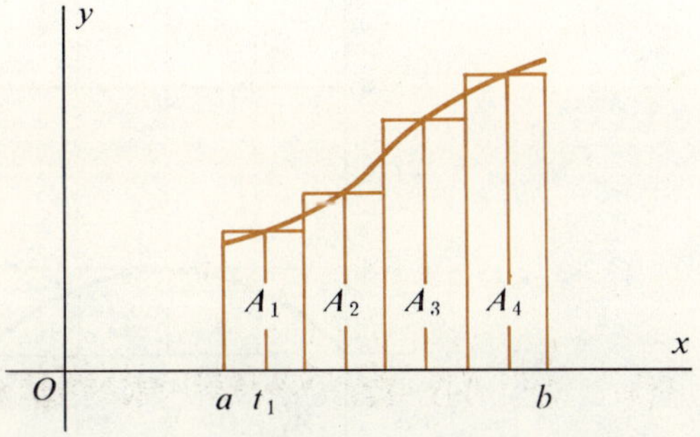

FIGURE 15-6

Similarly, $f(t_2)\,\Delta x$ gives the area, A_2, of the second rectangle, and so forth. Figure 15-6 suggests that $S_4 = A_1 + A_2 + A_3 + A_4$ gives an approximation for the area under the graph of $y = f(x)$ between $x = a$ and $x = b$.

In general, we have $S_n = A_1 + A_2 + \cdots + A_n \approx A$. Furthermore, this approximation of A by S_n gets better as n increases. In fact, the area under the graph of $y = f(x)$ between $x = a$ and $x = b$ is defined to be $A = \lim_{n\to\infty}(A_1 + A_2 + \cdots + A_n)$. This can be restated as

$$A = \lim_{n\to\infty} [f(t_1)\,\Delta x + f(t_2)\,\Delta x + \cdots + f(t_n)\,\Delta x]$$

$$A = \int_a^b f(x)\,dx \qquad \textit{by Definition 15.1}$$

Example 1. Consider the definite integral $\int_0^1 (x - x^2)\,dx$. Compute S_4 taking function values at the left endpoints of each subinterval and interpret each term in S_4 as the area of a rectangle.

Solution:

$$f(x) = x - x^2,\ \Delta x = 0.25$$

$$x_0 = 0,\ x_1 = \tfrac{1}{4},\ x_2 = \tfrac{1}{2},\ x_3 = \tfrac{3}{4},\ x_4 = 1$$

$$t_1 = 0,\ t_2 = \tfrac{1}{4},\ t_3 = \tfrac{1}{2},\ t_4 = \tfrac{3}{4}$$

$$S_4 = (0)0.25 + (\tfrac{1}{4} - \tfrac{1}{16})0.25 + (\tfrac{1}{2} - \tfrac{1}{4})0.25 + (\tfrac{3}{4} - \tfrac{9}{16})0.25$$

$$S_4 = (0 \cdot \tfrac{1}{4}) + (\tfrac{3}{16} \cdot \tfrac{1}{4}) + (\tfrac{1}{4} \cdot \tfrac{1}{4}) + (\tfrac{3}{16} \cdot \tfrac{1}{4})$$

$$S_4 = 0 + \tfrac{3}{64} + \tfrac{1}{16} + \tfrac{3}{64}$$

$$S_4 = A_1 + A_2 + A_3 + A_4$$

where $A_1 = 0$, $A_2 = \frac{3}{64}$, $A_3 = \frac{1}{16}$, and

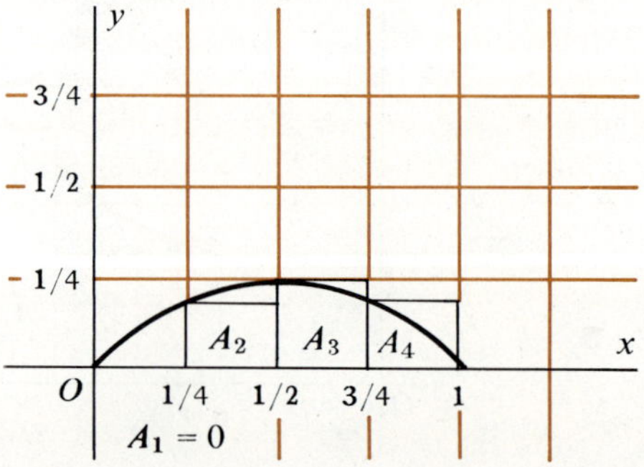

FIGURE 15-7

$$S_4 = \frac{0 + 3 + 4 + 3}{64} = \frac{10}{64} = \frac{5}{32} \approx A$$

Example 2. Draw the rectangles that correspond to the two terms of S_2 for $\int_0^1 x^2\,dx$, where $t_1 = \frac{1}{4}$ and $t_2 = \frac{3}{4}$.

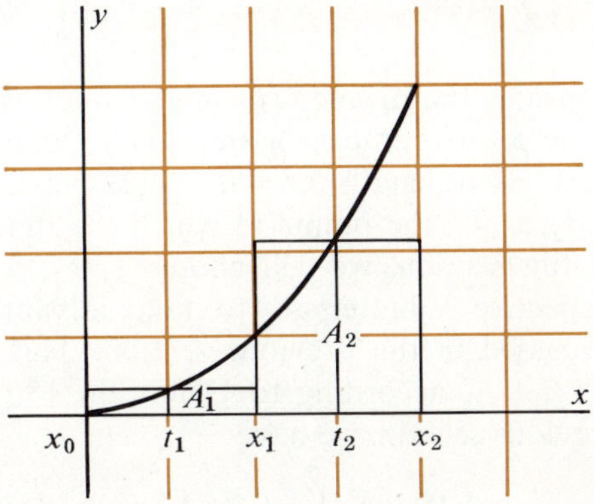

FIGURE 15-8

Solution:

$$A_1 = \tfrac{1}{16} \cdot \tfrac{1}{2} = \tfrac{1}{32}$$

$$A_2 = \tfrac{9}{16} \cdot \tfrac{1}{2} = \tfrac{9}{32}$$

$$A_1 + A_2 = \tfrac{10}{32} = \tfrac{5}{16}$$

The reader should observe that the rectangles in Example 2 *overestimate* the area under the curve between x_0 and t_1 and between x_1 and t_2. On the other hand, they *underestimate* the area between t_1 and x_1 and between t_2 and x_2. This *averaging* effect, which results from choosing the t_i's at the *midpoints* of the subintervals, is used in the next section to *approximate* some definite integrals.

PROBLEMS

1. Draw the rectangles that correspond to the terms of S_5 for the definite integral $\int_0^1 (x^2 + 2x)\, dx$. Take the function values at the left endpoints.
2. Repeat Problem 1 taking the function values at the right endpoints.
3. Compute S_5 in Problems 1 and 2.
4. Draw the graph of $y = x + e^{-x}$ from $x = 1$ to $x = 3$ and shade the area under the curve between $x = 1$ and $x = 3$.
5. Find an approximation for the area in Problem 4 using four rectangles.
6. Shade the area that is given by $\int_2^5 (2/x)\, dx$.
7. Draw the *two* rectangles that correspond to the terms of S_2 for the integral $\int_0^1 x^2\, dx$, where $t_1 = \tfrac{1}{2}$ and $t_2 = 1$.
8. Calculate S_2 for Problem 7.

~~15-3~~ INTEGRALS BY APPROXIMATION

OMIT

Consider the problem of approximating $\int_a^b f(x)\,dx$. First, we choose some positive integer n and divide the interval $a \leqslant x \leqslant b$ into n subintervals of length $\Delta x = (b-a)/n$. Next, we must make a choice for $t_1, t_2, \ldots, t_n$, the points at which the function $f(x)$ is to be evaluated. In this section, we will choose $t_1, t_2, \ldots, t_n$ at the midpoint of their respective subintervals to take advantage of the averaging effect discussed in the previous section. This is a permissible choice for $t_1, t_2, \ldots, t_n$ according to Definition 15.1. Using the notation of the previous section, we have

$$S_n = [f(t_1)\,\Delta x] + [f(t_2)\,\Delta x] + \cdots + [f(t_n)\,\Delta x]$$

where

$$\Delta x = \frac{b-a}{n} \qquad \text{and} \qquad t_k = x_{k-1} + \frac{1}{2}\Delta x \qquad (k = 1, 2, 3, \ldots, n)$$

By Definition 15.1,

$$\int_a^b f(x)\,dx = \lim_{n\to\infty} S_n$$

The method of approximating a definite integral by S_n, where S_n is formed as above, is called the *midvalue rule.*

Example 1. Use the midvalue rule with $n = 5$ to approximate $\int_0^1 x^2\,dx$.

Solution:

$a = 0, \; b = 1$

$b - a = 1 - 0 = 1$

$\Delta x = \frac{1}{5} = 0.2$

$t_1 = 0.1, \; t_2 = 0.3, \; t_3 = 0.5, \; t_4 = 0.7, \; t_5 = 0.9$

$$\begin{aligned} S_5 &= [(0.1)^2 \cdot 0.2] + [(0.3)^2 \cdot 0.2] + [(0.5)^2 \cdot 0.2] \\ &\quad + [(0.7)^2 \cdot 0.2] + [(0.9)^2 \cdot 0.2] \\ &= (0.01 + 0.09 + 0.25 + 0.49 + 0.81)0.2 \\ &= (1.65)(0.2) \end{aligned}$$

$S_5 = 0.33$

$$\int_0^1 x^2\,dx \approx 0.33 \qquad \textit{based on } S_5$$

Example 2. Use the midvalue rule with $n = 5$ to approximate $\int_0^1 e^{x^2}\,dx$.

Solution:

$$S_5 = [e^{(0.1)^2}0.2] + [e^{(0.3)^2}0.2] + [e^{(0.5)^2}0.2] + [e^{(0.7)^2}0.2] + [e^{(0.9)^2}0.2]$$

$$S_5 = 0.2(e^{0.01} + e^{0.09} + e^{0.25} + e^{0.49} + e^{0.81})$$

$$S_5 = 0.2(1.0101 + 1.0942 + 1.2840 + 1.6323 + 2.2479)$$

$$S_5 = 0.2(7.2685)$$

$$S_5 = 1.4537$$

$$\int_0^1 e^{x^2}\,dx \approx 1.454 \qquad \textit{based on } n = 5$$

The values of S_5, S_{10}, and S_{50} for the definite integral $\int_1^2 3x^2\,dx$ were found on a computer using the midvalue rule. Table 15-1 indicates these values.

TABLE 15-1

n	S_n	Δx
5	6.9899	0.20
10	6.9974	0.10
50	6.9998	0.02

These data suggest that $\int_1^2 3x^2\,dx = 7$. This will be found to be the case in the next section.

In the next section we will present the fundamental theorem of integral calculus. This theorem provides a means of finding the exact value of certain definite integrals.

PROBLEMS

1. Given the definite integral $\int_4^{10} f(x)\,dx$, find $t_1, t_2, \ldots, t_{12}$ and $x_0, x_1, \ldots, x_{12}$ for S_{12} using the midvalue rule.
2. Using the data in Problem 1, describe S_{12} as fully as possible.
3. Use the midvalue rule to approximate $\int_1^2 3x^2\,dx$ using $n = 5$, i.e., S_5. (Compare your result with the computer result given in Table 15-1.)
4. Repeat Problem 3 using $n = 10$ and compare your result with the computer result.
5. Use the midvalue rule with $n = 3$ to approximate $\int_0^6 2^x\,dx$.

6. What definite integral is approximated by the following?

$$S_5 = \{[(1.1)^2 + e^{1.1}] + [(1.3)^2 + e^{1.3}] + [(1.5)^2 + e^{1.5}] + [(1.7)^2 + e^{1.7}] + [(1.9)^2 + e^{1.9}]\} \, 0.2$$

7. Consider the definite integral $\int_1^2 x^2 \, dx$.
   (a) Complete the following table by calculating the appropriate S_n values

n	Area Using Left Endpoint	Area Using Right Endpoint	Area Using Midpoint
1			
2			
5			

   (b) S_n yields an approximation to $\int_1^2 x^2 \, dx$ as n approaches what value?

15-4 THE FUNDAMENTAL THEOREM OF INTEGRAL CALCULUS

A large class of definite integrals can be evaluated exactly and with very little work by making use of the fundamental theorem of integral calculus. This is one reason that this theorem is so important. The fundamental theorem of integral calculus is also important for theoretical reasons in that it involves a relationship between definite integrals and derivatives, the basic concepts of integral calculus and differential calculus, respectively.

THEOREM 15.1. *The following is a statement of the fundamental theorem of integral calculus. If $f(x)$ is continuous on the interval $a \leqslant x \leqslant b$ and $F(x)$ is a function whose derivative is $f(x)$, i.e., $F'(x) = f(x)$, then $\int_a^b f(x)\, dx = F(b) - F(a)$.*

We have not included a proof of this theorem.

Theorem 15.1 is applied in two steps:

Step 1: We must find a function $F(x)$ whose derivative is $f(x)$. $F(x)$ is called an *antiderivative* of $f(x)$.

Step 2: Next, evaluate $F(x)$ at $x = a$ and at $x = b$ and compute the difference $F(b) - F(a)$.

Example 1. Use the fundamental theorem of integral calculus to evaluate $\int_0^1 x^2 \, dx$.

Solution:

Step 1:

$$f(x) = x^2, \qquad a = 0 \qquad \text{and} \qquad b = 1$$

Recall that $(d/dx)(x^n) = nx^{n-1}$. We see that $F(x)$ must have the form $F(x) = kx^3$, where k is still undetermined. Since $F'(x)$ must equal $f(x) = x^2$, we have $3kx^2 = x^2$.

$$3k = 1$$

$$k = \tfrac{1}{3}$$

Therefore, the function $F(x) = \frac{1}{3}x^3$ has the desired property.

Step 2:

$$\int_0^1 x^2\,dx = F(1) - F(0)$$

$$= \tfrac{1}{3}(1)^3 - \tfrac{1}{3}(0)^3$$

$$= \tfrac{1}{3} - 0$$

$$\int_0^1 x^2\,dx = \tfrac{1}{3} = 0.333\cdots$$

The reader should recall the calculation in the previous section where it was found that $\int_0^1 x^2\,dx \approx 0.33$ (based on the midvalue rule with $n = 5$).

The following table of antiderivatives can be verified by differentiation.

TABLE 15-2

$f(x)$	$F(x)$
1. $cx^n (n \neq -1)$	$\dfrac{cx^{n+1}}{n+1}$
2. $cx^{-1} = \dfrac{c}{x}$	$c \ln x$
3. ce^{ax}	$\dfrac{c}{a}e^{ax}$
4. ca^x	$\dfrac{ca^x}{\ln a}$

Example 2. Use the fundamental theorem to evaluate $\int_1^2 3x^2\,dx$ and compare with the computer result given in Table 15-1.

Solution:

$$\int_1^2 3x^2\,dx, \qquad f(x) = 3x^2, \qquad F(x) = x^3$$

$$F(2) - F(1) = 8 - 1 = 7$$

$$\int_1^2 3x^2\,dx = 7$$

The computer results are as follows:

$$S_5 = 6.9899$$

$$S_{10} = 6.9974$$

$$S_{50} = 6.9998$$

Example 3. Use the fundamental theorem of integral calculus to evaluate

$$\int_1^3 (x^2 + 3x + 4)\,dx$$

Solution:

$$f(x) = x^2 + 3x + 4 = x^2 + 3x + 4x^0$$

$F(x) = (x^3/3) + (3x^2/2) + 4x$ is obtained by repeated application of item (1) of Table 15-2. Note that the derivative of a sum of functions is the sum of the derivatives of these functions.

Let $a = 1$, $b = 3$.

$$F(b) - F(a) = \frac{(3)^3}{3} + \frac{3(3)^2}{2} + 4(3) - \left(\frac{1}{3} + \frac{3}{2} + 4\right)$$

$$= \left(9 + \frac{27}{2} + 12\right) - \left(\frac{1}{3} + \frac{3}{2} + 4\right)$$

$$= \frac{54 + 81 + 72 - 2 - 9 - 24}{6}$$

$$= \frac{172}{6}$$

$$= 28\tfrac{2}{3}$$

Example 4. Evaluate $\int_1^2 (1/x)\,dx$.

Solution:

$$f(x) = \frac{1}{x}, \qquad F(x) = \ln x$$

$$\int_1^2 \frac{1}{x}\,dx = \ln(2) - \ln(1) = \ln(2) - 0 = \ln 2$$

Example 5. Evaluate $\int_2^4 (1/x^2)\,dx$.

Solution:

$$f(x) = \frac{1}{x^2} = x^{-2}$$

$$F(x) = \frac{x^{-1}}{-1}, \qquad a = 2, \qquad b = 4$$

$$F(b) - F(a) = -4^{-1} - (-2^{-1})$$

$$= \frac{-1}{4} + \frac{1}{2}$$

$$= \frac{1}{4}$$

Example 6. Evaluate $\int_1^4 \sqrt{x}\,dx$.

Solution:

$$f(x) = \sqrt{x}$$

$$f(x) = x^{1/2}$$

$$F(x) = \frac{x^{(1/2)+1}}{(1/2)+1} = \frac{x^{3/2}}{3/2}$$

$$F(x) = \frac{2}{3}x^{3/2}$$

$$\int_1^4 \sqrt{x}\,dx = \frac{2}{3}(4)^{3/2} - \frac{2}{3}(1)^{3/2}$$

$$= \frac{2}{3}[8-1] = \frac{14}{3}$$

Example 7. Use the fundamental theorem to evaluate $\int_{-1}^2 4e^{0.3x}\,dx$.

Solution:

$$f(x) = 4e^{0.3x}, \qquad a = -1, \qquad b = 2$$

$$F(x) = \frac{4}{0.3}e^{0.3x}$$

$$\int_{-1}^2 4e^{0.3x}\,dx = F(2) - F(-1)$$

$$= \frac{4}{0.3}e^{0.3(2)} - \frac{4}{3}e^{0.3(-1)}$$

$$= \frac{4}{0.3}[e^{0.6} - e^{-0.3}]$$

$$= \frac{40}{3}[1.8221 - 0.7408]$$

$$= \frac{40}{3}(1.0713)$$

$$\int_{-1}^{2} 4e^{0.3x}\,dx = 14.284$$

PROBLEMS

1. Find an antiderivative for each of the following functions:
   (a) $f(x) = x^2$ (c) $g(x) = e^{4x}$
   (b) $f(x) = 6/x$ (d) $g(x) = 5^x$
2. Find an antiderivative for each of the following functions:
   (a) $f(x) = 3x^3 + 4x^2$ (c) $f(x) = x^2 + 2/x$
   (b) $f(x) = \sqrt{x^3}$ (d) $f(x) = e^x + x^3$
3. Evaluate the following definite integrals by the fundamental theorem:
   (a) $\int_1^3 (x^2 + 2x + 3)\,dx$ (d) $\int_1^2 x^4\,dx$
   (b) $\int_0^4 (4x - x^3)\,dx$ (e) $\int_{-1}^2 (x^2 + 4)\,dx$
   (c) $\int_1^2 \frac{1}{x}\,dx$ (f) $\int_1^e (2/x)\,dx$
4. Evaluate the following definite integrals:
   (a) $\int_0^1 e^{2x}\,dx$ (c) $\int_{1/3}^1 e^{3x}\,dx$
   (b) $\int_0^1 2^x\,dx$ (d) $\int_0^2 3^x\,dx$
5. Verify by differentiation the antiderivatives given in Table 15-2.
6. Evaluate the following definite integrals:
   (a) $\int_0^2 (3x^2 + 2x + 3)\,dx$ (c) $\int_0^1 (e^{2x} + 3^x)\,dx$
   (b) $\int_1^2 (e^x + 2x)\,dx$ (d) $\int_1^{e^2} 3x^{-1}\,dx$
7. Evaluate
   (a) $\int_{-2}^2 x^3\,dx$ (b) $\int_{-2}^1 x^5\,dx$
8. State the fundamental theorem using $g(x)$ and $G(x)$.

15-5 APPLICATIONS OF THE DEFINITE INTEGRAL

The reader will recall from Section 15-2 that the areas under curves can be found using definite integrals. This fact will be illustrated further in this section.

Example 1. Use the midvalue rule to approximate $\int_0^4 x\,dx$ with $n = 4$.

Solution:

$$b = 4, \qquad a = 0, \qquad \Delta x = \frac{4}{4} = 1$$

$$S_4 = \frac{1}{2}(1) + \frac{3}{2}(1) + \frac{5}{2}(1) + \frac{7}{2}(1)$$

$$= \frac{1+3+5+7}{2}$$

$$= \frac{16}{2} = 8$$

This is the exact value because $f(x) = x$ is a linear function.

Example 2. Use the fundamental theorem to evaluate $\int_0^4 x\,dx$.

Solution:

$$f(x) = x, \qquad F(x) = \frac{x^2}{2}$$

$$\int_0^4 x\,dx = \frac{(4)^2}{2} - 0 = \frac{16}{2} = 8$$

Example 3. Use the fact that a definite integral gives the area under the curve and the formula for the area of a triangle to find $\int_0^4 x\,dx$.

Solution:

$$\int_0^4 x\,dx = \text{area under the graph of } y = x \text{ between } x = 0 \text{ and } x = 4$$

$$= \tfrac{1}{2}bh, \text{ i.e., } \tfrac{1}{2} \text{ base times height}$$

$$= \tfrac{1}{2}(4)(4)$$

Therefore,

$$\int_0^4 x\,dx = 8 \qquad \textit{as found in Examples 1 and 2}$$

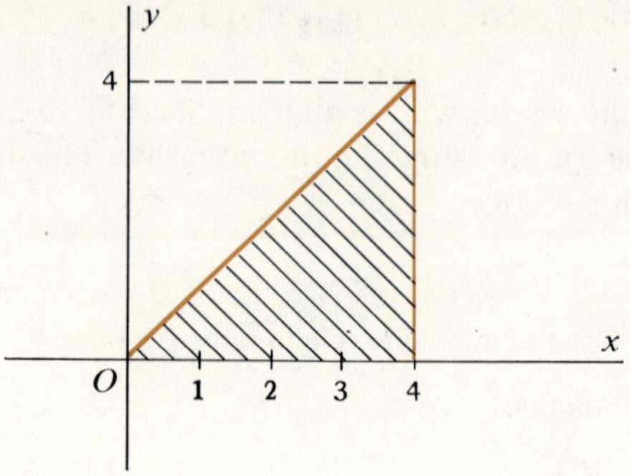

FIGURE 15-9

Example 4. Find the area under $y = x^2$ between $x = 1$ and $x = 3$.

Solution:

$$A = \int_1^3 x^2\,dx$$

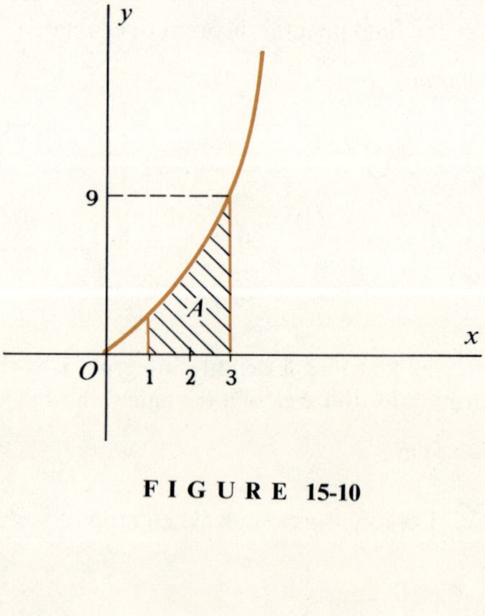

FIGURE 15-10

$$f(x) = x^2, \qquad F(x) = \frac{x^3}{3}$$

$$\int_1^3 x^2\,dx = \frac{27}{3} - \frac{1}{3} = \frac{26}{3} = 8\tfrac{2}{3}$$

Example 5. Find the area under $y = 4 - x^2$ between $x = -1$ and $x = 2$.

Solution:

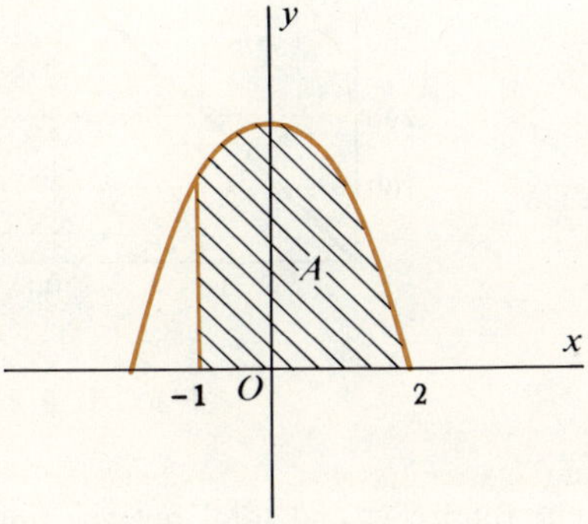

FIGURE 15-11

$$A = \int_{-1}^{2} (4 - x^2)\,dx$$

$$f(x) = 4 - x^2$$

$$F(x) = 4x - \frac{x^3}{3}$$

$$A = \left[4(2) - \frac{(2)^3}{3}\right] - \left[4(-1) - \frac{(-1)^3}{3}\right]$$

$$A = 8 - \tfrac{8}{3} + 4 - \tfrac{1}{3}$$

$$A = 12 - \tfrac{9}{3}$$

$$A = 9$$

Example 6. A certain firm sells its product for \$2.00 per unit. This gives the revenue function $R = 2x$. Find the area bounded by $R = 2x$, the vertical lines $x = 100$ and $x = 300$, and the x axis.

Solution: $A = \int_{100}^{300} 2x\,dx$. An antiderivative for this function is $F(x) = x^2$. Applying the fundamental theorem, we have $A = F(300) - F(100)$, which simplifies to $A = 80{,}000$.

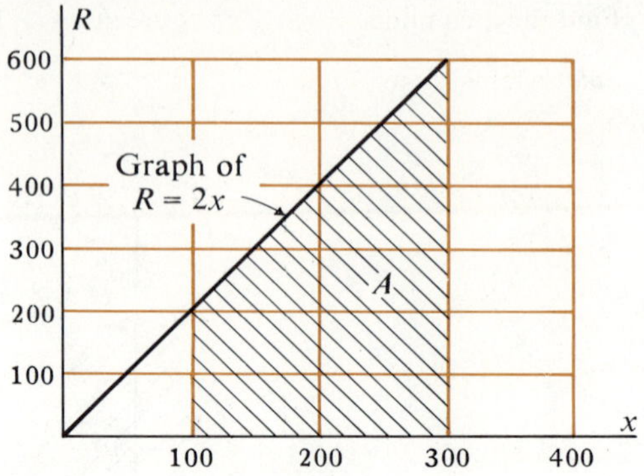

FIGURE 15-12

Example 7 Let the total cost and total revenue functions be given as $C = x + 4$ and $R = -2x^2 + 7x$. Find the area between the cost and revenue functions between $x = 1$ and $x = 2$, i.e., the x coordinates of the break-even points.

Solution: The area in question can be found by subtracting the area under the cost function from the area under the revenue function between these two points. Let A_2 denote the area under the revenue function, and let A_1 denote the area under the cost function. Then

$$A_1 = \int_1^2 (x+4)\,dx \qquad \text{and} \qquad A_2 = \int_1^2 (-2x^2 + 7x)\,dx$$

Therefore,

$$A_1 = 5\tfrac{1}{2} \qquad \text{and} \qquad A_2 = 5\tfrac{5}{6}$$

and the desired area is $\frac{1}{3}$.

PROBLEMS

1. Approximate $\int_0^2 x^2\,dx$ by the midvalue rule using $n = 5$.
2. Evaluate the integral in Problem 1 by the fundamental theorem and compare the results.
3. Shade the area given by the integral in Problem 1.
4. Approximate $\int_2^4 (2x + 3)\,dx$ using the midvalue rule with $n = 5$.
5. Evaluate the integral in Problem 4 in two ways, i.e., by the fundamental theorem and by plane geometry.
6. Explain why the answer in Problem 4 gives the exact value of the integral.
7. Use the midvalue rule with $n = 2$, 4, and 8 to approximate $\int_0^4 e^x\,dx$. Evaluate this integral by the fundamental theorem. Interpret these numerical results in light of the theory given in this chapter.

8. Approximate $\int_{-1}^{0} x^3\,dx$ by the midvalue rule with $n = 5$.

9. Evaluate the integral in Problem 8 by the fundamental theorem.

10. Evaluate A_1 and A_2 and draw the graph for Example 7 of this section.

11. **(a)** Shade the area under $y = e^{-x}$ between $x = 0$ and $x = 6$.
(b) Use the fundamental theorem to evaluate $\int_0^6 e^{-x}\,dx$.

~~15-6~~ AREAS UNDER A NORMAL CURVE

The function $y = (1/\sqrt{2\pi})e^{-x^2/2}$ is encountered frequently in the study of statistics. In particular, the area under the graph of this function between $x = 0$ and $x = z$ is of interest. The graph of the function under discussion is shown below.

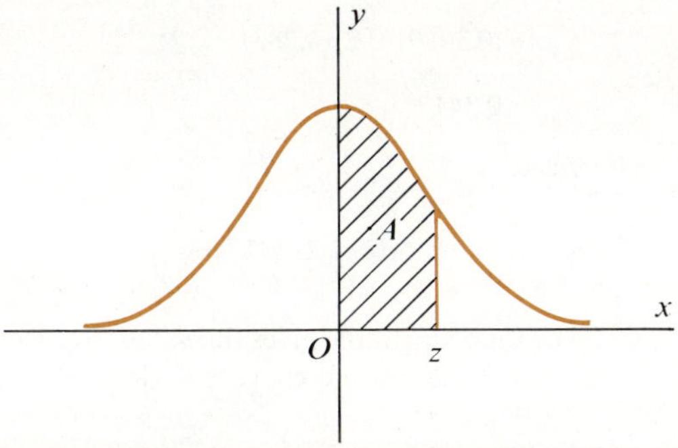

FIGURE 15-13

The area under the graph of $y = (1/\sqrt{2\pi})e^{-x^2/2}$ between $x = 0$ and $x = z$ is equal to the definite integral, $\int_2^z (1/\sqrt{2\pi})e^{-x^2/2}\,dx$. Now, $f(x) = (1/\sqrt{2\pi})e^{-x^2/2}$ is continuous between $x = 0$ and $x = z$ for any z. If we could find an antiderivative $F(x)$, then the desired area would be given by $A = F(z) - F(0)$. However, it has been shown by mathematicians that the function $f(x) = (1/\sqrt{2\pi})e^{-x^2/2}$ does not have an antiderivative in closed form, i.e., expressed as a formula with a finite number of terms. This means that some technique must be used to approximate $\int_0^z (1/\sqrt{2\pi})e^{-x^2/2}\,dx$ to a prescribed degree of accuracy.

The midvalue rule is used in the next example to obtain an approximation for $z = 1$.

Example 1. Approximate $\int_0^1 (1/\sqrt{2\pi})e^{-x^2\ 2}\,dx$ using the midvalue rule with $n = 5$.

Solution:

TABLE 15-3

x	$\frac{x^2}{2}$	$e^{-x^2/2}$
0.1	0.005	0.995
0.3	0.045	0.956
0.5	0.125	0.882
0.7	0.245	0.783
0.9	0.405	0.667

$\Delta x = \frac{1}{5} = 0.2$

$$\frac{1}{\sqrt{2\pi}} = 0.3989$$

$$S_5 = \frac{1}{\sqrt{2\pi}} 0.2[0.995 + 0.956 + 0.882 + 0.783 + 0.667]$$

$$S_5 = (0.3989)(0.2)(4.283)$$

$$S_5 = 0.3417$$

Therefore,

$$\int_0^1 \frac{1}{\sqrt{2\pi}} e^{-x^2/2}\, dx \approx 0.3417 = S_5$$

The following table gives the value to 4 decimal places for the area under the graph of the normal curve $y = (1/\sqrt{2\pi})e^{-x^2/2}$ between $x = 0$ and $x = z$.

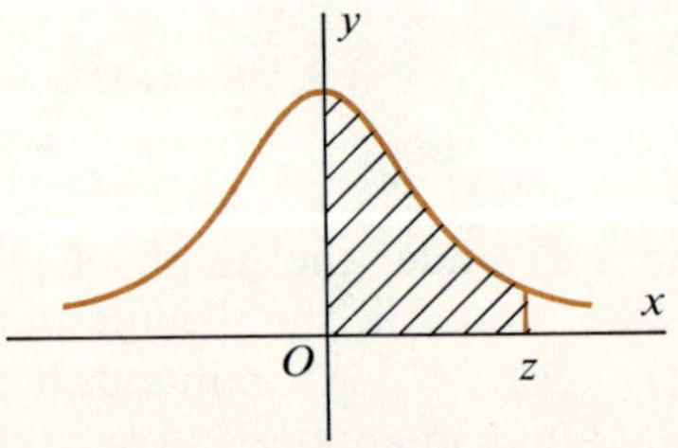

FIGURE 15-14

Example 2. Use Table 15-4 to find the value to 4 decimal places of $\int_0^1 (1/\sqrt{2\pi})e^{-x^2/2}\, dx$ and compare the result with Example 1. (See Figure 15-14.)

Solution: Since z is given as 1, from the table we obtain 0.3413 for the area under the curve between $x = 0$ and $x = 1$. Since the area under a curve is given by the definite integral, we have $\int_0^1 (1/\sqrt{2x})e^{-x^2/2}\, dx = 0.3413$ to 4 decimal places.

In Example 1, we obtained 0.3417 as an approximate value using the midvalue rule with $n = 5$.

TABLE 15-4[a]

z	.00	.01	.02	.03	.04	.05	.06	.07	.08	.09
0.0	.0000	.0040	.0080	.0120	.0160	.0199	.0239	.0279	.0319	.0359
0.1	.0398	.0438	.0478	.0517	.0557	.0596	.0636	.0675	.0714	.0753
0.2	.0793	.0832	.0871	.0910	.0948	.0987	.1026	.1064	.1103	.1141
0.3	.1179	.1217	.1255	.1293	.1331	.1368	.1406	.1443	.1480	.1517
0.4	.1554	.1591	.1628	.1664	.1700	.1736	.1772	.1808	.1844	.1879
0.5	.1915	.1950	.1985	.2019	.2054	.2088	.2123	.2157	.2190	.2224
0.6	.2257	.2291	.2324	.2357	.2389	.2422	.2454	.2486	.2518	.2549
0.7	.2580	.2612	.2642	.2673	.2704	.2734	.2764	.2794	.2823	.2852
0.8	.2881	.2910	.2939	.2967	.2995	.3023	.3051	.3078	.3106	.3133
0.9	.3159	.3186	.3212	.3238	.3264	.3289	.3315	.3340	.3365	.3389
1.0	.3413	.3438	.3461	.3485	.3508	.3531	.3554	.3577	.3599	.3621
1.1	.3643	.3665	.3686	.3708	.3729	.3749	.3770	.3790	.3810	.3830
1.2	.3849	.3869	.3888	.3907	.3925	.3944	.3962	.3980	.3997	.4015
1.3	.4032	.4049	.4066	.4082	.4099	.4115	.4131	.4147	.4162	.4177
1.4	.4192	.4207	.4222	.4236	.4251	.4265	.4279	.4292	.4306	.4319
1.5	.4332	.4345	.4357	.4370	.4382	.4394	.4406	.4418	.4429	.4441
1.6	.4452	.4463	.4474	.4484	.4495	.4505	.4515	.4525	.4535	.4545
1.7	.4554	.4564	.4573	.4582	.4591	.4599	.4608	.4616	.4625	.4633
1.8	.4641	.4649	.4656	.4664	.4671	.4678	.4686	.4693	.4699	.4706
1.9	.4713	.4719	.4726	.4732	.4738	.4744	.4750	.4756	.4761	.4767
2.0	.4772	.4778	.4783	.4788	.4793	.4798	.4803	.4808	.4812	.4817
2.1	.4821	.4826	.4830	.4834	.4838	.4842	.4846	.4850	.4854	.4857
2.2	.4861	.4864	.4868	.4871	.4875	.4878	.4881	.4884	.4887	.4890
2.3	.4893	.4896	.4898	.4901	.4904	.4906	.4909	.4911	.4913	.4916
2.4	.4918	.4920	.4922	.4925	.4927	.4929	.4931	.4932	.4934	.4936
2.5	.4938	.4940	.4941	.4943	.4945	.4946	.4948	.4949	.4951	.4952
2.6	.4953	.4955	.4956	.4957	.4959	.4960	.4961	.4962	.4963	.4964
2.7	.4965	.4966	.4967	.4968	.4969	.4970	.4971	.4972	.4973	.4974
2.8	.4974	.4975	.4976	.4977	.4977	.4978	.4979	.4979	.4980	.4981
2.9	.4981	.4982	.4982	.4983	.4984	.4984	.4985	.4985	.4986	.4986
3.0	.49865	.4987	.4987	.4988	.4988	.4989	.4989	.4989	.4990	.4990
4.0	.4999683									

[a] Adapted from Table A-1 in John Neter, William Wasserman, and G. A. Whitmore, *Fundamental Statistics for Business and Economics*, fourth ed. Boston: Allyn and Bacon, Inc., 1973.

Example 3. Use Table 15-4 to find $\int_1^2 (1/\sqrt{2x})\, e^{-x^2/2}\, dx$.

Solution: Since areas are additive, we have

$$\int_1^2 \frac{1}{\sqrt{2\pi}} e^{-x^2/2}\, dx = \int_0^2 \frac{1}{\sqrt{2\pi}} e^{-x^2/2}\, dx - \int_0^1 \frac{1}{\sqrt{2\pi}} e^{-x^2/2}\, dx$$

$$= 0.4772 - 0.3413$$

$$\int_1^2 \frac{1}{\sqrt{2\pi}} e^{-x^2/2}\, dx = 0.1359 \text{ to 4 decimal places}$$

PROBLEMS

1. Calculate the y coordinate for $x = 0$, $x = \pm 1$, $x = \pm 2$ for the function $y = e^{-x^2/2}$.
2. Use the results of Problem 1 to sketch the graph of $y = e^{-x^2/2}$.
3. Calculate the y coordinate for $x = 0$, $x = \pm 1$, $x = \pm 2$ for the function $y = (1/\sqrt{2\pi})e^{-x^2/2}$, where $1/\sqrt{2\pi} = 0.3989$.
4. Use the results of Problem 3 to sketch the graph of $y = (1/\sqrt{2\pi})e^{-x^2/2}$.
5. Use logs to calculate $1/\sqrt{2\pi}$; $\pi = 3.142$.
6. Use the midvalue rule with $n = 5$ to approximate $\int_0^z (1/\sqrt{2\pi})e^{-x^2/2}\,dx$; $z = \frac{1}{2}$.
7. Use Table 15-4 to find $\int_0^z (1/\sqrt{2\pi})e^{-x^2/2}\,dx$, where $z = \frac{1}{2}$. How does this result compare with the answer obtained in Problem 6?
8. Use Table 15-4 to find the following:

   (a) $\displaystyle\int_0^2 \frac{1}{\sqrt{2\pi}} e^{-x^2/2}\,dx$ **(c)** $\displaystyle\int_0^3 \frac{1}{\sqrt{2\pi}} e^{-x^2/2}\,dx$

   (b) $\displaystyle\int_2^3 \frac{1}{\sqrt{2\pi}} e^{-x^2/2}\,dx$ **(d)** $\displaystyle\int_0^{0.25} \frac{1}{\sqrt{2\pi}} e^{-x^2/2}\,dx$

9. Use Table 15-4 to find $\int_0^{0.8} e^{-x^2/2}\,dx$.

15-7 THE INDEFINITE INTEGRAL AND APPLICATIONS

In Section 15-4 we introduce the notion of an antiderivative in connection with the fundamental theorem of integral calculus. In this section we will discuss this concept further and introduce additional notation. Also, we will show some additional applications of antiderivatives.

We will begin by showing that a whole family of antiderivatives can be constructed for a given function once a single antiderivative has been found. To illustrate this, let $F_1(x)$ be an antiderivative of $f(x)$ and consider the function $F_2(x) = F_1(x) + C$, where C is an arbitrary constant. Since the derivative of a constant is zero and the derivative of a sum of functions is the sum of their derivatives, we have

$$\frac{d}{dx}F_2(x) = \frac{d}{dx}F_1(x) + C$$

$$= f(x) + 0 \qquad \textit{recall the definition of antiderivative}$$

$$= f(x)$$

Therefore, $F_2(x)$ is also an antiderivative of $f(x)$.

This motivates the following definition.

DEFINITION 15.2. *$\int f(x)\,dx = F(x) + C$, where $F(x)$ is an antiderivative of $f(x)$ and C is an arbitrary constant, is called the indefinite integral of $f(x)$.*

This arbitrary constant in this definition is called a *constant of integration.*

Two important properties of indefinite integrals are:

1. $\int cf(x)\,dx = c\int f(x)\,dx$
2. $\int [f(x) + g(x)]\,dx = \int f(x)\,dx + \int g(x)\,dx$

A very brief list of indefinite integrals is included below. Extensive tables of indefinite integrals can be found in mathematical handbooks.

I-1 $$\int x^n\,dx = \frac{x^{n+1}}{n+1} + C \qquad (n \neq -1)$$

I-2 $$\int \frac{1}{x}\,dx = \ln|x| + C$$

I-3 $$\int e^{bx}\,dx = \frac{1}{b}e^{bx} + C$$

I-4 $$\int a^{bx}\,dx = \frac{1}{b\ln a}a^{bx} + C$$

I-5 $$\int \frac{1}{x+b}\,dx = \ln|x+b| + C$$

I-6 $$\int \frac{x}{(x+b)^2}\,dx = \ln|x+b| + \frac{b}{x+b} + C$$

Example 1. Find the indefinite integral of $f(x) = 3x^2 + 2x$ and select the antiderivative of $f(x)$ that passes through point $A = (1, 4)$.

Solution:

$$\int (3x^2 + 2x)\,dx = x^3 + x^2 + C$$

To select the antiderivative that passes through point A, we let $y = x^3 + x^2 + C$.

$$4 = (1)^3 + (1)^2 + C$$

$$4 = 2 + C$$

$$C = 2$$

Therefore, $y = x^3 + x^2 + 2$ is the desired antiderivative.

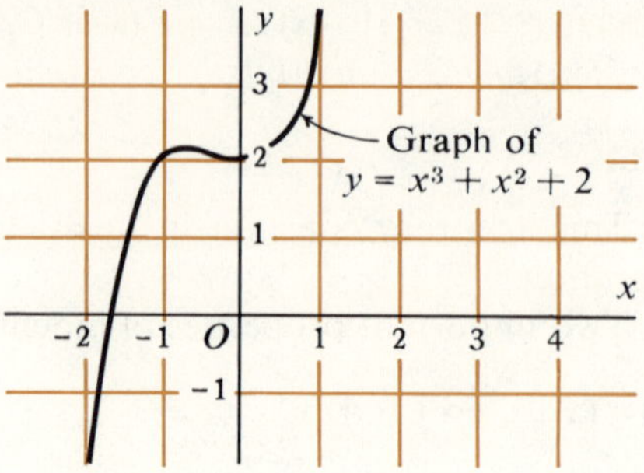

FIGURE 15-15

Example 2. Find the area bounded by $y = 2^{3x}$, the vertical lines $x = 1$ and $x = 2$, and the x axis.

Solution: Let A denote the desired area. Then $A = \int_1^2 2^{3x}\,dx$. Let $f(x) = 2^{3x}$. Then, by I-4, the indefinite integral of $f(x)$ is $(2^{3x}/3 \ln 2) + C$. By the fundamental theorem of integral calculus, we have

$$\int_1^2 2^{3x}\,dx = \left[\frac{2^{3(2)}}{3 \ln 2} + C\right] - \left[\frac{2^{3(1)}}{3 \ln 2} + C\right]$$

$$\int_1^2 2^{3x}\,dx = \frac{1}{3 \ln 2}(2^6 - 2^3)$$

Therefore, $A = 56/2.07944 = 26.93032$.

The reader will note that the arbitrary constant is retained in the solution of Example 2. This constant is cancelled when the fundamental theorem is applied. For this reason, it is customary to omit the constant of integration when the indefinite integral is used to find an antiderivative for use in the fundamental theorem.

Example 3. Given the marginal revenue function $MR = a/(x+b) - (ax/(x+b)^2) + k$, find the revenue function.

Solution: Since marginal revenue is the derivative of the revenue function, we have

$$R = \int \frac{a}{x+b}(dx) + \int \frac{-ax\,dx}{(x+b)^2} + \int k\,dx$$

$$R = a \ln|x+b| - a\left[\ln|x+b| + \frac{b}{x+b}\right] + kx + C$$

$$R = \frac{-ab}{x+b} + kx + C$$

Assume $R = 0$ at $x = 0$. We have $0 = -a + C$, or $C = a$. This gives

$$R = a + kx - \frac{ab}{x+b}$$

This can be simplified, algebraically, to give

$$R = \frac{ax}{x+b} + kx$$

PROBLEMS

1. Find the following indefinite integrals:

(a) $\int (x^3 + 2x + 4)\,dx$

(b) $\int e^{4x}\,dx$

(c) $\int \frac{1}{x+4}\,dx$

(d) $\int \frac{x}{(x-3)^2}\,dx$

2. Verify the following integration formulas by differentiation:

(a) $\int xe^{bx}\,dx = \frac{x}{b}e^{bx} - \frac{1}{b^2}e^{bx} + C$

(b) $\int \ln(bx)\,dx = x\ln(bx) - x + C$

3. Evaluate the following definite integrals:

(a) $\int_0^2 (x^3 - 2x)\,dx$

(b) $\int_2^4 \frac{1}{x-1}\,dx$

(c) $\int_0^1 e^{(1/2)x}\,dx$

(d) $\int_3^6 \frac{1}{x}\,dx$

4. In Example 3 of this section we showed that if

$$MR = \frac{a}{x+b} - \frac{ax}{(x+b)^2} + k$$

then $R = (ax/(x+b)) + kx$. Use this result to show that the demand law, which corresponds to this marginal revenue function, is $p = (a/(x+b)) + k$. [*Hint:* $p = R/x$.]

5. Given $MR = 2 - 6x$, where MR denotes marginal revenue, find the demand law. Assume revenue equals zero when x equals zero.

6. Repeat Problem 5 using $MR = ax + b$, where a and b are constants.

7. Find and shade the *area under the curve* $y = \ln 3x$ between $x = \frac{1}{3}$ and $x = 1$. Use the formula given in Problem 2(b).

8. Exponential growth. Assume that the rate of growth, dy/dx, of y is proportional to the value of y for all values of x

(a) Express the above statement in the form $dy/dx = ky$, where k is a constant.

(b) Rewrite $dy/dx = ky$ in the form $dy/y = k\,dx$, and integrate both sides of this equation. *Answer:* $\ln y = kx + C_1$.

(c) Let $C_1 = \ln C_2$ in (b) and show that $y = C_2 e^{kx}$.

9. Use the result of Problem 8 to determine y as a function of x if $dy/dx = 0.06y$ for all $x \geqslant 0$ and $y = 100$ at $x = 0$.

10. It was shown in Section 6-3 that if a principal is compounded continuously for n years at interest rate i, then the compound amount V_n is given by the formula $V_n = Pe^{ni}$. Use the result of Problem 8 to derive this result.

11. Complete Table 15-5.

TABLE 15-5

$f(x)$	x^2	e^{3x}	xe^{x^2}
$f'(x)$			
$f'(2)$			
$\int f(x)\,dx$			
$\int_1^3 f(x)\,dx$			
$\ln f(x)$			
$\frac{d}{dx}[\ln f(x)]$			

REFERENCES

Allen, R. G. D. *Mathematical Analysis for Economists.* New York: St. Martin's Press, Inc., 1967.

Andree, Richard V. *Introduction to Calculus with Analytic Geometry.* New York: McGraw-Hill Book Co., Inc., 1962.

Brabb, George J. *Introduction to Quantitative Management.* New York: Holt, Rinehart and Winston, Inc., 1968.

Daus, Paul H., and Whyburn, William M. *Introduction to Mathematical Analysis with Applications to Problems in Economics.* Reading, Mass.: Addison-Wesley Publishing Co., Inc., 1962.

Draper, Jean E., and Klingman, Jane S. *Mathematical Analysis Business and Economic Applications.* New York: Harper & Row, Publishers, 1972.

Johnson, R. E., Kiokemeister, F. L., and Wolk, E. S. *Calculus with Analytic Geometry*, fifth ed. Boston: Allyn and Bacon, Inc., 1974.

Leftwich, Richard H. *The Price System and Resource Allocation.* New York: Holt, Rinehart and Winston, 1966.

McBrien, V. O. *Introductory Analysis.* New York: Appleton-Century-Crofts, Inc., 1961.

McConnell, Campbell R. *Economics.* New York: McGraw-Hill Book Co., Inc., 1966.

Stockton, R. Stansburg. *Basic Inventory Systems: Concepts and Analysis.* Boston: Allyn and Bacon, Inc., 1965.

Tierney, John A. *Calculus and Analytic Geometry*, third ed. Boston: Allyn and Bacon, Inc., 1975.

Wasson, Chester R. *Understanding Quantitative Analysis.* New York: Appleton-Century-Crofts, 1969.

Youse, Bevon K., and Stalnaker, Ashford. *Calculus for Students of Business and Management.* Scranton, Pa.: International Textbook Co., Inc., 1967.

APPENDIX A

Tables

TABLE A-1 TABLE OF COMMON LOGARITHMS

N	L 0	1	2	3	4	5	6	7	8	9
0	$-\infty$	00 000	30 103	47 712	60 206	69 897	77 815	84 510	90 309	95 424
1	00 000	04 139	07 918	11 394	14 613	17 609	20 412	23 045	25 527	27 875
2	30 103	32 222	34 242	36 173	38 021	39 794	41 497	43 136	44 716	46 240
3	47 712	49 136	50 515	51 851	53 148	54 407	55 630	56 820	57 978	59 106
4	60 206	61 278	62 325	63 347	64 345	65 321	66 276	67 210	68 124	69 020
5	69 897	70 757	71 600	72 428	73 239	74 036	74 819	75 587	76 343	77 085
6	77 815	78 533	79 239	79 934	80 618	81 291	81 954	82 607	83 251	83 885
7	84 510	85 126	85 733	86 332	86 923	87 506	88 081	88 649	89 209	89 763
8	90 309	90 849	91 381	91 908	92 428	92 942	93 450	93 952	94 448	94 939
9	95 424	95 904	96 379	96 848	97 313	97 772	98 227	98 677	99 123	99 564
10	00 000	00 432	00 860	01 284	01 703	02 119	02 531	02 938	03 342	03 743
11	04 139	04 532	04 922	05 308	05 690	06 070	06 446	06 819	07 188	07 555
12	07 918	08 279	08 636	08 991	09 342	09 691	10 037	10 380	10 721	11 059
13	11 394	11 727	12 057	12 385	12 710	13 033	13 354	13 672	13 988	14 301
14	14 613	14 922	15 229	15 534	15 836	16 137	16 435	16 732	17 026	17 319
15	17 609	17 898	18 184	18 469	18 752	19 033	19 312	19 590	19 866	20 140
16	20 412	20 683	20 952	21 219	21 484	21 748	22 011	22 272	22 531	22 789
17	23 045	23 300	23 553	23 805	24 055	24 304	24 551	24 797	25 042	25 285
18	25 527	25 768	26 007	26 245	26 482	26 717	26 951	27 184	27 416	27 646
19	27 875	28 103	28 330	28 556	28 780	29 003	29 226	29 447	29 667	29 885
20	30 103	30 320	30 535	30 750	30 963	31 175	31 387	31 597	31 806	32 015
21	32 222	32 428	32 634	32 838	33 041	33 244	33 445	33 646	33 846	34 044
22	34 242	34 439	34 635	34 830	35 025	35 218	35 411	35 603	35 793	35 984
23	36 173	36 361	36 549	36 736	36 922	37 107	37 291	37 475	37 658	37 840
24	38 021	38 202	38 382	38 561	38 739	38 917	39 094	39 270	39 445	39 620
25	39 794	39 907	40 140	40 312	40 483	40 654	40 824	40 993	41 162	41 330
26	41 497	41 664	41 830	41 996	42 160	42 325	42 488	42 651	42 813	42 975
27	43 136	43 297	43 457	43 616	43 775	43 933	44 091	44 248	44 404	44 560
28	44 716	44 871	45 025	45 179	45 332	45 484	45 637	45 788	45 939	46 090
29	46 240	46 389	46 538	46 687	46 835	46 982	47 129	47 276	47 422	47 567
30	47 712	47 857	48 001	48 144	48 287	48 430	48 572	48 714	48 855	48 996
31	49 136	49 276	49 415	49 554	49 693	49 831	49 969	50 106	50 243	50 379
32	50 515	50 651	50 786	50 920	51 055	51 188	51 322	51 455	51 587	51 720
33	51 851	51 983	52 114	52 244	52 375	52 504	52 634	52 763	52 892	53 020
34	53 148	53 275	53 403	53 529	53 656	53 782	53 908	54 033	54 158	54 283
35	54 407	54 531	54 654	54 777	54 900	55 023	55 145	55 267	55 388	55 509
36	55 630	55 751	55 871	55 991	56 110	56 229	56 348	56 467	56 585	56 703
37	56 820	56 937	57 054	57 171	57 287	57 403	57 519	57 634	57 749	57 864
38	57 978	58 092	58 206	58 320	58 433	58 546	58 659	58 771	58 883	58 995
39	59 106	59 218	59 329	59 439	59 550	59 660	59 770	59 879	59 988	60 097
40	60 206	60 314	60 423	60 531	60 638	60 746	60 853	60 959	61 066	61 172
41	61 278	61 384	61 490	61 595	61 700	61 805	61 909	62 014	62 118	62 221
42	62 325	62 428	62 531	62 634	62 737	62 839	62 941	63 043	63 144	63 246
43	63 347	63 448	63 548	63 649	63 749	63 849	63 949	64 048	64 147	64 246
44	64 345	64 444	64 542	64 640	64 738	64 836	64 933	65 031	65 128	65 225
45	65 321	65 418	65 514	65 610	65 706	65 801	65 896	65 992	66 087	66 181
46	66 276	66 370	66 464	66 558	66 652	66 745	66 839	66 932	67 025	67 117
47	67 210	67 302	67 394	67 486	67 578	67 669	67 761	67 852	67 943	68 034
48	68 124	68 215	68 305	68 395	68 485	68 574	68 664	68 753	68 842	68 931
49	69 020	69 108	69 197	69 285	69 373	69 461	69 548	69 636	69 723	69 810
50	69 897	69 984	70 070	70 157	70 243	70 329	70 415	70 501	70 586	70 672
N	L 0	1	2	3	4	5	6	7	8	9

TABLE A-1 (*continued*)

N	L 0	1	2	3	4	5	6	7	8	9
50	69 897	69 984	70 070	70 157	70 243	70 329	70 415	70 501	70 586	70 672
51	70 757	70 842	70 927	71 012	71 096	71 181	71 265	71 349	71 433	71 517
52	71 600	71 684	71 767	71 850	71 933	72 016	72 099	72 181	72 263	72 346
53	72 428	72 509	72 591	72 673	72 754	72 835	72 916	72 997	73 078	73 159
54	73 239	73 320	73 400	73 480	73 560	73 640	73 719	73 799	73 878	73 957
55	74 036	74 115	74 194	74 273	74 351	74 429	74 507	74 586	74 663	74 741
56	74 819	74 896	74 974	75 051	75 128	75 205	75 282	75 358	75 435	75 511
57	75 587	75 664	75 740	75 815	75 891	75 967	76 042	76 118	76 193	76 268
58	76 343	76 418	76 492	76 567	76 641	76 716	76 790	76 864	76 938	77 012
59	77 085	77 159	77 232	77 305	77 379	77 452	77 525	77 597	77 670	77 743
60	77 815	77 887	77 960	78 032	78 104	78 176	78 247	78 319	78 390	78 462
61	78 533	78 604	78 675	78 746	78 817	78 888	78 958	79 029	79 099	79 169
62	79 239	79 309	79 379	79 449	79 518	79 588	79 657	79 727	79 796	79 865
63	79 934	80 003	80 072	80 140	80 209	80 277	80 346	80 414	80 482	80 550
64	80 618	80 686	80 754	80 821	80 889	80 956	81 023	81 090	81 158	81 224
65	81 291	81 358	81 425	81 491	81 558	81 624	81 690	81 757	81 823	81 889
66	81 954	82 020	82 086	82 151	82 217	82 282	82 347	82 413	82 478	82 543
67	82 607	82 672	82 737	82 802	82 866	82 930	82 995	83 059	83 123	83 187
68	83 251	83 315	83 378	83 442	83 506	83 569	83 632	83 696	83 759	83 822
69	83 885	83 948	84 011	84 073	84 136	84 198	84 261	84 323	84 386	84 448
70	84 510	84 572	84 634	84 696	84 757	84 819	84 880	84 942	85 003	85 065
71	85 126	85 187	85 248	85 309	85 370	85 431	85 491	85 552	85 612	85 673
72	85 733	85 794	85 854	85 914	85 974	86 034	86 094	86 153	86 213	86 273
73	86 332	86 392	86 451	86 510	86 570	86 629	86 688	86 747	86 806	86 864
74	86 923	86 982	87 040	87 099	87 157	87 216	87 274	87 332	87 390	87 448
75	87 506	87 564	87 622	87 679	87 737	87 795	87 852	87 910	87 967	88 024
76	88 081	88 138	88 195	88 252	88 309	88 366	88 423	88 480	88 536	88 593
77	88 649	88 705	88 762	88 818	88 874	88 930	88 986	89 042	89 098	89 154
78	89 209	89 265	89 321	89 376	89 432	89 487	89 542	89 597	89 653	89 708
79	89 763	89 818	89 873	89 927	89 982	90 037	90 091	90 146	90 200	90 255
80	90 309	90 363	90 417	90 472	90 526	90 580	90 634	90 687	90 741	90 795
81	90 849	90 902	90 956	91 009	91 062	91 116	91 169	91 222	91 275	91 328
82	91 381	91 434	91 487	91 540	91 593	91 645	91 698	91 751	91 803	91 855
83	91 908	91 960	92 012	92 065	92 117	92 169	92 221	92 273	92 324	92 376
84	92 428	92 480	92 531	92 583	92 634	92 686	92 737	92 788	92 840	92 891
85	92 942	92 993	93 044	93 095	93 146	93 197	93 247	93 298	93 349	93 399
86	93 450	93 500	93 551	93 601	93 651	93 702	93 752	93 802	93 852	93 902
87	93 952	94 002	94 052	94 101	94 151	94 201	94 250	94 300	94 349	94 399
88	94 448	94 498	94 547	94 596	94 645	94 694	94 743	94 792	94 841	94 890
89	94 939	94 988	95 036	95 085	95 134	95 182	95 231	95 279	95 328	95 376
90	95 424	95 472	95 521	95 569	95 617	95 665	95 713	95 761	95 809	95 856
91	95 904	95 952	95 999	96 047	96 095	96 142	96 190	96 237	96 284	96 332
92	96 379	96 426	96 473	96 520	96 567	96 614	96 661	96 708	96 755	96 802
93	96 848	96 895	96 942	96 988	97 035	97 081	97 128	97 174	97 220	97 267
94	97 313	97 359	97 405	97 451	97 497	97 543	97 589	97 635	97 681	97 727
95	97 772	97 818	97 864	97 909	97 955	98 000	98 046	98 091	98 137	98 182
96	98 227	98 272	98 318	98 363	98 408	98 453	98 498	98 543	98 588	98 632
97	98 677	98 722	98 767	98 811	98 856	98 900	98 945	98 989	99 034	99 078
98	99 123	99 167	99 211	99 255	99 300	99 344	99 388	99 432	99 476	99 520
99	99 564	99 607	99 651	99 695	99 739	99 782	99 826	99 870	99 913	99 957
100	00 000	00 043	00 087	00 130	00 173	00 217	00 260	00 303	00 346	00 389
N	L 0	1	2	3	4	5	6	7	8	9

TABLE A-1 (*continued*)

N	L 0	1	2	3	4	5	6	7	8	9	PP			
100	00 000	043	087	130	173	217	260	303	346	389		**44**	**43**	**42**
101	432	475	518	561	604	647	689	732	775	817	1	4.4	4.3	4.2
102	860	903	945	988	*030	*072	*115	*157	*199	*242	2	8.8	8.6	8.4
103	01 284	326	368	410	452	494	536	578	620	662	3	13.2	12.9	12.6
104	703	745	787	828	870	912	953	995	*036	*078	4	17.6	17.2	16.8
105	02 119	160	202	243	284	325	366	407	449	490	5	22.0	21.5	21.0
106	531	572	612	653	694	735	776	816	857	898	6	26.4	25.8	25.2
107	938	979	*019	*060	*100	*141	*181	*222	*262	*302	7	30.8	30.1	29.4
108	03 342	383	423	463	503	543	583	623	663	703	8	35.2	34.4	33.6
109	743	782	822	862	902	941	981	*021	*060	*100	9	39.6	38.7	37.8
110	04 139	179	218	258	297	336	376	415	454	493		**41**	**40**	**39**
111	532	571	610	650	689	727	766	805	844	883	1	4.1	4.0	3.9
112	922	961	999	*038	*077	*115	*154	*192	*231	*269	2	8.2	8.0	7.8
113	05 308	346	385	423	461	500	538	576	614	652	3	12.3	12.0	11.7
114	690	729	767	805	843	881	918	956	994	*032	4	16.4	16.0	15.6
115	06 070	108	145	183	221	258	296	333	371	408	5	20.5	20.0	19.5
116	446	483	521	558	595	633	670	707	744	781	6	24.6	24.0	23.4
117	819	856	893	930	967	*004	*041	*078	*115	*151	7	28.7	28.0	27.3
118	07 188	225	262	298	335	372	408	445	482	518	8	32.8	32.0	31.2
119	555	591	628	664	700	737	773	809	846	882	9	36.9	36.0	35.1
120	918	954	990	*027	*063	*099	*135	*171	*207	*243		**38**	**37**	**36**
121	08 279	314	350	386	422	458	493	529	565	600	1	3.8	3.7	3.6
122	636	672	707	743	778	814	849	884	920	955	2	7.6	7.4	7.2
123	991	*026	*061	*096	*132	*167	*202	*237	*272	*307	3	11.4	11.1	10.8
124	09 342	377	412	447	482	517	552	587	621	656	4	15.2	14.8	14.4
125	691	726	760	795	830	864	899	934	968	*003	5	19.0	18.5	18.0
126	10 037	072	106	140	175	209	243	278	312	346	6	22.8	22.2	21.6
127	380	415	449	483	517	551	585	619	653	687	7	26.6	25.9	25.2
128	721	755	789	823	857	890	924	958	992	*025	8	30.4	29.6	28.8
129	11 059	093	126	160	193	227	261	294	327	361	9	34.2	33.3	32.4
130	394	428	461	494	528	561	594	628	661	694		**35**	**34**	**33**
131	727	760	793	826	860	893	926	959	992	*024	1	3.5	3.4	3.3
132	12 057	090	123	156	189	222	254	287	320	352	2	7.0	6.8	6.6
133	385	418	450	483	516	548	581	613	646	678	3	10.5	10.2	9.9
134	710	743	775	808	840	872	905	937	969	*001	4	14.0	13.6	13.2
135	13 033	066	098	130	162	194	226	258	290	322	5	17.5	17.0	16.5
136	354	386	418	450	481	513	545	577	609	640	6	21.0	20.4	19.8
137	672	704	735	767	799	830	862	893	925	956	7	24.5	23.8	23.1
138	988	*019	*051	*082	*114	*145	*176	*208	*239	*270	8	28.0	27.2	26.4
139	14 301	333	364	395	426	457	489	520	551	582	9	31.5	30.6	29.7
140	613	644	675	706	737	768	799	829	860	891		**32**	**31**	**30**
141	922	953	983	*014	*045	*076	*106	*137	*168	*198	1	3.2	3.1	3.0
142	15 229	259	290	320	351	381	412	442	473	503	2	6.4	6.2	6.0
143	534	564	594	625	655	685	715	746	776	806	3	9.6	9.3	9.0
144	836	866	897	927	957	987	*017	*047	*077	*107	4	12.8	12.4	12.0
145	16 137	167	197	227	256	286	316	346	376	406	5	16.0	15.5	15.0
146	435	465	495	524	554	584	613	643	673	702	6	19.2	18.6	18.0
147	732	761	791	820	850	879	909	938	967	997	7	22.4	21.7	21.0
148	17 026	056	085	114	143	173	202	231	260	289	8	25.6	24.8	24.0
149	319	348	377	406	435	464	493	522	551	580	9	28.8	27.9	27.0
150	17 609	638	667	696	725	754	782	811	840	869				
N	**L 0**	**1**	**2**	**3**	**4**	**5**	**6**	**7**	**8**	**9**	**PP**			

TABLE A-1 (*continued*)

N	L 0	1	2	3	4	5	6	7	8	9
150	17 609	638	667	696	725	754	782	811	840	869
151	898	926	955	984	*013	*041	*070	*099	*127	*156
152	18 184	213	241	270	298	327	355	384	412	441
153	469	498	526	554	583	611	639	667	696	724
154	752	780	808	837	865	893	921	949	977	*005
155	19 033	061	089	117	145	173	201	229	257	285
156	312	340	368	396	424	451	479	507	535	562
157	590	618	645	673	700	728	756	783	811	838
158	866	893	921	948	976	*003	*030	*058	*085	*112
159	20 140	167	194	222	249	276	303	330	358	385
160	412	439	466	493	520	548	575	602	629	656
161	683	710	737	763	790	817	844	871	898	925
162	952	978	*005	*032	*059	*085	*112	*139	*165	*192
163	21 219	245	272	299	325	352	378	405	431	458
164	484	511	537	564	590	617	643	669	696	722
165	748	775	801	827	854	880	906	932	958	985
166	22 011	037	063	089	115	141	167	194	220	246
167	272	298	324	350	376	401	427	453	479	505
168	531	557	583	608	634	660	686	712	737	763
169	789	814	840	866	891	917	943	968	994	*019
170	23 045	070	096	121	147	172	198	223	249	274
171	300	325	350	376	401	426	452	477	502	528
172	553	578	603	629	654	679	704	729	754	779
173	805	830	855	880	905	930	955	980	*005	*030
174	24 055	080	105	130	155	180	204	229	254	279
175	304	329	353	378	403	428	452	477	502	527
176	551	576	601	625	650	674	699	724	748	773
177	797	822	846	871	895	920	944	969	993	*018
178	25 042	066	091	115	139	164	188	212	237	261
179	285	310	334	358	382	406	431	455	479	503
180	527	551	575	600	624	648	672	696	720	744
181	768	792	816	840	864	888	912	935	959	983
182	26 007	031	055	079	102	126	150	174	198	221
183	245	269	293	316	340	364	387	411	435	458
184	482	505	529	553	576	600	623	647	670	694
185	717	741	764	788	811	834	858	881	905	928
186	951	975	998	*021	*045	*068	*091	*114	*138	*161
187	27 184	207	231	254	277	300	323	346	370	393
188	416	439	462	485	508	531	554	577	600	623
189	646	669	692	715	738	761	784	807	830	852
190	875	898	921	944	967	989	*012	*035	*058	*081
191	28 103	126	149	171	194	217	240	262	285	307
192	330	353	375	398	421	443	466	488	511	533
193	556	578	601	623	646	668	691	713	735	758
194	780	803	825	847	870	892	914	937	959	981
195	29 003	026	048	070	092	115	137	159	181	203
196	226	248	270	292	314	336	358	380	403	425
197	447	469	491	513	535	557	579	601	623	645
198	667	688	710	732	754	776	798	820	842	863
199	885	907	929	951	973	994	*016	*038	*060	*081
200	30 103	125	146	168	190	211	233	255	276	298
N	**L 0**	**1**	**2**	**3**	**4**	**5**	**6**	**7**	**8**	**9**

P P

	29	**28**
1	2.9	2.8
2	5.8	5.6
3	8.7	8.4
4	11.6	11.2
5	14.5	14.0
6	17.4	16.8
7	20.3	19.6
8	23.2	22.4
9	26.1	25.2

	27	**26**
1	2.7	2.6
2	5.4	5.2
3	8.1	7.8
4	10.8	10.4
5	13.5	13.0
6	16.2	15.6
7	18.9	18.2
8	21.6	20.8
9	24.3	23.4

	25
1	2.5
2	5.0
3	7.5
4	10.0
5	12.5
6	15.0
7	17.5
8	20.0
9	22.5

	24	**23**
1	2.4	2.3
2	4.8	4.6
3	7.2	6.9
4	9.6	9.2
5	12.0	11.5
6	14.4	13.8
7	16.8	16.1
8	19.2	18.4
9	21.6	20.7

	22	**21**
1	2.2	2.1
2	4.4	4.2
3	6.6	6.3
4	8.8	8.4
5	11.0	10.5
6	13.2	12.6
7	15.4	14.7
8	17.6	16.8
9	19.8	18.9

TABLE A-1 (*continued*)

N	L 0	1	2	3	4	5	6	7	8	9	PP		
200	30 103	125	146	168	190	211	233	255	276	298			
201	320	341	363	384	406	428	449	471	492	514		**22**	**21**
202	535	557	578	600	621	643	664	685	707	728	1	2.2	2.1
203	750	771	792	814	835	856	878	899	920	942	2	4.4	4.2
204	963	984	*006	*027	*048	*069	*091	*112	*133	*154	3	6.6	6.3
205	31 175	197	218	239	260	281	302	323	345	366	4	8.8	8.4
206	387	408	429	450	471	492	513	534	555	576	5	11.0	10.5
207	597	618	639	660	681	702	723	744	765	785	6	13.2	12.6
208	806	827	848	869	890	911	931	952	973	994	7	15.4	14.7
209	32 015	035	056	077	098	118	139	160	181	201	8	17.6	16.8
210	222	243	263	284	305	325	346	366	387	408	9	19.8	18.9
211	428	449	469	490	510	531	552	572	593	613		**20**	
212	634	654	675	695	715	736	756	777	797	818	1	2.0	
213	838	858	879	899	919	940	960	980	*001	*021	2	4.0	
214	33 041	062	082	102	122	143	163	183	203	224	3	6.0	
215	244	264	284	304	325	345	365	385	405	425	4	8.0	
216	445	465	486	506	526	546	566	586	606	626	5	10.0	
217	646	666	686	706	726	746	766	786	806	826	6	12.0	
218	846	866	885	905	925	945	965	985	*005	*025	7	14.0	
219	34 044	064	084	104	124	143	163	183	203	223	8	16.0	
220	242	262	282	301	321	341	361	380	400	420	9	18.0	
221	439	459	479	498	518	537	557	577	596	616		**19**	
222	635	655	674	694	713	733	753	772	792	811	1	1.9	
223	830	850	869	889	908	928	947	967	986	*005	2	3.8	
224	35 025	044	064	083	102	122	141	160	180	199	3	5.7	
225	218	238	257	276	295	315	334	353	372	392	4	7.6	
226	411	430	449	468	488	507	526	545	564	583	5	9.5	
227	603	622	641	660	679	698	717	736	755	774	6	11.4	
228	793	813	832	851	870	889	908	927	946	965	7	13.3	
229	984	*003	*021	*040	*059	*078	*097	*116	*135	*154	8	15.2	
230	36 173	192	211	229	248	267	286	305	324	342	9	17.1	
231	361	380	399	418	436	455	474	493	511	530		**18**	
232	549	568	586	605	624	642	661	680	698	717	1	1.8	
233	736	754	773	791	810	829	847	866	884	903	2	3.6	
234	922	940	959	977	996	*014	*033	*051	*070	*088	3	5.4	
235	37 107	125	144	162	181	199	218	236	254	273	4	7.2	
236	291	310	328	346	365	383	401	420	438	457	5	9.0	
237	475	493	511	530	548	566	585	603	621	639	6	10.8	
238	658	676	694	712	731	749	767	785	803	822	7	12.6	
239	840	858	876	894	912	931	949	967	985	*003	8	14.4	
240	38 021	039	057	075	093	112	130	148	166	184	9	16.2	
241	202	220	238	256	274	292	310	328	346	364		**17**	
242	382	399	417	435	453	471	489	507	525	543	1	1.7	
243	561	578	596	614	632	650	668	686	703	721	2	3.4	
244	739	757	775	792	810	828	846	863	881	899	3	5.1	
245	917	934	952	970	987	*005	*023	*041	*058	*076	4	6.8	
246	39 094	111	129	146	164	182	199	217	235	252	5	8.5	
247	270	287	305	322	340	358	375	393	410	428	6	10.2	
248	445	463	480	498	515	533	550	568	585	602	7	11.9	
249	620	637	655	672	690	707	724	742	759	777	8	13.6	
250	794	811	829	846	863	881	898	915	933	950	9	15.3	
N	**L 0**	**1**	**2**	**3**	**4**	**5**	**6**	**7**	**8**	**9**	**P P**		

TABLE A-1 (*continued*)

N	**L 0**	**1**	**2**	**3**	**4**	**5**	**6**	**7**	**8**	**9**	**P P**	
250	39 794	811	829	846	863	881	898	915	933	950		**18**
251	967	985	*002	*019	*037	*054	*071	*088	*106	*123	1	1.8
252	40 140	157	175	192	209	226	243	261	278	295	2	3.6
253	312	329	346	364	381	398	415	432	449	466	3	5.4
254	483	500	518	535	552	569	586	603	620	637	4	7.2
255	654	671	688	705	722	739	756	773	790	807	5	9.0
256	824	841	858	875	892	909	926	943	960	976	6	10.8
257	993	*010	*027	*044	*061	*078	*095	*111	*128	*145	7	12.6
258	41 162	179	196	212	229	246	263	280	296	313	8	14.4
259	330	347	363	380	397	414	430	447	464	481	9	16.2
260	497	514	531	547	564	581	597	614	631	647		**17**
261	664	681	697	714	731	747	764	780	797	814	1	1.7
262	830	847	863	880	896	913	929	946	963	979	2	3.4
263	996	*012	*029	*045	*062	*078	*095	*111	*127	*144	3	5.1
264	42 160	177	193	210	226	243	259	275	292	308	4	6.8
265	325	341	357	374	390	406	423	439	455	472	5	8.5
266	488	504	521	537	553	570	586	602	619	635	6	10.2
267	651	667	684	700	716	732	749	765	781	797	7	11.9
268	813	830	846	862	878	894	911	927	943	959	8	13.6
269	975	991	*008	*024	*040	*056	*072	*088	*104	*120	9	15.3
270	43 136	152	169	185	201	217	233	249	265	281		**16**
271	297	313	329	345	361	377	393	409	425	441	1	1.6
272	457	473	489	505	521	537	553	569	584	600	2	3.2
273	616	632	648	664	680	696	712	727	743	759	3	4.8
274	775	791	807	823	838	854	870	886	902	917	4	6.4
275	933	949	965	981	996	*012	*028	*044	*059	*075	5	8.0
276	44 091	107	122	138	154	170	185	201	217	232	6	9.6
277	248	264	279	295	311	326	342	358	373	389	7	11.2
278	404	420	436	451	467	483	498	514	529	545	8	12.8
279	560	576	592	607	623	638	654	669	685	700	9	14.4
280	716	731	747	762	778	793	809	824	840	855		**15**
281	871	886	902	917	932	948	963	979	994	*010	1	1.5
282	45 025	040	056	071	086	102	117	133	148	163	2	3.0
283	179	194	209	225	240	255	271	286	301	317	3	4.5
284	332	347	362	378	393	408	423	439	454	469	4	6.0
285	484	500	515	530	545	561	576	591	606	621	5	7.5
286	637	652	667	682	697	712	728	743	758	773	6	9.0
287	788	803	818	834	849	864	879	894	909	924	7	10.5
288	939	954	969	984	*000	*015	*030	*045	*060	*075	8	12.0
289	46 090	105	120	135	150	165	180	195	210	225	9	13.5
290	240	255	270	285	300	315	330	345	359	374		**14**
291	389	404	419	434	449	464	479	494	509	523	1	1.4
292	538	553	568	583	598	613	627	642	657	672	2	2.8
293	687	702	716	731	746	761	776	790	805	820	3	4.2
294	835	850	864	879	894	909	923	938	953	967	4	5.6
295	982	997	*012	*026	*041	*056	*070	*085	*100	*114	5	7.0
296	47 129	144	159	173	188	202	217	232	246	261	6	8.4
297	276	290	305	319	334	349	363	378	392	407	7	9.8
298	422	436	451	465	480	494	509	524	538	553	8	11.2
299	567	582	596	611	625	640	654	669	683	698	9	12.6
300	712	727	741	756	770	784	799	813	828	842		
N	**L 0**	**1**	**2**	**3**	**4**	**5**	**6**	**7**	**8**	**9**	**P P**	

TABLE A-1 (*continued*)

N	L 0	1	2	3	4	5	6	7	8	9
300	47 712	727	741	756	770	784	799	813	828	842
301	857	871	885	900	914	929	943	958	972	986
302	48 001	015	029	044	058	073	087	101	116	130
303	144	159	173	187	202	216	230	244	259	273
304	287	302	316	330	344	359	373	387	401	416
305	430	444	458	473	487	501	515	530	544	558
306	572	586	601	615	629	643	657	671	686	700
307	714	728	742	756	770	785	799	813	827	841
308	855	869	883	897	911	926	940	954	968	982
309	996	*010	*024	*038	*052	*066	*080	*094	*108	*122
310	49 136	150	164	178	192	206	220	234	248	262
311	276	290	304	318	332	346	360	374	388	402
312	415	429	443	457	471	485	499	513	527	541
313	554	568	582	596	610	624	638	651	665	679
314	693	707	721	734	748	762	776	790	803	817
315	831	845	859	872	886	900	914	927	941	955
316	969	982	996	*010	*024	*037	*051	*065	*079	*092
317	50 106	120	133	147	161	174	188	202	215	229
318	243	256	270	284	297	311	325	338	352	365
319	379	393	406	420	433	447	461	474	488	501
320	515	529	542	556	569	583	596	610	623	637
321	651	664	678	691	705	718	732	745	759	772
322	786	799	813	826	840	853	866	880	893	907
323	920	934	947	961	974	987	*001	*014	*028	*041
324	51 055	068	081	095	108	121	135	148	162	175
325	188	202	215	228	242	255	268	282	295	308
326	322	335	348	362	375	388	402	415	428	441
327	455	468	481	495	508	521	534	548	561	574
328	587	601	614	627	640	654	667	680	693	706
329	720	733	746	759	772	786	799	812	825	838
330	851	865	878	891	904	917	930	943	957	970
331	983	996	*009	*022	*035	*048	*061	*075	*088	*101
332	52 114	127	140	153	166	179	192	205	218	231
333	244	257	270	284	297	310	323	336	349	362
334	375	388	401	414	427	440	453	466	479	492
335	504	517	530	543	556	569	582	595	608	621
336	634	647	660	673	686	699	711	724	737	750
337	763	776	789	802	815	827	840	853	866	879
338	892	905	917	930	943	956	969	982	994	*007
339	53 020	033	046	058	071	084	097	110	122	135
340	148	161	173	186	199	212	224	237	250	263
341	275	288	301	314	326	339	352	364	377	390
342	403	415	428	441	453	466	479	491	504	517
343	529	542	555	567	580	593	605	618	631	643
344	656	668	681	694	706	719	732	744	757	769
345	782	794	807	820	832	845	857	870	882	895
346	908	920	933	945	958	970	983	995	*008	*020
347	54 033	045	058	070	083	095	108	120	133	145
348	158	170	183	195	208	220	233	245	258	270
349	283	295	307	320	332	345	357	370	382	394
350	407	419	432	444	456	469	481	494	506	518
N	L 0	1	2	3	4	5	6	7	8	9

P P

	15
1	1.5
2	3.0
3	4.5
4	6.0
5	7.5
6	9.0
7	10.5
8	12.0
9	13.5

	14
1	1.4
2	2.8
3	4.2
4	5.6
5	7.0
6	8.4
7	9.8
8	11.2
9	12.6

	13
1	1.3
2	2.6
3	3.9
4	5.2
5	6.5
6	7.8
7	9.1
8	10.4
9	11.7

	12
1	1.2
2	2.4
3	3.6
4	4.8
5	6.0
6	7.2
7	8.4
8	9.6
9	10.8

TABLE A-1 (*continued*)

N	**L 0**	**1**	**2**	**3**	**4**	**5**	**6**	**7**	**8**	**9**
350	54 407	419	432	444	456	469	481	494	506	518
351	531	543	555	568	580	593	605	617	630	642
352	654	667	679	691	704	716	728	741	753	765
353	777	790	802	814	827	839	851	864	876	888
354	900	913	925	937	949	962	974	986	998	*011
355	55 023	035	047	060	072	084	096	108	121	133
356	145	157	169	182	194	206	218	230	242	255
357	267	279	291	303	315	328	340	352	364	376
358	388	400	413	425	437	449	461	473	485	497
359	509	522	534	546	558	570	582	594	606	618
360	630	642	654	666	678	691	703	715	727	739
361	751	763	775	787	799	811	823	835	847	859
362	871	883	895	907	919	931	943	955	967	979
363	991	*003	*015	*027	*038	*050	*062	*074	*086	*098
364	56 110	122	134	146	158	170	182	194	205	217
365	229	241	253	265	277	289	301	312	324	336
366	348	360	372	384	396	407	419	431	443	455
367	467	478	490	502	514	526	538	549	561	573
368	585	597	608	620	632	644	656	667	679	691
369	703	714	726	738	750	761	773	785	797	808
370	820	832	844	855	867	879	891	902	914	926
371	937	949	961	972	984	996	*008	*019	*031	*043
372	57 054	066	078	089	101	113	124	136	148	159
373	171	183	194	206	217	229	241	252	264	276
374	287	299	310	322	334	345	357	368	380	392
375	403	415	426	438	449	461	473	484	496	507
376	519	530	542	553	565	576	588	600	611	623
377	634	646	657	669	680	692	703	715	726	738
378	749	761	772	784	795	807	818	830	841	852
379	864	875	887	898	910	921	933	944	955	967
380	978	990	*001	*013	*024	*035	*047	*058	*070	*081
381	58 092	104	115	127	138	149	161	172	184	195
382	206	218	229	240	252	263	274	286	297	309
383	320	331	343	354	365	377	388	399	410	422
384	433	444	456	467	478	490	501	512	524	535
385	546	557	569	580	591	602	614	625	636	647
386	659	670	681	692	704	715	726	737	749	760
387	771	782	794	805	816	827	838	850	861	872
388	883	894	906	917	928	939	950	961	973	984
389	995	*006	*017	*028	*040	*051	*062	*073	*084	*095
390	59 106	118	129	140	151	162	173	184	195	207
391	218	229	240	251	262	273	284	295	306	318
392	329	340	351	362	373	384	395	406	417	428
393	439	450	461	472	483	494	506	517	528	539
394	550	561	572	583	594	605	616	627	638	649
395	660	671	682	693	704	715	726	737	748	759
396	770	780	791	802	813	824	835	846	857	868
397	879	890	901	912	923	934	945	956	966	977
398	988	999	*010	*021	*032	*043	*054	*065	*076	*086
399	60 097	108	119	130	141	152	163	173	184	195
400	206	217	228	239	249	260	271	282	293	304
N	**L 0**	**1**	**2**	**3**	**4**	**5**	**6**	**7**	**8**	**9**

P P

	13
1	1.3
2	2.6
3	3.9
4	5.2
5	6.5
6	7.8
7	9.1
8	10.4
9	11.7

	12
1	1.2
2	2.4
3	3.6
4	4.8
5	6.0
6	7.2
7	8.4
8	9.6
9	10.8

	11
1	1.1
2	2.2
3	3.3
4	4.4
5	5.5
6	6.6
7	7.7
8	8.8
9	9.9

	10
1	1.0
2	2.0
3	3.0
4	4.0
5	5.0
6	6.0
7	7.0
8	8.0
9	9.0

TABLE A-1 (*continued*)

N	L 0	1	2	3	4	5	6	7	8	9
400	60 206	217	228	239	249	260	271	282	293	304
401	314	325	336	347	358	369	379	390	401	412
402	423	433	444	455	466	477	487	498	509	520
403	531	541	552	563	574	584	595	606	617	627
404	638	649	660	670	681	692	703	713	724	735
405	746	756	767	778	788	799	810	821	831	842
406	853	863	874	885	895	906	917	927	938	949
407	959	970	981	991	*002	*013	*023	*034	*045	*055
408	61 066	077	087	098	109	119	130	140	151	162
409	172	183	194	204	215	225	236	247	257	268
410	278	289	300	310	321	331	342	352	363	374
411	384	395	405	416	426	437	448	458	469	479
412	490	500	511	521	532	542	553	563	574	584
413	595	606	616	627	637	648	658	669	679	690
414	700	711	721	731	742	752	763	773	784	794
415	805	815	826	836	847	857	868	878	888	899
416	909	920	930	941	951	962	972	982	993	*003
417	62 014	024	034	045	055	066	076	086	097	107
418	118	128	138	149	159	170	180	190	201	211
419	221	232	242	252	263	273	284	294	304	315
420	325	335	346	356	366	377	387	397	408	418
421	428	439	449	459	469	480	490	500	511	521
422	531	542	552	562	572	583	593	603	613	624
423	634	644	655	665	675	685	696	706	716	726
424	737	747	757	767	778	788	798	808	818	829
425	839	849	859	870	880	890	900	910	921	931
426	941	951	961	972	982	992	*002	*012	*022	*033
427	63 043	053	063	073	083	094	104	114	124	134
428	144	155	165	175	185	195	205	215	225	236
429	246	256	266	276	286	296	306	317	327	337
430	347	357	367	377	387	397	407	417	428	438
431	448	458	468	478	488	498	508	518	528	538
432	548	558	568	579	589	599	609	619	629	639
433	649	659	669	679	689	699	709	719	729	739
434	749	759	769	779	789	799	809	819	829	839
435	849	859	869	879	889	899	909	919	929	939
436	949	959	969	979	988	998	*008	*018	*028	*038
437	64 048	058	068	078	088	098	108	118	128	137
438	147	157	167	177	187	197	207	217	227	237
439	246	256	266	276	286	296	306	316	326	335
440	345	355	365	375	385	395	404	414	424	434
441	444	454	464	473	483	493	503	513	523	532
442	542	552	562	572	582	591	601	611	621	631
443	640	650	660	670	680	689	699	709	719	729
444	738	748	758	768	777	787	797	807	816	826
445	836	846	856	865	875	885	895	904	914	924
446	933	943	953	963	972	982	992	*002	*011	*021
447	65 031	040	050	060	070	079	089	099	108	118
448	128	137	147	157	167	176	186	196	205	215
449	225	234	244	254	263	273	283	292	302	312
450	321	331	341	350	360	369	379	389	398	408
N	L 0	1	2	3	4	5	6	7	8	9

P P

	11
1	1.1
2	2.2
3	3.3
4	4.4
5	5.5
6	6.6
7	7.7
8	8.8
9	9.9

	10
1	1.0
2	2.0
3	3.0
4	4.0
5	5.0
6	6.0
7	7.0
8	8.0
9	9.0

	9
1	0.9
2	1.8
3	2.7
4	3.6
5	4.5
6	5.4
7	6.3
8	7.2
9	8.1

TABLE A-1 (*continued*)

N	L 0	1	2	3	4	5	6	7	8	9
450	65 321	331	341	350	360	369	379	389	398	408
451	418	427	437	447	456	466	475	485	495	504
452	514	523	533	543	552	562	571	581	591	600
453	610	619	629	639	648	658	667	677	686	696
454	706	715	725	734	744	753	763	772	782	792
455	801	811	820	830	839	849	858	868	877	887
456	896	906	916	925	935	944	954	963	973	982
457	992	*001	*011	*020	*030	*039	*049	*058	*068	*077
458	66 087	096	106	115	124	134	143	153	162	172
459	181	191	200	210	219	229	238	247	257	266
460	276	285	295	304	314	323	332	342	351	361
461	370	380	389	398	408	417	427	436	445	455
462	464	474	483	492	502	511	521	530	539	549
463	558	567	577	586	596	605	614	624	633	642
464	652	661	671	680	689	699	708	717	727	736
465	745	755	764	773	783	792	801	811	820	829
466	839	848	857	867	876	885	894	904	913	922
467	932	941	950	960	969	978	987	997	*006	*015
468	67 025	034	043	052	062	071	080	089	099	108
469	117	127	136	145	154	164	173	182	191	201
470	210	219	228	237	247	256	265	274	284	293
471	302	311	321	330	339	348	357	367	376	385
472	394	403	413	422	431	440	449	459	468	477
473	486	495	504	514	523	532	541	550	560	569
474	578	587	596	605	614	624	633	642	651	660
475	669	679	688	697	706	715	724	733	742	752
476	761	770	779	788	797	806	815	825	834	843
477	852	861	870	879	888	897	906	916	925	934
478	943	952	961	970	979	988	997	*006	*015	*024
479	68 034	043	052	061	070	079	088	097	106	115
480	124	133	142	151	160	169	178	187	196	205
481	215	224	233	242	251	260	269	278	287	296
482	305	314	323	332	341	350	359	368	377	386
483	395	404	413	422	431	440	449	458	467	476
484	485	494	502	511	520	529	538	547	556	565
485	574	583	592	601	610	619	628	637	646	655
486	664	673	681	690	699	708	717	726	735	744
487	753	762	771	780	789	797	806	815	824	833
488	842	851	860	869	878	886	895	904	913	922
489	931	940	949	958	966	975	984	993	*002	*011
490	69 020	028	037	046	055	064	073	082	090	099
491	108	117	126	135	144	152	161	170	179	188
492	197	205	214	223	232	241	249	258	267	276
493	285	294	302	311	320	329	338	346	355	364
494	373	381	390	399	408	417	425	434	443	452
495	461	469	478	487	496	504	513	522	531	539
496	548	557	566	574	583	592	601	609	618	627
497	636	644	653	662	671	679	688	697	705	714
498	723	732	740	749	758	767	775	784	793	801
499	810	819	827	836	845	854	862	871	880	888
500	897	906	914	923	932	940	949	958	966	975

| N | L 0 | 1 | 2 | 3 | 4 | 5 | 6 | 7 | 8 | 9 | P P |

P P

	10
1	1.0
2	2.0
3	3.0
4	4.0
5	5.0
6	6.0
7	7.0
8	8.0
9	9.0

	9
1	0.9
2	1.8
3	2.7
4	3.6
5	4.5
6	5.4
7	6.3
8	7.2
9	8.1

	8
1	0.8
2	1.6
3	2.4
4	3.2
5	4.0
6	4.8
7	5.6
8	6.4
9	7.2

TABLE A-1 (*continued*)

N	L 0	1	2	3	4	5	6	7	8	9
500	69 897	906	914	923	932	940	949	958	966	975
501	984	992	*001	*010	*018	*027	*036	*044	*053	*062
502	70 070	079	088	096	105	114	122	131	140	148
503	157	165	174	183	191	200	209	217	226	234
504	243	252	260	269	278	286	295	303	312	321
505	329	338	346	355	364	372	381	389	398	406
506	415	424	432	441	449	458	467	475	484	492
507	501	509	518	526	535	544	552	561	569	578
508	586	595	603	612	621	629	638	646	655	663
509	672	680	689	697	706	714	723	731	740	749
510	757	766	774	783	791	800	808	817	825	834
511	842	851	859	868	876	885	893	902	910	919
512	927	935	944	952	961	969	978	986	995	*003
513	71 012	020	029	037	046	054	063	071	079	088
514	096	105	113	122	130	139	147	155	164	172
515	181	189	198	206	214	223	231	240	248	257
516	265	273	282	290	299	307	315	324	332	341
517	349	357	366	374	383	391	399	408	416	425
518	433	441	450	458	466	475	483	492	500	508
519	517	525	533	542	550	559	567	575	584	592
520	600	609	617	625	634	642	650	659	667	675
521	684	692	700	709	717	725	734	742	750	759
522	767	775	784	792	800	809	817	825	834	842
523	850	858	867	875	883	892	900	908	917	925
524	933	941	950	958	966	975	983	991	999	*008
525	72 016	024	032	041	049	057	066	074	082	090
526	099	107	115	123	132	140	148	156	165	173
527	181	189	198	206	214	222	230	239	247	255
528	263	272	280	288	296	304	313	321	329	337
529	346	354	362	370	378	387	395	403	411	419
530	428	436	444	452	460	469	477	485	493	501
531	509	518	526	534	542	550	558	567	575	583
532	591	509	607	616	624	632	640	648	656	665
533	673	681	689	697	705	713	722	730	738	746
534	754	762	770	779	787	795	803	811	819	827
535	835	843	852	860	868	876	884	892	900	908
536	916	925	933	941	949	957	965	973	981	989
537	997	*006	*014	*022	*030	*038	*046	*054	*062	*070
538	73 078	086	094	102	111	119	127	135	143	151
539	159	167	175	183	191	199	207	215	223	231
540	239	247	255	263	272	280	288	296	304	312
541	320	328	336	344	352	360	368	376	384	392
542	400	408	416	424	432	440	448	456	464	472
543	480	488	496	504	512	520	528	536	544	552
544	560	568	576	584	592	600	608	616	624	632
545	640	648	656	664	672	679	687	695	703	711
546	719	727	735	743	751	759	767	775	783	791
547	799	807	815	823	830	838	846	854	862	870
548	878	886	894	902	910	918	926	933	941	949
549	957	965	973	981	989	997	*005	*013	*020	*028
550	74 036	044	052	060	068	076	084	092	099	107
N	**L 0**	**1**	**2**	**3**	**4**	**5**	**6**	**7**	**8**	**9**

P P

	9
1	0.9
2	1.8
3	2.7
4	3.6
5	4.5
6	5.4
7	6.3
8	7.2
9	8.1

	8
1	0.8
2	1.6
3	2.4
4	3.2
5	4.0
6	4.8
7	5.6
8	6.4
9	7.2

	7
1	0.7
2	1.4
3	2.1
4	2.8
5	3.5
6	4.2
7	4.9
8	5.6
9	6.3

TABLE A-1 (*continued*)

N	L 0	1	2	3	4	5	6	7	8	9	P P
550	74 036	044	052	060	068	076	084	092	099	107	
551	115	123	131	139	147	155	162	170	178	186	
552	194	202	210	218	225	233	241	249	257	265	
553	273	280	288	296	304	312	320	327	335	343	
554	351	359	367	374	382	390	398	406	414	421	
555	429	437	445	453	461	468	476	484	492	500	
556	507	515	523	531	539	547	554	562	570	578	
557	586	593	601	609	617	624	632	640	648	656	
558	663	671	679	687	695	702	710	718	726	733	
559	741	749	757	764	772	780	788	796	803	811	
560	819	827	834	842	850	858	865	873	881	889	
561	896	904	912	920	927	935	943	950	958	966	**8**
562	974	981	989	997	*005	*012	*020	*028	*035	*043	1 0.8
563	75 051	059	066	074	082	089	097	105	113	120	2 1.6
564	128	136	143	151	159	166	174	182	189	197	3 2.4
565	205	213	220	228	236	243	251	259	266	274	4 3.2
566	282	289	297	305	312	320	328	335	343	351	5 4.0
567	358	366	374	381	389	397	404	412	420	427	6 4.8
568	435	442	450	458	465	473	481	488	496	504	7 5.6
569	511	519	526	534	542	549	557	565	572	580	8 6.4
570	587	595	603	610	618	626	633	641	648	656	9 7.2
571	664	671	679	686	694	702	709	717	724	732	
572	740	747	755	762	770	778	785	793	800	808	
573	815	823	831	838	846	853	861	868	876	884	
574	891	899	906	914	921	929	937	944	952	959	
575	967	974	982	989	997	*005	*012	*020	*027	*035	
576	76 042	050	057	065	072	080	087	095	103	110	
577	118	125	133	140	148	155	163	170	178	185	
578	193	200	208	215	223	230	238	245	253	260	
579	268	275	283	290	298	305	313	320	328	335	
580	343	350	358	365	373	380	388	395	403	410	**7**
581	418	425	433	440	448	455	462	470	477	485	1 0.7
582	492	500	507	515	522	530	537	545	552	559	2 1.4
583	567	574	582	589	597	604	612	619	626	634	3 2.1
584	641	649	656	664	671	678	686	693	701	708	4 2.8
585	716	723	730	738	745	753	760	768	775	782	5 3.5
586	790	797	805	812	819	827	834	842	849	856	6 4.2
587	864	871	879	886	893	901	908	916	923	930	7 4.9
588	938	945	953	960	967	975	982	989	997	*004	8 5.6
589	77 012	019	026	034	041	048	056	063	070	078	9 6.3
590	085	093	100	107	115	122	129	137	144	151	
591	159	166	173	181	188	195	203	210	217	225	
592	232	240	247	254	262	269	276	283	291	298	
593	305	313	320	327	335	342	349	357	364	371	
594	379	386	393	401	408	415	422	430	437	444	
595	452	459	466	474	481	488	495	503	510	517	
596	525	532	539	546	554	561	568	576	583	590	
597	597	605	612	619	627	634	641	648	656	663	
598	670	677	685	692	699	706	714	721	728	735	
599	743	750	757	764	772	779	786	793	801	808	
600	815	822	830	837	844	851	859	866	873	880	
N	**L 0**	**1**	**2**	**3**	**4**	**5**	**6**	**7**	**8**	**9**	**P P**

TABLE A-1 (*continued*)

N	L 0	1	2	3	4	5	6	7	8	9
600	77 815	822	830	837	844	851	859	866	873	880
601	887	895	902	909	916	924	931	938	945	952
602	960	967	974	981	988	996	*003	*010	*017	*025
603	78 032	039	046	053	061	068	075	082	089	097
604	104	111	118	125	132	140	147	154	161	168
605	176	183	190	197	204	211	219	226	233	240
606	247	254	262	269	276	283	290	297	305	312
507	319	326	333	340	347	355	362	369	376	383
608	390	398	405	412	419	426	433	440	447	455
609	462	469	476	483	490	497	504	512	519	526
610	533	540	547	554	561	569	576	583	590	597
611	604	611	618	625	633	640	647	654	661	668
612	675	682	689	696	704	711	718	725	732	739
613	746	753	760	767	774	781	789	796	803	810
614	817	824	831	838	845	852	859	866	873	880
615	888	895	902	909	916	923	930	937	944	951
616	958	965	972	979	986	993	*000	*007	*014	*021
617	79 029	036	043	050	057	064	071	078	085	092
618	099	106	113	120	127	134	141	148	155	162
619	169	176	183	190	197	204	211	218	225	232
620	239	246	253	260	267	274	281	288	295	302
621	309	316	323	330	337	344	351	358	365	372
622	379	386	393	400	407	414	421	428	435	442
623	449	456	463	470	477	484	491	498	505	511
624	518	525	532	539	546	553	560	567	574	581
625	588	595	602	609	616	623	630	637	644	650
626	657	664	671	678	685	692	699	706	713	720
627	727	734	741	748	754	761	768	775	782	789
628	796	803	810	817	824	831	837	844	851	858
629	865	872	879	886	893	900	906	913	920	927
630	934	941	948	955	962	969	975	982	989	996
631	80 003	010	017	024	030	037	044	051	058	065
632	072	079	085	092	099	106	113	120	127	134
633	140	147	154	161	168	175	182	188	195	202
634	209	216	223	229	236	243	250	257	264	271
635	277	284	291	298	305	312	318	325	332	339
636	346	353	359	366	373	380	387	393	400	407
637	414	421	428	434	441	448	455	462	468	475
638	482	489	496	502	509	516	523	530	536	543
639	550	557	564	570	577	584	591	598	604	611
640	618	625	632	638	645	652	659	665	672	679
641	686	693	690	706	713	720	726	733	740	747
642	754	760	767	774	781	787	794	801	808	814
643	821	828	835	841	848	855	862	868	875	882
644	889	895	902	909	916	922	929	936	943	949
645	956	963	969	976	983	990	996	*003	*010	*017
646	81 023	030	037	043	050	057	064	070	077	084
647	090	097	104	111	117	124	131	137	144	151
648	158	164	171	178	184	191	198	204	211	218
649	224	231	238	245	251	258	265	271	278	285
650	291	298	305	311	318	325	331	338	345	351
N	L 0	1	2	3	4	5	6	7	8	9

P P

	8
1	0.8
2	1.6
3	2.4
4	3.2
5	4.0
6	4.8
7	5.6
8	6.4
9	7.2

	7
1	0.7
2	1.4
3	2.1
4	2.8
5	3.5
6	4.2
7	4.9
8	5.6
9	6.3

	6
1	0.6
2	1.2
3	1.8
4	2.4
5	3.0
6	3.6
7	4.2
8	4.8
9	5.4

TABLE A-1 (*continued*)

N	L 0	1	2	3	4	5	6	7	8	9
650	81 291	298	305	311	318	325	331	338	345	351
651	358	365	371	378	385	391	398	405	411	418
652	425	431	438	445	451	458	465	471	478	485
653	491	498	505	511	518	525	531	538	544	551
654	558	564	571	578	584	591	598	604	611	617
655	624	631	637	644	651	657	664	671	677	684
656	690	697	704	710	717	723	730	737	743	750
657	757	763	770	776	783	790	796	803	809	816
658	823	829	836	842	849	856	862	869	875	882
659	889	895	902	908	915	921	928	935	941	948
660	954	961	968	974	981	987	994	*000	*007	*014
661	82 020	027	033	040	046	053	060	066	073	079
662	086	092	099	105	112	119	125	132	138	145
663	151	158	164	171	178	184	191	197	204	210
664	217	223	230	236	243	249	256	263	269	276
665	282	289	295	302	308	315	321	328	334	341
666	347	354	360	367	373	380	387	393	400	406
667	413	419	426	432	439	445	452	458	465	471
668	478	484	491	497	504	510	517	523	530	536
669	543	549	556	562	569	575	582	588	595	601
670	607	614	620	627	633	640	646	653	659	666
671	672	679	685	692	698	705	711	718	724	730
672	737	743	750	756	763	769	776	782	789	795
673	802	808	814	821	827	834	840	847	853	860
674	866	872	879	885	892	898	905	911	918	924
675	930	937	943	950	956	963	969	975	982	988
676	995	*001	*008	*014	*020	*027	*033	*040	*046	*052
677	83 059	065	072	078	085	091	097	104	110	117
678	123	129	136	142	149	155	161	168	174	181
679	187	193	200	206	213	219	226	232	238	245
680	251	257	264	270	276	283	289	296	302	308
681	315	321	327	334	340	347	353	359	366	372
682	378	385	391	398	404	410	417	423	429	436
683	442	448	455	461	467	474	480	487	493	499
684	506	512	518	525	531	537	544	550	556	563
685	569	575	582	588	594	601	607	613	620	626
686	632	639	645	651	658	664	670	677	683	689
687	696	702	708	715	721	727	734	740	746	753
688	759	765	771	778	784	790	797	803	809	816
689	822	828	835	841	847	853	860	866	872	879
690	885	891	897	904	910	916	923	929	935	942
691	948	954	960	967	973	979	985	992	998	*004
692	84 011	017	023	029	036	042	048	055	061	067
693	073	080	086	092	098	105	111	117	123	130
694	136	142	148	155	161	167	173	180	186	192
695	198	205	211	217	223	230	236	242	248	255
696	261	267	273	280	286	292	298	305	311	317
697	323	330	336	342	348	354	361	367	373	379
698	386	392	398	404	410	417	423	429	435	442
699	448	454	460	466	473	479	485	491	497	504
700	510	516	522	528	535	541	547	553	559	566
N	L 0	1	2	3	4	5	6	7	8	9

P P

	7
1	0.7
2	1.4
3	2.1
4	2.8
5	3.5
6	4.2
7	4.9
8	5.6
9	6.3

	6
1	0.6
2	1.2
3	1.8
4	2.4
5	3.0
6	3.6
7	4.2
8	4.8
9	5.4

TABLE A-1 (*continued*)

N	L 0	1	2	3	4	5	6	7	8	9
700	84 510	516	522	528	535	541	547	553	559	566
701	572	578	584	590	597	603	609	615	621	628
702	634	640	646	652	658	665	671	677	683	689
703	696	702	708	714	720	726	733	739	745	751
704	757	763	770	776	782	788	794	800	807	813
705	819	825	831	837	844	850	856	862	868	874
706	880	887	893	899	905	911	917	924	930	936
707	942	948	954	960	967	973	979	985	991	997
708	85 003	009	016	022	028	034	040	046	052	058
709	065	071	077	083	089	095	101	107	114	120
710	126	132	138	144	150	156	163	169	175	181
711	187	193	199	205	211	217	224	230	236	242
712	248	254	260	266	272	278	285	291	297	303
713	309	315	321	327	333	339	345	352	358	364
714	370	376	382	388	394	400	406	412	418	425
715	431	437	443	449	455	461	467	473	479	485
716	491	497	503	509	516	522	528	534	540	546
717	552	558	564	570	576	582	588	594	600	606
718	612	618	625	631	637	643	649	635	661	667
719	673	679	685	691	697	703	709	715	721	727
720	733	739	745	751	757	763	769	775	781	788
721	794	800	806	812	818	824	830	836	842	848
722	854	860	866	872	878	884	890	896	902	908
723	914	920	926	932	938	944	950	056	062	968
724	974	980	986	992	998	*004	*010	*016	*022	*028
725	86 034	040	046	052	058	064	070	076	082	088
726	094	100	106	112	118	124	130	136	141	147
727	153	159	165	171	177	183	189	195	201	207
728	213	219	225	231	237	243	249	255	261	267
729	273	279	285	291	297	303	308	314	320	326
730	332	338	344	350	356	362	368	374	380	386
731	392	398	404	410	415	421	427	433	439	445
732	451	457	463	469	475	481	487	493	499	504
733	510	516	522	528	534	540	546	552	558	564
734	570	576	581	587	593	599	605	611	617	623
735	629	635	641	646	652	658	664	670	676	682
736	688	694	700	705	711	717	723	729	735	741
737	747	753	759	764	770	776	782	788	794	800
738	806	812	817	823	829	835	841	847	853	859
739	864	870	876	882	888	894	900	906	911	917
740	923	929	935	941	947	953	958	964	970	976
741	982	988	994	999	*005	*011	*017	*023	*029	*035
742	87 040	046	052	058	064	070	075	081	087	093
743	099	105	111	116	122	128	134	140	146	151
744	157	163	169	175	181	186	192	198	204	210
745	216	221	227	233	239	245	251	256	262	268
746	274	280	286	291	297	303	309	315	320	326
747	332	338	344	349	355	361	367	373	379	384
748	390	396	402	408	413	419	425	431	437	442
749	448	454	460	466	471	477	483	489	495	500
750	506	512	518	523	529	535	541	547	552	558
N	**L 0**	**1**	**2**	**3**	**4**	**5**	**6**	**7**	**8**	**9**

P P

	7
1	0.7
2	1.4
3	2.1
4	2.8
5	3.5
6	4.2
7	4.9
8	5.6
9	6.3

	6
1	0.6
2	1.2
3	1.8
4	2.4
5	3.0
6	3.6
7	4.2
8	4.8
9	5.4

	5
1	0.5
2	1.0
3	1.5
4	2.0
5	2.5
6	3.0
7	3.5
8	4.0
9	4.5

P P

TABLE A-1 (*continued*)

N	L 0	1	2	3	4	5	6	7	8	9	P P
750	87 506	512	518	523	529	535	541	547	552	558	
751	564	570	576	581	587	593	599	604	610	616	
752	622	628	633	639	645	651	656	662	668	674	
753	679	685	691	697	703	708	714	720	726	731	
754	737	743	749	754	760	766	772	777	783	789	
755	795	800	806	812	818	823	829	835	841	846	
756	852	858	864	869	875	881	887	892	898	904	
757	910	915	921	927	933	938	944	950	955	961	
758	967	973	978	984	990	996	*001	*007	*013	*018	
759	88 024	030	036	041	047	053	058	064	070	076	
760	081	087	093	098	104	110	116	121	127	133	
761	138	144	150	156	161	167	173	178	184	190	
762	195	201	207	213	218	224	230	235	241	247	**6**
763	252	258	264	270	275	281	287	292	298	304	1 | 0.6
764	309	315	321	326	332	338	343	349	355	360	2 | 1.2
765	366	372	377	383	389	395	400	406	412	417	3 | 1.8
766	423	429	434	440	446	451	457	463	468	474	4 | 2.4
767	480	485	491	497	502	508	513	519	525	530	5 | 3.0
768	536	542	547	553	559	564	570	576	581	587	6 | 3.6
769	593	598	604	610	615	621	627	632	638	643	7 | 4.2
											8 | 4.8
770	649	655	660	666	672	677	683	689	694	700	9 | 5.4
771	705	711	717	722	728	734	739	745	750	756	
772	762	767	773	779	784	790	795	801	807	812	
773	818	824	829	835	840	846	852	857	863	868	
774	874	880	885	891	897	902	908	913	919	925	
775	930	936	941	947	953	958	964	969	975	981	
776	986	992	997	*003	*009	*014	*020	*025	*031	*037	
777	89 042	048	053	059	064	070	076	081	087	092	
778	098	104	109	115	120	126	131	137	143	148	
779	154	159	165	170	176	182	187	193	198	204	
780	209	215	221	226	232	237	243	248	254	260	**5**
											1 | 0.5
781	265	271	276	282	287	293	298	304	310	315	2 | 1.0
782	321	326	332	337	343	348	354	360	365	371	3 | 1.5
783	376	382	387	393	398	404	409	415	421	426	4 | 2.0
784	432	437	443	448	454	459	465	470	476	481	5 | 2.5
785	487	492	498	504	509	515	520	526	531	537	6 | 3.0
786	542	548	553	559	564	570	575	581	586	592	7 | 3.5
787	597	603	609	614	620	625	631	636	642	647	8 | 4.0
788	653	658	664	669	675	680	686	691	697	702	9 | 4.5
789	708	713	719	724	730	735	741	746	752	757	
790	763	768	774	779	785	790	796	801	807	812	
791	818	823	829	834	840	845	851	856	862	867	
792	873	878	883	889	894	900	905	911	916	922	
793	927	933	938	944	949	955	960	966	971	977	
794	982	988	993	998	*004	*009	*015	*020	*026	*031	
795	90 037	042	048	053	059	064	069	075	080	086	
796	091	097	102	108	113	119	124	129	135	140	
797	146	151	157	162	168	173	179	184	189	195	
798	200	206	211	217	222	227	233	238	244	249	
799	255	260	266	271	276	282	287	293	298	304	
800	309	314	320	325	331	336	342	347	352	358	
N	**L 0**	**1**	**2**	**3**	**4**	**5**	**6**	**7**	**8**	**7**	**P P**

TABLE A-1 (*continued*)

N	L 0	1	2	3	4	5	6	7	8	9	P P
800	90 309	314	320	325	331	336	342	347	352	358	
801	363	369	374	380	385	390	396	401	407	412	
802	417	423	428	434	439	445	450	455	461	466	
803	472	477	482	488	493	499	504	509	515	520	
804	526	531	536	542	547	553	558	563	569	574	
805	580	585	590	596	601	607	612	617	623	628	
806	634	639	644	650	655	660	666	671	677	682	
807	687	693	698	703	709	714	720	725	730	736	
808	741	747	752	757	763	768	773	779	784	789	
809	795	800	806	811	816	822	827	832	838	843	
810	849	854	859	865	870	875	881	886	891	897	
811	902	907	913	918	924	929	934	940	945	950	
812	956	961	966	972	977	982	988	993	998	*004	
813	91 009	014	020	025	030	036	041	046	052	057	
814	062	068	073	078	084	089	094	100	105	110	
815	116	121	126	132	137	142	148	153	158	164	
816	169	174	180	185	190	196	201	206	212	217	
817	222	228	233	238	243	249	254	259	265	270	
818	275	281	286	291	297	302	307	312	318	323	
819	328	334	339	344	350	355	360	365	371	376	
820	381	387	392	397	403	408	413	418	424	429	
821	434	440	445	450	455	461	466	471	477	482	
822	487	492	498	503	508	514	519	524	529	535	
823	540	545	551	556	561	566	572	577	582	587	
824	593	598	603	609	614	619	624	630	635	640	
825	645	651	656	661	666	672	677	682	687	693	
826	698	703	709	714	719	724	730	735	740	745	
827	751	756	761	766	772	777	782	787	793	798	
828	803	808	814	819	824	829	834	840	845	850	
829	855	861	866	871	876	882	887	892	897	903	
830	908	913	918	924	929	934	939	944	950	955	
831	960	965	971	976	981	986	991	997	*002	*007	
832	92 012	018	023	028	033	038	044	049	054	059	
833	065	070	075	080	085	091	096	101	106	111	
834	117	122	127	132	137	143	148	153	158	163	
835	169	174	179	184	189	195	200	205	210	215	
836	221	226	231	236	241	247	252	257	262	267	
837	273	278	283	288	293	298	304	309	314	319	
838	324	330	335	340	345	350	355	361	366	371	
839	376	381	387	392	397	402	407	412	418	423	
840	428	433	438	443	449	454	459	464	469	474	
841	480	485	490	495	500	505	511	516	521	526	
842	531	536	542	547	552	557	562	567	572	578	
843	583	588	593	598	603	609	614	619	624	629	
844	634	639	645	650	655	660	665	670	675	681	
845	686	691	696	701	706	711	716	722	727	732	
846	737	742	747	752	758	763	768	773	778	783	
847	788	793	799	804	809	814	819	824	829	834	
848	840	845	850	855	860	865	870	875	881	886	
849	891	896	901	906	911	916	921	927	932	937	
850	942	947	952	957	962	967	973	978	983	988	
N	**L 0**	**1**	**2**	**3**	**4**	**5**	**6**	**7**	**8**	**9**	**P P**

P P

	6
1	0.6
2	1.2
3	1.8
4	2.4
5	3.0
6	3.6
7	4.2
8	4.8
9	5.4

	5
1	0.5
2	1.0
3	1.5
4	2.0
5	2.5
6	3.0
7	3.5
8	4.0
9	4.5

Table A-1 (*continued*)

N	L 0	1	2	3	4	5	6	7	8	9
850	92 942	947	952	957	962	967	973	978	983	988
851	993	998	*003	*008	*013	*018	*024	*029	*034	*039
852	93 044	049	054	059	064	069	075	080	085	090
853	095	100	105	110	115	120	125	131	136	141
854	146	151	156	161	166	171	176	181	186	192
855	197	202	207	212	217	222	227	232	237	242
856	247	252	258	263	268	273	278	283	288	293
857	298	303	308	313	318	323	328	334	339	344
858	349	354	359	364	369	374	379	384	389	394
859	399	404	409	414	420	425	430	435	440	445
860	450	455	460	465	470	475	480	485	490	495
861	500	505	510	515	520	526	531	536	541	546
862	551	556	561	566	571	576	581	586	591	596
863	601	606	611	616	621	626	631	636	641	646
864	651	656	661	666	671	676	682	687	692	697
865	702	707	712	717	722	727	732	737	742	747
866	752	757	762	767	772	777	782	787	792	797
867	802	807	812	817	822	827	832	837	842	847
868	852	857	862	867	872	877	882	887	892	897
869	902	907	912	917	922	927	932	937	942	947
870	952	957	962	967	972	977	982	987	992	997
871	94 002	007	012	017	022	027	032	037	042	047
872	052	057	062	067	072	077	082	086	091	096
873	101	106	111	116	121	126	131	136	141	146
874	151	156	161	166	171	176	181	186	191	196
875	201	206	211	216	221	226	231	236	240	245
876	250	255	260	265	270	275	280	285	290	295
877	300	305	310	315	320	325	330	335	340	345
878	349	354	359	364	369	374	379	384	389	394
879	399	404	409	414	419	424	429	433	438	443
880	448	453	458	463	468	473	478	483	488	493
881	498	503	507	512	517	522	527	532	537	542
882	547	552	557	562	567	571	576	581	586	591
883	596	601	606	611	616	621	626	630	635	640
884	645	650	655	660	665	670	675	680	685	689
885	694	699	704	709	714	719	724	729	734	738
886	743	748	753	758	763	768	773	778	783	787
887	792	797	802	807	812	817	822	827	832	836
888	841	846	851	856	861	866	871	876	880	885
889	890	895	900	905	910	915	919	924	929	934
890	939	944	949	954	959	963	968	973	978	983
891	988	993	998	*002	*007	*012	*017	*022	*027	*032
892	95 036	041	046	051	056	061	066	071	075	080
893	085	090	095	100	105	109	114	119	124	129
894	134	139	143	148	153	158	163	168	173	177
895	182	187	192	197	202	207	211	216	221	226
896	231	236	240	245	250	255	260	265	270	274
897	279	284	289	294	299	303	308	313	318	323
898	328	332	337	342	347	352	357	361	366	371
899	376	381	386	390	395	400	405	410	415	419
900	424	429	434	439	444	448	453	458	463	468
N	L 0	1	2	3	4	5	6	7	8	9

P P

	6
1	0.6
2	1.2
3	1.8
4	2.4
5	3.0
6	3.6
7	4.2
8	4.5
9	4.8

	5
1	0.5
2	1.0
3	1.5
4	2.0
5	2.5
6	3.0
7	3.5
8	4.0
9	4.5

	4
1	0.4
2	0.8
3	1.2
4	1.6
5	2.0
6	2.4
7	2.8
8	3.2
9	3.6

TABLE A-1 *(continued)*

N	L 0	1	2	3	4	5	6	7	8	9
900	95 424	429	434	439	444	448	453	458	463	468
901	472	477	482	487	492	497	501	506	511	516
902	521	525	530	535	540	545	550	554	559	564
903	569	574	578	583	588	593	598	602	607	612
904	617	622	626	631	636	641	646	650	655	660
905	665	670	674	679	684	689	694	698	703	708
906	713	718	722	727	732	737	742	746	751	756
907	761	766	770	775	780	785	789	794	799	804
908	809	813	818	823	828	832	837	842	847	852
909	856	861	866	871	875	880	885	890	895	899
910	904	909	914	918	923	928	933	938	942	947
911	952	957	961	966	971	976	980	985	990	995
912	999	*004	*009	*014	*019	*023	*028	*033	*038	*042
913	96 047	052	057	061	066	071	076	080	085	090
914	095	099	104	109	114	118	123	128	133	137
915	142	147	152	156	161	166	171	175	180	185
916	190	194	199	204	209	213	218	223	227	232
917	237	242	246	251	256	261	265	270	275	280
918	284	289	294	298	303	308	313	317	322	327
919	332	336	341	346	350	355	360	365	369	374
920	379	384	388	393	398	402	407	412	417	421
921	426	431	435	440	445	450	454	459	464	468
922	473	478	483	487	492	497	501	506	511	515
923	520	525	530	534	539	544	548	553	558	562
924	567	572	577	581	586	591	595	600	605	609
925	614	619	624	628	633	638	642	647	652	656
926	661	666	670	675	680	685	689	694	699	703
927	708	713	717	722	727	731	736	741	745	750
928	755	759	764	769	774	778	783	788	792	797
929	802	806	811	816	820	825	830	834	839	844
930	848	853	858	862	867	872	876	881	886	890
931	895	900	904	909	914	918	923	928	932	937
932	942	946	951	956	960	965	970	974	979	984
933	988	993	997	*002	*007	*011	*016	*021	*025	*030
934	97 035	039	044	049	053	058	063	067	072	077
935	081	086	090	095	100	104	109	114	118	123
936	128	132	137	142	146	151	155	160	165	169
937	174	179	183	188	192	197	202	206	211	216
938	220	225	230	234	239	243	248	253	257	262
939	267	271	276	280	285	290	294	299	304	308
940	313	317	322	327	331	336	340	345	350	354
941	359	364	368	373	377	382	387	391	396	400
942	405	410	414	419	424	428	433	437	442	447
943	451	456	460	465	470	474	479	483	488	493
944	497	502	506	511	516	520	525	529	534	539
945	543	548	552	557	562	566	571	575	580	585
946	589	594	598	603	607	612	617	621	626	630
947	635	640	644	649	653	658	663	667	672	676
948	681	685	690	695	699	704	708	713	717	722
949	727	731	736	740	745	749	754	759	763	768
950	772	777	782	786	791	795	800	804	809	813
N	L 0	1	2	3	4	5	6	7	8	9

P P

	5
1	0.5
2	1.0
3	1.5
4	2.0
5	2.5
6	3.0
7	3.5
8	4.0
9	4.5

	4
1	0.4
2	0.8
3	1.2
4	1.6
5	2.0
6	2.4
7	2.8
8	3.2
9	3.6

TABLE A-1 (*continued*)

N	L 0	1	2	3	4	5	6	7	8	9	PP
950	97 772	777	782	786	791	795	800	804	809	813	
951	818	823	827	832	836	841	845	850	855	859	
952	864	868	873	877	882	886	891	896	900	905	
953	909	914	918	923	928	932	937	941	946	950	
954	955	959	964	968	973	978	982	987	991	996	
955	98 000	005	009	014	019	023	028	032	037	041	
956	046	050	055	059	064	068	073	078	082	087	
957	001	096	100	105	109	114	118	123	127	132	
958	137	141	146	150	155	159	164	168	173	177	
959	182	186	191	195	200	204	209	214	218	223	
960	227	232	236	241	245	250	254	259	263	268	**5**
961	272	277	281	286	290	295	299	304	308	313	1 | 0.5
962	318	322	327	331	336	340	345	349	354	358	2 | 1.0
963	363	367	372	376	381	385	390	394	399	403	3 | 1.5
964	408	412	417	421	426	430	435	439	444	448	4 | 2.0
965	453	457	462	466	471	475	480	484	489	493	5 | 2.5
966	498	502	507	511	516	520	525	529	534	538	6 | 3.0
967	543	547	552	556	561	565	570	574	579	583	7 | 3.5
968	588	592	597	601	605	610	614	619	623	628	8 | 4.0
969	632	637	641	646	650	655	659	664	668	673	9 | 4.5
970	677	682	686	691	695	700	704	709	713	717	
971	722	726	731	735	740	744	749	753	758	762	
972	767	771	776	780	784	789	793	798	802	807	
973	811	816	820	825	829	834	838	843	847	851	
974	856	860	865	869	874	878	883	887	892	896	
975	900	905	909	914	918	923	927	932	936	941	
976	945	949	954	958	963	967	972	976	981	985	
977	989	994	998	*003	*007	*012	*016	*021	*025	*029	
978	99 034	038	043	047	052	056	061	065	069	074	
979	078	083	087	092	096	100	105	109	114	118	
980	123	127	131	136	140	145	149	154	158	162	**4**
981	167	171	176	180	185	189	193	198	202	207	1 | 0.4
982	211	216	220	224	229	233	238	242	247	251	2 | 0.8
983	255	260	264	269	273	277	282	286	291	295	3 | 1.2
984	300	304	308	313	317	322	326	330	335	339	4 | 1.6
985	344	348	352	357	361	366	370	374	379	383	5 | 2.0
986	388	392	396	401	405	410	414	419	423	427	6 | 2.4
987	432	436	441	445	449	454	458	463	467	471	7 | 2.8
988	476	480	484	489	493	498	502	506	511	515	8 | 3.2
989	520	524	528	533	537	542	546	550	555	559	9 | 3.6
990	564	568	572	577	581	585	590	594	599	603	
991	607	612	616	621	625	629	634	638	642	647	
992	651	656	660	664	669	673	677	682	686	691	
993	695	699	704	708	712	717	721	726	730	734	
994	739	743	747	752	756	760	765	769	774	778	
995	782	787	791	795	800	804	808	813	817	822	
996	826	830	835	839	843	848	852	856	861	865	
997	870	874	878	883	887	891	896	900	904	909	
998	913	917	922	926	930	935	939	944	948	952	
999	957	961	965	970	974	978	983	987	991	996	
1000	00 000	004	009	013	017	022	026	030	035	039	
N	**L 0**	**1**	**2**	**3**	**4**	**5**	**6**	**7**	**8**	**9**	**PP**

Source: Table 5, pp. 84–103 in H. Arkin and R. Colton, *Tables for Statisticians*, New York: Barnes and Noble, Inc. 1968.

TABLE A-2 TABLE OF NATURAL LOGARITHMS

N	.0	.1	.2	.3	.4	.5	.6	.7	.8	.9
1	0.000	0.095	0.182	0.262	0.336	0.405	0.470	0.531	0.588	0.642
2	0.693	0.742	0.788	0.833	0.875	0.916	0.956	0.993	1.030	1.065
3	1.099	1.131	1.163	1.194	1.224	1.253	1.281	1.308	1.335	1.361
4	1.386	1.411	1.435	1.459	1.482	1.504	1.526	1.548	1.569	1.589
5	1.609	1.629	1.649	1.668	1.686	1.705	1.723	1.740	1.758	1.775
6	1.792	1.808	1.825	1.841	1.856	1.872	1.887	1.902	1.917	1.932
7	1.946	1.960	1.974	1.988	2.001	2.015	2.028	2.041	2.054	2.067
8	2.079	2.092	2.104	2.116	2.128	2.140	2.152	2.163	2.175	2.186
9	2.197	2.208	2.219	2.230	2.241	2.251	2.262	2.272	2.282	2.293
10	2.303	2.313	2.322	2.332	2.342	2.351	2.361	2.370	2.380	2.389

TABLE A-3 VALUES OF e^x AND e^{-x}

x	e^x	e^{-x}	x	e^x	e^{-x}
0.0	1.00	1.00	3.1	22.2	.045
0.1	1.11	.905	3.2	24.5	.041
0.2	1.22	.819	3.3	27.1	.037
0.3	1.35	.741	3.4	30.0	.033
0.4	1.49	.670	3.5	33.1	.030
0.5	1.65	.607	3.6	36.6	.027
0.6	1.82	.549	3.7	40.4	.025
0.7	2.01	.497	3.8	44.7	.022
0.8	2.23	.449	3.9	49.4	.020
0.9	2.46	.407	4.0	54.6	.018
1.0	2.72	.368	4.1	60.3	.017
1.1	3.00	.333	4.2	66.7	.015
1.2	3.32	.301	4.3	73.7	.014
1.3	3.67	.273	4.4	81.5	.012
1.4	4.06	.247	4.5	90.0	.011
1.5	4.48	.223	4.6	99.5	.010
1.6	4.95	.202	4.7	110	.0091
1.7	5.47	.183	4.8	122	.0082
1.8	6.05	.165	4.9	134	.0074
1.9	6.69	.150	5.0	148	.0067
2.0	7.39	.135	5.1	164	.0061
2.1	8.17	.122	5.2	181	.0055
2.2	9.02	.111	5.3	200	.0050
2.3	9.97	.100	5.4	221	.0045
2.4	11.0	.091	5.5	245	.0041
2.5	12.2	.082	5.6	270	.0037
2.6	13.5	.074	5.7	299	.0033
2.7	14.9	.067	5.8	330	.0030
2.8	16.4	.061	5.9	365	.0027
2.9	18.2	.055	6.0	403	.0025
3.0	20.1	.050			

Source: Table II and Table III on page 761 in R. E. Johnson, F. L. Kiokemeister, and E. S. Wolk, *Calculus with Analytic Geometry*, fifth ed., Boston: Allyn and Bacon, Inc., 1974.

TABLE A-4 COMPOUND INTEREST OF $1: $(1 + i)^n$

n	0.02	0.04	0.06	0.08	0.10	0.12	0.14	0.16	0.18	0.20
1	1.0200	1.0400	1.0600	1.0800	1.1000	1.1200	1.1400	1.1600	1.1800	1.2000
2	1.0404	1.0816	1.1236	1.1664	1.2100	1.2544	1.2996	1.3456	1.3924	1.4400
3	1.0612	1.1249	1.1910	1.2597	1.3310	1.4049	1.4815	1.5609	1.6430	1.7280
4	1.0824	1.1699	1.2625	1.3605	1.4641	1.5735	1.6890	1.8106	1.9388	2.0736
5	1.1041	1.2167	1.3382	1.4693	1.6105	1.7623	1.9254	2.1003	2.2878	2.4883
6	1.1262	1.2653	1.4185	1.5869	1.7716	1.9738	2.1950	2.4364	2.6996	2.9860
7	1.1487	1.3159	1.5036	1.7138	1.9487	2.2107	2.5023	2.8262	3.1855	3.5832
8	1.1717	1.3686	1.5938	1.8509	2.1436	2.4760	2.8526	3.2784	3.7589	4.2998
9	1.1951	1.4233	1.6895	1.9990	2.3579	2.7731	3.2519	3.8030	4.4355	5.1598
10	1.2190	1.4802	1.7908	2.1589	2.5937	3.1058	3.7072	4.4114	5.2338	6.1917
11	1.2434	1.5395	1.8983	2.3316	2.8531	3.4785	4.2262	5.1173	6.1759	7.4301
12	1.2682	1.6010	2.0122	2.5182	3.1384	3.8960	4.8179	5.9360	7.2876	8.9161
13	1.2936	1.6651	2.1329	2.7196	3.4523	4.3635	5.4924	6.8858	8.5994	10.6993
14	1.3195	1.7317	2.2609	2.9372	3.7975	4.8871	6.2613	7.9875	10.1472	12.8392
15	1.3459	1.8009	2.3966	3.1722	4.1772	5.4736	7.1379	9.2655	11.9737	15.4070
16	1.3728	1.8730	2.5404	3.4259	4.5950	6.1304	8.1372	10.7480	14.1290	18.4884
17	1.4002	1.9479	2.6928	3.7000	5.0545	6.8660	9.2765	12.4677	16.6722	22.1861
18	1.4282	2.0258	2.8543	3.9960	5.5599	7.6900	10.5752	14.4625	19.6733	26.6233
19	1.4568	2.1068	3.0256	4.3157	6.1159	8.6128	12.0557	16.7765	23.2144	31.9480
20	1.4859	2.1911	3.2071	4.6610	6.7275	9.6463	13.7435	19.4608	27.3930	38.3376
21	1.5157	2.2788	3.3996	5.0338	7.4002	10.8038	15.6676	22.5745	32.3238	46.0051
22	1.5460	2.3699	3.6035	5.4365	8.1403	12.1003	17.8610	26.1864	38.1421	55.2061
23	1.5769	2.4647	3.8197	5.8715	8.9543	13.5523	20.3616	30.3762	45.0076	66.2474
24	1.6084	2.5633	4.0489	6.3412	9.8497	15.1786	23.2122	35.2364	53.1090	79.4968
25	1.6406	2.6658	4.2919	6.8485	10.8347	17.0001	26.4619	40.8742	62.6686	95.3962

n	0.02	0.04	0.06	0.08	0.10	0.12	0.14	0.16	0.18	0.20
26	1.6734	2.7725	4.5494	7.3964	11.9182	19.0401	30.1666	47.4141	73.9490	114.4755
27	1.7069	2.8834	4.8223	7.9881	13.1100	21.3249	34.3899	55.0004	87.2598	137.3705
28	1.7410	2.9987	5.1117	8.6271	14.4210	23.8839	39.2045	63.8004	102.9666	164.8447
29	1.7758	3.1187	5.4184	9.3173	15.8631	26.7499	44.6931	74.0085	121.5005	197.8136
30	1.8114	3.2434	5.7435	10.0627	17.4494	29.9599	50.9502	85.8499	143.3706	237.3763
31	1.8476	3.3731	6.0881	10.8677	19.1943	33.5551	58.0832	99.5859	169.1774	284.8516
32	1.8845	3.5081	6.4534	11.7371	21.1138	37.5817	66.2148	115.5196	199.6293	341.8219
33	1.9222	3.6484	6.8406	12.6760	23.2252	42.0915	75.4849	134.0027	235.5625	410.1863
34	1.9607	3.7943	7.2510	13.6901	25.5477	47.1425	86.0528	155.4432	277.9638	492.2235
35	1.9999	3.9461	7.6861	14.7853	28.1024	52.7996	98.1002	180.3141	327.9973	590.6682
36	2.0399	4.1039	8.1473	15.9682	30.9127	59.1356	111.8342	209.1643	387.0368	708.8019
37	2.0807	4.2681	8.6361	17.2456	34.0039	66.2318	127.4910	242.6306	456.7034	850.5622
38	2.1223	4.4388	9.1543	18.6253	37.4043	74.1797	145.3397	281.4515	538.9100	1020.6747
39	2.1647	4.6164	9.7035	20.1153	41.1448	83.0812	165.6873	326.4837	635.9138	1224.8096
40	2.2080	4.8010	10.2857	21.7245	45.2593	93.0510	188.8835	378.7211	750.3783	1469.7715
41	2.2522	4.9931	10.9029	23.4625	49.7852	104.2171	215.3272	439.3165	885.4464	1763.7258
42	2.2972	5.1928	11.5570	25.3395	54.7637	116.7231	245.4730	509.6072	1044.8268	2116.4710
43	2.3432	5.4005	12.2505	27.3666	60.2401	130.7299	279.8392	591.1443	1232.8956	2539.7652
44	2.3901	5.6165	12.9855	29.5560	66.2641	146.4175	319.0167	685.7274	1454.8168	3047.7182
45	2.4379	5.8412	13.7646	31.9204	72.8905	163.9876	363.6791	795.4438	1716.6838	3657.2618
46	2.4866	6.0748	14.5905	34.4741	80.1795	183.6661	414.5941	922.7148	2025.6869	4388.7142
47	2.5363	6.3178	15.4659	37.2320	88.1975	205.7060	472.6373	1070.3492	2390.3106	5266.4570
48	2.5871	6.5705	16.3939	40.2106	97.0172	230.3908	538.8065	1241.6050	2820.5665	6319.7484
49	2.6388	6.8333	17.3775	43.4274	106.7190	258.0377	614.2395	1440.2618	3328.2684	7583.6981
50	2.6916	7.1067	18.4202	46.9016	117.3909	289.0022	700.2330	1670.7037	3927.3568	9100.4377

TABLE A-5 PRESENT VALUE OF \$1: $(1 + i)^{-n}$

n	0.02	0.04	0.06	0.08	0.10	0.12	0.14	0.16	0.18	0.20
1	0.9804	0.9615	0.9434	0.9259	0.9091	0.8929	0.8772	0.8621	0.8475	0.8333
2	0.9612	0.9246	0.8900	0.8573	0.8264	0.7972	0.7695	0.7432	0.7182	0.6944
3	0.9423	0.8890	0.8396	0.7938	0.7513	0.7118	0.6750	0.6407	0.6086	0.5787
4	0.9238	0.8548	0.7921	0.7350	0.6830	0.6355	0.5921	0.5523	0.5158	0.4823
5	0.9057	0.8219	0.7473	0.6806	0.6209	0.5674	0.5194	0.4761	0.4371	0.4019
6	0.8880	0.7903	0.7050	0.6302	0.5645	0.5066	0.4556	0.4104	0.3704	0.3349
7	0.8706	0.7599	0.6651	0.5835	0.5132	0.4523	0.3996	0.3538	0.3139	0.2791
8	0.8535	0.7307	0.6274	0.5403	0.4665	0.4039	0.3506	0.3050	0.2660	0.2326
9	0.8368	0.7026	0.5919	0.5002	0.4241	0.3606	0.3075	0.2630	0.2255	0.1938
10	0.8203	0.6756	0.5584	0.4632	0.3855	0.3220	0.2697	0.2267	0.1911	0.1615
11	0.8043	0.6496	0.5268	0.4289	0.3505	0.2875	0.2366	0.1954	0.1619	0.1346
12	0.7885	0.6246	0.4970	0.3971	0.3186	0.2567	0.2076	0.1685	0.1372	0.1122
13	0.7730	0.6006	0.4688	0.3677	0.2897	0.2292	0.1821	0.1452	0.1163	0.0935
14	0.7579	0.5775	0.4423	0.3405	0.2633	0.2046	0.1597	0.1252	0.0985	0.0779
15	0.7430	0.5553	0.4173	0.3152	0.2394	0.1827	0.1401	0.1079	0.0835	0.0649
16	0.7284	0.5339	0.3936	0.2919	0.2176	0.1631	0.1229	0.0930	0.0708	0.0541
17	0.7142	0.5134	0.3714	0.2703	0.1978	0.1456	0.1078	0.0802	0.0600	0.0451
18	0.7002	0.4936	0.3503	0.2502	0.1799	0.1300	0.0946	0.0691	0.0508	0.0376
19	0.6864	0.4746	0.3305	0.2317	0.1635	0.1161	0.0829	0.0596	0.0431	0.0313
20	0.6730	0.4564	0.3118	0.2145	0.1486	0.1037	0.0728	0.0514	0.0365	0.0261
21	0.6598	0.4388	0.2942	0.1987	0.1351	0.0926	0.0638	0.0443	0.0309	0.0217
22	0.6468	0.4220	0.2775	0.1839	0.1228	0.0826	0.0560	0.0382	0.0262	0.0181
23	0.6342	0.4057	0.2618	0.1703	0.1117	0.0738	0.0491	0.0329	0.0222	0.0151
24	0.6217	0.3901	0.2470	0.1577	0.1015	0.0659	0.0431	0.0284	0.0188	0.0126
25	0.6095	0.3751	0.2330	0.1460	0.0923	0.0588	0.0378	0.0245	0.0160	0.0105

n	0.02	0.04	0.06	0.08	0.10	0.12	0.14	0.16	0.18	[illegible]
26	0.5976	0.3607	0.2198	0.1352	0.0839	0.0525	0.0331	0.0211	0.0135	0.0087
27	0.5859	0.3468	0.2074	0.1252	0.0763	0.0469	0.0291	0.0182	0.0115	0.0073
28	0.5744	0.3335	0.1956	0.1159	0.0693	0.0419	0.0255	0.0157	0.0097	0.0061
29	0.5631	0.3207	0.1846	0.1073	0.0630	0.0374	0.0224	0.0135	0.0082	0.0051
30	0.5521	0.3083	0.1741	0.0994	0.0573	0.0334	0.0196	0.0116	0.0070	0.0042
31	0.5412	0.2965	0.1643	0.0920	0.0521	0.0298	0.0172	0.0100	0.0059	0.0035
32	0.5306	0.2851	0.1550	0.0852	0.0474	0.0266	0.0151	0.0087	0.0050	0.0029
33	0.5202	0.2741	0.1462	0.0789	0.0431	0.0238	0.0132	0.0075	0.0042	0.0024
34	0.5100	0.2636	0.1379	0.0730	0.0391	0.0212	0.0116	0.0064	0.0036	0.0020
35	0.5000	0.2534	0.1301	0.0676	0.0356	0.0189	0.0102	0.0055	0.0030	0.0017
36	0.4902	0.2437	0.1227	0.0626	0.0323	0.0169	0.0089	0.0048	0.0026	0.0014
37	0.4806	0.2343	0.1158	0.0580	0.0294	0.0151	0.0078	0.0041	0.0022	0.0012
38	0.4712	0.2253	0.1092	0.0537	0.0267	0.0135	0.0069	0.0036	0.0019	0.0010
39	0.4619	0.2166	0.1031	0.0497	0.0243	0.0120	0.0060	0.0031	0.0016	0.0008
40	0.4529	0.2083	0.0972	0.0460	0.0221	0.0107	0.0053	0.0026	0.0013	0.0007
41	0.4440	0.2003	0.0917	0.0426	0.0201	0.0096	0.0046	0.0023	0.0011	0.0006
42	0.4353	0.1926	0.0865	0.0395	0.0183	0.0086	0.0041	0.0020	0.0010	0.0005
43	0.4268	0.1852	0.0816	0.0365	0.0166	0.0076	0.0036	0.0017	0.0008	0.0004
44	0.4184	0.1780	0.0770	0.0338	0.0151	0.0068	0.0031	0.0015	0.0007	0.0003
45	0.4102	0.1712	0.0727	0.0313	0.0137	0.0061	0.0027	0.0013	0.0006	0.0003
46	0.4022	0.1646	0.0685	0.0290	0.0125	0.0054	0.0024	0.0011	0.0005	0.0002
47	0.3943	0.1583	0.0647	0.0269	0.0113	0.0049	0.0021	0.0009	0.0004	0.0002
48	0.3865	0.1522	0.0610	0.0249	0.0103	0.0043	0.0019	0.0008	0.0004	0.0002
49	0.3790	0.1463	0.0575	0.0230	0.0094	0.0039	0.0016	0.0007	0.0003	0.0001
50	0.3715	0.1407	0.0543	0.0213	0.0085	0.0035	0.0014	0.0006	0.0003	0.0001

TABLE A-6 AMOUNT OF AN ANNUITY OF \$1: $s_{\overline{n}|i} = \dfrac{(1+i)^n - 1}{i}$

n	0.02	0.04	0.06	0.08	0.10	0.12	0.14	0.16	0.18	0.20
1	1.0000	1.0000	1.0000	1.0000	1.0000	1.0000	1.0000	1.0000	1.0000	1.0000
2	2.0200	2.0400	2.0600	2.0800	2.1000	2.1200	2.1400	2.1600	2.1800	2.2000
3	3.0604	3.1216	3.1836	3.2464	3.3100	3.3744	3.4396	3.5056	3.5724	3.6400
4	4.1216	4.2465	4.3746	4.5061	4.6410	4.7793	4.9211	5.0665	5.2154	5.3680
5	5.2040	5.4163	5.6371	5.8666	6.1051	6.3528	6.6101	6.8771	7.1542	7.4416
6	6.3081	6.6330	6.9753	7.3359	7.7156	8.1152	8.5355	8.9775	9.4420	9.9299
7	7.4343	7.8983	8.3938	8.9228	9.4872	10.0890	10.7305	11.4139	12.1415	12.9159
8	8.5830	9.2142	9.8975	10.6366	11.4359	12.2997	13.2328	14.2401	15.3270	16.4991
9	9.7546	10.5828	11.4913	12.4876	13.5795	14.7757	16.0853	17.5185	19.0859	20.7989
10	10.9497	12.0061	13.1808	14.4866	15.9374	17.5487	19.3373	21.3215	23.5213	25.9587
11	12.1687	13.4864	14.9716	16.6455	18.5312	20.6546	23.0445	25.7329	28.7551	32.1504
12	13.4121	15.0258	16.8699	18.9771	21.3843	24.1331	27.2707	30.8502	34.9311	39.5805
13	14.6803	16.6268	18.8821	21.4953	24.5227	28.0291	32.0887	36.7862	42.2187	48.4966
14	15.9739	18.2919	21.0151	24.2149	27.9750	32.3926	37.5811	43.6720	50.8180	59.1959
15	17.2934	20.0236	23.2760	27.1521	31.7725	37.2797	43.8424	51.6595	60.9653	72.0351
16	18.6393	21.8245	25.6725	30.3243	35.9497	42.7533	50.9804	60.9250	72.9390	87.4421
17	20.0121	23.6975	28.2129	33.7502	40.5447	48.8837	59.1176	71.6730	87.0680	105.9306
18	21.4123	25.6454	30.9057	37.4502	45.5992	55.7497	68.3941	84.1407	103.7403	128.1167
19	22.8406	27.6712	33.7600	41.4463	51.1591	63.4397	78.9692	98.6032	123.4135	154.7400
20	24.2974	29.7781	36.7856	45.7620	57.2750	72.0524	91.0249	115.3797	146.6280	186.6880
21	25.7833	31.9692	39.9927	50.4229	64.0025	81.6987	104.7684	134.8405	174.0210	225.0256
22	27.2990	34.2480	43.3923	55.4568	71.4027	92.5026	120.4360	157.4150	206.3448	271.0307
23	28.8450	36.6179	46.9958	60.8933	79.5430	104.6029	138.2970	183.6014	244.4868	326.2369
24	30.4219	39.0826	50.8156	66.7648	88.4973	118.1552	158.6586	213.9776	289.4945	392.4842
25	32.0303	41.6459	54.8645	73.1059	98.3471	133.3339	181.8708	249.2140	342.6035	471.9811

n	0.02	0.04	0.06	0.08	0.10	0.12	0.14	0.16	0.18	0.20
26	33.6709	44.3117	59.1564	79.9544	109.1818	150.3339	208.3327	290.0883	405.2721	567.3773
27	35.3443	47.0842	63.7058	87.3508	121.0999	169.3740	238.4993	337.5024	479.2211	681.8527
28	37.0512	49.9676	68.5281	95.3388	134.2099	190.6989	272.8892	392.5028	566.4809	819.2233
29	38.7922	52.9663	73.6398	103.9659	148.6309	214.5827	312.0937	456.3032	669.4474	984.0679
30	40.5681	56.0849	79.0582	113.2832	164.4940	241.3327	356.7868	530.3117	790.9480	1181.8815
31	42.3794	59.3283	84.8017	123.3459	181.9434	271.2926	407.7370	616.1616	934.3186	1419.2578
32	44.2270	62.7015	90.8898	134.2135	201.1378	304.8477	465.8202	715.7474	1103.4960	1704.1094
33	46.1116	66.2095	97.3432	145.9506	222.2515	342.4294	532.0350	831.2670	1303.1252	2045.9313
34	48.0338	69.8579	104.1838	158.6267	245.4767	384.5210	607.5199	965.2698	1538.6878	2456.1175
35	49.9945	73.6522	111.4348	172.3168	271.0244	431.6635	693.5727	1120.7129	1816.6516	2948.3411
36	51.9944	77.5983	119.1209	187.1021	299.1268	484.4631	791.6729	1301.0270	2144.6489	3539.0093
37	54.0343	81.7022	127.2681	203.0703	330.0395	543.5987	903.5071	1510.1913	2531.6857	4247.8111
38	56.1149	85.9703	135.9042	220.3159	364.0434	609.8305	1030.9981	1752.8219	2988.3891	5098.3733
39	58.2372	90.4091	145.0585	238.9412	401.4478	684.0102	1176.3378	2034.2734	3527.2991	6119.0480
40	60.4020	95.0255	154.7620	259.0565	442.5925	767.0914	1342.0251	2360.7572	4163.2129	7343.8576
41	62.6100	99.8265	165.0477	280.7810	487.8518	860.1424	1530.9086	2739.4783	4913.5913	8813.6291
42	64.8622	104.8196	175.9505	304.2435	537.6370	964.3594	1746.2358	3178.7948	5799.0377	10577.3549
43	67.1595	110.0124	187.5076	329.5830	592.4007	1081.0826	1991.7088	3688.4020	6843.8645	12693.8259
44	69.5027	115.4129	199.7580	356.9496	652.6408	1211.8125	2271.5481	4279.5463	8076.7601	15233.5910
45	71.8927	121.0294	212.7435	386.5056	718.9048	1358.2300	2590.5648	4965.2737	9531.5769	18281.3092
46	74.3306	126.8706	226.5081	418.4261	791.7953	1522.2176	2954.2439	5760.7175	11248.2607	21938.5710
47	76.8172	132.9454	241.0986	452.9001	871.9748	1705.8837	3368.8380	6683.4323	13273.9477	26327.2852
48	79.3535	139.2632	256.5645	490.1321	960.1723	1911.5897	3841.4753	7753.7815	15664.2582	31593.7422
49	81.9406	145.8337	272.9584	530.3427	1057.1896	2141.9805	4380.2819	8995.3865	18484.8247	37913.4906
50	84.5794	152.6671	290.3359	573.7701	1163.9085	2400.0182	4994.5213	10435.6483	21813.0931	45497.1887

TABLE A-7 PRESENT VALUE OF AN ANNUITY OF \$1: $a_{\overline{n}|i} = \frac{1 - (1 + i)^{-n}}{i}$

n	0.02	0.04	0.06	0.08	0.10	0.12	0.14	0.16	0.18	0.20
1	0.9804	0.9615	0.9434	0.9259	0.9091	0.8929	0.8772	0.8621	0.8475	0.8333
2	1.9416	1.8861	1.8334	1.7833	1.7355	1.6901	1.6467	1.6052	1.5656	1.5278
3	2.8839	2.7751	2.6730	2.5771	2.4869	2.4018	2.3216	2.2459	2.1743	2.1065
4	3.8077	3.6299	3.4651	3.3121	3.1699	3.0373	2.9137	2.7982	2.6901	2.5887
5	4.7135	4.4518	4.2124	3.9927	3.7908	3.6048	3.4331	3.2743	3.1272	2.9906
6	5.6014	5.2421	4.9173	4.6229	4.3553	4.1114	3.8887	3.6847	3.4976	3.3255
7	6.4720	6.0021	5.5824	5.2064	4.8684	4.5638	4.2883	4.0386	3.8115	3.6046
8	7.3255	6.7327	6.2098	5.7466	5.3349	4.9676	4.6389	4.3436	4.0776	3.8372
9	8.1622	7.4353	6.8017	6.2469	5.7590	5.3282	4.9464	4.6065	4.3030	4.0310
10	8.9826	8.1109	7.3601	6.7101	6.1446	5.6502	5.2161	4.8332	4.4941	4.1925
11	9.7868	8.7605	7.8869	7.1390	6.4951	5.9377	5.4527	5.0286	4.6560	4.3271
12	10.5753	9.3851	8.3838	7.5361	6.8137	6.1944	5.6603	5.1971	4.7932	4.4392
13	11.3484	9.9856	8.8527	7.9038	7.1034	6.4235	5.8424	5.3423	4.9095	4.5327
14	12.1062	10.5631	9.2950	8.2442	7.3667	6.6282	6.0021	5.4675	5.0081	4.6106
15	12.8493	11.1184	9.7122	8.5595	7.6061	6.8109	6.1422	5.5755	5.0916	4.6755
16	13.5777	11.6523	10.1059	8.8514	7.8237	6.9740	6.2651	5.6685	5.1624	4.7296
17	14.2919	12.1657	10.4773	9.1216	8.0216	7.1196	6.3729	5.7487	5.2223	4.7746
18	14.9920	12.6593	10.8276	9.3719	8.2014	7.2497	6.4674	5.8178	5.2732	4.8122
19	15.6785	13.1339	11.1581	9.6036	8.3649	7.3658	6.5504	5.8775	5.3162	4.8435
20	16.3514	13.5903	11.4699	9.8181	8.5136	7.4694	6.6231	5.9288	5.3527	4.8696
21	17.0112	14.0292	11.7641	10.0168	8.6487	7.5620	6.6870	5.9731	5.3837	4.8913
22	17.6580	14.4511	12.0416	10.2007	8.7715	7.6446	6.7429	6.0113	5.4099	4.9094
23	18.2922	14.8568	12.3034	10.3711	8.8832	7.7184	6.7921	6.0442	5.4321	4.9245
24	18.9139	15.2470	12.5504	10.5288	8.9847	7.7843	6.8351	6.0726	5.4509	4.9371
25	19.5235	15.6221	12.7834	10.6748	9.0770	7.8431	6.8729	6.0971	5.4669	4.9476

Table A-7 Present Value of an Annuity of $1

n	0.02	0.04	0.06	0.08	0.10	0.12	0.14	0.16	0.18	0.20
26	20.1210	15.9828	13.0032	10.8100	9.1609	7.8957	6.9061	6.1182	5.4804	4.9563
27	20.7069	16.3296	13.2105	10.9352	9.2372	7.9426	6.9352	6.1364	5.4919	4.9636
28	21.2813	16.6631	13.4062	11.0511	9.3066	7.9844	6.9607	6.1520	5.5016	4.9697
29	21.8444	16.9837	13.5907	11.1584	9.3696	8.0218	6.9830	6.1656	5.5098	4.9747
30	22.3965	17.2920	13.7648	11.2578	9.4269	8.0552	7.0027	6.1772	5.5168	4.9789
31	22.9377	17.5885	13.9291	11.3498	9.4790	8.0850	7.0199	6.1872	5.5227	4.9824
32	23.4683	17.8736	14.0840	11.4350	9.5264	8.1116	7.0350	6.1959	5.5277	4.9854
33	23.9886	18.1476	14.2302	11.5139	9.5694	8.1354	7.0482	6.2034	5.5320	4.9878
34	24.4986	18.4112	14.3681	11.5869	9.6086	8.1566	7.0599	6.2098	5.5356	4.9898
35	24.9986	18.6646	14.4982	11.6546	9.6442	8.1755	7.0700	6.2153	5.5386	4.9915
36	25.4888	18.9083	14.6210	11.7172	9.6765	8.1924	7.0790	6.2201	5.5412	4.9929
37	25.9695	19.1426	14.7368	11.7752	9.7059	8.2075	7.0868	6.2242	5.5434	4.9941
38	26.4406	19.3679	14.8460	11.8289	9.7327	8.2210	7.0937	6.2278	5.5452	4.9951
39	26.9026	19.5845	14.9491	11.8786	9.7570	8.2330	7.0997	6.2309	5.5468	4.9959
40	27.3555	19.7928	15.0463	11.9246	9.7791	8.2438	7.1050	6.2335	5.5482	4.9966
41	27.7995	19.9931	15.1380	11.9672	9.7991	8.2534	7.1097	6.2358	5.5493	4.9972
42	28.2348	20.1856	15.2245	12.0067	9.8174	8.2619	7.1138	6.2377	5.5502	4.9976
43	28.6616	20.3708	15.3062	12.0432	9.8340	8.2696	7.1173	6.2394	5.5510	4.9980
44	29.0800	20.5488	15.3832	12.0771	9.8491	8.2764	7.1205	6.2409	5.5517	4.9984
45	29.4902	20.7200	15.4558	12.1084	9.8628	8.2825	7.1232	6.2421	5.5523	4.9986
46	29.8923	20.8847	15.5244	12.1374	9.8753	8.2880	7.1256	6.2432	5.5528	4.9989
47	30.2866	21.0429	15.5890	12.1643	9.8866	8.2928	7.1277	6.2442	5.5532	4.9991
48	30.6731	21.1951	15.6500	12.1891	9.8969	8.2972	7.1296	6.2450	5.5536	4.9992
49	31.0521	21.3415	15.7076	12.2122	9.9063	8.3010	7.1312	6.2457	5.5539	4.9993
50	31.4236	21.4822	15.7619	12.2335	9.9148	8.3045	7.1327	6.2463	5.5541	4.9995

TABLE A-8 Binomial Distribution

		p = .1		p = .2		p = .3		p = .4		p = .5	
n	x	f(x)	F(x)	f(x)	F(x)	f(x)	F(x)	f(x)	F(x)	f(x)	F(x)
1	0	0.9000	0.9000	0.8000	0.8000	0.7000	0.7000	0.6000	0.6000	0.5000	0.5000
1	1	0.1000	1.0000	0.2000	1.0000	0.3000	1.0000	0.4000	1.0000	0.5000	1.0000
2	0	0.8100	0.8100	0.6400	0.6400	0.4900	0.4900	0.3600	0.3600	0.2500	0.2500
2	1	0.1800	0.9900	0.3200	0.9600	0.4200	0.9100	0.4800	0.8400	0.5000	0.7500
2	2	0.0100	1.0000	0.0400	1.0000	0.0900	1.0000	0.1600	1.0000	0.2500	1.0000
3	0	0.7290	0.7290	0.5120	0.5120	0.3430	0.3430	0.2160	0.2160	0.1250	0.1250
3	1	0.2430	0.9720	0.3840	0.8960	0.4410	0.7840	0.4320	0.6480	0.3750	0.5000
3	2	0.0270	0.9990	0.0960	0.9920	0.1890	0.9730	0.2880	0.9360	0.3750	0.8750
3	3	0.0010	1.0000	0.0080	1.0000	0.0270	1.0000	0.0640	1.0000	0.1250	1.0000
4	0	0.6561	0.6561	0.4096	0.4096	0.2401	0.2401	0.1296	0.1296	0.0625	0.0625
4	1	0.2916	0.9477	0.4096	0.8192	0.4116	0.6517	0.3456	0.4752	0.2500	0.3125
4	2	0.0486	0.9963	0.1536	0.9728	0.2646	0.9163	0.3456	0.8208	0.3750	0.6875
4	3	0.0036	0.9999	0.0256	0.9984	0.0756	0.9919	0.1536	0.9744	0.2500	0.9375
4	4	0.0001	1.0000	0.0016	1.0000	0.0081	1.0000	0.0256	1.0000	0.0625	1.0000
5	0	0.5905	0.5905	0.3277	0.3277	0.1681	0.1681	0.0778	0.0778	0.0313	0.0313
5	1	0.3281	0.9185	0.4096	0.7373	0.3602	0.5282	0.2592	0.3370	0.1563	0.1875
5	2	0.0729	0.9914	0.2048	0.9421	0.3087	0.8369	0.3456	0.6826	0.3125	0.5000
5	3	0.0081	0.9995	0.0512	0.9933	0.1323	0.9692	0.2304	0.9130	0.3125	0.8125
5	4	0.0005	1.0000	0.0064	0.9997	0.0284	0.9976	0.0768	0.9898	0.1563	0.9688
5	5	0.0000	1.0000	0.0003	1.0000	0.0024	1.0000	0.0102	1.0000	0.0313	1.0000
6	0	0.5314	0.5314	0.2621	0.2621	0.1176	0.1176	0.0467	0.0467	0.0156	0.0156
6	1	0.3543	0.8857	0.3932	0.6554	0.3025	0.4202	0.1866	0.2333	0.0938	0.1094
6	2	0.0984	0.9842	0.2458	0.9011	0.3241	0.7443	0.3110	0.5443	0.2344	0.3438
6	3	0.0146	0.9987	0.0819	0.9830	0.1852	0.9295	0.2765	0.8208	0.3125	0.6563
6	4	0.0012	0.9999	0.0154	0.9984	0.0595	0.9891	0.1382	0.9590	0.2344	0.8906
6	5	0.0001	1.0000	0.0015	0.9999	0.0102	0.9993	0.0369	0.9959	0.0938	0.9844
6	6	0.0000	1.0000	0.0001	1.0000	0.0007	1.0000	0.0041	1.0000	0.0156	1.0000

TABLE A-8 (continued)

		p = .1		p = .2		p = .3		p = .4		p = .5	
n	x	f(x)	F(x)	f(x)	F(x)	f(x)	F(x)	f(x)	F(x)	f(x)	F(x)
7	0	0.4783	0.4783	0.2097	0.2097	0.0824	0.0824	0.0280	0.0280	0.0078	0.0078
7	1	0.3720	0.8503	0.3670	0.5767	0.2471	0.3294	0.1306	0.1586	0.0547	0.0625
7	2	0.1240	0.9743	0.2753	0.8520	0.3177	0.6471	0.2613	0.4199	0.1641	0.2266
7	3	0.0230	0.9973	0.1147	0.9667	0.2269	0.8740	0.2903	0.7102	0.2734	0.5000
7	4	0.0026	0.9998	0.0287	0.9953	0.0972	0.9712	0.1935	0.9037	0.2734	0.7734
7	5	0.0002	1.0000	0.0043	0.9996	0.0250	0.9962	0.0774	0.9812	0.1641	0.9375
7	6	0.0000	1.0000	0.0004	1.0000	0.0036	0.9998	0.0172	0.9984	0.0547	0.9922
7	7	0.0000	1.0000	0.0000	1.0000	0.0002	1.0000	0.0016	1.0000	0.0078	1.0000
8	0	0.4305	0.4305	0.1678	0.1678	0.0576	0.0576	0.0168	0.0168	0.0039	0.0039
8	1	0.3826	0.8131	0.3355	0.5033	0.1977	0.2553	0.0896	0.1064	0.0313	0.0352
8	2	0.1488	0.9619	0.2936	0.7969	0.2965	0.5518	0.2090	0.3154	0.1094	0.1445
8	3	0.0331	0.9950	0.1468	0.9437	0.2541	0.8059	0.2787	0.5941	0.2188	0.3633
8	4	0.0046	0.9996	0.0459	0.9896	0.1361	0.9420	0.2322	0.8263	0.2734	0.6367
8	5	0.0004	1.0000	0.0092	0.9988	0.0467	0.9887	0.1239	0.9502	0.2188	0.8555
8	6	0.0000	1.0000	0.0011	0.9999	0.0100	0.9987	0.0413	0.9915	0.1094	0.9648
8	7	0.0000	1.0000	0.0001	1.0000	0.0012	0.9999	0.0079	0.9993	0.0313	0.9961
8	8	0.0000	1.0000	0.0000	1.0000	0.0001	1.0000	0.0007	1.0000	0.0039	1.0000
9	0	0.3874	0.3874	0.1342	0.1342	0.0404	0.0404	0.0101	0.0101	0.0020	0.0020
9	1	0.3874	0.7748	0.3020	0.4362	0.1556	0.1960	0.0605	0.0705	0.0176	0.0195
9	2	0.1722	0.9470	0.3020	0.7382	0.2668	0.4628	0.1612	0.2318	0.0703	0.0898
9	3	0.0446	0.9917	0.1762	0.9144	0.2668	0.7297	0.2508	0.4826	0.1641	0.2539
9	4	0.0074	0.9991	0.0661	0.9804	0.1715	0.9012	0.2508	0.7334	0.2461	0.5000
9	5	0.0008	0.9999	0.0165	0.9969	0.0735	0.9747	0.1672	0.9006	0.2461	0.7461
9	6	0.0001	1.0000	0.0028	0.9997	0.0210	0.9957	0.0743	0.9750	0.1641	0.9102
9	7	0.0000	1.0000	0.0003	1.0000	0.0039	0.9996	0.0212	0.9962	0.0703	0.9805
9	8	0.0000	1.0000	0.0000	1.0000	0.0004	1.0000	0.0035	0.9997	0.0176	0.9980
9	9	0.0000	1.0000	0.0000	1.0000	0.0000	1.0000	0.0003	1.0000	0.0020	1.0000

TABLE A-8 (*continued*)

		p = .1		p = .2		p = .3		p = .4		p = .5	
n	x	f(x)	F(x)	f(x)	F(x)	f(x)	F(x)	f(x)	F(x)	f(x)	F(x)
10	0	0.3487	0.3487	0.1074	0.1074	0.0282	0.0282	0.0060	0.0060	0.0010	0.0010
10	1	0.3874	0.7361	0.2684	0.3758	0.1211	0.1493	0.0403	0.0464	0.0098	0.0107
10	2	0.1937	0.9298	0.3020	0.6778	0.2335	0.3828	0.1209	0.1673	0.0439	0.0547
10	3	0.0574	0.9872	0.2013	0.8791	0.2668	0.6496	0.2150	0.3823	0.1172	0.1719
10	4	0.0112	0.9984	0.0881	0.9672	0.2001	0.8497	0.2508	0.6331	0.2051	0.3770
10	5	0.0015	0.9999	0.0264	0.9936	0.1029	0.9527	0.2007	0.8338	0.2461	0.6230
10	6	0.0001	1.0000	0.0055	0.9991	0.0368	0.9894	0.1115	0.9452	0.2051	0.8281
10	7	0.0000	1.0000	0.0008	0.9999	0.0090	0.9984	0.0425	0.9877	0.1172	0.9453
10	8	0.0000	1.0000	0.0001	1.0000	0.0014	0.9999	0.0106	0.9983	0.0439	0.9893
10	9	0.0000	1.0000	0.0000	1.0000	0.0001	1.0000	0.0016	0.9999	0.0098	0.9990
10	10	0.0000	1.0000	0.0000	1.0000	0.0000	1.0000	0.0001	1.0000	0.0010	1.0000
11	0	0.3138	0.3138	0.0859	0.0859	0.0198	0.0198	0.0036	0.0036	0.0005	0.0005
11	1	0.3835	0.6974	0.2362	0.3221	0.0932	0.1130	0.0266	0.0302	0.0054	0.0059
11	2	0.2131	0.9104	0.2953	0.6174	0.1998	0.3127	0.0887	0.1189	0.0269	0.0327
11	3	0.0710	0.9815	0.2215	0.8389	0.2568	0.5696	0.1774	0.2963	0.0806	0.1133
11	4	0.0158	0.9972	0.1107	0.9496	0.2201	0.7897	0.2365	0.5328	0.1611	0.2744
11	5	0.0025	0.9997	0.0388	0.9883	0.1321	0.9218	0.2207	0.7535	0.2256	0.5000
11	6	0.0003	1.0000	0.0097	0.9980	0.0566	0.9784	0.1471	0.9006	0.2256	0.7256
11	7	0.0000	1.0000	0.0017	0.9998	0.0173	0.9957	0.0701	0.9707	0.1611	0.8867
11	8	0.0000	1.0000	0.0002	1.0000	0.0037	0.9994	0.0234	0.9941	0.0806	0.9673
11	9	0.0000	1.0000	0.0000	1.0000	0.0005	1.0000	0.0052	0.9993	0.0269	0.9941
11	10	0.0000	1.0000	0.0000	1.0000	0.0000	1.0000	0.0007	1.0000	0.0054	0.9995
11	11	0.0000	1.0000	0.0000	1.0000	0.0000	1.0000	0.0000	1.0000	0.0005	1.0000

TABLE A-8 (*continued*)

		p = .1		p = .2		p = .3		p = .4		p = .5	
n	x	f(x)	F(x)	f(x)	F(x)	f(x)	F(x)	f(x)	F(x)	f(x)	F(x)
12	0	0.2824	0.2824	0.0687	0.0687	0.0138	0.0138	0.0022	0.0022	0.0002	0.0002
12	1	0.3766	0.6590	0.2062	0.2749	0.0712	0.0850	0.0174	0.0196	0.0029	0.0032
12	2	0.2301	0.8891	0.2835	0.5583	0.1678	0.2528	0.0639	0.0834	0.0161	0.0193
12	3	0.0852	0.9744	0.2362	0.7946	0.2397	0.4925	0.1419	0.2253	0.0537	0.0730
12	4	0.0213	0.9957	0.1329	0.9274	0.2311	0.7237	0.2128	0.4382	0.1208	0.1938
12	5	0.0038	0.9995	0.0532	0.9806	0.1585	0.8822	0.2270	0.6652	0.1934	0.3872
12	6	0.0005	0.9999	0.0155	0.9961	0.0792	0.9614	0.1766	0.8418	0.2256	0.6128
12	7	0.0000	1.0000	0.0033	0.9994	0.0291	0.9905	0.1009	0.9427	0.1934	0.8062
12	8	0.0000	1.0000	0.0005	0.9999	0.0078	0.9983	0.0420	0.9847	0.1208	0.9270
12	9	0.0000	1.0000	0.0001	1.0000	0.0015	0.9998	0.0125	0.9972	0.0537	0.9807
12	10	0.0000	1.0000	0.0000	1.0000	0.0002	1.0000	0.0025	0.9997	0.0161	0.9968
12	11	0.0000	1.0000	0.0000	1.0000	0.0000	1.0000	0.0003	1.0000	0.0029	0.9998
12	12	0.0000	1.0000	0.0000	1.0000	0.0000	1.0000	0.0000	1.0000	0.0002	1.0000

APPENDIX B

Answers to selected problems

SECTION P-1

1. $\{2, 4, 6, \ldots\}$

3. $\frac{2}{3}, \frac{4}{7}, -\frac{1}{3}, \frac{3}{5}, 1\frac{7}{8}, \frac{99}{100}, \frac{499}{500}, \frac{-8}{3}$

5. $0 > -5,\ 3 > -5,\ 10 > -5$
$3 > 0,\ 10 > 0,\ 10 > 3$

7. 20

9. 24

11. 50

13. 5

SECTION P-2

1. 9

3. -15

5. -1

7. 12

9. 24

11. 20

13. 3

15. 24

17. 35

19. $-5x + 5y$ or $5(y - x)$

SECTION P-3

1. 1

3. $\frac{8}{15}$

5. $\frac{5}{6}$

7. $\frac{7}{12}$

9. $\dfrac{a+3}{7}$

11. $\frac{5}{21}$

13. $\frac{9}{1225}$

15. $\frac{61}{14}$

17. 0

SECTION P-4

1. 1.65
3. 2.57%
5.

	.25	25
	.20	20
$\frac{2}{5}$		40
$\frac{7}{20}$		35
$\frac{1}{10}$	.10	
$\frac{1}{8}$	.125	

7. **(a)** 0.794% **(b)** 0.787% **(c)** No
9. **(a)** \$27.50 **(b)** \$24.75
11. \$95.04
13. **(a)** Yes
    (b) 2.776% 2.523% 2.164% 2.180% 2.084%
    (c) Amount is increasing, but percentage of GNP is decreasing.

SECTION P-5

1. x^9
3. x^2
5. $\frac{1}{b}$
7. $\frac{1}{8}$
9. 12
11. $\sqrt[4]{x^3}$
13. $\sqrt[4]{q}$
15. $\frac{1}{\sqrt[3]{x-y}}$
17. $x^{1/3}$
19. x^2y^2
21. 3
23. -1
25. 64
27. 2
29. 3
31. -2

SECTION P-6

1. $2, 4, 3, 0, -1, -3, -2$
3. $2, -3, 5, \frac{1}{3}, \frac{1}{2}, \frac{1}{3}, \frac{1}{3}, \frac{2}{3}$
5. $6 = 6$
7. Use $b^x = M$ and $b^y = N$
9. **(a)** 3 **(b)** 0 **(c)** -1 **(d)** $2 + \log_b(ab+1)$
11. *Hint:* Let $x = \log_a N$.
    Then $a^x = N$ (1)
    Take log of both sides of (1).

SECTION P-7

1. $2x + 2y$
3. $x^3 - x^2y$
5. $x^2 + x - 6$
7. $x^2 + 4x + 4$
9. $x^2 - 24x + 144$
11. $x(y+z)$
13. $3x^2y(1-2y^2)$
15. $(x+5)^2$
17. $(x+11)(x-11)$
19. $(x+5)(x-2)$

SECTION P-8

1. 3
3. 5
5. $-a/b$
7. $-2, 3$
9. $-3, -4$
11. $0, -3$
13. $-2, a$
15. 6
17. $P = \dfrac{a}{x+b} - c$
19. 9.9 million bales

SECTION P-9

1. $(1, -1)$
3. $(\frac{3}{2}, \frac{5}{2})$
5. No solution
7. $(-\frac{1}{2}, 0)$
9. $(3, -\frac{2}{3})$
11. $(-11, 9)$
13. No solution
15. $(2, 1, -3)$
17. $(\frac{14}{41}, \frac{51}{41}, \frac{6}{41})$
19. $(\frac{9}{4}, \frac{-19}{4}, -\frac{1}{4})$
21. $(-1, 1, 0)$

SECTION P-10

1. 15
3. $\frac{77}{30}$
5. 76
7. 60
9. **(a)** $c_{13} = a_{11}b_{13} + a_{12}b_{23} + a_{13}b_{33}$
(b) $c_{31} = a_{31}b_{11} + a_{32}b_{21} + a_{33}b_{31}$
13. **(a)** 516 **(b)** 39
15. $1395
21. 50, 69, 540, 730, 2500

CHAPTER 1
SECTION 1-1

1. **(a)** 1 **(b)** 1 **(c)** 16 **(d)** $h^2 + 6h + 9$
3. Domain: All real numbers between 4 and 9, inclusive.
Range: All real numbers between 10 and 25, inclusive.
5. Domain: All real numbers between -1 and 2, inclusive.
Range: All real numbers between -1 and 3, inclusive.
7. A real variable u is said to be a function of the real variable t if one and only one value of u corresponds to each value of t.
9. Domain: All real numbers between 2,000 and 4,000, inclusive.
Range: 350,000.
11. **(a)** Tax = .20 (Taxable Income), if we let x represent Taxable Income and y Tax then we have: $y = .20x$, x between 0 and 20,000, inclusive.
(b) Tax (y) and Taxable Income (x); Tax is the dependent variable.
(c) Domain: All real numbers between 0 and 20,000 inclusive.
Range: All real numbers between 0 and 4000 inclusive.

SECTION 1-2

1. Lines AB, AC, and AD all have slope -3.
2. **(a)** 2 **(b)** 3 **(c)** -2 **(d)** 2 **(e)** $\frac{1}{2}$
3. **(a)** $-\frac{3}{2}, 3$ **(b)** $1, -3$ **(c)** $\frac{5}{2}, 5$ **(d)** $-3, 6$ **(e)** $3, -\frac{3}{2}$
7. $y = -3x + 8$
8. **(a)** $y = 3x - 5$ **(b)** $y = 2x + 1$ **(c)** $y = -\frac{1}{2}x + 2$
(d) $y = 4x - 12$ **(e)** $y = -5x + 3$ **(f)** $y = -x + 7$

9. **(c)** $\text{Slope} = \dfrac{y_2 - y_1}{x_2 - x_1} = \dfrac{-6-4}{3-(-2)} = \dfrac{-10}{5} = -2$

(d)
$$\begin{aligned} y - y_1 &= m(x - x_1) \\ y - 4 &= -2(x - (-2)) \\ y - 4 &= -2x - 4 \\ y &= -2x \end{aligned}$$

10. **(a)** $y = 5x + 10$ **(b)** $y = -2x - 2$
11. **(a)** parallel **(b)** perpendicular **(c)** neither **(d)** neither
(e) perpendicular **(f)** parallel
13. **(a)** 5 **(b)** 5 **(c)** 5 **(d)** $\sqrt{37} = 6.0828$ **(e)** $\sqrt{52} = 7.211$
(f) $\sqrt{a^2 + b^2}$
14. **(c)** yes **(d)** $\sqrt{200} = \sqrt{180 + 20}$
15. **(b)** no **(c)** $\sqrt{17} \neq \sqrt{2} + \sqrt{5}$

SECTION 1-3

1. Yes **3.** No, linear **5.** No **7.** $(-4, -14)$ **9.** $(\frac{1}{3}, \frac{2}{3})$

11.

x	y
-3	7
-2	5
$-\frac{3}{2}$	$\frac{19}{4}$
-1	5
0	7

13.

x	y
5	4
4	1
3	0
2	1
1	4

15.

x	y
-1	-2
2	5
-3	-6
-4	-5
-5	-2

17. 7 **19.** -28

SECTION 1-4

1. **(a)** 1.50 **(b)** $R = 1.50x$
3. **(a)** $p = -.00005x + 45$
5. Parabola opening downward with vertex at $x = 30$.
7. **(a)** R is positive for $0 \leqslant x \leqslant 30$.
(b) Revenue is maximized at $x = 15$
9. A real variable y is said to be a function of the real variable x if one and only one value of y corresponds to each value of x. In this case for a given p value there are many x values.

SECTION 1-5

1. (a) $C = 0.50x + 2500$
   (b) The vertical intercept is the fixed cost.
   (c) The slope is the variable cost per unit.
   (d) In some cases the assumption of a linear relationship is a good one. In other cases this assumption of linearity may be relevant only for certain values of the independent variable. A typical example might read as follows. The cost function $C = f(x)$ is nearly linear for $100 \leqslant x \leqslant 300$.
3. \$400
5. $R = 3.00x$ and $C = 2.00x + 300$
7. $C = \$4000 + 8.00x$
9. No. In the earlier stages of production one should realize economies of scale and in the later stages diseconomies of scale.
14. One lot of 200 shares, \$202.00; two lots of 100 shares, \$130.00; savings = \$72.00.
15. One lot of 200 shares, \$58.00; two lots of 100 shares, \$76.00; loss = \$18.00.
19. (b) Yes, they both cost \$14.40.

SECTION 1-6

1. $(\frac{3}{2}, 6)$
3. 750 units
5. $C = 1.50x + 2000$; $R = 2.00x$
7. If you are using the car more than 2000 miles per month, choose Plan 1. Including the cost of gas has no effect on the indifference point.
8. (a) Option 2
   (b) Use Option 1 if less than 33 copies are required.
   (c) Use Option 1 if less than 37 copies are required.
11. (b) P max at $x = 22.5$ (c) No
13. d
15. c

SECTION 1-7

1.

x	$y = 2/x$
-4	-0.5
-2	-1
-1	-2
$-\frac{1}{2}$	-4
$\frac{1}{2}$	4
1	2
2	1
4	0.5

3.

x	$y = 2/x + 4$
-5	3.8
-4	3.5
-2	3
-1	2
1	6
2	5
4	4.5
5	4.2

5. (a) $y = 200/x$

7. **(a)** $y = \dfrac{100{,}000}{x} + 10$

(c) Total cost

(d) $c = \left(\dfrac{100{,}000}{x} + 10\right)x$

(e) No matter how many units are produced unit costs will not drop below $10.

SECTION 1-8

1. *d* **2.** *a* **3.** *a* **4.** *b* **5.** *c* **6.** *b*

CHAPTER 2
SECTION 2-1

1. **(a)** equal
(b) not equal; one is a row, the other is a column
(c) not equal; different number of components
(d) not equal; first components differ

2. **(a)** (3, 7, 9)
(b) one is a row, the other is a column
(c) the numbers of components differ
(d) (a, b, c)

5. The point corresponding to the sum is
(a) (3, 8) **(b)** (2, −2) **(c)** (0, 3) **(d)** (2, 5)

7. An example is (7, −2, 3, 1) and $\begin{pmatrix} a \\ b \\ c \\ d \\ e \end{pmatrix}$

9. (2, 1, 1)

10. $\begin{pmatrix} 100{,}000 \\ 150{,}000 \\ 125{,}000 \end{pmatrix}$ **11.** $\begin{pmatrix} 1200 \\ 1000 \\ 800 \end{pmatrix}$

12. $\begin{pmatrix} 30 \\ 15 \\ 25 \\ 30 \end{pmatrix}$

SECTION 2-2

1. **(a)** $\begin{pmatrix} 2 \\ 3 \\ 1 \end{pmatrix}$ **(c)** $\begin{pmatrix} x \\ 4x \\ 2x \end{pmatrix}$

(b) (12, 24, 21) **(d)** $\begin{pmatrix} 20 \\ 10 \\ 5 \end{pmatrix}$

2. **(a)** 17 **(b)** 58 **(c)** 57 **(d)** 107

6. $x\begin{pmatrix} 4 \\ 5 \\ 5 \end{pmatrix} + y\begin{pmatrix} 6 \\ 4 \\ 6 \end{pmatrix} + z\begin{pmatrix} 3 \\ 1 \\ 0 \end{pmatrix} = \begin{pmatrix} 2 \\ 1 \\ 4 \end{pmatrix}$.

7. $4w + 2x + 3y = 4$
$9w + x + 2y = 9$
$3w + 2x + 4y = 3$

10. (1, 2, 3, 4, 5), (4, −2, 3, −6, 8), (−1, −4, 3, 6 10)

11. *Hint:* Use closure under multiplication of real number system.

12. *Hint:* Use closure under addition of real number system.

13. **3a** **(a)** (6, −9) **(b)** (6, 3, −18) **(c)** (−3, 6, −9, 12, −9)
−b **(a)** (1, −4) **(b)** (−1, −3, −4) **(c)** (−2, −4, −3, −6, −1)
a + b **(a)** (1, 1) **(b)** (3, 4, −2) **(c)** (1, 6, 0, 10, −2)
a − b **(a)** (3, −7) **(b)** (1, −2, −10) **(c)** (−3, −2, −6, −2, −4)

14. **(a)** $\sqrt{58} = 7.616$ **(b)** $\sqrt{105} = 10.247$ **(c)** $\sqrt{69} = 8.307$

SECTION 2-3

1. **(a)** not equal; $1 \neq 2$ **(b)** not equal; the sizes are not the same

3. **(a)** $\begin{pmatrix} 9 & 9 \\ 9 & 6 \end{pmatrix}$ **(b)** $\begin{pmatrix} 6 & 5 & 11 \\ 5 & 5 & 7 \end{pmatrix}$ **(c)** $\begin{pmatrix} 16 & 12 & 12 \\ 10 & 10 & 5 \\ 7 & 3 & 5 \end{pmatrix}$

5. $a = 3,\ b = 5$

8.

	W_1	W_2	W_3	W_4
R_1	5	7	9	16
R_2	10	5	11	10

9. **(a)** $C = \begin{pmatrix} 5 & 7 & 9 & 16 \\ 10 & 5 & 11 & 10 \\ 8 & 7 & 4 & 8 \end{pmatrix}$

(b) $c_{11} = 5$

(c) c_{21} corresponds to the route from Region 2 to Warehouse 1.
c_{12} corresponds to the route from Region 1 to Warehouse 2.

10. **(a)** $V = \begin{pmatrix} 20 & 10 \\ 15 & 5 \\ 10 & 3 \\ 5 & 2 \end{pmatrix}$

(b) v_{11} gives the material cost of model 1.
v_{32} gives the variable overhead for model 2.

(c) total variable cost for model 1.

11.

		buy	lease
Demand	low	10,000	12,000
	high	20,000	5,000

12. $P = \begin{pmatrix} 10{,}000 & 12{,}000 \\ 20{,}000 & 5{,}000 \end{pmatrix}$

13. $x = \frac{1}{5}$, $y = \frac{3}{5}$

SECTION 2-4

1. **(a)** $\begin{pmatrix} 48 & 72 \\ 16 & 8 \end{pmatrix}$ **(c)** (1.2, 2.4)

(b) $\begin{pmatrix} 6 & 3 & 12 \\ 15 & 18 & 24 \\ 21 & 9 & 6 \end{pmatrix}$ **(d)** $\begin{pmatrix} 12.6 \\ 2.8 \\ 1.4 \end{pmatrix}$

2. **(a)** $\begin{pmatrix} 20 & 11 \\ 48 & 29 \end{pmatrix}$ **(c)** 43

(b) $\begin{pmatrix} 31 & 23 \\ 70 & 56 \end{pmatrix}$ **(d)** $\begin{pmatrix} 56 & 68 & 36 \\ 28 & 22 & 30 \\ 34 & 54 & 22 \end{pmatrix}$

5. **(a)** $\begin{pmatrix} -4 & -3 \\ 0 & -13 \end{pmatrix}$ **(d)** $\begin{pmatrix} 126 & 68 & 40 \\ 118 & 81 & 49 \\ 85 & 56 & 46 \end{pmatrix}$

(b) $\begin{pmatrix} 60 & 50 \\ 66 & 70 \end{pmatrix}$ **(e)** $\begin{pmatrix} 7 & 20 & 18 & 13 & 12 \\ 17 & 42 & 38 & 27 & 28 \end{pmatrix}$

(c) not defined

11. *a*

SECTION 2-5

1. $\begin{pmatrix} 1 & -2 & 1 \\ 2 & 1 & 2 \\ 1 & 1 & 3 \end{pmatrix}\begin{pmatrix} x \\ y \\ z \end{pmatrix} = \begin{pmatrix} 9 \\ -1 \\ 2 \end{pmatrix}$

4. $$\begin{aligned} 4x_1 + 2x_2 + 6x_3 &= 4 \\ 9x_1 + x_2 &= 9 \\ 3x_1 + 2x_2 + 4x_3 &= 3 \end{aligned}$$

SECTION 2-6

3. \$56.75

6. (25, 52); 25 units of *A* and 52 units of *B* are required

8. \$206

9.

	I	II	III
P_1	4	2	3
P_2	2	0	1

11.

	061	072
P_1	22	31
P_2	8	11

13. \$822

15. Three hundred units of part *a* are needed to fill the order.

CHAPTER 3
SECTION 3-1

1. $\begin{pmatrix} 2 & -4 \\ 1 & 7 \end{pmatrix}$ **3.** $\begin{pmatrix} 1 & 0 \\ 0 & 1 \end{pmatrix}$

5. $\begin{pmatrix} -3 & 3 & -2 \\ -1 & -5 & -5 \\ -4 & -3 & -8 \end{pmatrix}$ **7.** $\begin{pmatrix} 0 & 0 \\ 0 & 0 \end{pmatrix}$

9. $\begin{pmatrix} 0 & 0 \\ 0 & 0 \end{pmatrix}$

11. **(a)** $\begin{pmatrix} 0 & 0 & 0 & 0 \\ 0 & 0 & 0 & 0 \\ 0 & 0 & 0 & 0 \\ 0 & 0 & 0 & 0 \end{pmatrix}$ **(b)** $\begin{pmatrix} 0 & 0 \\ 0 & 0 \end{pmatrix}$

12. **(a)** $\begin{pmatrix} 1 & 0 & 0 & 0 \\ 0 & 1 & 0 & 0 \\ 0 & 0 & 1 & 0 \\ 0 & 0 & 0 & 1 \end{pmatrix}$ **(b)** $\begin{pmatrix} 1 & 0 \\ 0 & 1 \end{pmatrix}$

SECTION 3-2

1. $\begin{pmatrix} 14 & 20 \\ 10 & 7 \end{pmatrix} = \begin{pmatrix} 14 & 20 \\ 10 & 7 \end{pmatrix}$

3. $A(B+C) = AB + AC$

$$\begin{pmatrix} 2 & 4 \\ 6 & 3 \end{pmatrix}\begin{pmatrix} 12 & 16 \\ 4 & 4 \end{pmatrix} = \begin{pmatrix} 24 & 18 \\ 54 & 45 \end{pmatrix} + \begin{pmatrix} 16 & 30 \\ 30 & 63 \end{pmatrix}$$

$$\begin{pmatrix} 40 & 48 \\ 84 & 108 \end{pmatrix} = \begin{pmatrix} 40 & 48 \\ 84 & 108 \end{pmatrix}$$

5. $A + B = B + A$

$$\begin{pmatrix} 10 & 11 \\ 8 & 4 \end{pmatrix} = \begin{pmatrix} 10 & 11 \\ 8 & 4 \end{pmatrix}$$

7. No

9. **(a)** No **(b)** No **(c)** No

11. **(a)** Yes **(b)** Yes

14. $\begin{pmatrix} 15 & 8 & 9 \\ 17 & 8 & 10 \\ 13 & 8 & 8 \end{pmatrix}$ **15.** $\begin{pmatrix} 13 & 7 & 22 \\ 13 & 6 & 17 \\ 8 & 4 & 12 \end{pmatrix}$

16. No. It appears that the commutative law of multiplication does not hold for matrices.

SECTION 3-3

1. $AB = \begin{pmatrix} 19 & 13 \\ 30 & 22 \end{pmatrix}, \; BA = \begin{pmatrix} 13 & 42 \\ 8 & 28 \end{pmatrix}$

3. $\begin{pmatrix} 0 & 27 \\ -26 & 24 \end{pmatrix}$

5. $x = 1$

SECTION 3-4

1. **(a)** $\begin{pmatrix} -2 & -1 \\ -5 & -6 \end{pmatrix}$ **(b)** $\begin{pmatrix} -3 & -2 & -8 \\ -4 & -5 & -2 \\ -1 & -7 & -6 \end{pmatrix}$

2. $\begin{pmatrix} -\frac{1}{7} & +\frac{2}{7} \\ +\frac{6}{7} & -\frac{5}{7} \end{pmatrix}$

5. a, c, and d do not have a multiplicative inverse.

9. $A^{-1} = \begin{pmatrix} 1/a & 0 & 0 \\ 0 & 1/b & 0 \\ 0 & 0 & 1/c \end{pmatrix}$

10. $A^{-1} = \begin{pmatrix} 1/2 & 0 & 0 \\ 0 & 1/3 & 0 \\ 0 & 0 & -1/4 \end{pmatrix}$

SECTION 3-5

1. **(a)** $\begin{pmatrix} 2 & 3 \\ 1 & -2 \end{pmatrix}\begin{pmatrix} x \\ y \end{pmatrix} = \begin{pmatrix} 4 \\ -5 \end{pmatrix}$

(b) $\begin{pmatrix} 2 & 3 \\ 1 & -2 \end{pmatrix}\begin{pmatrix} \frac{2}{7} & \frac{3}{7} \\ \frac{1}{7} & -\frac{2}{7} \end{pmatrix} = \begin{pmatrix} 1 & 0 \\ 0 & 1 \end{pmatrix}$

(c) $x = -1, \; y = 2$

3. **(a)** $\begin{pmatrix} 1 & 1 \\ 3 & -1 \end{pmatrix}\begin{pmatrix} x \\ y \end{pmatrix} = \begin{pmatrix} 4 \\ 2 \end{pmatrix}$

(b) $\begin{pmatrix} +\frac{1}{4} & +\frac{1}{4} \\ +\frac{3}{4} & -\frac{1}{4} \end{pmatrix}$

(c) $x = 1.5, \; y = 2.5$

5. **(a)** $\begin{pmatrix} 3 & 2 & 2 \\ 0 & 1 & 0 \\ 2 & 2 & 2 \end{pmatrix}\begin{pmatrix} x_1 \\ x_2 \\ x_3 \end{pmatrix} = \begin{pmatrix} 4 \\ 5 \\ 6 \end{pmatrix}$

(b) $\begin{pmatrix} 3 & 2 & 2 \\ 0 & 1 & 0 \\ 2 & 2 & 2 \end{pmatrix}\begin{pmatrix} 1 & 0 & -1 \\ 0 & 1 & 0 \\ -1 & -1 & \frac{3}{2} \end{pmatrix} = \begin{pmatrix} 1 & 0 & 0 \\ 0 & 1 & 0 \\ 0 & 0 & 1 \end{pmatrix}$

(c) $x_1 = -2, \; x_2 = 5, \; x_3 = 0$

7. *Hint:* Use $\mathbf{x} = I\mathbf{x}$.

SECTION 3-6

1. **(a)** $\left(\begin{array}{cc|cc} 2 & 0 & 1 & 0 \\ 1 & 1 & 0 & 1 \end{array}\right)$ **(b)** $\left(\begin{array}{cc|cc} 1 & 0 & 1 & 0 \\ 2 & 2 & 0 & 1 \end{array}\right)$

3. **(a)** $\begin{pmatrix} -\frac{1}{7} & \frac{2}{7} \\ \frac{6}{7} & -\frac{5}{7} \end{pmatrix}$ **(b)** $\begin{pmatrix} \frac{2}{14} & +\frac{4}{14} \\ \frac{5}{14} & -\frac{3}{14} \end{pmatrix}$

5. **(a)** $\begin{pmatrix} \frac{2}{7} & \frac{3}{7} \\ \frac{1}{7} & -\frac{2}{7} \end{pmatrix}$ **(b)** $\begin{pmatrix} 1 & -3 & 7 \\ 0 & 1 & -2 \\ 0 & 0 & 0.5 \end{pmatrix}$

7. $\begin{pmatrix} 0 & \frac{1}{3} & 0 \\ 0.5 & -\frac{2}{3} & 0 \\ -0.5 & -1\frac{1}{3} & 1 \end{pmatrix}$

SECTION 3-7

4. $x_1 = 3230,\ x_2 = 3711$ **8.** $x_1 = 200,\ x_2 = 100,\ x_3 = 100$

9. $x_1 = 211.88,\ x_2 = 100.78,\ x_3 = 100.75$ **10.** $(I - A)^{-1} = \begin{pmatrix} \frac{3}{2} & \frac{2}{3} \\ \frac{1}{2} & \frac{4}{3} \end{pmatrix}$

11. $x_1 = 850,\ x_2 = 950$ **12.** $x_1 = 895,\ x_2 = 965$

SECTION 3-8

1. *a* **2.** *d* **3.** *a* **4.** *b* **5.** *c* **6.** *b* **7.** \$58,000

CHAPTER 4
SECTION 4-1

1. **(a)** (4)(3), (1)(2)
(b) $c_{11}c_{22},\ c_{21}c_{22}$
(c) (2)(6)(4), (3)(2)(1), (4)(5)(7), (1)(6)(4), (7)(2)(2), (4)(5)(3)
(d) $x8,\ 2y$

3. **(a)** 2 **(b)** 3 **(c)** 8 or some other even number **(d)** 1
6. odd **8.** **(a)** one **(b)** three **9.** minus, minus
11. **(a)** plus **(b)** minus **(c)** minus **(d)** minus
13. **(a)** 0 **(b)** -100

SECTION 4-2

1. -9 **3.** -110 **4.** 99 **7.** $x = 3$

SECTION 4-3

3. 8, -1, -10 **4.** $|A| = 3$ **5.** 25 **7.** **(a)** -18 **(b)** -9
11. 76 **12.** **(a)** 4 **(b)** -1

SECTION 4-4

1. **(a)** $x = 2.5,\ y = -1.5$ **3.** $x = 1$
5. $a = 2,\ b = -1,\ c = -3$

SECTION 4-5

1. **(a)** $A^{-1} = \begin{pmatrix} -\frac{1}{6} & \frac{2}{3} \\ \frac{1}{2} & -1 \end{pmatrix}$ **(b)** $A^{-1} = \begin{pmatrix} \frac{3}{19} & -\frac{2}{19} \\ -\frac{4}{19} & \frac{9}{19} \end{pmatrix}$

3. **(a)** $\begin{pmatrix} 6 & 2 & 6 \\ 5 & 3 & 8 \\ 4 & 1 & 9 \end{pmatrix}$ **(c)** $(2 \quad 6 \quad 1)$

(b) $\begin{pmatrix} c_{11} & c_{21} & c_{31} \\ c_{12} & c_{22} & c_{32} \end{pmatrix}$ **(d)** $\begin{pmatrix} 2 & 4 & 6 \\ 3 & 5 & 7 \end{pmatrix}$

5. **(a)** 8 **(c)** $\begin{pmatrix} 6 & -2 \\ -8 & 4 \end{pmatrix}$

(b) $\begin{pmatrix} 6 & -8 \\ -2 & 4 \end{pmatrix}$ **(d)** $B^{-1} = \begin{pmatrix} \frac{3}{4} & -\frac{1}{4} \\ -1 & \frac{1}{2} \end{pmatrix}$

7. **(a)** Determinant = 0 and inverse does not exist.

(b) $\begin{pmatrix} 0 & \frac{1}{3} & 0 \\ \frac{1}{2} & -\frac{2}{3} & 0 \\ -\frac{1}{2} & -\frac{4}{3} & 1 \end{pmatrix}$

13. Revenue = tr $AB^t = 21{,}700$ **14.** yes, $B^t = B,\ \therefore\ AB^t = AB$

SECTION 4-6

1. $x = 3.5,\ y = -1.5$ **3.** $x = 3,\ y = -1,\ z = -1$
5. $x = 1,\ y = 2,\ z = -1$ **7.** $x_1 = 2,\ x_2 = -1,\ x_3 = -3$

CHAPTER 5
SECTION 5-1

2. **(a)** $h = 1$ **(b)** $h = 6$ **(c)** $h = 3$ **(d)** $h = 3$ **4.** $5 < 10$
6. **(a)** $-3 \leqslant x$ **(b)** $x \leqslant 1$ **(c)** $3 \leqslant x$
8. **(a)** $x \leqslant -\frac{5}{3}$ **(b)** $x < \frac{10}{3}$ **(c)** $x \leqslant -2$
9. Special case of Problem 10. See graph.
10. $a = 0.375,\ b = 0.125,\ c = 0.5$

15. **(b)** $f(x) = \begin{cases} 1926x, & 2 \leqslant x \leqslant 10 \\ 1712x, & 11 \leqslant x \leqslant 20 \\ 1498x, & 21 \leqslant x \leqslant 30 \\ 1284x, & 31 \leqslant x \end{cases}$

SECTION 5-2

5. The entire first quadrant.
7. The corners of the solution space are $(-\frac{24}{5}, \frac{18}{5})$, $(\frac{6}{7}, \frac{12}{7})$, and $(4, -3)$.
8. The corners of the solution space are $(0, 0)$, $(0, 3)$, $(2, 2)$, and $(4, 0)$.
11. $S \geqslant 1{,}000{,}000$, $CGS \leqslant 650{,}000$, $NI > 100{,}000$, $RR \geqslant 11\%$
13. $D \leqslant 20$, $S \geqslant 10{,}000$, $V \geqslant 2$

SECTION 5-3

1. $$\begin{array}{ll} 4x + y \leqslant 36 & P = 6x + 5y \\ 2x + 4y \leqslant 40 & x \geqslant 0 \\ & y \geqslant 0 \end{array}$$
2. $$\begin{array}{ll} 4x + y \leqslant 36 & P = 6x + 5y \\ 2x + 4y \leqslant 40 & x \geqslant 0 \\ 2x + y \leqslant 20 & y \geqslant 0 \end{array}$$
5. **(a)** *c* **(b)** *e* **(c)** *e* **(d)** *d* **(e)** *d*
7. **(a)** *a* **(b)** *c*

SECTION 5-4

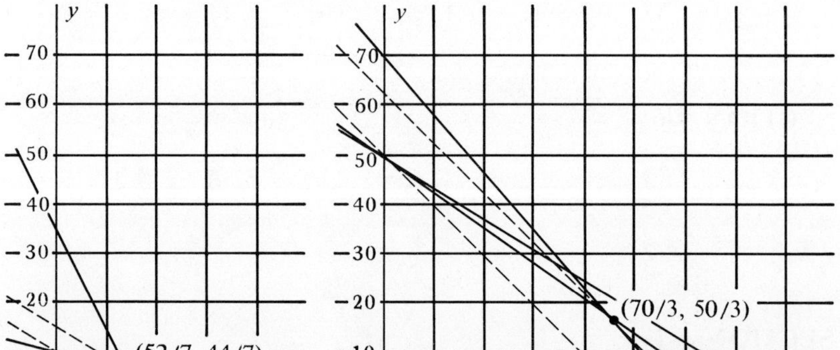

5. Corners: $(0, 0)$, $(0, 10)$, $(9, 0)$, $(\frac{52}{7}, \frac{44}{7})$
   $P = 76$ at $(\frac{52}{7}, \frac{44}{7})$ or $P = 72$ at $(7, 6)$
7. Corners: $(0, 1)$, $(0, 8)$, $(7, 1)$
   $P = 22$ at $(7, 1)$

9. & 10. All lines have slope $= -\frac{2}{3}$ and L_2 is farthest from origin.

SECTION 5-5

1. $$\begin{pmatrix} 1 & 0 & 4 & 1 \\ 0 & 1 & 2 & 4 \end{pmatrix} \begin{pmatrix} w \\ z \\ x \\ y \end{pmatrix} = \begin{pmatrix} 36 \\ 40 \end{pmatrix}$$

2. $$\begin{matrix} 1 & 0 & 0 & 4 & 1 \\ 0 & 1 & 0 & 2 & 4 \\ 0 & 0 & 1 & 2 & 1 \end{matrix} \begin{pmatrix} w \\ z \\ t \\ x \\ y \end{pmatrix} = \begin{pmatrix} 36 \\ 40 \\ 20 \end{pmatrix}$$

5.

	w	z	x	y	Constant
w	1	0	4	1	36
z	0	1	2	4	40
C	0	0	6	5	
Z			0	0	
NET			6	5	

6.

	w	z	t	x	y	Constant
w	1	0	0	4	1	36
z	0	1	0	2	4	40
t	0	0	1	2	1	20
C	0	0	0	6	5	
Z				0	0	
NET				6	5	

9.

	x	y	w	z	Constant
x	1	0	$\frac{4}{14}$	$\frac{1}{14}$	$\frac{52}{7}$
y	0	1	$-\frac{2}{14}$	$\frac{4}{14}$	$\frac{44}{7}$
C	6	5	0	0	
Z			1	1	
NET			-1	-1	

11.

	y	x	z	w	Constant
y	1	0	-0.5	0.5	25
x	0	1	.4	-0.2	20
C	1.00	2.00	0	0	
Z			0.30	0.10	
NET			-0.30	-0.10	

13. **(a)** *d* **(b)** *c* **(c)** *d* **(d)** *b* **(e)** *b*

SECTION 5-6

1. $$\begin{pmatrix} 3 & 3 & 1 & 0 & 0 \\ 1 & 5 & 0 & 1 & 0 \\ 4 & 4 & 0 & 0 & 1 \end{pmatrix} \begin{pmatrix} x \\ y \\ w \\ z \\ t \end{pmatrix} \quad \begin{pmatrix} 2400 \\ 1600 \\ 1600 \end{pmatrix}$$

3. TABLE 1

	w	z	t	x	y	Constant
w	1	0	0	4	1	36
z	0	1	0	2	4	40
t	0	0	1	2	1	20
C				6	5	
Z				0	0	
NET				6	5	

TABLE 2

	x	z	t	w	y	Constant
x	1	0	0	$\frac{1}{4}$	$\frac{1}{4}$	9
z	0	1	0	$-\frac{1}{2}$	$\frac{7}{2}$	22
t	0	0	1	$-\frac{1}{2}$	$\frac{1}{2}$	2
C	6	0	0	0	5	
Z				$\frac{6}{4}$	$\frac{6}{4}$	
NET				$-\frac{1}{4}$	$\frac{7}{2}$	

TABLE 3

	x	z	y	w	t	Constant
x	1	0	0	$\frac{1}{2}$	$-\frac{1}{2}$	8
z	0	1	0	3	-7	8
y	0	0	1	-1	2	4
C	6	0	5	0	0	
Z				-2	7	
NET				$+2$	-7	

TABLE 4

	x	w	y	z	t	Constant
x	1	0	0	$-\frac{1}{6}$	$\frac{2}{3}$	$6\frac{2}{3}$
w	0	1	0	$\frac{1}{3}$	$-\frac{7}{3}$	$2\frac{2}{3}$
y	0	0	1	$\frac{1}{3}$	$-\frac{1}{3}$	$6\frac{2}{3}$
C	6	0	5	0	0	
Z				$\frac{2}{3}$	$2\frac{1}{3}$	
NET				$-\frac{2}{3}$	$-2\frac{1}{3}$	

7. Using x and y as your variables and w and z as slack variables the solution is $x = 12$, $y = 4$.

FINAL TABLE

	x	y	w	z	Constant
x	1	0	0.2	-0.25	12
y	0	1	-0.1	0.25	4
C	3	4	0	0	
Z			0.2	0.25	
NET					

8. One unit of x contributes \$6 to profit rather than \$3 (Problem 9); hence we should expect more x in the solution. Using x and y for your variables and w and z as slack variables, the solution is $x = 16$ and $z = 16$.

9. **(a)** *a* **(b)** *b* **(c)** *c* **(d)** *b* **(e)** *d* **(f)** *b* **(g)** *a* **(h)** *a* **(i)** *d* **(j)** *d* **(k)** *d* **(l)** *e*

CHAPTER 6
SECTION 6-1

1. **(a)** 6, 4, 3.333, ..., 3, 2.8
(b) 1, 4, 9, 16, 25
(c) 1, 1.414, ..., 1.732, ..., 2, 2.236, ...
(d) 0, 0.301, 0.477, 0.602, 0.699
(e) 0, 0.693, 1.099, 1.386, 1.609
(f) 2, 4, 8, 16, 32
(g) 2.718, 7.389, 20.086, 54.598, 148.413
(h) 0.368, 0.135, 0.050, 0.018, 0.007

4. 1, 2, 6, 24, 120 **5.** 1, 1, 2, 6, 24 **6.** 2, 5, 8, 11, 14

7. $a, a + d, a + 2d, a + 3d, a + 4d$

8. **(a)** 2, 7, 15, 26, 40
(b) $a, 2a + d, 3a + 3d, 4a + 6d, 5a + 10d$

11. 2, 6, 18, 54, 162 $a_5 = 2(3)^4 = 2(81) = 162$

15. $1, \frac{1}{1.06}, \frac{1}{1.06^2}$, i.e., 1, 0.943, 0.890

SECTION 6-2

1. **(a)** 3 **(b)** $a_4 = 13,\ a_{10} = 31,\ a_{20} = 61$ **(c)** 650
3. 5050
4. **(a)** $r = 3$ **(b)** $a_2 = 6,\ a_4 = 54,\ a_6 = 486$ **(c)** 728
5. $r = 2$ **6.** 255 **9.** $\frac{1}{2}(x_1 + x_n)$ **10.** 2
11. **(a)** 1, 1.08, 1.1664, 1.2597
(b) $100, 100(1+i), 100(1+i)^2, 100(1+i)^3$
12. 1, 0.9259, 0.8573, 0.7938 **13.** 4374.61

SECTION 6-3

1. \$3649.96
4. **(a)** 200 **(b)** 1.04 **(c)** 0.04
5. \$5126.60 **9.** \$1.0824

SECTION 6-4

1. \$79.38
4. **(a)** 1.0000 **(b)** 0.9434 **(c)** 0.06
5. \$7903

SECTION 6-5

2. \$27,931.20 **3.** \$758.68 **5.** \$13,621.80
6. \$2774.08 **7.** \$34,211

SECTION 6-6

1.

	Year 1	*Year 2*	*Year 3*
(a)	5,500	5,500	5,500
(b)	10,000	9,000	8,000

3.

	Year 1	*Year 2*	*Year 3*	*Year 4*	*Year 5*
(a)	15,000	15,000	15,000	15,000	15,000
(b)	25,000	20,000	15,000	10,000	5,000

5. **(a)** 44,000 **(b)** $(\frac{34}{55})(110{,}000) = 68{,}000$
7.

Income	Depreciation	Taxable Income	Income After Taxes Plus Depreciation	Present Value of \$1.00	Discounted Cash Flow
26,000,000	10,000,000	16,000,000	18,000,000	.9091	16,363,800
26,000,000	8,000,000	18,000,000	17,000,000	.8264	14,048,800
26,000,000	6,000,000	20,000,000	16,000,000	.7513	12,020,8000
28,000,000	4,000,000	24,000,000	16,000,000	.6830	10,928,000
28,000,000	2,000,000	26,000,000	15,000,000	.6209	9,313,500
	30,000,000		82,000,000		62,674,900

9. $n+\dfrac{n(n-1)}{2}$

11. $s_{40} = 820$

13. *Hint:* $\sum_{i=1}^{n} i = n+\sum_{i=1}^{n-1} i$, in general, and $n-1$ is even if n is odd.

SECTION 6-7

1. \$490,905 game

3. Worth = \$42,570.32; Income = \$3,405.63

5. **(a)** arithmetic, $a=-2$, $d=2$, $a_5=6$, $S_{10}=70$, $S_3=0$

(b) geometric, 1.00, 1.06, $a_4=1.2625$, $S_4=4.375$

7. **(a)** d **(b)** a **(c)** c **(d)** c

9. All the options are acceptable because the net present value of all three proposals is positive. The construction of special-purpose equipment option has the largest net present value (\$17,974).

11.

Payment Number	Payment on Interest	Payment on Principal	Balance
1	6.67	80.23	919.77
2	6.13	80.87	838.90
3	5.59	81.41	757.49
4	5.05	81.95	675.54
5	4.50	82.50	593.04
6	3.95	83.05	509.99
7	3.40	83.60	426.39
8	2.84	84.16	342.23
9	2.28	84.72	257.51
10	1.72	85.28	172.23
11	1.15	85.85	
12	.58	86.38	0.00

CHAPTER 7

SECTION 7-1

3. **(a)** Yes **(b)** Yes **(c)** No **(d)** Yes **(e)** Yes **(f)** No

5. a, b, d, g, h

9. $U' = \varnothing$, $\varnothing' = U$

11. 8

13. 16

SECTION 7-2

1. **(a)** $A' = \{1, 3, 5, 7, 8, 9, 10\}$

(b) $B' = \{2, 3, 7, 8, 9, 10\}$

(c) $A \cap B = \{4, 6\}$

(d) $A \cup B = \{1, 2, 4, 5, 6\}$

3. $A' \cap B' = (A \cup B)'$ and $A' \cup B' = (A \cap B)'$
5. Assume $x(A \cap B)'$ and show $x' \in A' \cap B'$. Then assume $x \in A' \cap B'$ and show $x \in (A \cap B)'$.
7. **(a)** $\varnothing$
   (b) {product 1, product 2, product 3, plant 1, plant 2, plant 3, plant 4}
9. **(a)** $p_1: (x-2)(x-3)(x-6) = 0$
   $p_2: (x-1)(x-3)(x-6)(x-8) = 0$
   (b) $A_1 = \{2, 3, 6\}$ and $A_2 = \{1, 3, 6, 8\}$
   (c) $A_1 \cap A_2 = \{3, 6\}$ and $A_1 \cup A_2 = \{1, 2, 3, 6, 8\}$
   (d) $A_1 \cap A_2 = \{x: (x-3)(x-6) = 0\}$
   (e) $A_1 \cup A_2 = \{x: (x-1)(x-2)(x-3)(x-6)(x-8) = 0\}$
11. **(a)** $a+b$ **(d)** $A \cup B = \varnothing$
   (b) $\min(a, b)$ **(e)** $B \cap A = \varnothing$
   (c) $A \cap B = \varnothing$ **(f)** $B \subset A$
13. 1000 have both types
17. They are the same.

SECTION 7-3

1. $\{H, T\}$
3. $\{(H, H), (H, T), (T, H), (T, T)\}$
5. No
7. 0.85

SECTION 7-4

1. $\{(H, H), (H, T), (T, H), (T, T)\}$
3. $\{K, O, l, m, g, r, v\}$
5. Any event with 2 or more sample points
7. 10
9. **(a)** 26 **(c)** 1 **(d)** 2
11. **(a)** Finite **(b)** Infinite **(c)** Finite
13. No

SECTION 7-5

1. $P(x_i) = \frac{1}{6}, \quad i = 1, 2, \ldots, 6$
5. **(a)** $f(x_1) + f(x_2) + f(x_3) + f(x_4) + f(x_5)$
   (b) $x_1^2 + x_2^2 + x_3^2 + x_4^2$
   (c) $(x_1 + x_2 + x_3 + x_4)^2$
   (d) $x_1 f(x_1) + x_2 f(x_2) + x_3 f(x_3) + x_4 f(x_4) + x_5 f(x_5)$
7. $P(x_i) = \frac{1}{12}, \quad i = 1, 2, \ldots, 12$
9. We obtained $P = \frac{13}{20}$ based on 20 tosses
11. **(a)** 4 **(b)** 3 **(c)** 2
13. $\{(H, H), (H, T), (T, H), (T, T)\}$
15. $\{A, B, C, D, F\}$
17. $\{(1, 1), (1, 2), \ldots, (6, 5), (6, 6)\}$
   36 sample points
19. Infinite
21. $\{1, 2, \ldots, n, \ldots\}$

23. The technical answer is:
18—continuous sample space
21—discrete sample space
Both sample spaces are infinite, however.

SECTION 7-6

1. A and B

3. **(a)** $\frac{1}{13}$ **(b)** $\frac{1}{13}$ **(c)** $\frac{3}{13}$

5. **(a)** A and K are mutually exclusive.
(b) A and K are not exhaustive.

7. **(a)** 0.5 **(d)** 0.7
(b) 1 **(e)** 0.8
(c) 0 **(f)** 0.4

SECTION 7-7

1.

\$16	\$61	\$ 83	\$125	\$163
22	67	90	135	176
35	74	104	146	186
48	79	115	159	191

3. 0.2, 0.3, 0.25, 0.25

5. 67: $\frac{3}{15}$
71: $\frac{2}{15}$
73: $\frac{2}{15}$
each of the other 8 heights: $\frac{1}{15}$

7. $\frac{4}{15}, \frac{5}{15}, \frac{6}{15}$

SECTION 7-8

1. **(a)** 720 **(c)** 5040
(b) 10 **(d)** 21

3. **(a)** 10 **(c)** 4
(b) 1 **(d)** 16

5. 120

7. $\dfrac{1}{4{,}845}$

9. $\dfrac{1}{11{,}880}$

11. 210

13. 635, 013, 559, 600

15. 151, 200

SECTION 7-9

1. $\frac{27}{56}$

3. 0.14005

5. 2, 598, 960

7. **(a)** $\frac{1}{16}$ **(d)** $\frac{1}{4}$
(b) $\frac{1}{4}$ **(e)** $\frac{1}{16}$
(c) $\frac{3}{8}$
9. 0.059535
11. **(a)** 10
(b) {100, 160, 120}, {100, 160, 150}, etc.
(c) 126.67, 136.67, etc.
13. 495
17. 12

CHAPTER 8

SECTION 8-1

1. **(a)** $P(E_1) = 0.15$, $P(E_2) = 0.20$, $P(E_3) = 0.30$, $P(E_4) = 0.20$, $P(E_5) = 0.15$
(b) 0.20 **(c)** 0.35
(d) The probability that an emp.oyee chosen at random earns between \$1.50 and \$3.49 an hour.

4. **(a)**

Type of Work Sex	Males	Females	
Production	6000	1000	7000
Sales	2000	1500	3500
	8000	2500	10,500

(b) 8000 **(c)** 2500 **(d)** 8500

5. **(a)**

	A_1	A_2	A_3	A_4	Total
B_1	30	40	60	70	200
B_2	270	210	190	130	800
Total	300	250	250	200	1000

(b) 300 **(e)** 0.3
(c) 800 **(f)** 0.8
(d) 1000

SECTION 8-2

1.

	D_1	D_2	D_3	D_4	D_5	Total
E_1	$\frac{9}{52}$	$\frac{1}{52}$	$\frac{1}{52}$	$\frac{1}{52}$	$\frac{1}{52}$	$\frac{13}{52}$
E_2	$\frac{9}{52}$	$\frac{1}{52}$	$\frac{1}{52}$	$\frac{1}{52}$	$\frac{1}{52}$	$\frac{13}{52}$
E_3	$\frac{9}{52}$	$\frac{1}{52}$	$\frac{1}{52}$	$\frac{1}{52}$	$\frac{1}{52}$	$\frac{13}{52}$
E_4	$\frac{9}{52}$	$\frac{1}{52}$	$\frac{1}{52}$	$\frac{1}{52}$	$\frac{1}{52}$	$\frac{13}{52}$
Total	$\frac{9}{13}$	$\frac{1}{13}$	$\frac{1}{13}$	$\frac{1}{13}$	$\frac{1}{13}$	1

3. (a)

	1	2	3	4	5	6	Total
H	$\frac{1}{12}$	$\frac{1}{12}$	$\frac{1}{12}$	$\frac{1}{12}$	$\frac{1}{12}$	$\frac{1}{12}$	$\frac{6}{12}$
T	$\frac{1}{12}$	$\frac{1}{12}$	$\frac{1}{12}$	$\frac{1}{12}$	$\frac{1}{12}$	$\frac{1}{12}$	$\frac{6}{12}$
Total	$\frac{1}{6}$	$\frac{1}{6}$	$\frac{1}{6}$	$\frac{1}{6}$	$\frac{1}{6}$	$\frac{1}{6}$	1

(b) $\frac{1}{12}$ (c) $\frac{1}{12}$ (d) $\frac{7}{12}$

5. (a) 0.5 (b) 0.4 (c) 0.06 (d) 0.93

7-8. (a) $P(B_1) = 0.1$ (b) $P(A_1) = 0.5$ (c) $P(B_3 \cup A_1) = 0.75$ (d) $P(B_2 \cap A_1) = 0.1$

SECTION 8-3

1. (a) $\frac{1}{52}$ (b) $\frac{2}{13}$ (c) $\frac{1}{4}$ (d) $\frac{1}{4}$ (e) $\frac{1}{13}$

3. (a) joint (c) marginal (d) conditional (e) conditional

5.

	A_1	A_2	
B_1	0.28	0.22	0.5
B_2	0.42	0.08	0.5
	0.7	0.3	1

7. (a) 0.80 (b) 0.25 (c) 0.13 (d) 0.20 (e) 0.90 (f) 0.87

10. (a) 0.7 (b) 0.8 (c) 0.5 (d) 1 (e) $\frac{1}{3}$

11. (a) 0.12 (b) 0.32 (c) 0.68 (d) 0.6 (e) $\frac{3}{8}$

SECTION 8-4

1. y_1, x_1 and y_2, x_1 are independent events.

7.

	A_1	A_2	
B_1	0.12	0.18	0.30
B_2	0.28	0.42	0.70
	0.40	0.60	1

	A_1	A_2
B_1	0.3	0.3
B_2	0.7	0.7
	1	1

8. The variables A and B are not independent.
9. **(a)** $0.7^5 = 0.16807$ **(d)** $10(0.7)^3(0.3)^2 = 0.3087$
(b) $0.3^5 = 0.00243$ **(e)** 0.96922
(c) $10(0.7)^2(0.3)^3 = 0.1323$ **(f)** 0.83692
10. **(a)** 0.0156 **(c)** 0.0156
(b) 0.3125 **(d)** 0.65625
11. Note that the probability of x successes is equal to the number of possible ways x successes can occur multiplied by the probability of x successes.
12. See answers for Problems 9 and 10.
14. d

SECTION 8-5

1. $P(B_1/A_1) = \frac{2}{3}$ $P(B_2/A_1) = \frac{1}{3}$
$P(B_1/A_2) = \frac{3}{4}$ $P(B_2/A_2) = \frac{1}{4}$
$P(B_1/A_3) = \frac{1}{3}$ $P(B_2/A_3) = \frac{2}{3}$

3.

	x_1	x_2	
y_1	0.06	0.48	0.54
y_2	0.14	0.32	0.46
	0.20	0.80	1

	x_1	x_2
y_1	0.3	0.6
y_2	0.7	0.4
	1	1

7. $P(C/U) = \frac{1}{3}$

SECTION 8-6

1. **(a)** yes, each outcome is a real number
(b) x could equal 1, 2, 3, 4, 5 or 6
(c) $\frac{1}{6}, \frac{1}{6}, \frac{1}{2}, \frac{2}{3},$
(d, e)

x:	1	2	3	4	5	6
$f(x)$:	$\frac{1}{6}$	$\frac{1}{6}$	$\frac{1}{6}$	$\frac{1}{6}$	$\frac{1}{6}$	$\frac{1}{6}$
$F(x)$:	$\frac{1}{6}$	$\frac{1}{3}$	$\frac{1}{2}$	$\frac{2}{3}$	$\frac{5}{6}$	1

3. **(a)** $E(X) = 1.6$
(b) $p = 0.4$
5. $S = \{(H, H, H), (H, T, T), \ldots\}$, 8 sample points
7. **(a)** $a^2 + 2ab + b^2$
(b) $a^3 + 3a^2b + 3ab^2 + b^3$
(c) $a^5 + 5a^4b + 10a^3b^2 + 10a^2b^3 + 5ab^4 + b^5$

SECTION 8-7

3. **(a)** $P(x = 7) = \binom{10}{7} .5^7 \ .5^3 = .1172$

(b) $P(x \geqslant 6) = .2051 + .1172 + .0439 + .0098 + .0010 = .3770$
or
$P(x \geqslant 6) = 1 - P(x \leqslant 5) = 1 - .6230 = .3770$

5. **(a) & (b)**

x	0	1	2	3	4	5
$f(x)$	.3277	.4096	.2048	.0512	.0064	.0003

(c) 0, 1, 2

7. $\binom{10{,}000}{5}(.0005)^5\,(.9995)^{9995}$

9. **(a)** $P(x = 2) = \binom{10}{2}\ .1^2\ \ .9^8 = .1937$

(b) $P(x \leq 2) = .9298$

11. **(a)** $\binom{100}{4}(.05)^4\,(.95)^{96}$ **(b)** $1 - F(4)$

SECTION 8-8

1. **(a)**

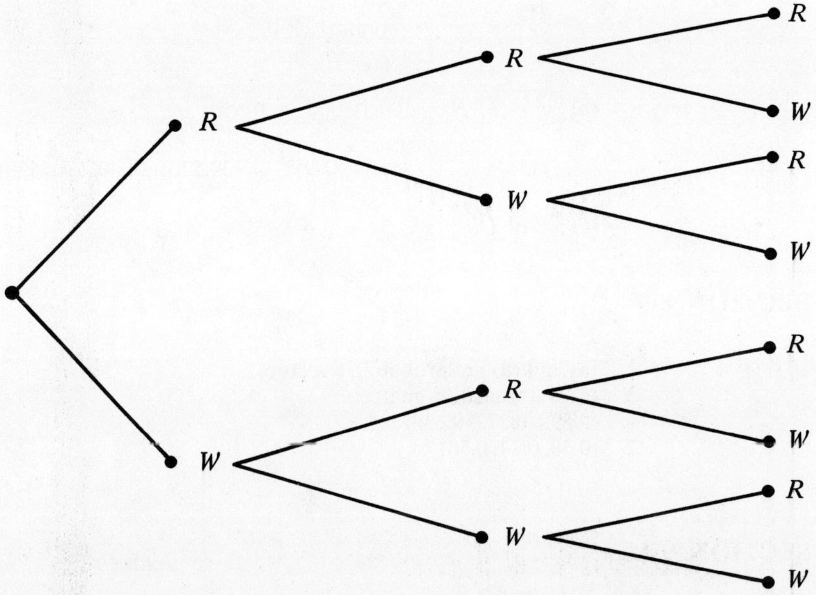

(b) $(\frac{6}{10}) \cdot (\frac{5}{9}) \cdot (\frac{4}{8}) = \frac{1}{6}$

(c) $N = 10,\ M = 6,\ x = 3,\ n = 3$

$$f(3) = \frac{\binom{10-6}{3-3}\binom{6}{3}}{\binom{10}{3}} = \frac{\binom{4}{0}\binom{6}{3}}{\binom{10}{3}} = \frac{\dfrac{6\cdot5\cdot4}{3\cdot2\cdot1}}{\dfrac{10\cdot9\cdot8}{3\cdot2\cdot1}} = \frac{6\cdot5\cdot4}{10\cdot9\cdot8}$$

$$= \frac{120}{720}$$

$$= .1667$$

3. Whether one is sampling with or without replacement, i.e., binomial when sampling with replacement, hypergeometric when sampling without replacement.
5. **(a)** $N = 1000,\ M = 50,\ n = 50,\ x = 0$

$$f(0) = \frac{\binom{950}{50}\binom{50}{0}}{\binom{1000}{50}}$$

(b) $$p = \frac{M}{N} = \frac{50}{1000} = .05$$

$$f(0) = \binom{50}{0}(.05)^0(.95)^{50} = (1)(1)(.0769) = .0769$$

7. **(a)** $N = 10{,}000,\ M = 100,\ n = 60,\ x = 3$

$$f(3) = \frac{\binom{9900}{57}\binom{100}{3}}{\binom{10000}{60}}$$

(b) $p = M/N = 100/10{,}000 = .01$

$$f(3) = \binom{60}{3}(.01)^3(.99)^{57} = (34{,}220)(.000001)(.5639) = .0192$$

(c) No. $1 - p(x \leqslant 3) = 1 - .9967 = .0033$

SECTION 8-9

1. **(a)** and **(d)** are probability vectors.
3. **(a)** is a transition matrix.
5. (0.592, 0.272, 0.136)
7. (0.38, 0.24, 0.38)

CHAPTER 9
SECTION 9-1

1. That the classes be mutually inclusive and exhaustive. That there be neither too few nor too many classes (i.e., 5 to 10). That the class intervals be equal. That one takes care in choosing the end points of each class.
3. We can no longer compute the range within which observations fall. We will not be able to compute the mid-point of the open-ended interval which will be important in calculations required in future sections.
5.

Interval	*Number of observations*
00 and under 100	14
100 and under 200	2
200 and under 300	8
300 and under 400	0
400 and under 500	7

SECTION 9-2

1. $(350+450)/2 = \mathit{400}$
$(540+560+600+610+630+670)/6 = \mathit{601.66}$
$(740+760+760+780+800+810+830+840+890)/9 = \mathit{801.11}$
$(910+930+960+970+980+1010+1030+1030+1050+1060+1080)/11$
$= \mathit{1000.909}$
$(1110+1150+1170+1200+1220+1240+1290)/7 = \mathit{1197.14}$
$(1350+1420+1440)/3 = \mathit{1403.33}$
$(1550+1630)/2 = \mathit{1590}$

3.

Observation	*Frequency*	
18	2	36
19	7	133
20	3	60
21	2	42
22	1	22
23	1	23
24	2	48
40	1	40
	19	404

$\bar{x} = 404/19 = 21.26$

5. **(a)** Mean = 4.29 rounded to 2 decimal places
(b) Median = 3
(c) Mode = 3

7. **(a)** $\bar{x} = 6{,}150/31 = 198.39$
(b) $31/2 = 15.5$th item
(c) $100 + \left(\dfrac{15.5-14}{2}\right)100 = 175$

9. $\bar{x} = \$20.61/28 = \$.74$
Item that is median = 14.5th item
Median = $(\$.58+\$.60)/2 = \$.59$
Mode = $0.00

11. That the observations in the median class are evenly distributed within this class.

SECTION 9-3

1. $\sum_{i=1}^{n} x_i = 38{,}400$

$\sum_{i=1}^{n} x_i^2 = 40{,}377{,}800$

$\left(\sum_{i=1}^{n} x_i\right)^2 = 1{,}474{,}560{,}000$

$s^2 = (40{,}377{,}800 - 36{,}864{,}000)/39 = 90{,}097.44$

3. $\sum_{j=1}^{k} m_j^2 f_j = 40{,}320{,}000$

$$\left(\sum_{j=1}^{k} m_j f_j\right)^2 = (38{,}400)^2 = 1{,}474{,}560{,}000$$

$$s^2 = (40{,}320{,}000 - 36{,}864{,}000)/390$$

$$s^2 = 3{,}456{,}000/39 = 88{,}615.38$$

5. $\sum_{j=1}^{4} m_j^2 f_j = 1{,}735{,}000$

$$\left(\sum_{j=1}^{4} m_j f_j\right)^2 = (4700)^2 = 22{,}090{,}000$$

$$s^2 = \frac{1{,}735{,}000 - \dfrac{22{,}090{,}000}{14}}{13} = 12{,}087.92$$

$$s = 109.945$$

7.

	Accounting	Management	Marketing	Total
Mean	\$9400.00	\$8900.00	\$9150.00	\$9066.66
Standard Deviation	1049.29	1085.02	965.48	1055.50
Coefficient of Variation	.1116	.1219	.1055	.1164

SECTION 9-4

1. **(a) & (b)**

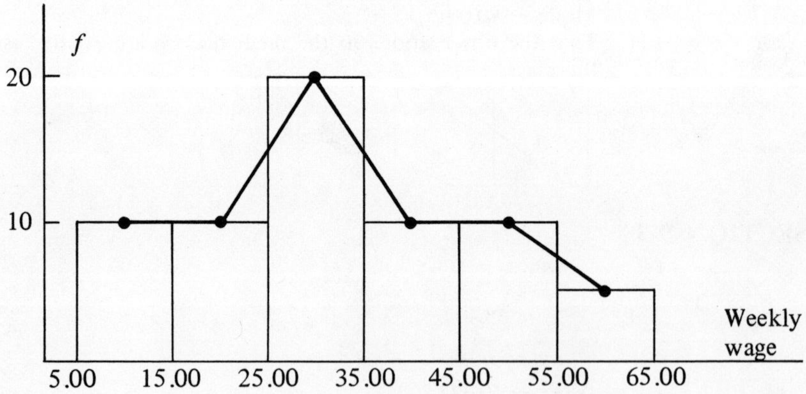

(c)

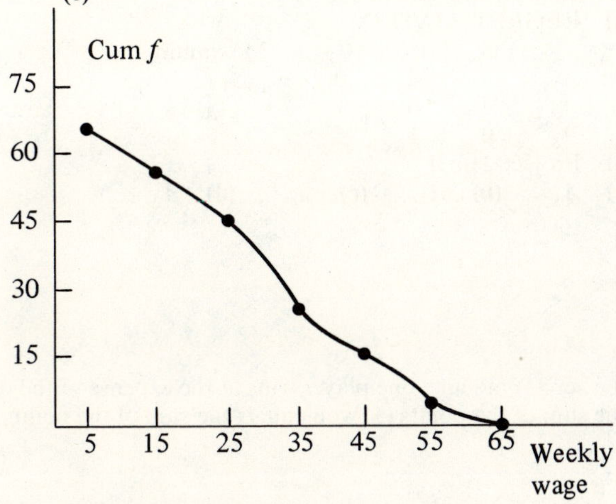

Weekly wage	Cum f
\$ 5.00	65
15.00	55
25.00	45
35.00	25
45.00	15
55.00	5
65.00	0

(d) positive by observation
Mean = \$32.31
Median = \$31.25
Standard deviation = \$14.87
Skewness = 3(32.31 − 31.25)/14.87 = +0.2138

3.

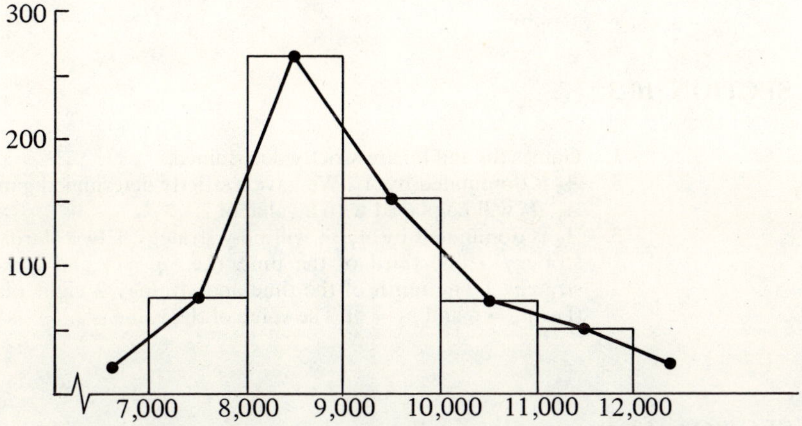

5. Mean = \$9066.66, Median = \$9000, Standard Deviation = \$1055.50
Skewness = 3(9066.66 − 9000)/1055.50 = +0.063

CHAPTER 10
SECTION 10-1

1. Maximax, Maximin, Minimax Regret
3. In a risk situation one knows all of the outcome states and the probability associated with each outcome. Under uncertainty one does not know the probability of the outcome states.

5. Certainty, Risk, Uncertainty
7. (a) REGRET MATRIX

	0_1	0_2	0_3	Maximum
A_1	7	2	0	7
A_2	3	0	1	③
A_3	0	1	5	5

(b) Project 2 or A_2.
9. (a) A_1 (b) A_2 (c) A_1 (d) A_2

SECTION 10-2

1. In a zero-sum game one player wins at the expense of the other. The sum of the points ($) won equals the sum of the points ($) lost.
3.

	Y_1	Y_2
X_1	-5	-2
X_2	2	-7

5. Yes
7. (a) A_1, B_2 with value of 15
(b) A_2, B_1 with value of -5
(c) A_1, B_2 with value of 4
(d) A_2, B_3 with value of -2

SECTION 10-3

1. Games (b) and (c) are strictly determined.
3. A_3 is dominated by A_1. We have a strictly determined game. A_1, B_2 will be played with a value of 2
5. A_2 is dominated by A_1. A will play strategy 1 two thirds of the time and strategy 3 one third of the time. (I.e., $q_1 = \frac{2}{3}$ and $q_2 = \frac{1}{3}$). B will play strategy 1 one ninth of the time and strategy 2 eight ninths of the time. (I.e., $p_1 = \frac{1}{9}$ and $p_2 = \frac{8}{9}$). The value of the game is $\frac{7}{3}$.

SECTION 10-4

2. $\begin{pmatrix} 0.0003 & 0.4652 \\ -0.0003 & 0.5348 \end{pmatrix}$
5. $EV(\text{Act } 1) = \$14{,}000$ $EV(\text{Act } 2) = \$11{,}000$
6. (a)

	Demand low	Demand high
Option 1	61,446	122,892
Option 2	75,816	94,770

(b)

$$\begin{matrix}\text{Option 1}\\ \text{Option 2}\end{matrix}\begin{pmatrix} 11{,}446 & 72{,}892 \\ 25{,}816 & 44{,}770 \end{pmatrix}\begin{pmatrix} 0.5 \\ 0.5 \end{pmatrix} = \begin{pmatrix} 42{,}169 \\ 35{,}293 \end{pmatrix}$$

		Year 1	Year 2	Year 3
7. **(a)**	Total probable sales	\$800,000	\$1,860,000	\$435,000
(b)	Probable Net Income	42,000	324,000	24,800
(c)	Net cash flow	417,000	574,000	149,800
(d)	Net present value of cash flow	396,150	493,640	116,844

8. Order 30,000 and have maximum expected value of \$4,000.

CHAPTER 11

SECTION 11-1

1. **(a)** slope $= 2$, y intercept $= -3$
(b) slope $= \frac{1}{2}$, y intercept $= 4$
(c) slope $= -\frac{2}{3}$, y intercept $= 2$
3. $y = -2x^2 + 2x + 3$
5. $y = 2x^3 - x^2 - 3x + 2$
7. **(a)** $y = \dfrac{100{,}000}{x} + 10$ **(c)** Total Cost
(d) $C = \left(\dfrac{100{,}000}{x} + 10\right)x$
(e) No matter how many units are produced unit costs will not drop below \$10.00.
9. The coefficient b must be negative.
11. $p = -2x + 120$
13. $p = -3x + 90$
15. Revenue is maximized at $x = 15$

SECTION 11-2

1. 2, 4, 3, 0, -1, -3, -2
3. 2, -3, 5, $\frac{1}{3}$, $\frac{1}{2}$, $\frac{1}{3}$, $\frac{1}{3}$, $\frac{2}{3}$
8. $\log_b x = 6$
9. **(a)** 3 **(c)** -1
(b) 0 **(d)** $2 + \log_b(ab+1)$
12. Common value is 2. **13.** $\log_{10}(\frac{5}{3})$
14. **(a)** $\log_3 5$ **(b)** $\log_9 81 = 2$ **15.** $\log V_n = \log P + N\log(1+i)$
16. **(a)** $x^{1/2}$ **(d)** x^2y^2
(b) $x^{1/3}$ **(e)** $x^{3/2}$
(c) $x^{3/2}$ **(f)** $(x+y)^{1/3}$
18. **(a)** 27 **(c)** $\frac{1}{8}$
(b) x^3 **(d)** $\frac{1}{9}$
20. **(a)** $a(b+c)^2$ **(c)** $\frac{5}{7}$
(b) $\frac{3}{4}$ **(d)** $\sqrt{\frac{13}{3}}$
22. **(a)** $x = 2$ **(d)** $x = -1$
(b) $x = 0$ **(e)** $x = -2$
(c) $x = 3$ **(f)** $x = 4$

SECTION 11-3

1. **(a)** 1 and 2 **(b)** 1 and 2 **(c)** 2 and 3 **(d)** 2 and 3 **(e)** 3 and 4 **(f)** 3 and 4 **(g)** -1 and 0 **(h)** -3 and -2

3. **(a)** 4.82×10^2 **(b)** 5.372×10^3 **(c)** 4.3×10^{-2} **(d)** 8.7×10^{-1} **(e)** 3.52×10^6 **(f)** 4.8×10^3 **(g)** 5.73×10^5 **(h)** 4.3×10^{-4}

5. **(a)** 3 **(b)** 3 **(c)** 1 **(d)** 2 **(e)** -2 **(f)** -1 **(g)** 5 **(h)** -3

7. **(a)** 4.68320 **(b)** $0.92000 - 2$ **(c)** 1.37495 **(d)** 3.92246 **(e)** $0.09160 - 1$ **(f)** 5.86607

8. $b = 1.47403$

SECTION 11-4

1. **(a)**

x	y
1	0
2	1
4	2
8	3

(b)

x	y
1	0
4	1
8	1.5
16	2

(c)

x	y
1	0
2	1.386
e	2
4	2.772

(d)

x	y
1	0
4	1.20412
$\sqrt{10}$	1
10	2

4. $y = 2 \log x + 2$

6. $y = \log x + 5$

8. **(a)** Common value is 0.81291. **(b)** Common value is 0.99123.

SECTION 11-5

1. **(a)**

x	y
-2	$\frac{1}{9}$
-1	$\frac{1}{3}$
0	1
1	3
2	9

(c)

x	y
-2	$\frac{3}{16}$
-1	$\frac{3}{4}$
0	3
1	12
2	48

(b)

x	y
-2	0.27
-1	0.736
0	2
1	5.44
2	14.78

(d)

x	y
-2	8
-1	4
0	2
1	1
2	$\frac{1}{2}$

3. $y = 3(2^x)$

5. **(a)** \$35,246,000 **(b)** \$49,520,000

6. **(a)**

x	1	2	3	4	5
y	1.100	1.2100	1.3310	1.4641	1.6105

(b) $y = 1{,}000{,}000(1.10)^x$

SECTION 11-6

1. **(a)**

x	y
-2	-8
-1	-1
0	0
1	1
2	8

(c)

x	y
0	0
1	$\frac{1}{2}$
4	16
9	121.5

(b)

x	y
0	0
1	3
2	4.24
4	6
9	9

(d)

x	y
-8	-4
-1	-2
0	0
1	2
8	4

3. $y = x$

SECTION 11-7

1. x axis in log scale

(a)

x	y
1	2
2	2.602
5	3.397
10	4

(c)

x	y
1	0.301
5	1
10	1.301
50	2

(b)

x	y
1	2
2	2.903
5	4.094
10	5

(d)

x	y
1	3.301
5	4
10	4.301
50	5

3. $y = \log x + 5$
5. $y = 3(2^x)$
8. $y = x$
9. $y = 1(2^x)$
12. **(a)** exponential **(b)** $y = ab^x$ **(c)** semilog where y axis is logarithmic
13, 14. Data will plot as a straight line with $a = 1{,}000{,}000$ and $b = 1.10$.
15. exponential
16. $a = 160$, $\log b = (\log 180 - \log 160)/3 = 0.01705$, $b = 1.04$

APPENDIX TO CHAPTER 11

1. **(a)** 10,000 **(b)** 0.01 **(c)** 1 **(d)** 468 **(e)** 0.005133 **(f)** 4.572
3. **(a)** 4 **(b)** 7.5595 **(c)** 7.1138 **(d)** 0.3979
5. **(a)** 21,019 **(b)** 0.047575 **(c)** 0.00066665 **(d)** 8.3094
7. **(a)** 3581.6 **(b)** 1.4693 **(c)** 2191.1
9. 2
10. **(a)** 2 **(b)** 125 **(c)** b **(d)** b^a

CHAPTER 12

SECTION 12-1

1. 8
3. 6.001
5. 0.31
7. 3
9. 0.4624752956
11. 3
13. 0.6

SECTION 12-2

1. 1
3. 1
5. 2
7. *Ans.* 0
fact: $(0.4)^{247} = 5.11 \times 10^{-99}$
9. -12
11. $\ln e^2 = 2$
13. 3
15. 0

SECTION 12-3

1. 8.41, 8.9401, 8.994001
3. $x - 3$: 0.1, 0.01, 0.001
$f(x) - 9$: 0.61, 0.0601, 0.006001
5. $f(x) - 3$: 0.41, 0.0401, 0.004001
7.

x	x^2
4.1	16.81
4.01	16.0801
4.001	16.008001
3.99999	15.99992

11. 3

SECTION 12-4

1. $y_1 = 19$
y_2: 32, 23.72, 19.2208, 19.011002
$\dfrac{\Delta y}{\Delta x}$: 13, 11.8, 11.04, 11.002

3. $\dfrac{\Delta y}{\Delta x} = 2\,\Delta x + 11,\ f'(2) = 11$

5. $g'(3) = 29$

7. $h'(4) = 11,\ h'(a) = 2a + 3$

9. $f'(4) = 0.25$ is the correct limit.

11. $2 + \dfrac{1}{n}$

13. $R_f(A, B_n)$ approaches 2.

SECTION 12-5

1. slopes: 6, 5, 4.5, 4.1

3. slopes: 4, 3, 2.5, 2.1

5. slopes: 6, 6, 6
(d) when the function is of the form $y = mx + b$ (i.e., linear)

7. 5

9. **(a)** 20% **(b)** 10% **(c)** 2 **(d)** Yes

11. $e_d = \dfrac{(\Delta x/x)(100)}{(\Delta P/P)(100)}$ etc.

13. $e_d = \dfrac{P}{x} \cdot \dfrac{\Delta x}{\Delta P}$

$$= \frac{P}{x} \cdot \frac{f(P + \Delta P - f(P))}{\Delta P}$$

$$= \frac{P}{x} \cdot R_f(A, B)$$

SECTION 12-6

1. **(a)** $2x + 3$ **(b)** $2a + 3$ **(c)** 11

3. $y = 5x + 3$

5. **(a)** $2u$ **(b)** 1 **(c)** $2u + 2$ **(d)** $-2u$

7. $g'(u)$ and $\dfrac{dw}{du}$

9. $e_d = \dfrac{P}{x} f'(P)$

11. 2

SECTION 12-7

1. $y' = 6u - 2$

3. $y = f(u) + g(u)$
$y' = f'(u) + g'(u)$

5. **(a)** 0 **(b)** 1 **(c)** $2u$ **(d)** $4u^3$ **(e)** $5u^4$

(f) $\frac{1}{3}u^{-2/3} = \dfrac{1}{3u^{2/3}} = \dfrac{1}{3\sqrt[3]{u^2}}$

7. $\dfrac{dw}{dz} = kz^{k-1}$

9. **(a)** $24u^3$ **(b)** 5 **(c)** $8u - 8$

(d) $4u + 2 + \frac{1}{4}u^{-3/4}$

Can also be written as

$$4u + 2 + \frac{1}{4\sqrt[4]{u^3}}$$

11. $\frac{8}{3}(2x+3)^{1/3}$

13. $3x(x^2+9)^{-1/2} = \dfrac{3x}{\sqrt{x^2+9}}$

SECTION 12-8

1. $2x + 2$

3. $\frac{3}{2}x^{-1/2} - 2x^{-3/2}$

5. $12x^2 - 16x^3$

7. $4(2x+3)$

9. $\frac{5}{2}x^{3/2} - 9x^{1/2} + \frac{9}{2}x^{-1/2}$

11. $\frac{1}{5}x^{-4/5}$

13. 0

15. $-20x^{-3} = \dfrac{-20}{x^3}$

17. $10x^{99}$

19. $11 + \frac{3}{2}x^{-1/4} + \frac{1}{2}x^{-1/2}$

SECTION 12-9

1. **(a)** $\dfrac{1}{u}\log_3(e)$

(b) $\dfrac{1}{u}\log_5(e)$

(c) $2^u \ln 2 = 2^u(0.69315)$

(d) $10^u \ln 10 = 10^u(2.30259)$

3. **(a)** $h\log(1+h)$

(b) $\dfrac{1}{h}\ln\left(1+\dfrac{1}{h}\right)$

(c) $\dfrac{1}{h}\log_a\left(1+\dfrac{1}{h}\right)$

(d) $\dfrac{u}{\Delta u}\log_a\left(1+\dfrac{\Delta u}{u}\right)$

5. **(a)** $2.7183 - 2.25 = 0.4683$

(b) $2.7183 - 2.5936 = 0.1247$

(c, d) $2.7183 - 2.704 = 0.0143$

7. **(a)** $2^x \ln 2 + (1/x)\log_2(e)$

(b) $3x^2 + 3^x \ln 3$

(c) $-3x^{-4} + (1/x)$

(d) $-5x^{-7/2} - \frac{3}{2}x^{1/2} + 4^x \ln 4$

9. $V_n = p[(1+h)^{1/h}]^{ni}$ since $h = i/k$ and $k/i = 1/h$.

$\lim_{k\to\infty} h = \lim_{k\to\infty}(1/k) = 0$

11. **(a)** $5.82689(1.06)^u$
(b) $3.84805(1+0.08)^u$
(c) $a\ln(1+r)(1+r)^u$

13. $4(2^{4x})\ln 2 + 2x(e^{x^2})$

15. $\frac{7}{x}\log_5(e) + \frac{6}{x}\log_7(e)$

or by D-3′ (*alternative*)

$$\frac{7}{x\ln 5} + \frac{6}{x\ln 7} = \frac{7}{1.60944x} + \frac{6}{1.94591x}$$

17. $\frac{5}{x}\log_7(e) = \frac{5}{x\ln 7}$

19. $\frac{3}{x}\log_{1/2}(e) + \frac{3}{x\ln(0.5)}$

SECTION 12-10

1. $y = 12x - 18$

3. $y = x - 1$

5. $MC = 3x^2$, $MC = 48$ when $x = 4$

9. (1) $P = a - bx$ (2) $P = \frac{a}{x+b} - c$ (3) $P = a - bx^2$

10. Graph $P = 10 - x$ (1).

	x	2	4	6	8	10
(1)	P	8	6	4	2	0

11. (1) $R = ax - bx^2$ (2) $R = \frac{ax}{x+b} - cx$ (3) $R = ax - bx^3$

12. (1) $a - 2bx$ (2) $\frac{ab}{(x+b)^2} - c$

13. $c = x^3 - 3x^2 + 3x + 1$ (*cubic cost function*)
$MC = 3x^2 - 6x + 3$ (*quadratic marginal cost function*)

14. (1) a (2) $2ax + b$ (3) $3ax^2 - 2bx + c$

15. $\frac{bx-a}{bx}$

CHAPTER 13

SECTION 13-1

1. **(a)** 0.331 **(b)** 0.030301 **(c)** 0.003003001 **(d)** 0.000300030001

3. Yes, Δy approaches 0 as Δx approaches 0 because $y = x^3$ is a differentiable function.

5. **(a)** e^x **(b)** e^{x^2} **(c)** e^{x^2+4} **(d)** e^{-x^2}

7. Continuous except at $x = 1$.

9. $y' = 3x^2 - 4x + 3$ is valid for all real x.

11. $y = \ln x$ is not defined (as a real-valued function) for $x \leqslant 0$. Therefore, it is not continuous there. On the other hand $d(\ln x)/dx = 1/x$ for $x > 0$. Therefore, $y = \ln x$ is differentiable and hence continuous for all $x > 0$.

SECTION 13-2

1. (a) $12x^{11}$ (b) $30x^{29}$ (c) $3x^2$ (d) $3x^2$
3. (a) $2e^{2x}$ (b) $5e^{5x}$ (c) $4(2^{4x})\ln 2$ (d) $5(3)^{5x}\ln 3$
5. (a) $2xe^{x^2}$ (c) $3x^2(3)^{x^3}\ln 3$
   (b) $-2xe^{-x^2}$ (d) $\dfrac{2^{\sqrt{x}}\ln 2}{2\sqrt{x}}$
7. (a) $(2x+3)e^{x^2+3x}$ (c) $9x^2(x^3+2)^2$
   (b) $\dfrac{2x+4}{x^2+4x+7}$ (d) $3e^{3x}+\dfrac{x}{\sqrt{x^2+4}}$
9. (a) $\Delta x = 2$ (b) $\Delta u = 6.38906$ (c) $\Delta y = 53.59815$

SECTION 13-3

2. (a) $18x^2$ (d) $6(2)^x\ln 2$
   (b) $\dfrac{3}{x}\log e$ (e) $5e^x$
   (c) $\dfrac{5}{x}$ (f) $10(3)^x\ln 3$
3. (a) $2h(x)^{h-1}$ (b) $ch(x)^{h-1}$ (c) he^x (d) $c(a)^x\ln a$
5. (a) $18x^2$ (b) $x^2e^x(x+3)$ (c) $60x$ (d) $1+\ln x$
   (e) $x\log e + 2x\log x$ (f) $5^x x^4(x\ln 5+5)$
7. (a) $\dfrac{3x+3}{2}$ (b) $\frac{1}{2}$ (c) $\frac{1}{4}e^x$ (d) $\dfrac{1-\ln x}{3x^2}$
   (e) $\dfrac{\log e - 2\log x}{x^3}$ (f) $\dfrac{5^x(x\ln 5-3)}{6x^4}$
9.

$f(x)$	$f'(x)$	$f'(1)$
$4\sqrt{x}$	$2/\sqrt{x}$	2
$2e^{3x}$	$6e^{3x}$	120.5132
$e^x + e^{-x}$	$e^x - e^{-x}$	2.3504
$e^x + e^{-2x}$	$e^x - 2e^{-2x}$	2.4476
$\dfrac{\ln x}{e^x}$	$e^{-x}\left[\dfrac{1}{x} - \ln x\right]$	0.3679

SECTION 13-4

1. $2x+6$
3. $3x^2$
5. $2xe^{x^2}+4$
7. $3/x$
9. $\dfrac{6^{\sqrt{2x^2+5}}(2x)\ln 6}{\sqrt{2x^2+5}}$
11. $24x^3+72x$
13. $1/(1+2x)^2$
15. $\dfrac{-5}{2(x-3)^2}\left(\dfrac{x+2}{x-3}\right)^{-1/2}$
17. $2ax+b$
19. $2f(x)f'(x)$

21.

Function	$\frac{dy}{dx}$	$\left.\frac{dy}{dx}\right\vert_{x=1}$	$\left.\frac{dy}{dx}\right\vert_{x=2}$
$y=3$	0	0	0
$y=x^2/2$	x	1	2
$y=x^{1/3}$	$\frac{1}{3}x^{-2/3}$	$\frac{1}{3}$	$\frac{1}{3}\sqrt[3]{4}$
$y=2^x$	$2x \ln 2$	1.386	2.772
$y=e^x$	e^x	2.72	7.398
$y=xe^x$	$e^x(1+x)$	5.44	22.17
$y=\ln x$	$1/x$	1	$\frac{1}{2}$
$y=\log x^2$	$2/x \log e$	$.868$	$.434$
$y=x^2 \ln 2x$	$x(1+2\ln 2x)$	2.386	7.544
$y=2+3e^{-4x}$	$-12e^{-4x}$	-0.21923	-0.004

SECTION 13-5

1. $y = -x + 2$
3. $MC = 0.122$ at $x = 60$ and $MC = 0.244$ at $x = 80$
5. $MC = e^x + \dfrac{1}{x}$
7. $x = 2.5$
9. $P = 0.21$ at $x = 2.3$, $P = 0.25$ at $x = 2.5$ and $P = 0.24$ at $x = 2.6$
11. $R = ax - bx^2$ and $MR = a - 2bx$

SECTION 13-6

1. (a) 1.44 (b) 1.4
3. 0.2
5. $dw = g'(z)\,dz$, $\Delta w = g(z+\Delta z) - g(z)$, $dz = \Delta z$ and $\Delta z = z_2 - z_1$
6. (a) $dy = (2x+3)\,dx$ (d) $dy = \dfrac{dx}{x}$

   (b) $dy = \dfrac{dx}{2\sqrt{x}}$ (e) $dy = -\frac{1}{2}x^{-3/2}\,dx$

   (c) $dy = 3e^{3x}\,dx$ (f) $dy = 2xe^{x^2}\,dx$

CHAPTER 14

SECTION 14-1

1. (a) $x = \frac{3}{2}$ (d) $x = \sqrt{2}, -\sqrt{2}$

   (b) $x = -\frac{3}{2}$ (e) $x = 2, -3$

   (c) $x = 0, -\frac{4}{3}$ (f) $x = \dfrac{-4+\sqrt{52}}{6}, \dfrac{-4-\sqrt{52}}{6}$
3. (b) $f'(-3) = -2$, $f'(-1) = 2$ (c) minimum

5. **(a)** minimum at $(-\frac{3}{2}, \frac{7}{4})$
   (b) maximum at $x = \dfrac{-2-2\sqrt{28}}{3}$, minimum at $x = \dfrac{-2+2\sqrt{28}}{3}$
   (c) minimum at $(0, 1)$
7. **(a)** yes
   (c) no; $f'(x)$ fails to be continuous at $x = 1$. $f(x)$ is not differentiable at $x = 1$.
   (d) maximum at $(1, 1)$, minimum at $(2, -2)$

SECTION 14-2

1. **(a)** $y'' = 2$
   (b) $y'' = -\frac{1}{4}x^{-3/2} + \frac{5}{16}x^{-9/4}$
   (c) $z'' = 6x$
   (d) $y'' = (-3-4\log e)/x^2$
3. y'' is negative at $x = -2, -1$; y'' is positive at $x = 0, 2$
7. **(a)** inflection point at $(0, 0)$
   (b) inflection point at $(0, 0)$
   (c) inflection point at $(-1, 2)$
   (d) no inflection point

SECTION 14-3

3. **(a)** maximum at $x = -\frac{2}{3}-\frac{4}{3}\sqrt{7}$
   minimum at $x = -\frac{2}{3}+\frac{4}{3}\sqrt{7}$
   (b) minimum at $(0, 1)$
5. minimum at $(e^{-1}, -e^{-1})$
7. maximum at $(3, 0)$, minimum at $(\frac{11}{3}, -\frac{4}{27})$

SECTION 14-4

1. maximum at $(0, 0)$
3. maximum at $x = 2-(\frac{1}{3})\sqrt{30}$, minimum at $x = 2+(\frac{1}{3})\sqrt{30}$, inflection point at $x = 2$
5. maximum at $x = -3$, minimum at $x = 2$, inflection point at $x = -\frac{1}{2}$

SECTION 14-5

1. $q = \frac{110}{32}$
3. maximum profit at $x = 50$, maximum loss at $x = \sqrt{10}$
5. minimum cost $= 100$ at $x = 100$

SECTION 14-6

1. **(b)** 13 units
3. Yes, no
5. Assume $y = \dfrac{a}{x-b} + c$

partial answer:
Has asymptotes   No max and no min   No inflection point

7. Assume $y = e^{kx}$
No max and no min   No inflection point

9. Has a minimum loss, i.e., maximum profit, but profit is still negative even at its maximum value.

CHAPTER 15
SECTION 15-1

1. **(a)** 2.04 **(b)** 2.64 **(c)** 2.33

3. 1.716

5. See Definition 15.1.

7. The integral is the limit of S_n as n approaches infinity. I.e.,

$$\int_a^b f(x)\,dx = \lim_{n\to\infty} S_n$$

SECTION 15-2

3. 1.04 and 1.64

5. 4.319 based on $t_1 = 1.2$, $t_2 = 1.8$, $t_3 = 2.2$, and $t_4 = 2.8$

8. $\frac{5}{16}$

SECTION 15-3

1. $x_0 = 4$, $x = 4.5, \ldots, x_{12} = 10$
$t_1 = 4\frac{1}{4}$, $t_2 = 4\frac{3}{4}, \ldots, t_{12} = 9\frac{3}{4}$

3. 6.99

5. 84

SECTION 15-4

1. **(a)** $x^3/3$ **(b)** $6 \ln x$ **(c)** $e^{4x}/4$ **(d)** $5^x/\ln 5$

3. **(a)** $22\frac{2}{3}$ **(b)** -32 **(c)** 0.693 **(d)** 6.2 **(e)** 15 **(f)** 2

4. **(a)** 3.195 **(b)** $1/\ln 2$ **(c)** 5.79 **(d)** $8/\ln 3$

7. **(a)** 0 **(b)** -10.5

SECTION 15-5

1. 2.64 exact area is $2\frac{2}{3}$

2. $\frac{8}{3}$

5. 18

7. $S_2 = 45.64$
$S_4 = 51.43$ exact value: 53.6

9. $-\frac{1}{4}$

SECTION 15-6

1. 1, 0.607, 0.135
3. 0.3989, 0.24213, 0.05385
5. 0.39892
7. 0.1915
9. 0.7222

SECTION 15-7

1. **(a)** $(x^4/4) + x^2 + 4x + C$
(b) $(e^{4x}/4) + C$
(c) $\ln|x+4| + C$
(d) $\ln|x-3| - \dfrac{3}{x-3} + C$

3. **(a)** 0 **(b)** 1.099 **(c)** 1.3 **(d)** 0.693
5. $p = 2 - 3x$
6. $p = (ax/2) + b$

Index